ADVANCES IN EQUINE NUTRITION II

ACKNOWLEDGEMENTS

The editors are sincerely grateful to Catherine Bishop, Dennis Christenson and Mark Llewellyn for contributions that helped bring this volume to life.

Advances in Equine Nutrition II

Edited by Joe D. Pagan Ph.D
and Ray J. Geor BVSc, MVSc, Ph.D, Dipl.ACVIM
Kentucky Equine Research Inc., Versailles, Kentucky, USA

NOTTINGHAM
University Press

Nottingham University Press
Manor Farm, Main Street, Thrumpton
Nottingham, NG11 0AX, United Kingdom

NOTTINGHAM

First published 2001
© Kentucky Equine Research, Inc. 2001

British Library Cataloguing in Publication Data
A catalogue record for this book is available from the British Library

ISBN   1-897676-78-6

Typeset by Kentucky Equine Research, Inc.
Printed and bound by Hobbs the Printers Ltd., Hampshire, UK

CONTENTS

General Nutrition

Growth and Development

Nutrient Requirements

Pathological Conditions

Feeding Practices

The Performance Horse

GENERAL NUTRITION

A PRACTICAL METHOD FOR RATION EVALUATION AND DIET FORMULATION: AN INTRODUCTION TO SENSITIVITY ANALYSIS

DAVID KRONFELD

Virginia Polytechnic Institute and State University, Blacksburg, VA

Summary

The evaluation of rations and formulation of diets for horses customarily uses mean values for nutritional requirements, compositions of ingredients, and intakes of forages and feeds to yield a single solution, that is, one ration, one diet or one supplement. More realistic is the representation of these variables as ranges (for example, as lower, middle and upper values), and testing the effects of these ranges to yield three or more solutions--*sensitivity analysis.* This method tests the flexibility and robustness of the design of a diet or supplement, and greatly increases the probability of detecting weaknesses in a ration.

Introduction

An orientation of nutritional practice (Figure 1) has been successfully used by the author and associates for many species, including the horse. The focus is the setting of goals for intakes of energy and nutrients. These goals are not specified as single numbers or requirements, but rather as optimal or target ranges, with upper and lower limits as well as middle values (Figure 2). This paper deals with the interactions between nutritional goals and the evaluation of a ration (daily intake) or the design of a diet or supplement. It introduces the use of *sensitivity analysis*, which is borrowed from economics and epidemiology (Anderson, 1974; Martin et al., 1987).

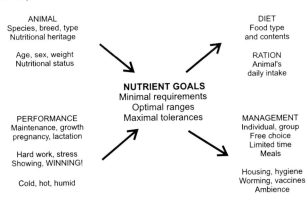

Figure 1. An orientation of nutritional practice focuses on setting nutritional goals for energy and nutrients in keeping with an animal's nature, desired performance, and state of health. The goals are optimal ranges (Figure 2), which are usually specified as lower, middle and upper values.

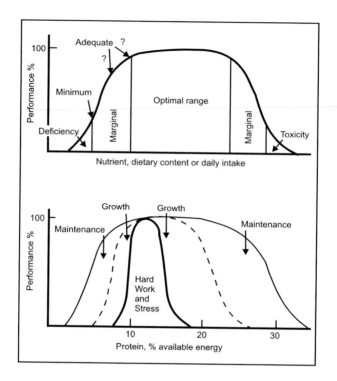

Figure 2. The influence of dietary content or nutrient intake on a specified measure of performance rises to a plateau, the optimal range, then declines. Minimum requirements are usually set to avoid an adverse effect in 50% of the population, and optimal function may be reduced before such an adverse effect becomes evident. For practical purposes, nutritional goals or target zones should be in optimal ranges rather than be restricted to the minimum requirement.

Sensitivity analysis tests the effects of varying parameter values of a model through a defined range and observing the resultant changes in the outcome. A simple example would be the prediction of dry matter intake (DMI):

$$\text{DMI} = a\text{X} + b\text{Y} + c\text{Z}$$
$$\text{intake} \quad \text{age} \quad \text{weight} \quad \text{activity}$$

where the measured variables are X, Y and Z; the relating parameters are *a*, *b* and *c*. Sensitivity analysis enables us to cope with variation, including estimating error, in DMI or digestible energy (DE) intake, nutrient content and availability, and DE and nutrient needs of an equine population. It requires that nutritional goals should be represented by a set of optimal ranges rather than a set of single values for DE and each nutrient.

Achieving nutritional goals is simple for those animals fed a complete and balanced diet, for example, dogs and cats, or a total mixed ration, for example, feedlot cattle. Legend has it that the first complete and balanced diet for a domestic animal, Purina Horse Chow, was made by John Danforth in the 1890s to "take the worry out" of feeding horses. This concept has prospered for pet foods more than horse feeds, which commonly are supplements for forages. Most horses are offered a maintenance level of forages (pasture and hay) supplemented with concentrates (mainly grain).

The traditional approach to designing a diet or supplement leads to a single solution, for example, one concentrate formula for each forage analysis, to achieve nutritional goals expressed as requirements (NRC, 1989). In contrast, the procedure summarized here specifies nutritional goals as target ranges and tests the effects of ranges of nutrients in ingredients and proportions of concentrate:forage consumed. It acknowledges and deals with the variations that exist in animals and feedstuffs and, for pasture-fed animals, the huge errors in estimating pasture intake (Figure 3).

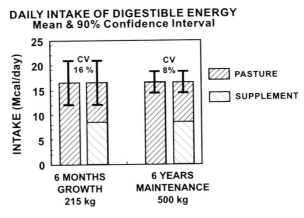

Figure 3. The common proximate analysis of carbohydrates determines the insoluble fibers of plant cell walls as neutral detergent fiber (NDF) and includes the soluble fibers and other non-hydrolyzable, soluble carbohydrates in the nonstructural carbohydrates (NSC), which are calculated by difference: NSC = total carbohydrates - NDF. In view of the digestive physiology and metabolism of hindgut fermenters, such as the horse, hydrolyzable carbohydrates need to be split from fermented carbohydrates, and the latter should be further divided into those fibers that yield lactic acid and those that yield acetic, propionic and butyric acids.

In practice, the diet formulation program is run three or more times instead of just once; this does not require three or more times the work, however, because only one number needs to be changed from one run to the next. By visualizing the effects of ranges, this method generates more general formulas capable of sustaining wider ranges of uses. Engineers would call these more flexible and robust products.

Similarly, for evaluating the ration of a horse, the traditional approach is to assume a single, middle value for pasture intake (Lewis, 1995; Pagan et al., 1996). This approach neglects the huge variation or error in estimating pasture intake, especially when a supplement is fed and all of the variation in daily intake is confined to the pasture intake (Figure 3). The present method addresses the practical importance of this variation or error by running the ration calculations at least three times, using lower and upper limits as well as the middle values for pasture intake.

Nutritional Goals

Nutritional science usually is applied to the determination of minimal nutrient requirements, whereas practical nutrition really needs optimal ranges for specified purposes (Figure 2). The performance traditionally evaluated in farm animal nutrition is energetic efficiency, but other measures of interest would include reproductive efficiency, conformation, tractability, low disease incidence and competitive athletic ability---winning! For horses, optimal ranges of vitamin A have been determined for growth and reproduction (Donoghue et al., 1981; Greiwe-Crandell et al., 1995). Also, parabolic curves have been used to determine the optimal dietary protein for growth in horses (Thiers and Kronfeld, unpublished data) and the optimal dietary fat with respect to muscle glycogen concentration (Kronfeld et al., 1994). An optimal range for dietary calcium for growth has been determined for large dogs (Kronfeld et al., 1994); it may apply equally well to the horse, because the minimum requirements are the same and the upper limit appears to depend on the risk of osteochondrosis in both species. Much remains to be done in equine nutrition to establish nutritional goals in the form of optimal ranges with scientific rigor. Meanwhile tentative optimal ranges or target zones need to be used as nutritional goals; practical experience, as well as science, contributes to craft and technology.

The most widely used nutritional standards for animals are the NRC series for many species. These requirements are performance-oriented for production animals, but they are mean minimums for companion animals, prudently viewed as sufficient to prevent lesions or growth retardation in 50% of animals. Failure to recognize this crucial difference in the NRC series may lead to the improper use of the NRC minimum requirements for dogs, cats and horses in the same way as NRC standards for cattle, swine and chickens (or the recommended dietary allowances, RDAs, for humans). Recognizing that the nutrient requirements for dogs and cats have little practical value, the Association of American Feed Control Officials created nutrient profiles for dogs and cats that were about 1.3- to 2-times corresponding NRC values. The human RDAs are two standard deviations above mean minimum requirements, thereby being sufficient for 98% of the population (Food and Nutrition Board, 1989). When the standard deviation is not known for a nutrient, the CV of 15% for energy is used; thus many RDAs are 1.3-times the mean minimum. Applying the same approach to the horse, a set of equine RDAs would be about 1.2-times the NRC requirements for maintenance and up to 1.5-times the NRC requirement for rapid growth (Kronfeld et al., 1994). Such equine RDAs would be more likely than the NRC requirements to reach the lower regions of the optimal ranges for nutrients (Figure 2), and hence be more useful guides for practical nutrition.

Researchers have recommended that horses should be fed 2- to 5-times the vitamin A requirement specified by the NRC for growth and reproduction (Donoghue et al., 1981; Greiwe-Crandell, 1996). Others have recommended 2- to 3-times the Cu requirement specified by the NRC (Knight et al., 1985). Both vitamin A and Cu interact with several other vitamins and minerals, so elevating these two nutrients but not the other vitamins and minerals is likely to induce imbalances. In practice, our goals for vitamins and minerals are about 1.5 to 3 times the NRC minimums, a range based partly on the literature and partly on our own experiments with vitamin A, phosphorus, calcium, zinc and selenium (Kronfeld et al., 1996). Target ranges are also modified in line with likely availabilities of nutrients in various ingredients.

Goals for dietary protein are historically the most contentious in nutritional science. Applying a quadratic curve to data from three good growth studies in the literature indicates peak weight gain at a crude protein (CP) level of 160 g/kg of DM, with 1.0 SE below the peak at 130 and 190 g/kg, which may be taken to be an optimal range (Thiers and Kronfeld, unpublished). The NRC's minimum requirement for maintenance, 8.0% CP (80 g/kg), seems like a sub-subsistence for wasting tissue in 50% or more horses, especially for old horses (Ralston and Breuer, 1997). Examination of the NRC's Tables 5-2a and 5-2b reveals a range of CP contents for concentrates from 10.8% of DM for light work to 17.3% for rapid growth (NRC, 1989). Most manufacturers make a series of concentrates with CP contents of not less than 12, 14 and 16% as fed (13.3 to 17.8% DM). This series neglects the competitive athlete, which must balance protein benefits against disadvantages. More protein may be needed for hypertrophy and repair, for stress and, especially in the horse, to compensate for nitrogen losses in sweat. On the other hand, the athlete needs less protein to minimize the production of acid, heat and urea. To minimize protein quantity without compromising protein needs, protein quality should be as high as possible in a diet for top athletes. High quality protein is needed also to minimize pasture contamination with nitrogen.

Goals for dietary carbohydrates are currently the most contentious in equine nutrition. The traditional guideline of forage intake equal to 1% body weight (NRC, 1989) needs refinement for various purposes; for example, we use entirely different design objectives for our athlete's diet versus our pasture supplement. The proximate analysis of carbohydrates designed for ruminants is inappropriate for the horse (Figure 4, upper part). For a hindgut fermenter, the two main physiological groups should be hydrolyzed carbohydrates (CHO-H) and fermented carbohydrates (CHO-F), since CHO-H yields mainly glucose, whereas CHO-F yields mainly acetate, propionate and butyrate. Digestive and metabolic processes are much more efficient for CHO-H than for CHO-F. Also, the CHO-F should probably be divided into slowly fermented fibers (CHO-S) and rapidly fermented fibers (CHO-R), because the latter tend to give rise to lactate rather than acetate, thus raising risks of several disorders.

NONSTRUCTURAL
CARBOHYDRATE

- *HYDROLYZABLE*
  Monosaccharides
  Disaccharides
  Some starches

- *FERMENTED*
  Oligosaccharides
  Resistant starches
  Gums, mucilage, pectins

NEUTRAL DETERGENT
FIBER

Hemicelluloses

ACID DETERGENT
FIBER

Celluloses
Lignocelluloses
(Lignin)

- *HYDROLYZABLE* → glucose
  Sugars and some starches

- *FERMENTED RAPIDLY* → lactic acid → acetic acid
  Resistant starches, oligosaccharides
  Gums, mucilages, pectins
  Hemicelluloses

- *FERMENTED SLOWLY* → acetic acid
  Cellulose
  Lignocellulose

Figure 4. The daily intake of DM and DE is highly variable, with coefficients of variation (CV) of about 8% for maintenance and about 16% for rapid growth. For a mean intake of 16.4 Mcal DE, the 90% confidence interval (±1.7 CV) is 14.2 to 18.6 Mcal for the mature horse and 11.9 to 20.9 Mcal for the weanling. Thus, if 20 weanlings are group fed, one will consume nearly twice as much DE as another one. If half the intake is provided as a highly palatable supplement, then all of the variation will be compressed into the pasture intake. Now the mean intake of pasture is 8.2 Mcal, and the range is 6.0 to 10.4 Mcal for the horse and 3.7 to 12.7 Mcal for the weanling. If 20 yearlings are group fed, one will consume 3.4-times as much pasture as another one. The impact of this huge variation in pasture intake should be assessed by sensitivity analysis.

We have enzymatically assayed CHO-H in 130 forages and 30 concentrates and found that CHO-H, hence also CHO-R, can be predicted approximately from nonstructural carbohydrates, NSC (Hoffman and Kronfeld, unpublished data):

CHO-H = 0.3 x NSC in forages
CHO-H = 0.6 x NSC in concentrates
CHO-F = (NSC - CHO-H)

Of course, any overload of CHO-H that escapes hydrolysis in the small intestine will be fermented rapidly in the large bowel. Thus, we are now able to evaluate rations and design diets and supplements with goals specified for physiologically different carbohydrates (Figure 4, lower part).

Most clients of professional nutritionists are seeking help to achieve a high level of performance of various kinds. Each nutritionally competent person or corporation in equine practice has a set of nutritional goals or tentative target ranges, and these goals are usually regarded as proprietary and valuable intellectual properties.

Ration Evaluation

Ration evaluation is needed most often to identify and describe the role of the food and feeding management in poor performance or disease. The traditional approach is to determine amounts of concentrates and hays offered, which is usually fairly precise, then how much remains unconsumed, which is often zero for concentrates but sometimes more difficult to determine for hays (Kronfeld, 1978; Lewis, 1995; Pagan et al., 1996). The cumulative errors are probably less than ±10% for stall-fed horses, but much larger for horses with access to pasture, up to ±20% for horses at maintenance and up to ±40% for young horses (Figure 4). These huge errors are seldom acknowledged by nutritionists and epidemiologists. The procedures described herein, however, are designed to cope with variation and to provide a realistic picture of the lack of precision in ration evaluation.

In practice, one visits the farm or stable and interviews the staff to determine the volumes of hays, concentrates and other supplements that are offered and, if possible, any amounts not consumed. At least two, preferably five, of these volume measures (cans or cups of concentrates, flakes or bales of hay) are weighed. At least two samples of each concentrate and five samples of each hay are taken. Pastures are usually sampled by walking the two diagonals and clipping the grasses and legumes, neglecting the weeds, every 10 paces. For each forage or feed, samples are combined and thoroughly mixed, then duplicate subsamples are submitted for proximate and mineral analyses.

The subsamples are sent to a laboratory that analyzes forages and feeds. The routine profile is likely to be suitable for ruminants (for dairy cattle in the US, reflecting the importance of the Dairy Herd Improvement Association). Routine profiles usually include dry matter (DM), crude protein (CP), acid detergent fiber (ADF), neutral detergent fiber (NDF), nonstructural carbohydrates (NSC, by difference), ether extract (EE or crude fat), ash, Ca, Mg, P, Na and K. Trace element assays are usually available, and Cu, Zn and Se are often requested for the horse, together with Fe, Mn, Mb, I and S, depending partly on expense. The main inadequacies of ruminant nutrient profiles for the horse are the proximate assays for carbohydrates (Figure 4), as discussed above.

The ration and diet can now be calculated as weighted averages (Figure 5). The energy and nutrient content of each dietary component (e.g., g/kg food) are multiplied by the component's daily intake (kg/day/animal) to obtain the daily intakes of energy and nutrients from that component. Then the sums of these component intakes of energy and nutrients comprise the ration. These intake totals for energy and nutrients are divided by the sum of the intake weights to give the overall diet. The example (Figure 5) is for a 215 kg (473 lb) weanling,

6 months old in October. The supplement is a typical, high quality concentrate with an energy density of 3.3 Mcal/kg DM and a guaranteed analysis for CP >15% (170 g/kg of DM). The pasture was browned off in a hot August, so some hay is provided. A single weighted average for energy and all measured nutrients is where the traditional ration evaluation stops.

WEIGHTED AVERAGES
WEANLING, 6 MONTHS, 215 KG

VARIABLE	CONC[a]	HAY	PASTURE	RATION[b]	DIET[c]
DM, kg/d	3	2	1	6 (2.8% W)	
DE, Mcal/kg	3.8	2.0	1.8		2.62
Mcal/kd	9.9	4.0	1.8	15.7	
CP, g/kg	170	100	80		132
g/kd	510	200	80	790	

[a] Concentrate.   [b] Ration = Conc + Hay + Pasture.   [c] Diet = Ration/5.91.

Figure 5. Rations and diets are calculated as weighted averages. For example, the concentrate intake of 3 kg/d of dry matter (DM) is multiplied by the concentrate's content of digestible energy (DE), 3.3 Mcal/kg DM, to give the intake of 9.9 Mcal/d of DE from concentrate. The same calculation is performed for hay (4 Mcal/d) and for pasture (1.8 Mcal/d), then the three DE intakes are added to give the total intake, 15.7 Mcal/d, which is the energy provided by the ration. This number is divided by the total DM intake, 6 kg, to yield the energy density of the diet, 2.62 Mcal/kg. The procedure is repeated for crude protein (CP). In practice, weighted averages are obtained for energy and all measured nutrients to describe the ration and diet.

Sensitivity analysis may now be applied to the variation in DM intake, hence the error in estimating pasture intake (Figure 6). The DM intake of weanlings has a range of 2% to 3.5% of body weight (NRC, 1989). In this example, the midpoint is taken as 2.8% or 6.0 kg, and the intakes of concentrate and hay are 3 and 2 kg, respectively (from Figure 5), so that the intake of pasture is 1.0 kg. The upper limit is 7.5 kg, and the corresponding pasture intake is 2.5 kg. The lower limit should be 4.3 kg (2% of 215 kg), but this youngster is already consuming 5 kg of DM, so its pasture intake is estimated at -0.7 kg, which we reduce to zero. If we inspect the diets, the one with the lowest pasture content is the best. This is misleading, however, because consuming too little of an admittedly better diet is still underfeeding. In practice, full attention should be given to the ration. Clearly the yearling should consume DM between the middle value and upper limits, 6 and 7.5 kg/day, to obtain the energy and nutrients that it needs--at least 16.1 Mcal DE, 800 g CP, and 32 g Ca (NRC, 1989). This example (Figure 6) supports the need for sensitivity analysis, and emphasizes why the ration should be evaluated, rather than the diet.

SENSITIVITY ANALYSIS
WEANLING, 6 MONTHS, 215 KG

	DM INTAKE, %W		
	2	2.8	3.5
DM, intake, kg/d	5[a]	6.0	7.5
DE, Mcal/day	13.4	15.7	18.4
Mcal/kg DM	2.68	2.62	2.45
CP, gl/day	710	783	912
g/kg DM	124	132	122
Ca, gl/day	38	41	46
g/kg DM	7.6	6.8	6.1

[a] 2% of 215 = 4.3 kg; Conc 3 kg, Hay 2 kg,
so pasture intake is - *0.7 kg!!!*

Figure 6. Sensitivity analysis is applied to the full range of dry matter (DM) intake for a weanling. In this example, solutions are found for the lower limit, 2% of body weight (W), the middle value, 2.8%, and the upper limit, 3.5%. The best diet is found for the lowest intake, but an adequate ration is provided between the middle and upper intakes. The calculation of a negative pasture intake is not uncommon although clearly an indication of lack of precision and accuracy in estimates of intakes. This example emphasizes the need to evaluate the ration rather than the diet of an animal.

The calculation of diets and rations by means of weighted averages is simple but tedious, time consuming and subject to errors that are hard to find and usually repeated. For these reasons, ration evaluations were usually limited to energy, protein, calcium and phosphorus before the advent of the personal computer. The mathematical drudgery of ration evaluation has been eliminated by the availability of inexpensive, user-friendly computer programs. Commonly available spreadsheets, such as Lotus and Excel, readily perform weighted averages.

Specifically designed dietetic programs calculate the total diet and the ration or daily intakes of energy and nutrients, then compare these sets of values with nutritional goals, such as the nutrient requirements of horses (NRC, 1989). In addition, some programs will calculate diets and rations from combinations of feed ingredients, either from specified amounts or on a least-cost basis if the cost of each ingredient is entered into the program.

The relative merits of rival programs are arguable both objectively and subjectively. I became familiar with *Animal Nutritionist* (Version 2.5, 1987, N-Squared Computing, Silverton, OR) a decade ago and continue to use it. This software can be used to evaluate a ration, calculate a diet from specified amounts of ingredients (stored in an adjustable data bank), or calculate a least cost diet from a stipulated set of ingredients. Much simpler to use and inexpensive is *Spartan* (Cooperative Extension, Michigan State University, East Lansing, MI). A recent program for ration evaluation is *Microsteed* ™ (Kentucky Equine Research Inc., Versailles, KY).

This software has several advantages: it is designed solely for the horse; it is simple to use; it is the subject of detailed explanatory articles (Pagan et al., 1996); it is likely to be updated and expanded; and, most of all, it offers an option of using the minimum requirements of the NRC (1989) or Kentucky Equine Research's own set of nutritional allowances.

Diet or Supplement Formulation

A maintenance level of nutrition is usually provided by forage-pasture or pasture conserved as hay, preferably at peak energy and nutrient contents. For above maintenance purposes, such as growth, reproduction and hard work, forage intakes are usually supplemented with concentrates, which have higher contents of energy and essential nutrients. One might expect, perhaps, that a single forage supplement could be designed to meet all of these purposes by appropriate feeding management, that is, by varying the concentrate:forage ratio. This ratio may be varied from 0:100 at maintenance to 70:30 for rapid growth, as recommended by the NRC, changing the energy density from 2.0 to 2.9 Mcal/kg.

Pasture varies in nutritional quality through the seasons, so presents wide ranges of contents of most nutrients. In north-central Virginia, we found 90% confidence intervals for 130 samples as for the first 33 samples, so believe that these ranges are reasonably well defined for this region (Hoffman et al., 1996; Wilson et al., 1997). Thus, we had the information needed for the statistical exercise of designing a robust, flexible pasture supplement capable of reaching target zones for all nutrients of interest when used to provide 25 to 50% of the ration. The concentrate would be used to increase energy and nutrient densities above the maintenance level, that is, for growth, reproduction and hard work.

Copper is given as an example of the use of sensitivity analysis in designing a pasture supplement (Figure 7), because it was often marginal or deficient in our survey, and because of its potential importance in developmental orthopedic disease. The minimum requirement is 10 mg/kg DM (NRC, 1989). Our target zone is set by lower, middle and upper values of 15, 20 and 30 mg/DM, respectively. Virginia forages have a mean of 8 mg/kg and a 90% confidence interval of 6-12 mg/kg. We start with the middle values, 9 mg/kg Cu in forage, 20 mg Cu target and 33% concentrate (concentrate:forage, 1:2). A Pearson square is used to calculate that the concentrate should contain 42 mg/kg of Cu (Figure 7). Next we practice sensitivity analysis, using the 6-12 mg/kg Cu range in the forages and the range of 25-50% concentrate (concentrate:forage, 1:3 to 1:1). The resulting range of Cu in the mixtures is 15-27 mg/kg, right on our target of 15-30 mg/kg. If the initial estimate of concentrate Cu had given a value outside the target range, further iterations would be explored to find a best fit.

Similar calculations were performed for iron, manganese, zinc, molybdenum, selenium, iodine, calcium, phosphorus, magnesium, potassium, and sodium. We could determine mineral specifications for a single pasture supplement for all grass forages and all grass:legume mixtures that contained up to about 35% legumes. Another formula would be needed for alfalfa hay.

SENSITIVITY ANALYSIS in DIET FORMULATION

GIVEN: VIRGINIA FORAGES, COPPER 6 - 12 mg/kg DM
TARGET: MIDDLE 20, LIMITS 15 AND 30 mg/kg
CONCENTRATE FORAGE 25:75 to 50:50.
TASK: SELECT CONCENTRATE COPPER.

1. Calculate for middle values, Cu 9 mg/kg, C:F 33:67.

9   ↘   ↗22   67%
        20
42 ↗   ↘11   33%

2. Do sensitivity analysis for ranges.

Cu	CONCENTRATE:FORAGE		
mg/kg	25:75	33:67	50:50
6	15	18	24
9	17.2	20	25.5
12	19.5	22	27

Figure 7. Sensitivity analysis is applied to determine the copper (Cu) content of a pasture supplement. First, middle values for the nutritional goal (20 mg/kg), forage Cu content (9 mg/kg), and concentrate:forage ratio (33:67) are used in a Pearson square to calculate a first estimate of Cu content (42 mg/kg) in the concentrate. Second, sensitivity analysis is applied to the lower, middle and upper values for forage Cu content and concentrate:forage ratio. In this case, there was no need to attempt further iterations of the concentrate's Cu content.

Conclusion

The availability of computer software for ration evaluation and diet formulation enables the practical use of sensitivity analysis to explore the effects of reasonable ranges of pasture intakes, ingredient composition and nutritional goals. Balancing rations and diets by means of ranges with specified lower, middle and upper values, instead of single, middle numbers, gives a truer picture of the real world. It improves the chances of detecting weakness in rations, and it enables the design of more flexible and robust diets and pasture supplements for horses.

References

Anderson JR. 1974. Simulation methodology and application in agricultural economics. Rev Markt Agric Econ 42: 3-25.

Donoghue S, Kronfeld DS et al. 1981. Vitamin A nutrition of the equine: growth, serum biochemistry and hematology. J Nutr 111: 365-374.

Greiwe-Crandell KM, Kronfeld DS et al. 1995. Seasonal variation in vitamin A status of grazing horses is assessed better by the relative dose response test than by serum retinol concentration. J Nutr 125: 2711-2716.

Hoffman RM, Kronfeld DS et al. 1996. Dietary starch and sugar versus fat and fiber: growth and development of foals. Pferdeheilkunde 12: 312-316.

Knight DA, Gabel AA et al. 1985. Correlation of dietary mineral to incidence and severity of metabolic bone disease in Ohio and Kentucky. Proc Am Ass Equine Practnr 31: 445-461.

Kronfeld DS. 1978. Feeding on horse breeding farms. Proc Am Ass Equine Practnr 24: 461-464.

Kronfeld DS, Cooper WL et al. 1996. Supplementation of pasture for growth. Pferdeheilkunde 12: 317-319.

Kronfeld DS, Ferrante PL, Grandjean D. 1994. Optimal nutrition for athletic performance, with emphasis on fat adaptation in dogs and horses. J Nutr 124: 2745S-2753S.

Martin SW, Meek AH, Willeberg P. 1987. Veterinary Epidemiology: Principles and Methods. Iowa State University Press, Ames.

National Research Council. 1989. Nutrient Requirements of Horses, ed. 5. National Academy Press, Washington, DC.

Pagan JD, Jackson S, Duren S. 1996. Computing horse nutrition: How to properly conduct an equine nutrition evaluation using Microsteed™ equine ration evaluation software. World Equine Vet Rev 1(2): 10B1-7.

Ralston SL, Breuer LH. 1996. Field evaluation of a feed formulated for geriatric horses. J Equine Vet Sci 16: 334-338.

Wilson JA, Kronfeld DS et al. 1997. Seasonal variation in nutrient composition of northern Virginia forages. Proc Equine Nutr Physiol Soc

FACTORS AFFECTING MINERAL DIGESTIBILITY IN HORSES

JOE D. PAGAN
Kentucky Equine Research, Inc., Versailles, KY

Mineral requirements for horses are usually calculated using information about the horse's body weight, age, weight gain, physiological status (pregnancy, lactation, etc.) and level of activity. Rarely are other factors that affect digestibility taken into consideration. In most cases, other substances in the ration don't significantly alter mineral utilization, but there are cases where the presence of other minerals or inhibitory compounds should be considered when formulating a horse ration. This paper will review some of the most important factors affecting the digestibility of minerals in horses and will use data from a series of digestion trials conducted by Kentucky Equine Research to illustrate how relevant nutrient interactions are in typical horse rations.

Studying Mineral Digestibility

Kentucky Equine Research has conducted a large number of digestibility studies with a wide variety of feedstuffs. A summary of 30 of these studies has been published (Pagan, 1998) and these data will be further evaluated in the present paper. The diets evaluated ranged from pure alfalfa hay to a combination of sweet feed and fescue hay to pelleted concentrates fed with timothy hay. Table 1 lists the average concentration of each nutrient measured along with the standard deviation, maximum and minimum ranges. As Table 2 illustrates, these combined experiments represent a wide range of nutrient intakes for mature horses.

The majority of horses used in these studies have been Thoroughbreds, although Quarter Horses, Appaloosas, and warmbloods have also been included. Most of these horses have averaged between 500-600 kg of body weight.

Mineral Digestibility

Mineral digestibility can be expressed in two different ways. One way is as apparent digestibility. Using this calculation, the amount of a specific mineral that is recovered in the feces is subtracted from the total daily intake of that mineral. The amount that disappeared (intake - feces) is divided by the total daily intake to produce a percentage of intake. Apparent digestibility is a fairly crude way to evaluate digestibility since it only measures the total amount of a particular nutrient in the feces. There are two possible sources of these fecal nutrients. Some of the nutrient could be the undigested residue left from the feed, but some may have actually been excreted into the digestive tract from the horse's system or it might have sloughed off the intestinal wall. The fecal substances that originate from inside the horse are considered *endogenous* in nature and they result in an underestimation of *true* nutrient digestibility.

13

Table 1. Average nutrient concentrations of diets studied (100% dry matter basis).

Nutrient	Average concentration	Standard deviation	Maximum	Minimum
Crude protein	13.1 %	2.6 %	20.4 %	9.6 %
ADF	28.8 %	4.6 %	40.6 %	20.6 %
NDF	46.9 %	5.4 %	57.4 %	38.3 %
Hemicellulose	18.1 %	4.7 %	24.1 %	6.0 %
Crude fiber	22.8 %	3.9 %	31.8 %	15.4 %
Soluble CHO	28.9 %	5.0 %	36.9 %	18.3 %
Fat	3.6 %	0.8 %	5.5 %	2.1 %
Calcium	0.89 %	0.24 %	1.50 %	0.55 %
Phosphorus	0.39 %	0.09 %	0.58 %	0.20 %
Magnesium	0.22 %	0.03 %	0.29 %	0.17 %
Potassium	1.63 %	0.54 %	3.29 %	0.98 %
Iron	287 ppm	119 ppm	753 ppm	127 ppm
Zinc	84 ppm	38 ppm	147 ppm	20 ppm
Copper	22 ppm	8 ppm	38 ppm	7 ppm
Manganese	83 ppm	29 ppm	127 ppm	29 ppm
Ash	7.45 %	1.53 %	11.62 %	5.99 %

Table 2. Average nutrient intakes of diets studied.

Nutrient	Average concentration	Standard deviation	Maximum	Minimum
Dry matter	7,119 grams	1,464 grams	10,541 grams	4,777 grams
Crude protein	946 grams	326 grams	1,808 grams	572 grams
ADF	2,097 grams	709 grams	4,266 grams	984 grams
NDF	3,371 grams	927 grams	5,427 grams	1,968 grams
Hemicellulose	1,275 grams	411 grams	2,191 grams	393 grams
Soluble CHO	2,007 grams	285 grams	2,627 grams	1,195 grams
Fat	254 grams	66 grams	469 grams	144 grams
Calcium	64 grams	27 grams	158 grams	33 grams
Phosphorus	28 grams	9 grams	57 grams	13 grams
Magnesium	16 grams	5 grams	27 grams	9 grams
Potassium	119 grams	53 grams	269 grams	47 grams
Iron	2,059 mg	1,250 mg	7,912 mg	773 mg
Zinc	567 mg	216 mg	885 mg	131 mg
Copper	149 mg	57 mg	282 mg	46 mg
Manganese	572 mg	201 mg	1,179 mg	210 mg
Ash	540 grams	199 grams	1,136 grams	299 grams

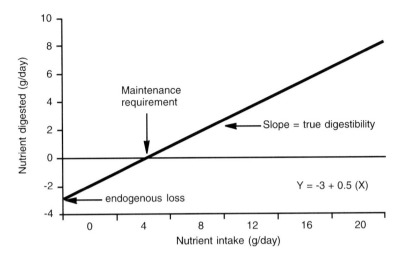

Figure 1. Lucas test for determining true digestibility and endogenous fecal losses.

To overcome the interference of endogenous losses in the determination of digestibility, a statistical procedure called a Lucas test can be utilized. In this test, a range of nutrient intakes are studied. The amount of nutrient that is digested is regressed against its corresponding level of intake. This procedure is illustrated in Figure 1. If there are real endogenous losses associated with a particular nutrient, then the calculated level of nutrient digested at a nutrient intake of zero will be a negative number. The slope of the regression line represents the *estimated true digestibility* for the nutrient. If the regression line intersects the vertical axis at or above zero, then there are no endogenous losses for that particular substance.

Table 3 shows the estimated true digestibilities of several minerals along with their calculated endogenous losses. These endogenous losses can in turn be used to calculate the true digestibility of each mineral in an individual digestibility trial.

Table 3. Estimated true digestibility and endogenous losses for each mineral measured.

Nutrient	*True digestibility*	*Endogenous loss*	*Endogenous loss real[a]*	*R^2*
Calcium	74.7 %	17.4 g/d	yes	0.94
Phosphorus	25.2 %	4.7 g/d	yes	0.33
Magnesium	51.8%	2.2 g/d	yes	0.76
Zinc	20.8 %	54 mg/d	yes	0.57
Copper	40.0 %	38 mg/d	yes	0.71
Manganese	28.5 %	110 mg/d	yes	0.40

[a]intercept of regression equation significantly different from zero (p<0.05)

Calcium

Requirement

Horses are generally very efficient at digesting and absorbing calcium. The upper half of the small intestine is the major site of calcium absorption. In calculating the maintenance requirement of calcium, the 1989 NRC assumes an endogenous loss of 20 mg of calcium/kg of body weight/day based on metabolism studies at Cornell University (Schryver et al., 1970, 1971). The efficiency of absorption of calcium was assumed to be 50%, leading to a daily requirement of 22 grams of calcium for a 550 kg BW horse.

The relationship between calcium intake and digested calcium is shown in Figure 2. The true digestibility of calcium from these rations was 75% and the daily endogenous loss was 17.4 grams, or 29-34 mg calcium/kg BW/day. Using these data, the daily requirement for a 550 kg horse would be around 23 grams/day, a figure almost identical to the NRC estimate.

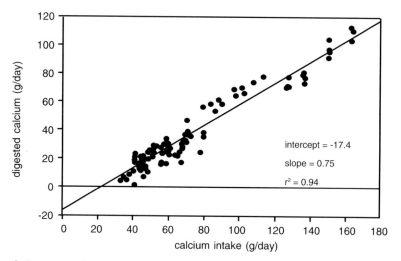

Figure 2. Lucas test for calcium digestibilty.

Factors Affecting Calcium Digestibility

To evaluate whether the concentration of other substances in the horse's ration affects calcium digestibility, linear regressions were calculated between estimated true calcium digestibility and the concentration of several other nutrients in the rations. The coefficients generated from these regressions are listed in Table 4. If an interaction exists between a nutrient and calcium digestibility, then the slope of the regression equation should be significantly different from zero. There were no interactions between calcium digestibility and the concentration of protein, fat, magnesium, iron, zinc, copper, or manganese.

Figure 3 shows the regression between calcium digestibility and iron. There was no relationship between the level of iron in the diets and the true digestibility of calcium. On the other hand, calcium digestibility was negatively correlated with NDF and phosphorus concentration. The relationship between dietary phosphorus and calcium digestibility is shown in Figure 4.

Table 4. Regression coefficients between true calcium digestibility and other substances in the ration.

Nutrient	Intercept	Slope	Slope real[a]	R^2
Crude protein	.67	0.006	no	0.02
NDF	1.08	-0.007	yes	0.13
Soluble CHO	.60	0.005	yes	0.07
Fat	.81	-0.02	no	0.02
Calcium	.68	0.07	yes	0.04
Phosphorus	1.00	-.65	yes	0.35
Ca:P ratio	.68	0.03	yes	0.13
Magnesium	.71	0.13	no	0.00
Iron	.70	0.0002	no	0.02
Zinc	.72	0.0004	no	0.02
Copper	.77	-.001	no	0.01
Manganese	.73	0.0002	no	0.01

[a] slope of regression equation significantly different from zero ($p<0.05$)

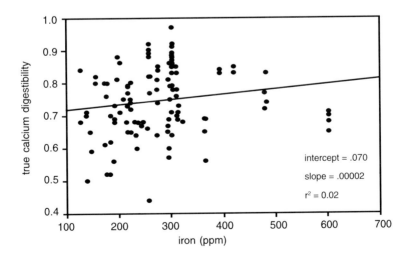

intercept = .070

slope = .00002

$r^2 = 0.02$

Figure 3. Relationship between iron concentration and true calcium digestibility. The slope of the regression is not significantly different from zero.

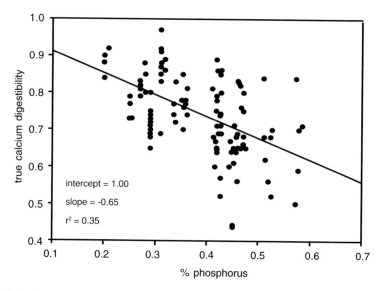

Figure 4. Relationship between phosphorus concentration and true calcium digestibility. The slope of the regression is significantly different from zero (p<.05).

Phosphorus

Requirement

The NRC estimates that mature idle horses absorb phosphorus with an efficiency of 35 percent and that endogenous losses equal 10 mg/kg BW/day. Therefore, a 550 kg horse would require 15.7 grams of phosphorus/day. In the present study, the horses digested phosphorus with an estimated efficiency of 25% and had endogenous losses equal to 4.7 g/day (~8.5 mg/kg BW/day)(Figure 5). Using these data, a 550 kg horse's maintenance phosphorus requirement would equal 18.7 grams/day, only slightly higher than the NRC estimates.

Factors Affecting Phosphorus Digestibility

Coefficients generated from linear regressions of true phosphorus digestibility against other nutrients are shown in Table 5. Phosphorus digestibility, as a function of intake, was much more variable than calcium digestibility and several nutrients in the diet appeared to influence digestibility. There was no relationship between protein, calcium or calcium:phosphorus ratio and phosphorus digestibility. True phosphorus digestibility regressed against Ca:P ratio is shown in Figure 6. True phosphorus digestibility was negatively correlated with fiber content. This relationship is presented graphically in Figure 7.

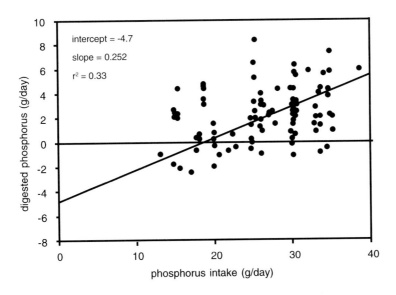

Figure 5. Lucas test for phosphorus digestibility.

Table 5. Regression coefficients between true phosphorus digestibility and other substances in the ration.

Nutrient	Intercept	Slope	Slope real[a]	R^2
Crude protein	.34	-0.005	no	0.02
NDF	.72	-0.01	yes	0.23
Soluble CHO	-.57	0.012	yes	0.36
Fat	.14	0.04	yes	0.08
Calcium	.31	-0.04	no	0.01
Phosphorus	.38	-0.28	yes	0.07
Ca:P ratio	.28	-0.002	no	0.001
Magnesium	.57	-1.29	yes	0.14
Iron	.22	0.0002	yes	0.04
Zinc	.16	0.0015	yes	0.31
Copper	.22	0.002	yes	0.04
Manganese	.18	0.001	yes	0.15

[a]slope of regression equation significantly different from zero (p<0.05)

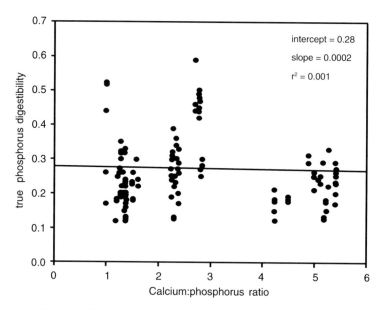

Figure 6. Relationship between Ca:P and true phosphorus digestibility. The slope regression is not significantly different from zero.

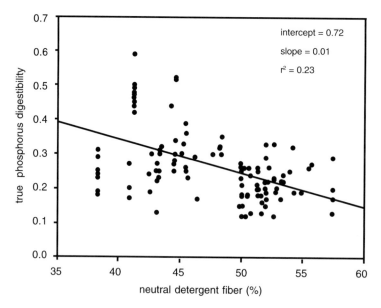

Figure 7. Relationship between neutral detergent fiber (NDF) and true digestibility. The slope of the regression is significantly different from zero (p<.05).

Magnesium

Requirement

The NRC uses a true absorption efficiency of 40 percent and endogenous losses of 6 mg/kg BW/day to calculate the magnesium requirement of the adult horse leading to a daily requirement of 8.25 grams of magnesium for a 550 kg animal. Our data suggest an efficiency of absorption of 52% and endogenous losses of only 2.2 grams/day (~4 mg/kg BW/day)(Figure 8). Using these figures, a mature 550 kg horse would require only about 4.2 grams of magnesium per day.

Factors Affecting Magnesium Digestibility

Coefficients generated from linear regressions of true magnesium digestibility against other nutrients are shown in Table 6. There was no relationship between protein, fat, magnesium, iron, copper, or manganese and true magnesium digestibility. There was a significant negative correlation between fiber and phosphorus levels in the diet and magnesium digestibility. Calcium content and calcium:phosphorus ratio were positively correlated with magnesium digestibility.

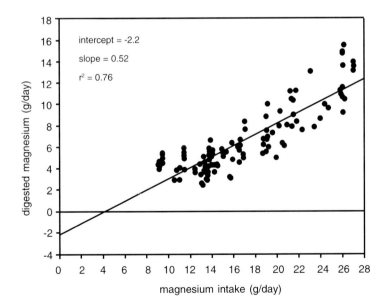

Figure 8. Lucas test for magnesium digestibility.

Table 6. Regression coefficients between true magnesium digestibility and other substances in the ration.

Nutrient	Intercept	Slope	Slope real[a]	R^2
Crude protein	.44	0.006	no	0.03
NDF	.85	-0.007	yes	0.14
Soluble CHO	.39	0.005	yes	0.07
Fat	.58	-0.02	no	0.02
Calcium	.43	0.106	yes	0.10
Phosphorus	.73	-0.54	yes	0.28
Ca:P ratio	.45	0.027	yes	0.16
Magnesium	.59	-0.29	no	0.01
Iron	.51	0.00004	no	0.002
Zinc	.49	0.0004	no	0.03
Copper	.53	-0.0006	no	0.003
Manganese	.51	0.0002	no	0.005

[a]slope of regression equation significantly different from zero ($p<0.05$)

Zinc

Requirement

The NRC estimates that zinc absorption is around 5-10% and that mature horses require 40 ppm zinc in their diets. A 550 kg horse would therefore require around 330 mg zinc/day. In our study, the horses digested zinc with an average efficiency of 21% with endogenous losses of 54 mg/day (.1 mg/kg BW/day)(Figure 9). Using these data, a 550 kg horse would require about 260 mg zinc per day.

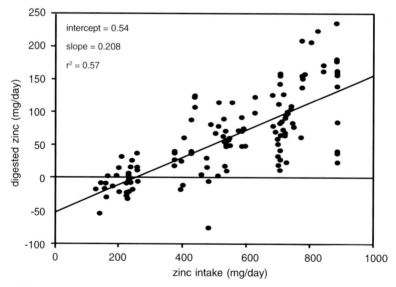

Figure 9. Lucas test for zinc digestibility.

Factors Affecting Zinc Digestibility

Coefficients generated from linear regressions of true zinc digestibility against other nutrients are presented in Table 7. The only nutrient that was significantly correlated to zinc digestibility was magnesium (Figure 10). None of the trace minerals, including iron (Figure 11), affected zinc digestibility.

Table 7. Regression coefficients between true zinc digestibility and other substances in the ration.

Nutrient	Intercept	Slope	Slope real[a]	R^2
Crude protein	.14	0.005	no	0.03
NDF	.22	-0.0002	no	0.0002
Soluble CHO	.27	-0.002	no	0.02
Fat	.20	0.003	no	0.001
Calcium	.18	0.04	no	0.02
Phosphorus	.21	-0.007	no	0.0001
Ca:P ratio	.20	0.006	no	0.01
Magnesium	.05	0.73	yes	0.08
Iron	.23	0.00008	no	0.01
Zinc	.23	-0.0003	no	0.02
Copper	.23	-0.001	no	0.02
Manganese	.23	-0.0002	no	0.009

[a]slope of regression equation significantly different from zero ($p<0.05$)

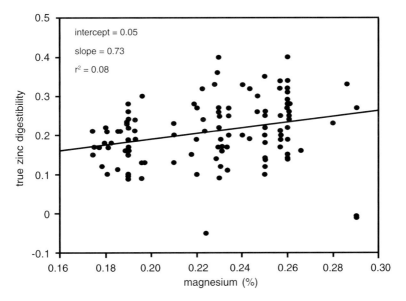

Figure 10. Relationship between magnesium content and zinc digestibility. The slope regression is significantly different from zero ($p >.05$).

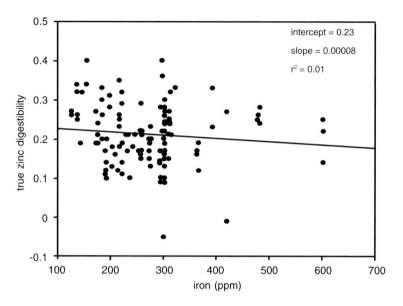

Figure 11. Relationship between iron and zinc digestibility. The slope regression is not significantly different from zero.

Copper

Requirement

The NRC recommends that mature horses should receive 10 ppm copper in their diets or around 85 mg copper/day. In the present study, true copper digestibility was estimated to be 40% and endogenous losses equaled 38 mg/day (~.07 mg/kg BW/ day). Using these figures, daily maintenance copper requirements for a 550 kg horse would equal around 95 mg/day.

Factors Affecting Copper Digestibility

Coefficients generated from linear regressions of true copper digestibility against other nutrients are presented in Table 8. Protein and calcium were negatively correlated to copper digestibility. The relationship between calcium and copper digestibility is shown in Figure 12. Iron level in the diet did not affect copper digestibility (Figure 13).

Table 8. Regression coefficients between true copper digestibility and other substances in the ratio.

Nutrient	Intercept	Slope	Slope real[a]	R^2
Crude protein	.62	-0.02	yes	0.15
NDF	.30	0.002	no	0.008
Soluble CHO	.27	0.004	yes	0.04
Fat	.40	-0.0001	no	0.000
Calcium	.51	-0.13	yes	0.09
Phosphorus	.36	0.10	no	0.007
Ca:P ratio	.45	-0.02	yes	0.05
Magnesium	.38	-0.06	no	0.0002
Iron	.37	0.00009	no	0.006
Zinc	.32	0.0009	yes	0.09
Copper	.36	0.001	no	0.01
Manganese	.28	0.001	yes	0.15

[a] slope of regression equation significantly different from zero (p<0.05)

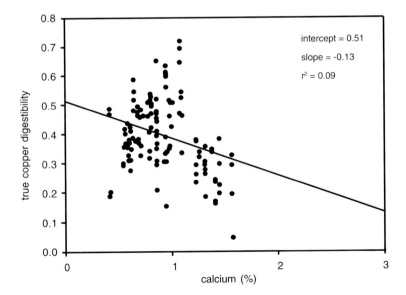

Figure 12. Relationship between calcium and copper digestibility. The slope of the regression is significantly different from zero (p<.05).

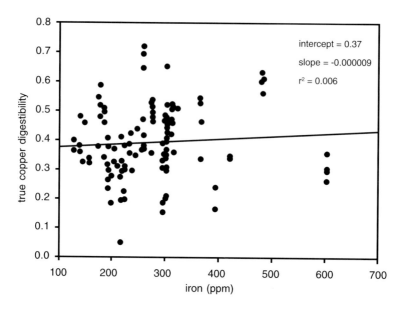

Figure 13. Relationship between iron and copper digestibility. The slope of the
regression is not significantly different from zero.

Conclusions

The mineral requirements calculated for mature horses in the present study
generally agree with the requirements suggested by the 1989 NRC. True
calcium digestibility was negatively correlated with the phosphorus content of
the diet. Phosphorus digestibility, as a function of intake, was much more variable
than calcium digestibility and several nutrients in the diet appeared to influence
digestibility. Magnesium digestibility was negatively correlated to dietary phosphorus content. Zinc digestibility was unaffected by all other nutrients in the diet,
other than magnesium, which was positively related to zinc digestibility. Calcium
was negatively correlated with copper digestibility. Surprisingly, iron content did
not affect the digestibility of any of the minerals in this study, even though the
content of iron in the diets was fairly high (127 ppm-753 ppm). Most of this iron
was not supplemented and was probably in the form of iron oxide, so it remains
to be determined whether supplemental sources of iron have a greater effect.

References

National Research Council. 1989. Nutrient Requirements of Horses (5th Ed.). National Academy Press, Washington, D.C.

Pagan, J.D. 1998. Nutrient digestibility in horses. In: Advances in Equine Nutrition, J.D. Pagan, ed., Nottingham University Press, Nottingham, UK.

Schryver, H.F., P.H. Craig, and H.F. Hintz. 1970. Calcium metabolism in ponies fed varying levels of calcium. J. Nutr. 100:955.

Schryver, H.F., H.F. Hintz, and P.H. Craig. 1971. Calcium metabolism in ponies fed high phosphorus diets. J. Nutr. 101:259.

NUTRACEUTICALS:
WHAT ARE THEY AND DO THEY WORK?

KATHLEEN CRANDELL AND STEPHEN DUREN
Kentucky Equine Research, Inc.,Versailles, KY

Introduction

In the past five years, the world has witnessed the explosive growth of a multi-billion dollar industry known as nutraceuticals. The term "nutraceutical" combines the word "nutrient" (a nourishing food or food component) with "pharmaceutical" (a medical drug). The word "nutraceutical" has been used to describe a broad list of products sold under the premise of being dietary supplements (i.e. a food), but for the expressed intent of treatment or prevention of disease. What is the legal definition of a nutraceutical? How do they differ from either a nutrient or a drug? What rules govern their safety and efficacy? What nutraceuticals have found their way into the horse industry? These topics will be addressed in the following paper.

By Definition

Several terms need to be defined in order to gain an understanding of nutraceuticals.

Nutrient: As defined by AAFCO (1996), "a feed constituent in a form and at a level that will help support the life of an animal." The chief classes of feed nutrients are proteins, fats, carbohydrates, minerals and vitamins.

Feed: As defined by AAFCO (1996), "edible materials which are consumed by animals and contribute energy and/or nutrients to the animal's diet."

Food: As defined by the Food, Drug and Cosmetic Act (1968), "an article that provides taste, aroma or nutritive value. Food and Drug Administration (FDA) considers food as 'generally recognized as safe' (GRAS)."

Drug: As defined by AAFCO (1996), "a substance intended for use in the diagnosis, cure, mitigation, treatment or prevention of disease in man or other animals. A substance other than food intended to affect the structure or any function of the body of man or other animals."

Dietary Supplement: As defined by the Dietary Supplement Health and Education Act (DSHEA, 1994), "a product that contains one or more of the following dietary ingredients: vitamin, mineral, herb, or other botanical, and amino acid (protein). Includes any possible component of the diet as well as concentrates, constituents, extracts or metabolites of these compounds."

Nutraceutical: As commonly defined by the dietary supplement industry, "any nontoxic food component that has scientifically proven health benefits, including disease treatment and prevention."

Veterinary Nutraceutical: As defined by the newly created North American Veterinarian Nutraceutical Council, Inc. (NAVNC), "a substance which is produced in a purified or extracted form and administered orally to patients to provide agents required for normal body structure and function and administered with the intent of improving the health and well-being of animals."

Food or Drug

Using the above definitions, it is still difficult to determine what is and what isn't a nutraceutical. Are nutraceuticals considered food or feeds? According to definition, a feed is an edible substance that contributes energy or nutrients to an animal's diet. Feeds can make claims only about the nutrients they contain and the scientific functions of those nutrients. Both of the definitions presented in this paper for nutraceuticals either include the word "food" or state they are "required for normal body structure and function." A potential difference between a feed and a nutraceutical is that a nutraceutical is unlikely to have an established nutritive value (Boothe, 1997). Feeds are required to have nutritive value and are accountable, via labeling, for these values. Another difference between a feed (food) and a nutraceutical is that feed is generally recognized as safe (GRAS). Nutraceuticals may contain substances that are "natural" but may not be generally recognized as safe.

The other component of our definition of nutraceutical includes the statements "for disease treatment and prevention" and "administered with the intent of improving the health and well-being of animals." When a dietary supplement, nutraceutical or other feed is intended to be used for the treatment or prevention of disease, in essence it "becomes" a drug (Dzanis, 1998). Drugs are subject to an approval process prior to marketing. To be approved, a drug must demonstrate safety and efficacy for its intended use (Dzanis, 1998). Drugs that are not properly approved are subject to regulatory action. Nutraceuticals are not drugs simply because they have not gone through an approval process (Boothe, 1997).

From this discussion, it seems nutraceuticals fall somewhere in between food and drug. They have many advantages over either food or drug since they are not required to list nutrient profiles as required by feeds, and in many cases are intended to treat or prevent disease without first undergoing proper drug approval. Determining if a product is a food, or is subject to regulation as a drug, is a function of the manufacturer's claims that establish intent. Boothe (1997) cites the example of vitamin E. When vitamin E is added to the diet as an essential nutrient it is considered a feed component. However, when vitamin E is claimed to treat or prevent azoturia (tying-up) in horses, it is a drug.

Regulations

The primary set of rules governing the human nutraceutical market is the Dietary Supplement Health and Education Act (DSHEA) passed in 1994. This act does not permit FDA to consider a new product a "drug" or "food additive" if it falls under the definition of a "dietary supplement," which includes among other substances any possible component of the diet as well as concentrates, constituents, extracts or metabolites of these components (Dzanis, 1998). This gives human nutraceutical manufacturers a wide range of substances that may be able to satisfy these requirements. The other major component of this act shifts the burden of safety. The FDA now has to prove a substance is unsafe rather than the manufacturer proving the substance safe (Dzanis, 1998).

The DSHEA rules do not apply to nutraceuticals intended for animals. In a nutshell, the federal government has cited differences in metabolism of substances between humans and animals and potential safety issues with nutraceuticals used in food producing animals as reasons to exclude animals from provisions of the DSHEA. Therefore, expressed or implied claims relating use of a product with the treatment or prevention of disease, or with an effect on the structure or function of the body in a manner distinct from what would be normally ascribed to "food" (e.g. that it does something other than provide known essential nutrients), could cause a product to be subject to regulation as an unapproved "drug" (Dzanis, 1998).

Safety and Efficacy

Many nutraceuticals are being used as alternatives for both nutrition and medicine. A substantial number of these products make illegal drug claims without regulation and proper data to support their safety and efficacy. As such, consumers need assurance that a product is safe and hopefully able to do what it say it does.

Above anything else, nutraceuticals should be safe. Stock should not be taken in the old adages "if a little is good, a lot is better" or "it can't hurt." Nutraceuticals, like many substances, may cause problems due to direct toxic effects, or by delay of more appropriate treatment (Dzanis, 1998). Safety of a nutraceutical product is often easier to establish than efficacy. Studies that test doses of nutraceutical several fold greater than the intended (recommended) dose help to establish toxicity data. These studies must test animal reaction to the product both short- and long-term. Finally, a lack of reported toxicity problems with any nutraceutical should not be interpreted as evidence of safety (Boothe, 1998).

Does the nutraceutical do what it says it can do? Is the product effective? Evidence of efficacy is generally provided by studies that document the pharmaceutical, pharmacokinetic, and pharmacodynamic characteristics of a compound (Boothe, 1998). Pharmaceutical data are an evaluation of quality of manufacturing, purity of product and accuracy of labeling. Pharmacokinetic data consists of tracking the compound through the animal's body. It also answers questions about absorption, tissue distribution, metabolism and excretion. Pharmacodynamic evaluation describes how the animal responds to the compound. This step is the most difficult to define for nutraceuticals since most of these compounds are involved in a cascade of different reactions throughout the body.

Since the market for nutraceuticals is booming, many products are available that have not been tested for either safety or efficacy. A simple test of a quality nutraceutical product may be to ask for research data (peer reviewed and published) which support the product. This will go a long way in limiting quackery and the ever present danger of parting you from your money.

Nutraceuticals and the Horse

The theory behind the mode of action of nutraceuticals is to provide functional benefits by increasing the supply of natural building blocks in the body. Replacement of these building blocks can work in two ways: to diminish disease signs or to improve performance. The use of nutraceuticals as performance enhancers is much more common than treatment of disease.

Much of the data used to promote nutraceuticals to the public come from human research. There are only a limited number of nutraceuticals in which research has been done in the horse. For this reason, we will discuss only products that have at least had preliminary testing in the horse. The following is a brief discussion of the theorized mode of action, a summary of human studies and results of equine research.

Carnitine
Carnitine is an amino acid found in abundance in cardiac and skeletal muscle. Since carnitine is involved with the utilization of fatty acids for energy in the muscle cell, it has been hypothesized that supplementation of carnitine would be glycogen sparing and reduce lactic acid production. The end result would be improved muscle function and endurance. While human studies have not been able to show an increase in muscle carnitine from supplementation or improvement in performance with healthy individuals, improvement in humans with impaired oxygen supply was seen in heart and skeletal muscle from carnitine supplementation (Cerretelli and Marconi, 1990). Endogenous synthesis, primarily in the liver, is probably adequate in the normal healthy adult but may be deficient at times of stress and in certain disease states.

The carnivorous human diet is typically very high in carnitine, while the herbivore diet is very low and the horse will have to produce the majority of its carnitine supply endogenously. Taking this into consideration, researchers have investigated responsiveness of yearlings and adult horses to oral carnitine supplementation in adult and yearlings. Results are somewhat inconclusive in that supplementation increased plasma free carnitine in adult and most yearling horses, while long term feeding did not increase muscle carnitine concentrations. Since an increase in muscle carnitine is what would improve performance, there is no evidence in the horse that supplementation really helps.

Coenzyme Q$_{10}$
Ubiquinone, more commonly known as coenzyme Q$_{10}$, is a substance found in the body as a component of the mitochondrial respiratory chain. It works in concert with other substances to regenerate ATP (energy) in a cell. Coenzyme Q$_{10}$ also functions as a powerful antioxidant and free radical scavenger. An antioxidant is a substance that gives up electrons easily and can act to neutralize

harmful oxidants and free radicals. In humans, the levels of coenzyme Q_{10} have been found to be below normal in patients with cardiovascular disease and periodontal disease. Whether low levels are a cause or effect is not clear, but coenzyme Q_{10} supplementation has been reported to have been used successfully in the treatment of heart problems, muscular dystrophy, myopathies and periodontal disease (Greenburg and Fishman, 1990; Nishikawa et al., 1989). Use of coenzyme Q_{10} is not very well researched in the horse. One study found that coenzyme Q_{10} may have an indirect effect on the utilization of oxygen within the tissues, but had no effect on lactate metabolism or heart rates (Rathgeber-Lawrence et al., 1991). Perhaps the use of coenzyme Q_{10} in the horse has potential in treatment of heart and muscle disorders, but it needs to be investigated further.

Creatine
The idea behind creatine supplementation is that increasing the creatine content of muscle will increase the corresponding creatine phosphate (PCr) concentration. A simple reaction involving creatine phosphate is the muscle cell's first and fastest source of energy for contraction. It is present in limited amounts in the muscle cell and has a fairly rapid turnover rate. The availability of PCr has been proposed as one of the most likely limitations to muscle performance during intense, fatiguing, short-term exercise. Scientific investigations in humans indicated that the PCr content of muscle was increased by taking creatine supplements, and that exercise performance is improved by ingestion of creatine over a period of days prior to the exercise test (Greenhaff et al., 1993; Harris et al., 1993). However, in another study 30% of subjects ingesting creatine failed to show an increase in muscle creatine or retain substantial quantities (Greenhaff et al., 1994). One major problem with creatine supplementation is the frequency that the supplement has to be taken (4-6 x per day) which makes it impractical for the horse. One study done in racing Thoroughbred horses failed to show marked increases in muscle creatine after supplementation and no improvement in performance (Sewell and Harris, 1995).

DMG
Dimethylglycine (DMG), a derivative of the amino acid glycine, is a normal intermediate in choline metabolism and has been proposed to enhance creatine phosphate stores in muscles. Many of the claims of the benefits of DMG supplementation are to increase oxygen utilization, reduce lactic acid accumulation in the muscles, strengthen the horse's natural immune response system, prevent tying-up, increase a horse's tolerance to vigorous physical activity and improve overall performance. DMG first came to the forefront when it was found that the Russian athletes were using the "super drug" that they called vitamin B15 or pangamic acid as a performance enhancer. There has been work done in humans receiving DMG orally which suggests that it will boost the immune system (Sellnow, 1987). The use of DMG in the horse has been studied more than most nutraceuticals, but still far from extensively, with mixed results. Studies on supplementation of DMG to Standardbreds and Quarter horses found reduced

blood lactate, while a third study with Thoroughbreds found no benefit (Levine et at., 1982; Moffit et al., 1985; Rose et al., 1989). The fervor over DMG has died down and little work has been done on the subject in recent years.

HMB

ß-hydroxy-ß-methylbutyrate or more simply HMB is a product made in the muscle tissue from the amino acid leucine. Once HMB is formed it serves as a building block for intramuscular cholesterol synthesis. During stressful situations like heavy training and exercise, it is theorized that the muscle cell may not be able to make enough cholesterol for maximal growth or function. Supplying HMB in the diet supposedly would keep blood cholesterol at an optimal level. Work in humans has indicated that aerobic exercise performance and muscular strength can both be improved with HMB supplementation. Very recent studies in Thoroughbreds show promise for the nutraceutical. A treadmill study found lower muscle tissue breakdown in HMB supplemented horses with higher blood glucose during exercise than controls (Nissen et al., 1997). One study with horses in actual race training and racing conditions found a lower amount of muscle enzymes (indicative of muscle damage) in the HMB supplemented horses after a race (Miller and Fuller, 1998). The general impression of HMB supplementation was that it allowed the horses to condition faster.

MSM

Methylsulfonylmethane (MSM) is an odorless and tasteless derivative of the pungent dimethylsulfoxide (DMSO). Its main action is to supply bioavailable sulfur to the horse, and it has been proclaimed to have numerous beneficial effects: moderating allergic reactions and gastrointestinal tract upset, correcting malabsorption of other nutrients (in particular minerals related to developmental orthopedic disease), relieving pain and inflammation, acting as a natural antimicrobial, antioxidant and antiparasitic. Exactly where MSM goes in the body after ingestion has been studied intensively (Metcalf, 1983). With the help of tracer studies, it appears that MSM given orally will eventually end up in every cell in organosulfur molecules. Effectiveness of MSM in treating all of the above stated conditions to this date is mostly anecdotal.

Oral Joint Supplements

The intent of oral joint supplements is to 1) work as an anti-inflammatory agent and/or 2) supply additional building blocks for the formation and maintenance of normal joint cartilage. Ultimately, the idea is to make movement in the joint less painful for the individual. Most oral joint supplements contain chondroitin sulfate and/or glucosamine. Joint supplementation is one area of research where the work done in the horse stimulated interest for use of the products in the human. The most intriguing question about oral joint supplements is whether the product gets to the joint in order to help. The fact that oral supplementation does at least get the product into the bloodstream has been established (Baici et al., 1992). There is a study currently underway looking at the arrival of the products from the bloodstream to the joint being done by researchers at Marion DuPont Equine Medical Center. Experimental evidence of effectiveness of oral joint supplements on improvement of lameness in horses would indicate that at least some of the product is getting to the joint (Hanson et al., 1996). Certainly, this type of nutraceutical is one area where a flourish of research can be expected in the coming years.

Conclusion

It becomes blatantly clear from the brief summaries above that product testing in the horse is sparse for the number of nutraceuticals available on the market. Since efficacy and safety testing are not required in order to market a product (as long as the product does not have medicinal or performance enhancing claims on the label or in the literature), it is difficult to say whether the testing will ever be done. Certainly, if testing ever became required for nutraceuticals, the resulting increase in price of the products may make them prohibitively expensive. Because of the lack of regulation for these products, horse owners themselves become the researchers and their beloved horses the subjects in their own fact finding missions on the truth and efficacy of the nutraceutical.

References

AAFCO. 1996. Association of American Feed Control Officials Incorp. - 1996 Official Publication. p. 175-267.

Baici, A. 1992. Analysis of glycosaminoglycans in human serum after oral administration of chondriotin sulfate. Rheumatol. Int. 12:81-88.

Boothe, D.M. 1997. Nutraceuticals in Veterinary Medicine. Part I. Definitions and Regulations. Comp Cont Ed Vol 19, No. 11 p. 1248-1255.

Boothe, D.M. 1998. Nutraceuticals in Veterinary Medicine. Part II. Safety and Efficacy. Comp Cont Ed Vol 20, No. 1, p. 15-21.

Cerretelli, P. and Marconi, C. 1990. L-carnitine supplementation in humans. The effects of physical performance. J. Sports Med. 11:1-14.

Dzanis, D.A. 1998. Nutraceuticals: Food or Drug? TNAVC Proceedings. p. 430-431.

Greenburg, S. and Frishman, W.H. 1990. Coenzyme Q_{10}: a new drug for cardiovascular disease. J. Clin Pharmacol. 30:596-608.

Greenhaff, P.L., Bodin, K., Soderlund, K. and Hultman, E. 1994. The effect of oral creatine supplementation on skeletal muscle phosphocreatine resynthesis. Am. J. Physiol. 266:E725-E730.

Greenhaff, P.L., Casey, A. Short, A.H., Harris, R.C., Soderlund, K. and Hultman, E. 1993. Influence of oral creatine supplementation on muscle torque during repeated bouts of maximal voluntary exercise in man. Clin. Sci. 84:565-71.

Hanson, R., Smalley, L.R., Huff, F.K., White, S. and Hammand, T. 1996. Oral treatment with a glucosamine-chondroitin sulfate compound for degenerative joint disease in horse: 25 cases. Equine Practice 19/9:16-22.

Harris, R.C., Viru, M., Greenhaff, P.L. and Hultman, E. 1993. The effect of oral creatine supplementation on running performance during maximal short term exercise in man. J. Physiol. 476:74.

Hishikawa, Y., Takahashe, M., Yorifuji, S. 1989. Long term coenzyme Q_{10} therapy for mitochondrial encephalomyopathy with cytochrome C oxidase deficiency. Neurology 39:399-403.

Levine, S.B., Myhre,G.B., Smith, G.L., Burns, J.G. and Erb, H. 1982. Effect of a nutritional supplement containing N,N-dimethylglycine (DMG) on the racing Standardbred. Equine Practice 4:17-21.

Metcalf, J.W. 1983. MSM A dietary derivative of DMSO. Equine Vet. J. 3/5:148.

Miller, P. and Fuller, J.C. 1998. The effects of supplemental ß-hydroxy-ß-methylbutyrate (HMB) on training and racing Thoroughbreds. Abstract from the 17th Annual Meeting AESM, Leesburg, VA, p.13.

Moffit, P.G., Potter, G.D., Kreider, J.L. and Moritani, T.M. 1985. Venous lactic acid levels in exercising horses fed N,N-dimethylglycine. The 9th Equine Nutrition and Physiology Symposium, East Lansing, Michigan, pp. 248-252.

NAVNC. 1996. Nutraceutical Council. J. Eq. Vet . Sci. Vol. 16, No. 11 p. 486.

Nissen, S., Fuller, J. and Rathmacher, J. 1997. ß-hydroxy ß-methylbutyrate (HMB) supplementation in training horses. Metabolic Technologies Bulletin, Ames, Iowa.

Rathgeber-Lawrence, R.A., Ratzlaff, M.H., Grant, B.D. and Grimes, K.L. 1991. The effects of coenzyme Q_{10} as a nutritional supplement on cardiovascular and musculoskeletal fitness in the exercising horse. Proc. 10th Annual Meeting of Assoc. Equine Sports Med., Reno NV, pp. 30-34.

Rose, R.J., Schlierf, H.A., Knight, P.K., Plummer, C., Davis, M. and Ray, S.P. 1989. Effects of N, N-dimethylglycine on cardiorespiratory function and lactate production in Thoroughbred horses performing incremental treadmill exercise. Vet. Record 125:268-271.

Sewell, D.A., Harris, R.C. 1995. Effects of creatine supplementation in the Throughbred horse. In: Equine Exercise Physiology 4, suppl. Eq. Vet. J., pp. 239-242.

FEED MANUFACTURING TECHNOLOGY: CURRENT ISSUES AND CHALLENGES

KEITH C. BEHNKE
Kansas State University, Manhattan, KS

Animal agriculture has been changing rapidly for the last two decades. One of the most significant changes has been in the area of feed requirements for optimum performance. Genetic improvements are placing constant pressure on feed manufacturers to produce quality feeds that match the increased requirements of improved breeds without creating additional physiological or health stress. Increased understanding of nutrition, the environmental impact on animal performance, and even philosophical differences between nutritionists have caused feed manufacturers to alter "business as usual." Additional factors, such as the increased demand for high value specialty feeds for pets, equine, aquaculture, ratites, early weaning diets for nursery pigs and the like, have contributed to the need for improved feed manufacturing techniques.

This paper will focus primarily on three areas of feed manufacturing that are under active research attention. These areas are:

1) Grain particle size and its effect on animal performance;
2) Feed (nutrient) uniformity and its effect on animal performance;
3) Pellet quality issues.

Additional topics briefly discussed will be in-line pellet moisture and quality control, hygienic treatment of finished feeds, odor control, and trends toward liquid ingredients.

Grain Particle Size and its Effect On Animal Performance

Grain is ground prior to mixing to increase surface area for improved rate of digestion, decreased segregation and mixing problems, and to facilitate further processes such as extrusion or pelleting. Extensive grinding requires more energy; however, even small improvements in feed efficiency will often justify the added cost (Ensminger, 1985).

A study to evaluate the effects of grain sorghum particle size on nutrient digestibility was conducted by Owsley et al. (1981). Using ileal cannulated pigs, they found that upper tract digestibilities of N, DM, GE, starch, and most amino acids were increased as particle size was reduced. Total tract digestibility trends were similar but the differences were not as great. For example, ileal starch digestion was 19% greater in pigs fed grain sorghum ground through a 3.2 mm screen vs. those fed a coarsely rolled grain sorghum. Total tract difference was only 3%. It is recognized that starch is fermented to VFA in the large intestine and absorbed by the pig (Argenzio and Southworth, 1975). However, synthesis of VFA into glucose is less efficient than release and absorption of glucose from starch in the small intestine (Black, 1971). It can then be argued that decreased particle size of grain is one technique useful in improving its utilization in growing-finishing pigs.

Reducing the particle size of barley by 14% (789μ vs. 676μ) improved ADG and G/F by 5% for starter pigs (Goodband and Hines, 1988). The finely ground barley gave G/F equal to that of grain sorghum ground to 753μ. They concluded that the nutritional value of barley could be improved more than grain sorghum by fine grinding.

Ohh et al. (1983) found improved G/F and DM, GE, and N digestibility when the particle size of corn and grain sorghum was reduced. Roller milling and hammermill grinding gave similar results. Wu (1985) found a response to fine grinding of corn in finishing pigs but not in weanling pigs.

Performance of pigs fed wheat may not be improved by extensive particle size reduction. Average daily gain and G/F were similar in starter pigs fed diets containing wheat ground to 860μ or 1710μ average particle size (Seerley et al., 1988). Growing pigs fed wheat ground to the same particle size were 8% more efficient than pigs fed diets with 1710μ wheat but ADG was not affected. Finishing pigs gained 9% slower and were 6% less efficient when fed diets with wheat ground to 860μ vs. 1710μ. These studies would indicate that optimum particle size of wheat increases for older pigs.

Fine grinding has been implicated in the development of ulcers in the esophageal region of pigs (Reimann et al., 1968; Hedde et al., 1985). This is still an area of active investigation. Cabrera et al. (1993) found that stomach morphology was negatively affected by fine grinding (<600μ) of corn and two grain sorghum genotypes; however, improved performance would likely make fine grinding an acceptable compromise. They also found a dramatic decrease in daily DM and N excretion as grain particle size was reduced.

Wondra et al. (1993) investigated the influence of mill type (hammermill vs. roller mill) on finishing pig performance and stomach morphology. Mill type did not affect growth performance, but pigs fed corn ground by a roller mill had greater digestibilities of DM, N and GE and excreted 18% less DM and 13% less N than pigs fed hammermill ground corn. Pigs fed the roller mill ground corn had slightly better ulcer scores than those fed hammermilled grain.

Information regarding the effect of grain particle size on broiler performance is limited. Reese et al. (1986) reported improved G/F for broilers fed diets with corn ground to 910μ vs. 1024μ but only when the diets were pelleted. The authors implied that uniformity of particle size might also affect broiler performance.

Healy et al. (1994) investigated the effect of particle size of corn and hard and soft grain sorghum on growth performance and nutrient utilization in broiler chicks. Starting with seven-day-old chicks, they conducted a 21-day growth assay using diets made from each grain ground to 900, 700, 500, or 300μ. Gains for day 0 to 7 were maximized at 500μ for corn, 700μ for the hard endosperm grain sorghum, and 300μ for the soft grain sorghum. Maximum G/F for corn and soft grain sorghum was 300μ and 500μ for hard grain sorghum. For days 0 to 21 gain was optimum for corn at 700μ, for hard grain sorghum at 500μ, and for soft grain sorghum at 300μ. The authors concluded that reduced particle size improved growth performance and that, if properly processed, grain sorghum has a nutritional value equal to corn.

The fineness of grind found to be optimum in the above investigation is much finer than typically found in the U.S. broiler industry. In general, hammermills are equipped with 10/64" (4 mm) or 12/64" (4.8 mm) screens, usually resulting in an average particle size in the 800μ to 1000μ range. An interesting finding in the Healy et al. (1994) study is that gizzard weights were less in chicks fed the 300μ treatment. It is apparent that if the organ is not utilized for its intended function, development can be inhibited. This may have a significant economic impact by reducing the proportion of lower value parts.

Current Issues - Grinding

There is a great deal of interest in evaluating roller mills as an alternative to hammermills. The largest feed mill in the world has but one small hammermill used to grind "overs" of the ground grain scalper. All other grinding is done with roller mills.

Some U.S. manufacturers are looking at the common European practice of post-mix grinding to improve pellet quality. In most cases, minor improvements are noted but at a substantial cost in energy and capital. A relatively new concept in mill design has been proposed by a large engineering firm (Ibberson, Minneapolis, MN). In effect, this is a hybrid between pre-grind and post-grind and allows management to select the operation of choice depending upon the operation and feed quality desired.

Air-assisted hammermill systems continue to gain acceptance. In addition to increasing throughput by up to 25%, moisture shrink and product temperature rise are substantially reduced.

A new hammermill design has been developed by Buhler, Inc. (Minneapolis, MN). The design is novel in that the shaft is vertical rather than horizontal. Screen changing is easily accomplished and particle uniformity and grinding efficiency is said to be improved over conventional hammermills.

Evaluating the effect of grind fineness on animal performance continues to be an active area of research. Much needs to be learned regarding other cereals as well as protein meal in this regard. It is known that fermentation rate is increased as particle size is reduced (for ruminants). Little is known of the effects of fine grinding on hindgut fermentation, as in horses.

Feed (Nutrient) Uniformity And Its Effect On Animal Performance

Mixing is considered to be one of the most critical and essential operations in feed manufacturing regardless of whether it's on-farm or in a commercial facility. Lack of proper mixing can lead to reduced diet uniformity, affecting not only animal performance but regulatory compliance as well.

Few feed manufacturers really know how their mixers are performing or what mix time is required for a given diet. An ongoing survey of commercial mixers found that over 50% did not meet the "de facto" industry standard of a coefficient of variation (CV) of less than 10% when using methionine and lysine as the tracer (Wicker and Poole, 1991). A survey of farm feed mixers found very similar results (Stark et al., 1991). In their study, 42% of the participants had CVs of less than 10%, 47% were between 10 and 20% CV, and 11% had CVs greater than 20%. Salt was used as the tracer in this study.

Even though the importance of diet uniformity is intuitive, there is very little credible research that relates diet uniformity to animal performance. Numerous authors have cited uniformity as one of the most important aspects in feed production (Beumer, 1991; Ensminger et al., 1991; Wilcox and Balding, 1986). However, credible animal studies relating factors such as mix time, diet uniformity, and ingredient segregation are not available in abundance. Duncan (1973) provided insight into the effect of ingredient nutrient variability on animal performance and quality control. If nutrient density is highly variable, it would have the same effect as nutrient variation because of poor mixing or segregation.

McCoy et al. (1994) conducted two experiments to investigate the effects of diet nonuniformity, caused by inadequate mixing, on the performance of broiler chicks. In both experiments, a common diet formulation was mixed for different times to represent poor, intermediate, and adequate uniformity. In the first study, only G/F was improved as diet uniformity increased. In the second experiment (a 28-day growth assay), improvements were noted in ADG, ADFI, and G/F as diet uniformity increased from poor to intermediate. No further enhanced performance was noted as diet uniformity was increased to adequate.

In a recent study, Traylor et al. (1994) examined the effect of mix time on diet uniformity and subsequent growth performance of nursery and finishing pigs. For Experiment 1, 120 weanling pigs were used in a 27-day growth assay. Treatments were mix times of 0, 0.5, 2.0, and 4.0 minutes for a Phase II nursery diet. They found that ADG improved by 49% and F/G was decreased by 16% as diet uniformity was increased. In the finishing pig study, 128 animals were fed from 124 lbs to 260 lbs final body weight. Treatments were the same (mix times) as used in the nursery study. Increasing mix time reduced diet variability (CV) from 54% to less than 10% but did not significantly affect ADG, ADFI, or F/G. However, pigs fed the most nonuniform diet had the poorest rates and efficiencies of gain.

Conclusions reached in both of these studies (McCoy et al., 1994; Traylor et al., 1994) were that body size and, therefore, daily food intake, of young animals is a critical aspect relating diet uniformity to animal performance. Older, more mature animals that consume larger meals and retain digesta longer are less sensitive to variation in diet uniformity than young animals.

Current Issues - Diet Uniformity

There is obviously a great deal of research needed to document more fully the level of uniformity needed to optimize performance of target animals. It is disturbing to contemplate the fact that the basis for all formulation, nutrition, and regulatory control is the assumption of nutrient or additive uniformity, yet very little real data exist to support this assumption.

There is a distinct possibility that regulatory agencies such as FDA/CVM will require that all mixers used to manufacture medicated feed will have to be validated as to blending ability and required mix time. At present, at least one Canadian province requires biannual testing of medicated feed mixers.

The recent introduction of short cycle mixers will have a great impact on mill design and operation. If the mixing cycle is reduced from 5-8 minutes to 1-3 minutes, it is obvious that other operations, such as hand weighment, batching, and liquid additions, will have to be streamlined. Management will become a greater challenge.

Perhaps the most needed item is the development of a dependable mixer testing procedure that is acceptable to both the industry and regulatory agencies. At present, there is no "standard" procedure available that meets the requirements of accuracy, safety, expense, and utility needed for acceptance. This is fertile ground for creative research.

Pellet Quality Issues

Since its introduction in the 1930s, pelleting has become an important process to the feed industry. Estimates vary on the percentage of annual feed production that is pelleted but it is likely in the 60% range.

In recent years there has been a dramatic increase in pelleted feed tonnage because of the rapid growth in hog integration. This industry, like the broiler industry, has concluded that the cost of pelleting is more than offset by improved animal performance.

Pelleted diets can affect animal performance in a variety of ways. The following is a partial list of pelleting attributes that might contribute to improved performance (Behnke, 1994):

1. Decreased feed wastage
2. Reduced selective feeding
3. Decreased ingredient segregation
4. Less time and energy expended for prehension
5. Destruction of pathogenic organisms
6. Thermal modification of starch and protein
7. Improved palatability

The above factors are critical for feeding food producing animals. In the case of pelleted feed for horses, the criteria to be applied are somewhat different. Perhaps most important, however, is the ability to prevent selective feeding. Nearly as important, however, is the value of clean, bright feed to the horse owner. These owners typically place a high value on packaging and "presentation," and pellets do have a role to play in this regard.

In addition, pelleting allows the use of a wider variety of ingredients without obvious changes in the physical properties of the diet. This often will allow diet costs to be reduced with little or no affect on performance.

Research conducted in Europe and North America has shown that pelleted pig nursery diets will increase ADG and G/F by 9 to 10%. Pelleted grower-finisher diets result in an increase of 3 to 5% in ADG and 7 to 10% in G/F (Hanke et al., 1972; Tribble et. al., 1979; Harris et al., 1979; Skoch et al., 1983; Walker et al., 1989; Wondra et al., 1994).

It is accepted that pelleting will improve performance; however, the effect of pellet quality and/or diet fines content on animal performance is poorly understood. Stark (1994) conducted two nursery experiments and a finishing experiment to determine the effects of meal vs. pellets and the effects of fines on pelleted diet performance in pigs. In the nursery studies, Phase II diets were fed for 28 days. In the first study, diets were fed as meal, screened pellets, or screened pellets with 25% fines to reflect a low quality pellet. Pelleting improved gain to feed by 12% compared to the meal diet. There was a trend toward reduced G/F ($P < .07$) if fines were included in the pellet diet.

In the second nursery study, similar treatments were used except fines were included at 15 or 30% rather than 25%. Pelleting increased ADG and G/F by 8% and 15%, respectively, compared to the meal diet. The addition of fines did not affect ADG or F/G ($P < .04$).

In the finishing study, 80 gilts were used to evaluate the effect of diet form and pellet fines on growth performance. The treatments were 1) meal; 2) screened pellets; 3) pellets with 20% fines; 4) pellets with 40% fines; or 5) pellets with 60% fines. Animal performance was not affected by diet form (meal vs. pellet). The presence of fines in the pellets tended to decrease feed efficiency ($P < 0.09$) but did not influence ADG.

In another recent finishing pig study (Wondra et al., 1994), pelleting improved ADG by 4% and G/F by 6%. Feed intake was not affected.

Nearly all broiler and turkey feeds are pelleted. It is well established that pelleting improved both ADG and F/G compared to meal diets. However, recent interest in the broiler industry centers on the effect of pellet quality and dietary fines on bird performance. Scheideler (1991) found that broilers fed 75% pellets (25% fines) had substantially better feed conversion than birds fed 25% pellets (75% fines). Selective feeding and nutrient density changes because of selective feeding were noted during the study.

Turkeys appear to be more sensitive to pellet quality and fines. Several studies have indicated that pellet fines decrease turkey performance (Proudfoot and Hulan, 1982; Salmon, 1985; Moran, 1989; Waibel et al., 1992). Moran (1989) showed a decrease in growth and performance when reground pellets were fed, compared to whole pellets. Proudfoot and Hulan (1982) found decreased turkey performance as fines were increased from 0 to 60%.

Current Issues - Pelleting And Pellet Quality

Perhaps the most challenging issue is the development of the high shear conditioning concept (expanders). Many European manufacturers have incorporated expanders in their pelleting systems over the past decade. Over the past two or three years, the U.S. industry has begun to investigate high shear conditioning as a way to improve pellet quality, mill throughput, feed sanitation, and, ultimately, animal performance. Peisker (1993a,b; 1994) has presented data on the applicability of high shear conditioning. In a broiler feeding study (1993b), the author reported that ME was increased by expander conditioning by nearly 5% in a broiler diet. Significant improvements were found in fat and starch digestion. In an earlier report (Peisker, 1993a), the effect of expansion on starch gelatinization was demonstrated. In general, expansion conditioning doubled the level of gelatinization in feed.

The use of high shear conditioning may have significant application in some animal applications; however, because of capital and operating and maintenance costs, the industry must use judgement in identifying those applications. At present, there are as many as one hundred of these systems on order, under installation, or installed in the U.S. Most are in the poultry industry. There is concern that many of the decisions being made are coming from a "crowd mentality" rather than being justified by factual information under domestic conditions. This is certainly an area that needs further unbiased research.

Another contemporary research topic is the elucidation and control of factors that dictate pellet quality. Formulation has a tremendous impact on pellet quality; however, it is not yet possible to include "pelletability" factors in least cost formulation programs.

Pellet quality issues can be partitioned into several individual components and their contribution of each component. These are: formulation (40%); grind (20%); conditioning (20%); die selection (15%); and cooling/drying (5%). It should be noted that 60% of pellet quality is determined before the feed reaches the pellet mill.

There is significant ongoing disagreement within both swine and poultry integrator companies concerning the affect of pellet fines on animal performance. Traditionally, most broiler integrators have held that there is no advantage of having more than 60% pellets (40% fines) in the feeder pans. Recent privately funded studies, however, have shown that contemporary broilers need at least 80% pellets for optimum performance. In general, swine integrators are finding similar results.

There are actually two issues that must be addressed if a high percentage of pellets is to be presented to the animal. First of all is pellet quality. If a high quality, durable pellet is not produced, no amount of gentle handling will assure a high percentage of pellets at the feeder. If the earlier quality allocation is accurate, we must begin to formulate diets with pellet quality in mind. To date, reliable attributes are not available to use least cost formulation to predict pellet quality. However, there are several charts available that address the relative "pelletability" of most common ingredients.

The second issue is screening operations after cooling. The trend in the U.S. for the last 10 to 15 years has been to build feed mills with no pellet screening equipment. Even in older feed mills, the screens are often by-passed or blanked with sheet metal. The net result is that all fines generated within the feed mill are not re-pelleted but are sent to loadout or bagging and carried to the farm.

Given the size and capacity of today's feed mills, fines generation is unavoidable. Fines contribute to pellet cooling problems, fat or molasses coating problems, segregation, and selective feeding.

One of the easiest and most effective ways to reduce fines delivered to the feeder is to prevent them from leaving the feed mill by screening. Screening operations are not very expensive in terms of capital equipment or operating cost, but if recycle is significant, a reduction in plant throughput equal to the recycle tonnage can be expected. This brings the discussion back to initial pellet quality. If high quality pellets are produced, recycle will be minimized.

Beyond the expander, there have been several recent innovations in pelleting. The first is known as the compactor. This device replaces the standard conditioner and serves a purpose similar to the expander. The compactor will not result in the high levels of gelatinization found with extrusion or expansion but does result in substantial improvements in pellet quality and feed hygiene.

Another novel approach is the Universal Pellet Cooker (UPC) introduced by the Wenger Corporation (Sabetha, KS). This device incorporates long-term conditioning with short-term/high temperature extrusion. The product from the machine is similar to standard pellets in appearance but of higher typical durability. This device may find wide application in specialty feed production such as horse feeds.

Future Trends

Odor Control

Given current trends in dust emission control, the next step may be odor control for feed mills. This is a huge issue with swine production units and is becoming so with many other types of production units. Technology is presently available to reduce odor emission to near zero. However, it is costly and adds no value to the product. If adopted, it will be done only under regulatory enforcement.

Hygienic Treatment of Finished Feeds

Feed manufacturers are under pressure to provide clean, pathogen-free feeds that do not contribute to the health problems of the target animal. Many manufacturers are investigating or initiating Hazard Analysis and Critical Control Point programs to accomplish that goal. Present technology in the form of hydrothermal treatment (i.e., expanders, compactors, UPCs, etc.) can do a great deal; however, it will take a great deal in the form of facility design and management skill to be successful.

In-Line Pellet Moisture and Quality Control

Several prototype and production models of in-line pellet moisture monitors are available. For successful adoption, these monitors must be given feedback control capabilities to be able to control residence time and air flow in coolers. Both temperatures and moisture content are important to shelf-life and must be controlled.

In addition to controlling coolers and dryers, in-line moisture monitoring and control is being adopted in the pre-pellet area. The concept is to refine management of feed moisture to improve pellet quality and production rate while controlling moisture shrink in the final product. At least two suppliers are actively promoting this technology in the feed industry.

Liquid Ingredients

Several major vitamin and animal drug manufacturers are pursuing developments of liquid forms of their products. In addition, several enzyme preparations, designed to improve nutrient availability, are entering the market (e.g., phytase). Often these have a use level of a few ounces per ton. There are several reasons for this development, not the least of which is a reduced cost of manufacturing and improved nutrition that may be shared with the final user.

With the development of mass flow meters and other highly reliable liquid proportioning systems, accurate application rates are feasible down to less than one-half kilogram per ton. When this is coupled with the fact that liquids, when applied properly, do not segregate, it is easy to understand the interest.

Summary

Feed cost represents the major item in the cost of animal production. Without doubt, efforts will continue to refine feed manufacturing techniques to reduce the cost of feed and to increase its value to the target animal. The possibilities for improvements in feed manufacturing are endless; however, the cost of each innovation must be carefully weighed against demonstrated improvements in animal performance.

In some cases, changes in feed manufacturing technology will be dictated, not by animal response, but by other motivations such as regulatory guidelines or health concerns. A case in point is the use of hydrothermal processes, such as pelleting, extrusion, or roasting, to reduce the microbiological load in the feed. The concern has little to do with animal health but will add significantly to the cost of feed.

References

Argenzio, R.A., and Southworth, J.S. 1975. Sites of organic acid production and absorption in gastrointestinal tract of the pig. Amer. J. Physiol. 228:454.
Behnke, K.C. 1994. Factors affecting pellet quality. Proc. Maryland Nutrition

Conference. Dept. of Poultry Science and Animal Science, College of Agriculture, Univ. of Maryland, College Park.

Beumer, I.H. 1991. Quality assurance as a tool to reduce losses in animal feed production. Adv. Feed Technol. 6:6-23.

Black, J.L. 1971. A theoretical consideration of the effect of preventing rumen fermentation on the efficiency of utilization of dietary energy and protein in the lamb. Br. J. Nutr. 25:31.

Cabrera, M.R., Hancock, J.D., Behnke, K.C., Bramel-Cox, P.J., and Hines, R.H. 1993. Sorghum genotype and particle size affect growth performance, nutrient digestibility, and stomach morphology in finishing pigs. Swine Day Report-93, Kansas St. University, Manhattan. p 129.

Duncan, M.S. 1973. Nutrient variation: Effect on quality control and animal performance. Ph.D. Dissertation. Kansas St. University, Manhattan.

Ensminger, M.E. 1985. Processing effects on nutrition. Page 529 in R.R. McEllhiney, ed. Feed Manufacturing Technology III. American Feed Industry Association, Inc., Arlington, VA.

Ensminger, M.E., Oldfield, J.E., and Heinmen, W.W. 1991. Feeds and nutrition. 2nd Ed Ensminger Pub. Co. Clovis, CA.

Goodband, R.D., and Hines, R.H. 1988. An evaluation of barley in starter diets for swine. J. Anim. Sci. 66:3086.

Hanke, H.E., Rust, J.W., Meade, R.J., and Hanson, L.E. 1972. Influence of source of soybean protein and of pelleting on rate of gain and gain/feed of growing swine. J. Anim. Sci. 25:958.

Harris, D.D., Tribble, L.F., and Orr, Jr, D.E. 1979. The effect of meal versus different size pelleted forms of sorghum-soybean meal diets for finishing swine. Proc. 27th Annual Swine Short Course, Texas Tech. Univ.

Healy, B.J., Hancock, J.D., Bramel-Cox, P.J., Behnke, K.C., and Kennedy, G.A. 1991. Optimum particle size of corn and hard and soft sorghum for broiler chicks. Swine Day Report-91, Kansas St. University, Manhattan. pp 56-62.

Hedde, R.D., Lindsey, T.O., Parish, R.C., Daniels, H.D., Morgenthien, E.A., and Lewis, H.B. 1985. Effect of diet particle size and feeding of H2-receptor antagonists on gastric ulcers in swine. J. Anim. Sci. 61:179.

McCoy, R.A., Behnke, K.C., Hancock, J.D., and McEllhiney, R.R. 1994. Effect of mixing uniformity on broiler chick performance. Poult. Sci. 73:443-451.

Moran, E.T. 1989. Effect of pellet quality on the performance of meat birds. Recent Advances in Animal Nutrition. Butterworths, London.

Ohh, S.J., Allee, G.L., Behnke, K.C., and Deyoe, C.W. 1983. Effects of particle size of corn and sorghum grain on performance and digestibility of nutrients for weaned pigs. J. Anim. Sci. 57 (Suppl.1):260 (Abstr.).

Owsley, W.F., Knabe, D.A., and Tanksley, Jr., T.D. 1981. Effect of sorghum particle size on digestibility of nutrients at the terminal ileum and over the total digestive tract of growing- finishing pigs. J. Anim. Sci. 52:557.

Peisker, M. 1993a. New conditioners for pelleting. Feed Management 44(1). p 16-21.

Peisker, M. 1993b. Conditioning for better animal performance. Feed Management 44(7). p 22-24.

Peisker, M. 1994. An expander's effect on wheat bran in piglet rations. Extrusion Communique 7(2). p 18-19.

Proudfoot, F.G., and Hulan, H.W. 1982. Feed texture effects on the performance of turkey broilers. Poult. Sci. 61:408.

Reese, F.N., Elliot, B.D., and Deaton, J.W. 1986. The effects of hammermill screen size on ground corn particle size, pellet durability, and broiler performance. Poult. Sci. 65:1257.

Reimann, E.M., Maxwell, C.V., Kawalczyk, T., Benevenga, N.J., Grummer, R.H., and Hoekstra, W.G. 1968. Effect of fineness of grind of corn on gastric lesions and contents of swine. J. Anim. Sci. 27:992.

Salmon, R.E. 1985. Effects of pelleting, added sodium bentonite and fat in a wheat based diet on performance and carcass characteristics of small white turkeys. Anim. Feed Sci. Tech. 12:223.

Scheideler, S.E. 1991. Pelleting is important for broilers. Proc. Carolina Poultry Nutrition Conf., Carolina Feed Industry Assn., Sanford, NC.

Seerley, R.W., Vandergrift, W.L., and Hale, O.M. 1988. Effect of particle size of wheat on performance of nursery, growing, and finishing pigs. J. Anim. Sci. 66:2484.

Skoch, E.R., Binder, S.F., Deyoe, C.W., Allee, G.L., and Behnke, K.C. 1983. Effects of pelleting conditions on performance of pigs fed a corn-soybean meal diet. J. Anim. Sci. 57:922.

Stark, C.R. 1994. Pellet quality I. Pellet quality and its effects on swine performance. Ph.D. Dissertation. Kansas St. University. Manhattan.

Stark, C.R., Behnke, K.C., Goodband, R.D. and Hansen, J.A. 1991. On-farm feed uniformity survey. Swine Day Report-1991. Kansas St. University. Manhattan. p 144.

Traylor, S.L., Hancock, J.D., Behnke, K.C., Stark, C.R. and Hines, R.H. 1994. Mix time affect diet uniformity and growth performance of nursery and finishing pigs. Swine Day Report-1994. Kansas Agric. Experiment Station. Kansas St. University. Manhattan.

Tribble, L.F., Harris, D.D., and Orr, Jr., D.E. 1979. Effect of pellet size (diameter) on performance of finishing swine. Proc. 27th Annual Swine Short Course. p 59. Texas Tech Univ.

Waibel, P.E., Noll, S.L., Hoffbeck, S., Vickers, Z.M., and Salmon, R.E. 1992. Canola meal in diets for market turkeys. Poult. Sci. 71:1059.

Walker, W.R., Myer, R.O, Brendemuhl, J.H., and DeGregorio, R.M. 1989. The use of pelleted or meal type prestarter diets for sow or milk replacer rearer pigs. University of Florida Swine Field Day. Dept. Anim. Sci. Res. Rep. MA-1989-5.

Wicker, D.L., and Poole, D.R. 1991. How is your mixer performing. Feed Management 42(9). p 40.

Wilcox, R.A., and Balding, J.L. 1986. Feed manufacturing problems: Incomplete mixing and segregation. Bulletin C-555 Revised. Kansas St. University Extension Service. Manhattan.

Wondra, K.J., Hancock, J.D., Behnke, K.C., Hines, R.H., and Stark, C.R. 1993. Effects of hammermills and rollermills on growth performance, nutrient digestibility and stomach morphology in finishing pigs. Swine Day Report-1993. Kansas St. University. Manhattan. p 135.

Wondra, K.J., McCoy, R.A., Hancock, J.D., Behnke, K.C., Hines, R.H., Fahrenholz, C.H., and Kennedy, G.A. 1994. Effect of particle size and pelleting on growth performance, nutrient digestibility, and stomach morphology in finishing pigs. J. Anim. Sci. 73:757-763.

Wu, J.F. 1985. Effects of particle size of corn, sorghum grain, and wheat on pig performance and nutrient digestibility. Ph.D. Dissertation. Kansas St. University, Manhattan.

MOLASSES IN FEEDS

DAVE CALDWELL
Westway Trading Corporation, Cedar Lake, IN

Molasses is the liquid residue left after condensing the sap of sugar cane or sugar beets until sugar crystals precipitate. When all the crystals that can be formed have been centrifuged off, the syrup can be as high as 85% dry matter and 90° Brix. It is too viscous to be handled by ordinary mill equipment and is referred to as high Brix molasses. The molasses is shipped in the concentrated high Brix form to reduce freight costs. Upon arrival at a terminal distribution point, it will be diluted down to (in the U.S.) a standard of 79.5° Brix, which is still quite thick, but which is possible to handle in many mills.

The ASFCO (American Society of Feed Control Officials) definition of cane molasses is:

> "A63.7 Cane molasses is a by-product of the manufacture of
> sucrose from sugar cane. It must contain not less than 43%
> total sugars expressed as invert. If its moisture content
> exceeds 27%, its density determined by double dilution must
> not be less than 79.5° Brix."

Measurements of Molasses

Brix is a measure of the **specific gravity** of a liquid. At 79.5° Brix the specific gravity is 1.41, which is equivalent to 11.75 **lb/gallon**.

Brix is tested by diluting a representative sample of the molasses with an equal weight of water, letting the mixture stand for about 20 minutes to allow entrained air to escape, and observing the level of the liquid on the stem of the hydrometer. The reading is then doubled to allow for the dilution with water.

Volume. Most liquid meters measure volume which is then converted to a weight based on the density of the liquid. This can lead to **errors in metering.** If one is using 79.5° Brix molasses, it would be logical to set the mill meter at 11.75 lb/gallon. However, the result will be incorrect. **Entrained air** from the turbulence of pumping and mixing will always be present in varying amounts, causing the effective density of the molasses to be less than indicated by the Brix. Variance can be 1/2 lb or more/gallon, or about 5%. The result will be less molasses applied than intended. It is a good practice to adjust the meter by filling a container of known volume. This should be done at least quarterly.

49

Temperature effects on molasses volume changes are not significant. Molasses does expand with increasing temperature, but the coefficient of expansion is small (about 0.0007 lb/gallon/°F), and is overwhelmed by the effect of entrained air. In fact, the increased viscosity of colder liquids will tend to hold more entrained air causing an increase in volume, the opposite effect of contraction due to the cold.

Viscosity is the measure of resistance to flow. At room temperature, one extensive industry survey found molasses viscosities ranging from 1,100 - 7,150 cps (centipoids/second) depending on the source. Cane variety and processing can affect viscosity. Heating molasses reduces viscosity, which is why it may be desirable to have a preheat tank in a mill to help improve the mixability of the liquid product. As for the effect of cold, we in the molasses business could retire if we got a nickel every time we heard the term "as thick as molasses in January," but it is true that very cold molasses can be so viscous that it can appear frozen. Viscosities can increase by 10 or more times with a 50 degree drop in temperature.

Chemical Assay of Molasses

Dry Matter — The most important chemical assay of molasses is its dry matter content. Obviously, nutrients can be found only in the dry matter fraction. Dry matter is measured by several means:

> Refractometer — Light passing through a liquid is bent, or refracted. The amount of refraction increases with density of the liquid. Increasing density is directly related to increasing dry matter content. The refractometer gives the least accurate of the estimates of dry matter in a liquid, but is the quickest and least affected by technique.

> Brix — Brix is a measure of density, and is therefore related to dry matter content. However, different solutes have different densities, so two liquids at the same Brix can have slightly different dry matter content. This method gives a slightly more accurate estimate of dry matter content than refraction, but is more easily affected by technique.

> Drying — This is the most common laboratory method of determining dry matter. It usually involves use of an oven, but some infrared heater/balance instruments are being used. This method will produce drastically incorrect results on molasses and other liquids if the oven or heater is set at too high a temperature. The oven should be at no more than 65°C because excessive temperature can drive off non-water components and can cause reactions that liberate compounds other than water. I have seen commercial

laboratories report dry matters as much as 15% too low when excessive oven heat was used. Correctly done, this method is very accurate.

Karl-Fischer — This method measures the reaction of the water in a liquid with a chemical mixture. The extent of the reaction is directly related to the moisture content. The instrument is much more expensive than that used in any other method, but it gives rapid results and is very accurate.

Sugars — Total Sugars as Invert (TSAI) — Even though sugar is the primary material extracted from the condensed syrup, a great deal of unextractable sugar is still left in molasses. The sugar level remaining in molasses is generally expressed as TSAI.

What does TSAI mean? Most naturally-occuring sugars exist as single sugar molecules (monosaccharides), or as disaccharides (two sugar molecules joined together). The ratio of monosaccharides and disaccharides in molasses and other liquids varies. In order to compare molasses with different proportions of monosaccharides and disaccharides, all the disaccharides are hydrolyzed to monosaccharides [by the addition of a molecule of water] is added to replace the bond between the two sugars.

$$C_{12}H_{22}O_{11} + H_2O \longrightarrow C_6H_{12}O_6 + C_6H_{12}O_6$$

Atomic	342 sugar+	360 sugar
Weights	18 water	

The term "invert" refers to the fact that the plane of polarized light is rotated when passed through a disaccharide sugar solution. When the disaccharides are hydrolized to monosaccharides, the direction of rotation is reversed or "inverted," thus the term Total Sugar as Invert.

Several methods are used to measure sugar content. The most widely used is the Lane-Eynon, a volumetric/colorimetric method. Coming into more widespread use is individual sugar analysis by HPLC (high pressure liquid chromatography).

Ash — The mineral matter in molasses can be measured by direct incineration. It commonly amounts to about 6-10% of the dry matter.

Protein — There is a small amount of natural protein in cane molasses (about 3%) and a slightly larger amount in beet molasses (8%). In some cases, depending on origin and processing, cane molasses has contained as much as 10.6% protein. Protein content is projected from the nitrogen content obtained by the same wet chemistry means used for dry feedstuffs. NIR has not been used for liquid assay to my knowledge.

Fiber — There is essentially no measureable fiber (crude, ADF, NDF) in molasses.

Fat — Low levels of fat may be found in molasses, on the order of 1-2%. These assays are normally done by acid hydrolysis.

Feed Energy — Some early digestion trials with high levels of molasses resulted in low digestibility values. This was due to the laxative effects of feeding impractical levels of molasses (more than 20%) and led to incorrect energy values for molasses which persist even today. More recent metabolism work in ruminants has shown that at normal inclusion rates (4-10%), the organic matter in molasses is almost completely utilized. This is in accordance with the fact that there is no structural fiber to impede microbial or enzymatic access to the organic compounds in the non-ash fraction. Thus, recent reviewers have found the energy content of molasses for ruminants to be 98% of the dry matter minus the ash content. Equine studies indicate that nonstructural carbohydrates are well utilized by the horse, but there is very little information on the energy value of molasses as such in the horse.

Other nutrients — Wet chemistry methods are reliable for the macro- and microminerals.

Typical Composition

The following is compiled from extensive corporate molasses databases.

Typical Values for Cane Molasses at 79.5° Brix
(As fed basis)

Nutrient	Unit	Low	High	Average
Dry Matter	%	67.5	74.6	71.5
TSAI	%	44.9	50.2	47.2
Protein	%	1.2	10.6	3.5
Ash	%	5.9	13.4	8.5
Est. TDN - Cattle	%	58.2	66.0	62.9
Calcium	%	0.02	0.84	0.57
Chlorine	%	0.33	2.27	1.19
Magnesium	%	0.12	0.58	0.33
Phosphorus	%	0.05	0.48	0.10
Potassium	%	1.24	2.64	3.98
Sodium	%	0.05	0.48	0.21
Sulfur	%	0.34	1.31	0.86
Cobalt	ppm	1.5	4.23	2.45
Copper	ppm	6.6	68.4	14.0
Iodine	ppm	-	-	-
Iron	ppm	145	640	297
Manganese	ppm	2.1	67.1	28.3
Zinc	ppm	7.5	37.3	13.1

Storing and Handling Liquids

Most practical situations will have a mixture of best, intermediate, and worst conditions. The closer one can get to the "best" side of the chart, the better the results.

	Best		**Worst**
Tank Orientation	Vertical		Horizontal
Tank Shape	Cylinder w. cone	Cylinder	Box
Tank Diameter	<1/2 liquid depth		>liquid depth
Tank Material	Poly	Steel	Concrete
Tank Location	Inside, near mixer		Outside, distant
Liquid Storage Time	< 2 weeks		> 3 months
Min. Storage Temperature	55° F		> 10°F
Max. Storage Temperature	80° F		> 90° F
Agitation	Mechanical	Recirculation Air	None
Pump Type	2@ + gear		1" Centrifugal
Pump Location	Beside/Below Tank		Beside Mixer
Pump Maintenance	Monthly		Huh?
Meter	Checked Monthly		Never
Application Rate Control	Automated + Eyeball	Automated	Eyeball
Lines	2@ + Insulated		1" Bare
Line Bends	45° Elbows		90° Elbows
Line Shape	Straight		Many bends
Cleanout	Annual		Huh?

A setup that sits mostly not on the "best" side of the chart can still function, but should be limited to simple liquid blends that are low viscosity at all temperatures. Operations on the "best" side will be able to use suspensions, high fat blends, etc.

Liquid Blends

Molasses is the primary liquid ingredient used in feed mill liquids, but there are many blends available. In fact, the majority of mills use blends instead of straight cane molasses.

Other liquid ingredients that are available as part of mill blends are:

Corn steepwater
Condensed molasses solubles
Soy solubles
Distillers solubles
Brewers solubles
Lignin sulfonate
Condensed whey

Inclusion of these ingredients has become popular because the blends are lower viscosity and easier to handle. Generally, the blends are sold on the basis of dry matter content, ranging from 62% to 70%. They require less energy for mixing, and result in less buildup on equipment than straight cane molasses.

Lignin sulfonate and whey are used in blends to aid pellet binding. Corn steepwater adds some natural protein to the mix. Brewers and distillers solubles and some of the condensed molasses solubles are derived from fermentation processes and are high in B-vitamins and amino acid precursors.

Mold inhibitors can be included in the molasses blends, though they should be considered to be only part of the mold inhibition program.

Flavors, vitamins, phosphorus and trace minerals can be added if desired.

Fats are being added to some mill blends more for the effect they have on the appearance and handling characteristics of finished feed than for nutritional benefit. Texturized feeds retain a moist feel and appearance when treated with a molasses blend containing as little as 3% fat. Cold weather handling of texturized feed is dramatically improved by using a fat-containing blend. Fat-containing products are more demanding and are best used by mills with equipment fitting the left side of the Storing and Handling chart.

Why Use Liquids

Molasses and liquid blends:

- Improve palatability
- Reduce dust
- Reduce sorting
- Aid pelleting
- Help maintain moistness
- Help extend shelf life
- Improve mixing integrity
- Improve winter handling
- Carry additional nutrients

References

Official Publication - Association of American Feed Control Officials - 1997

The Analysis of Molasses - Pacific Molasses Company - 1986

Westway Trading Corporation - internal files - National Molasses Company survey

Nutritional Requirements of Dairy Cattle - 1989. National Research Council, National Academy Press. Washington, D.C.

Berger, L. - Determining the Energy Value of Liquid Feed Ingredients - 1997 Liquid Feed Symposium - American Feed Industry Association

Lofgreen, G.P and K.K. Otagaki, 1960, The net energy of blackstrap molasses for fattening steers as determined by a comparative slaughter technique. J. Anim. Sci. 19:392.

Morrison, F.B., 1956, Feeds and Feeding (22nd. ed.) The Morrison Publishing Company

Nofziger, J. - Determining the Nutritive Energy of Molasses -1995 Liquid Feed Symposium - American Feed Industry Association

Scott, et. al. - Use of Chemical Composition or Near Infrared Reflectance Spectros copy to Predict the Gross Energy Content of Cane Molasses - 1990 Animal Sciences Research Report - Oklahoma State University - pp 147-153

GRAIN PROCESSING FOR HORSES: DOES IT PAY?

R.J. COLEMAN
University of Kentucky, Lexington, KY

The question of whether grain for horses should be fed whole or rolled has been an issue for horse owners for at least 100 years. Horse owners believe that horses digest processed grain better than whole grain because they readily observe hulls in the manure of horses fed whole grain. However, this observation is not a good indication of grain digestibility. When the grain is processed, there is still undigested material in the manure but it is just not as easy to identify.

Processed grain, for most horsemen, means grain that has been dry rolled, crimped, steam rolled or steam flaked. For the purpose of this discussion, the focus will be on these processes and how they affect the horse's ability to extract nutrients from the grain.

The reason grain is processed is to alter the physical form of the grain to improve the availability of nutrients. With processes such as rolling or crimping, the particle size is changed, thereby increasing the surface area to allow for greater exposure of the feedstuff to digestive enzymes. When the process includes heat, the starch in the grain is gelatinized and this may increase starch digestion.

Researchers in the early 1900s, evaluating rolled grain versus whole grain, reported improvements in growth performance of horses fed rolled grain. Morrison et al. (1919) suggested a 5-6% improvement with working horses fed crushed oats. The horses on crushed oat diets were fed at 95% of the grain intake of the whole grain fed horses yet still had better weight gains, and were observed to be in better condition. Caine (1931) reported 24% faster gains on 21% less feed with horses fed crushed versus whole grain. The result of this research has been the foundation of the recommendation that rolling or crushing oats will result in a 5-7% improvement while processing hard grains such as barley and wheat will result in even greater improvement. These early reports suggest that processing grain will result in faster growth and better utilization of the processed grain. However, more recent research that has looked at either growth performance or nutrient digestibility suggests that processing of grains fed to horses does not result in improvements in performance or nutrient digestion that justify the cost of processing.

Hintz et al. (1972) reported the effects on digestion of crimped oats versus whole oats. The results (Table 1) indicate that there is no improvement in digestibility of dry matter, crude protein, or neutral detergent fiber. French researchers comparing oats, corn or wheat fed either whole or crushed indicated no differences due to processing in the parameters measured (Table 2) (Wolter et al., 1982). Meyer et al. (1993) compared processing methods and their effects on prececal starch digestibility of oats and corn. This research indicated that in order to increase starch digestibility in the small intestine, a significant increase in surface area or greater disruption of the starch granules by heat is required, when feeding corn. Dr. Meyer's work shows that the processing of oats by rolling does not improve prececal starch digestion.

Table 1. Comparison of digestibility of whole and crimped oats.

	Digestibility		
	Dry Matter	Crude Protein	Neutral Detergent Fiber
Oats			
Whole	73.2 ± 4.	85.6 ± 2.7	36.4 ± 4.0
Crimped	75.8 ± 2.2	84.7 ± 3.0	39.2 ± 4.5

Hintz et al. (1972)

Table 2. Apparent digestibility of diets containing oats, corn, wheat*.

	Oats		Corn		Wheat	
	Whole	Rolled	Whole	Crushed	Whole	Rolled
Dry Matter, %	72.7 ± 3.0	74.0 ± 1.8	76.9 ± 1.8	76.6 ± 2.6	75.4 ± 5.2	74.4 ± 7.1
Crude Protein, %	75.0 ± 4.1	77.8 ± 3.2	59.8 ± 10	58.3 ± 10.6	60.0 ± 2.4	58.0 ± 2.35
Organic Matter, %	74.0 ± 3.4	75.4 ± 1.9	78.8 ± 1.8	80.0 ± 3.4	77.0 ± 5.1	77.0 ± 6.4
Starch, %	99.3	99.3	99.6	99.7	99.7	99.0

*Values are means ± standard deviations.

Wolter et al. (1982)

Research in Alberta compared whole oats to dry rolled oats, and whole barley to dry rolled barley fed to mature horses at maintenance (Coleman et al., 1985). In addition to processing, the oat diets were fed at two different levels of intake. The processing of the oats did not improve dry matter or energy digestibilities (Table 3). As well, dry rolled barley had similar digestibilities for dry matter and energy as whole barley (Table 3).

Table 3. Dry matter and energy digestibilities for whole oats, rolled oats, whole barley or rolled barley.

Grain	Daily Grain Intake	DM Digestibility %	Energy Digestion %
Whole Oats	2.25	55.4[a]	57.5[a]
Rolled Oats	2.25	53.9[a]	57.9[a]
Whole Oats	4.5	63.5[a]	63.0[a]
Rolled Oats	4.5	62.9[a]	62.1[a]
Whole Barley	4.5	69.3[b]	67.3[b]
Rolled Barley	4.5	70.5[b]	69.8[b]

All diets included alfalfa cubes as forage component. Coleman et al. (1985)
[ab]Values in each column with different superscripts are significantly different (P>0.05)

Processing grain for young growing horses has been reported to improve gains and feed utilization. In a growth trial, 60 yearling colts of mixed genetic background were fed diets based on oats (whole or rolled) or barley (whole or rolled). Grain was processed by dry rolling for both grain treatments. The diets were 80% concentrate, 20% forage and the horses were fed to appetite. The horses on the oat-based diets gained 0.93 and 0.92 kg/day for the whole and rolled groups respectively (Table 4). These differences were not significant. The horses fed the barley diets had gains of 0.70 kg/day for the whole barley and 0.86 kg/day for the rolled barley. The difference in gain due to processing of the barley diets was significant (P<0.05) (Coleman, unpublished data). Feed:gain ratios for the treatments in this trial were not significantly different (Table 4).

Information on the effects of processing cereal grains for horses is limited. For oats, it appears that processing does not improve growth performance or improve nutrient availability. Frape (1986) suggested that if processing increased the cost of the oat grain by more than 10%, it could not be justified. The research available would support this and the information presented here supports that processing oats for the reasons of improving nutrient utilization is not justified.

Table 4. The effect of rolling oats or barley on the daily gains and feed: gain ratio of yearling horses[1].

	Oats Whole	Rolled	Barley Whole	Rolled	SEM[2]	Probability
Daily Gain	0.92[a]	0.93[a]	0.70[b]	0.86[a]	0.05	0.04
Feed:Gain Ratio	10.0	9.5	11.90	10.5	0.62	0.11

[1]Values are least square means.
[2]Pooled standard error of the mean.
[ab]Values within a row with unlike superscripts are significantly different

With other grains such as barley, corn, and wheat, the use of a processing method such as dry rolling or crimping does not result in significant improvements in nutrient utilization. In order to improve starch availability in corn, Meyer et al. (1993) suggests that grinding or the addition of heat are effective in increasing starch digestion prececally while only cracking the grain does not. This may also be true for barley in that a greater disruption of the starch by using a heat processing method may be required for improved digestion.

Conclusions

The reason for processing cereal grains for horses is to improve nutrient availability. Of particular importance is the availability of energy because grain is fed primarily to supply energy.

Therefore, should you process grain for horses?

1) The processing of oats has little effect on nutrient availability and is not recommended from a nutritional point of view.
2) Dry rolling barley did not improve energy or dry matter digestibility in mature horses but did improve daily gain in yearlings. Use of a heat process such as steam flaking may improve barley even further and would be justified, particularly if the feed is used for young horses.
3) Grinding or heat processing of corn significantly increases the availability of the starch fraction and should be considered. Processing corn by just cracking it does not improve nutrient availability enough to justify the additional cost.
4) Feeds for young horses under 1 year of age should include processed grain. (If the feed is a creep feed, a pelleted product would be the best). Processing of grain for older horses with teeth problems is recommended.

The use of rolled, crimped or steam flaked grain will continue in feeding horses. The reasons for its use will not reflect advantages in nutrient availability but will be due to marketing and production of commercial feeds. There will be the continued concern by horse owners when whole grains, particularly oats, are included in the concentrate mixture, because hulls will be visible in the manure. It is important to remember that most of these visible hulls are just that, only hulls which are poorly digested. Even with rolled or crimped oats, hulls are in the manure; however, they are not easily distinguished from the rest of the manure. The process selected and the cost of processing grain must be evaluated in relation to improvements in performance. For most feeding situations, these additional costs are not justified. Therefore, the challenge when supplying horse feeds is to provide sound, nutritionally correct feeds in a form the consumer wants to buy.

References

Caine, AB (1931) Rolled vs whole oats as a feed for draft colts. In: Proceedings of the American Society of Animal Production 31:200.

Coleman, RJ, JD Milligan and RJ Christopherson (1985) Energy and dry matter digestibility of processed grain for horses. In: Proceedings of the 9th Equine Nutrition and Physiology Symposium, East Lansing, Michigan.

Frape, DL (1986) Equine Nutrition and Feeding. Published by Longman Group, UK Limited.

Hintz, HF, HF Schryver and JE Lowe (1972). Digestion in the horse in feedstuffs, July 2, 1972.

Meyer, H, S Radicke, E Kienzle, S Wilke and D Kleffken (1993). Investigations on Preileal Digestion of Oats, Corn and Barley Starch in Relation to Grain Processing. In: Proceedings of the 12th Equine Nutrition and Physiology Symposium, p. 92. University of Florida, Gainsville,Florida.

Morrison, FB, JG Fuller and G Bohstedt (1919). Crushed versus whole oats for work horses. Wis. Agric. Exp. Stn. Bull. 302, p. 63.

Ott, EA (1973). Effects of Processing Feeds in Their Nutritional Value for Horses in the Effect of Processing on the Nutritional Value of Feeds, National Academy of Sciences.

Wolter, R, JP Valette, A Durix, JC Letourneau, and M Carcelen (1982). Compared digestibility of four cereals (oats, barley, maize and wheat) according to physical form in ponies. Ann. Zootech 31(4)p. 445.

PARTITIONING DIGESTION IN HORSES AND PONIES

DEREK CUDDEFORD

Edinburgh University, Scotland

Introduction

The only certain way to partition digestion between different parts of the digestive tract is to cannulate each compartment and then measure the exact rate of degradation of a feed therein. The late Frank Alexander used cannulated ponies (Alexander and Donald, 1949) to determine the digestive processes operating within the large intestine and revived interest in the digestive physiology of horses. An alternative approach to cannulation was to slaughter animals and then tie off segments of the gut (Hintz et al., 1971). However, this drastic approach has limited value in terms of understanding the dynamics of the digestive process. Ileal cannulation, together with the use of chromic oxide as a marker, was the method employed to examine digestion of starch in the small and large intestine during the late 1970s and 1980s (Potter et al., 1992). Also cannulation of the terminal jejunum of horses was undertaken to determine preileal starch digestion of whole, rolled or ground oats and maize (Kienzle et al., 1992). More recently, both cecal and colonic cannulations have been performed in the same animal (Drogoul et al., 1995) to simultaneously assess feed degradation in different compartments of the horse's large intestine. Easy access to the rumen of both sheep and cattle has allowed ruminant nutritionists to characterize the digestive process in these animals in considerable detail.

The relative ease of cannulating ruminants and the durability of the animal model have enabled research groups all over the world to develop and refine techniques that allow significant advances to be made in terms of feed evaluation and in defining animal requirements. The relative paucity of equine experimental models and the apparent reluctance to transfer ruminant methodologies to the equine are in part responsible for the lack of progress in the field of equine nutrition. Only recently have techniques that have been developed in pigs, cattle and sheep been adapted for use with equids. These fundamental techniques are the *in situ* incubation method (Mehrez and Orskov, 1977), the mobile nylon bag method (Sauer et al., 1983) and the use of markers, singly and in combination.

In Situ Methods

These methods were developed to measure the degradation of feeds that were incubated in the rumen. Mehrez and Orskov (1977) evaluated the artificial fiber bag technique for assessing the proportions of dietary dry matter and nitrogen which "disappeared" in the rumen. They found that the technique was satisfactory as a simple and rapid guide for measuring nutrient disappearance from Dacron bags containing feeds that were suspended in the rumen. These original bags were hand-sewn, using material from old, discarded parachutes. This method was developed by Orskov and McDonald (1979) in order to estimate protein degradability in the rumen. Importantly, they weighted the incubation measurements according to rate of passage. This was achieved by conducting a separate experiment in which the rate of passage of the protein source was measured using a marker, chromium. Potential degradability, 'p', was related to incubation time, 't', by the following equation:

$$p = a + b(1 - e^{-ct}) \qquad (1)$$

where 'a' is the rapidly degradable (soluble) fraction, 'b' is the slowly degraded fraction, and 'c' is the rate at which 'b' is degraded. Orskov and McDonald (1979) showed that, by measuring the rate of passage from the rumen to the abomasum with chromium, a rate constant, 'k', could be derived. Thus, allowing for rate of passage, the percentage degradation, 'P', was shown to be:

$$P = a + [bc/(c+k)](1 - e^{-(c+k)t}) \qquad (2)$$

by time, 't', after feeding. As t increases, this tends to the asymptotic value a+bc/(c+k), providing an estimate of degradability under the prevailing feeding conditions. Equation 1 provides an empirical fit to incubation data and a, b and c are constants that can be fitted by an iterative, least-squares procedure. There are two important underlying assumptions to the above and these are:

1. that compartment volume remains constant;
2. that rate of passage of untreated food particles is the same as that of chromium-treated particles.

While comminution during ingestion and rumination will have the same effect on treated or untreated particles, chromium-mordanting renders particles completely indigestible and thus they will not be fragmented by microbial action.

The *in situ* technique has been modified for use in the horse cecum as described by a number of authors (Applegate and Hershberger, 1969; Miraglia et al., 1988; Drogoul et al., 1995). While the different procedural aspects that influence the results obtained using this technique in ruminants have been extensively studied and were recently reviewed by Huntington and Givens (1995), these aspects have not been fully evaluated in equids. However, Hyslop et al. (1999a) have recently reported a methodology based on the use of 6.5 x 20 cm, *in situ* bags, containing 16mg of feed per cm^2, incubated in the ceca of mature pony geldings (mean weight 285 kg). A complete exchange method (Paine et al., 1982) was used to evaluate two different incubation sequences, forward (3, 5, 16, 8, 24, 48h) and reverse (48, 24, 8, 16, 5, 3h). Degradation profiles were shown to be sensitive to incubation sequence in contrast to the findings of Huntington and Givens (1995), who showed no effect of incubation sequence on the degradability of hay, soya or fishmeal in cattle and sheep.

Digesta reaching the cecum of the horse may be more homogeneous than that entering the rumen because it will have been thoroughly chewed and exposed to the digestive influences of both the stomach and small intestine. However, the small size of the cecal pool means that variation within it, due to outflow to the ventral colon and inflow from the terminal ileum, could account for some of the sensitivity measured by Hyslop et al. (1999a). The equine cecum accounts for only about 16% of the total gastrointestinal tract in contrast to the rumen, which may represent up to 70% of the total tract (Argenzio, 1993). Additionally, digesta outflow from the cecum has been measured at 20% per hour (k= 0.20) (Hintz, 1990) which is much faster than that measured from the rumen; the latter has been assessed at between 2 and 8% per hour (k= 0.02 to 0.08) (Agricultural and Food Research Council, 1992). Rapid outflow from the cecum may lead to a less stable microbial environment. Furthermore, the microbial ecosystem within the cecum has to cope with two (Goodson et al., 1988) to fourfold (Argenzio et al., 1974) changes in cecal volume which may arise from changing the nature of the diet or in relation to the time of feeding. Clearly, the foregoing factors could have a major impact on gut activity, digesta mixing, microbial populations and their activity, thereby affecting the *in situ* degradation of foodstuffs.

Mobile Nylon Bag Methods

The concept of enclosing feed in a container and then allowing the container to pass through the digestive tract is not new. In 1782, Spallanzani enclosed bread and meat in linen bags and then swallowed them; the bags were recovered in the feces within 24 h and the contents had "disappeared" (cited by Sauer et al., 1983). Petry and Handlos (1978) orally administered small nylon bags containing feed to pigs, but unfortunately they were retained in the stomach. This problem of gastric retention was overcome by Sauer et al. (1983) who inserted 25 x 40 mm monofilament nylon bags (50 µm pores) through a cannula directly into the duodenum. These authors concluded that the technique could be used for the rapid determination of protein digestibility. Hvelplund (1985) used 60 x 60 mm polyamide bags (9 and 22 µm pores) and introduced them via a cannula into the duodenum of dairy cows. Some bags were recovered from an ileal cannula and the rest from the feces. Hvelplund (1985) concluded that the technique showed promise as a tool to predict digestibility in the small intestine of ruminants. Independently, Macheboeuf et al. (1996) and Hyslop and Cuddeford (1996) used mobile nylon bags to study nutrient disappearance throughout the digestive tract of horses and ponies respectively. The latter authors tested a range of bag sizes and showed that large bag sizes (19 x 110 or 19 x 55 or 45 x 45 mm) resulted in transit times in excess of 100 h and as a result, there were large feed constituent disappearances. Bags which gave transit times and feed disappearances in accordance with expectation measured 10 x 60 mm; the pore size was 41 µm. This was the size of bag used by Macheboeuf et al. (1996), Moore-Colyer et al. (1997), Hyslop et al. (1998) and McLean et al. (1999). Moore-Colyer et al. (1997) used the mobile nylon bag technique to partition fiber degradation in the digestive tract of cecally fistulated ponies. Bags were filled with 350 mg of a dietary fiber source (unmolassed sugar beet pulp or hay cubes or soya hulls or an oat hull:naked oat mixture-50:50) ground to pass through a 1mm screen. On two consecutive mornings, 20 bags were administered directly into the stomach of each pony using a nasogastric tube. A magnetic capture device was placed in the cannula just posterior to the ileo-cecal valve. Each mobile bag contained 2 x 100 mg steel washers so that as the bags entered the cecum, they were captured. Ten to 16 bags were captured in this manner and the balance allowed to continue through the gut and were collected in the feces. Bags that were recovered entering the cecum, 1 to 8 h after dosing, provided data on disappearance following a range of incubation times in the prececal part of the gastrointestinal tract. Degradation profiles were fitted to the DM losses from the mobile bags using the same models that were applied to *in situ* disappearances (Orskov and McDonald, 1979; Dhanoa, 1985). This experiment also allowed the calculation of both prececal and total tract losses of non-starch polysaccharide (NSP).

The results of this study contradicted the widely held view that dietary fiber is only degraded in the large intestine of the horse. Prececal losses of NSP were 84, 111, 127 and 164 g/kg of the total tract NSP disappearance for the mixture of oat products, hay cubes, soya hulls and beet pulp. Allowing mobile nylon bags to pass through the entire length of the digestive tract of the horse and then recovering them from the feces over an extended period of time yields data that reflect a range of incubation times in the whole tract (Hyslop et al., 1998). Degradation profiles can be fitted to the losses from these bags using the same models as before.

Marker Methods

Continuously dosed markers can be used to measure digesta flow while the residence time of food particles in the tract from a discrete meal can be estimated by using a "pulse dose" of marker. Owens and Hanson (1992) have reviewed the use of markers for determining the site and extent of digestion in ruminants; no comparable information is available for equids.

External markers are preferred because they will be unique to the "pulse dose;" they should remain associated with the undigested nutrient of interest and the flow and digestion of marked fragments should be the same as that for unmarked fragments. Rare earths such as ytterbium, samarium, lanthanum and europium are alternatives to the historically popular chromium. While they are appropriate flow markers for ruminants, rare earths can be displaced from their foodstuff-binding sites by protons at low pH. This displacement is of little consequence in ruminant studies because ruminal digesta is the primary source of variation in the flow of particles; any effect of passage of rare earth-treated foods through the stomach of the horse on marker displacement is unknown. Bertone et al. (1989) assumed that a high proportion of ytterbium would be unbound in the stomach of the horse. However, they considered that this was not a "big concern" (*sic*) because, provided the marker was delivered to the cecum, it would bind to particulate matter with high affinity at the higher cecal pH.

Internal markers are ideal flow markers but since they are not unique to a given meal, flux and compartmental mass must be independently determined. Acid–detergent lignin (ADL), indigestible neutral detergent fiber (INDF) and alkanes are examples of internal markers that have been used in different species.

Markers can be dosed *per os,* directly into the stomach, using a stomach/nasogastric tube or via a cannula, which can be located in any of the different segments of the gastrointestinal tract. Liquid and solid phase markers may be administered together *per os* in whole animals in order to measure digesta passage rates; feeds can be marked using the methods of Uden et al. (1980). The major advantage of this approach is that there is no need to surgically interfere with animals. Following a pulse dose of marked feed, fecal output must be sampled over an extended period of time (about 100 h). Subsequent drying of fecal samples and determination of marker concentration yields a data set that can then have a model fitted to it in order to estimate mean compartmental residence times. The purpose of using a compartmental model is to subdivide total gastrointestinal time into residence times within those particular segments or compartments of the total gastrointestinal tract that are associated with distinct modes of food breakdown. For example, in the ruminant gastric fermentation is followed by hydrolytic digestion and then post-gastric fermentation. In contrast, in equids hydrolytic digestion is followed by a significant, post-gastric fermentation although it is acknowledged that some fermentation occurs within the stomach. To reliably determine passage rate through sub-segments of the gastrointestinal tract requires that the region be cannulated so that markers can be introduced directly. Hyslop et al. (1999b) pulse dosed chromium-mordanted feeds into the cecum of cecally-fistulated ponies and withdrew cecal digesta samples by suction at regular intervals over a 10 h period. The chromium concentration in cecal digesta samples was measured and cecal outflow rate (k) was determined by fitting simple exponential relationships of the form:

$$[Cr] = A \exp^{-kt} \qquad\qquad (3)$$

to the chromium concentration data. Cecal outflow rate (k) varied between 0.240 and 0.387 depending on the type of feed marked while the R^2 of the exponential relationship ranged from 0.717 to 0.948.

Models: the Mathematician's Dream and the Nutritionist's Nightmare!

The 1980s saw the development of a number of models that could be used to fit ruminant data based on the use of markers. The classic two-compartment model of Grovum and Williams (1973) was followed by the development of a multicompartmental model proposed by Dhanoa et al. (1988). Pond et al. (1988) discussed the applications and limitations of several models in expressing and interpreting digesta flow, and more recently, France et al. (1993) have incorporated diffusion and viscosity concepts into compartmental models. Models have increased in complexity although it is questionable whether they explain the biology of the animal any more clearly than before (for example, see Holland et al., 1998).

Models can be formulated deterministically or stochastically. The former relies on fixed input parameters and no account is taken of uncertainty, predominantly random variation, whereas the latter describes this variation so that the outcomes occur with a probability (Thrusfield, 1997). Deterministic models are preferred (J France – personal communication) because it can be argued that the variation in stochastic models could hide the inadequacies of data acquired through poor experimentation.

One of the questions that has to be addressed when deciding which model to apply to horse data is whether or not the flow of digesta is time-dependent or time-independent. Briefly, the simplest situation is the time-independent paradigm where:

1. there is complete and instantaneous mixing of influxing particles with those resident in the compartment;
2. there is an equal opportunity for escape of all particles from within the compartment;
3. there is constant inflow, outflow and compartmental mass (Ellis et al., 1994).

The above could describe the situation in the stomach, cecum, ventral and dorsal colon of the horse. However, laminar flow probably occurs in the small intestine, small colon and rectum of the horse with little mixing of particles taking place. It could thus be argued that a time-dependent model would be more appropriate for these regions of the horse's gastrointestinal tract, since the probability of passage will be greater, the longer the particles reside in these particular segments of the gut.

Time-independent passage in the horse's gut is best illustrated by the outflow of chromium-mordanted particles from the cecum which follow an exponential distribution (Hyslop et al., 1999b). Excretion of ytterbium-labelled particles in the feces of horses followed a unimodal distribution when the animals were given an oral pulse dose of marked feed (Hyslop, 1998); the data were fitted using the models of Grovum & Williams (1973) and Dhanoa et al. (1985), which rely on an exponential relationship. Hyslop (1998) has proposed that small intestinal passage rate in the horse is time-dependent and to this end, McLean et al. (1999) have modelled mobile bag data assuming a Gamma 2 time dependency (Ellis et al., 1994).

Application of Models

Hyslop (1998) has suggested that degradation profiles obtained from *in situ* studies can be combined with those from mobile nylon bag studies to provide an overall impression of feed degradation in the horse. Data from marker experiments can be fitted using time-independent models which enable the calculation of k, the rate constant for the exponential rate of digesta passage. With this information, ED can be measured as follows (Orskov and McDonald 1979):

$$ED = a + \frac{(bc)}{c+k}$$

Time-dependent rates of passage can be accounted for in the calculation of ED by using l values or rate constants appropriate to this type of passage. Ellis et al. (1994) describe these time-dependent Gamma functions which vary from Gamma 2 to Gamma 6, the Gamma 1 function representing the time-independent, exponential relationship. ED can be calculated using the Gamma 2 function as follows:

$$ED = a + \frac{(bc)}{c + [(2/MRT)0.59635]}$$

Hyslop (1998) has combined knowledge of degradation profiles with estimates of digesta passage rate using different models to partition digestion throughout the digestive tract of the horse. Assuming MRT of 3, 4 and 41 h in the small intestine, cecum and colon respectively, he calculated the loss of dry matter from sugar beet pulp (SBP) and hay cubes in different parts of the gastrointestinal tract by using the above ED equations. These losses are shown in Table 1.

Table 1. Proposed partition of dry matter (DM) degradation (%) of two feeds throughout the digestive tract of ponies.

	SBP	Hay Cubes
ED in:		
Small intestine[1]	17	31
Cecum[2]	41	13
Colon[2]	12	8

[1] time-dependent passage; [2] time-independent passage.

Conclusion

The techniques reviewed above provide a means by which feed degradation in the horse can be described. A major weakness appears to be that of accurately quantifying digesta passage rates in order to be able to weight degradation data appropriately; these rates will be affected by both the nature of the diet and the level of feeding. A further problem is that of knowing whether passage rate in different segments of the digestive tract is time-dependent or time-independent. However, the magnitude of the MRT will affect the relevance of this distinction. In order to define the partition of digestion of food entities between different parts of the horse's gastrointestinal tract, more work must be done with cannulated animals. In contrast to this approach, compartmental analysis, based on total tract marker studies, can dispense with the need for surgically modifying animals although the interpretation of this analysis, together with choice of model, are still issues that demand further examination.

Acknowledgments

Recent work at Edinburgh, in collaboration with the Welsh Institute of Rural Studies and the Institute of Grassland and Environmental Research, was funded by the Horserace Betting Levy Board (Project 612) and Dodson and Horrell Ltd.

References

Agricultural and Food Research Council (AFRC) (1992). Technical Committee on Response to Nutrients: Nutritive requirements of ruminant animals. Report No 9: Protein. Nutrition Abstracts and Reviews (Series B) 62, 787-835.

Alexander, F, and Donald, D E (1949) Cecostomy in the Horse. Journal of Comparative Pathology 59, 127-132.

Applegate, C S, and Hershberger, T V (1969) Evaluation of *in vitro* and *in vivo* cecal fermentation techniques for estimating the nutritive value of forages for equines. Journal of Animal Science 28, 18-22.

Argenzio, R A (1993) Digestion, absorption and metabolism. In: Duke's Physiology of Domestic Animals, 11th edition (ed. M J Swenson and W O Reece) pp 325-335. Comstock Publishing Associates, Ithaca, USA.

Argenzio, R A, Lowe, J E, Pickard, D W and Stevens, C E (1974) Digesta passage and water exchange in the equine large intestine. American Journal of Physiology 226, 1035-1042.

Bertone, A L, Van Soest, P J, Johnson, D, Ralston, S L and Stashak, T S (1989) Large intestinal capacity, retention time and turnover rates of particulate ingesta associated with extensive large-colon resection in horses. American Journal of Veterinary Research 50, 1621-1629.

Dhanoa, M S, Siddons, R C, France, J and Gale, DL (1985). A multicompartmental model to describe marker excretion patterns in ruminant feces. British Journal of Nutrition 53, 663-671.

Drogoul C, Faurie, F and Tisserand, J L (1995) Estimation of the contribution of the pony's colon in fiber digestion: a methodological approach. Annales de Zootechnie 44, 182 (Supplement).

Ellis, W C, Matis, J H, Hill, T M and Murphy, M R (1994). Methodology for estimating digestion and passage kinetics of forages. In: Forage quality, evaluation and utilization, Ed G C Fahey Jnr., American Society of Agronomy: Crop Science Society of America: Soil Science Society of America. Madison, Wisconsin, USA.

France, J, Thornley, H M, Siddons, R C and Dhanoa, M S (1993). On incorporating diffusion and viscosity concepts into compartment models for analyzing fecal marker excretion patterns in ruminants. British Journal of Nutrition 70, 369-378.

Goodson, J, Tyznik, W J, Cline, J H and Dehority, B A (1988) Effects of an abrupt diet change from hay to concentrate on microbial numbers and physical environment in the cecum of the pony. Applied and Environmental Biology 58, 1946-1950.

Grovum, W L and Williams, V J (1973) Rate of passage of digesta in sheep 4. Passage of marker through the alimentary tract and the biological relevance of rate-constants derived from the changes in concentration of marker in feces. British Journal of Nutrition 30, 313-329.

Hintz, H F (1990) Digestive Physiology. In: The Horse 2nd Edition (ed. J W Evans, A Borton, H F Hintz and L D Van Vleck). p 189-207. Freeman and Company, New York, USA.

Hintz, H F, Hogue, D E, Walker, E F, Lowe, J E and Schryver, H F (1971) Apparent digestion in various segments of the digestive tract of ponies fed diets with varying roughage-grain ratios. Journal of Animal Science 32, 245-248.

Holland, J L, Kronfeld, D S, Sklan, D and Harris, P A (1998) Calculation of fecal kinetics in horses fed hay or hay and concentrate. Journal Animal Science 76, 1937-1944.

Huntington, J A and Givens, D I (1995) The *in situ* technique for studying the rumen degradation of feeds: a review of the procedure. Nutrition Abstracts and Reviews (Series B) 65, 63-93

Hvelplund, T (1985) Digestibility of rumen microbial protein and undegraded dietary protein estimated in the small intestine of sheep and by *in sacco* procedure. Acta Agricultural Scandinavia (Supplement) 25, 132-144.

Hyslop J J and Cuddeford, D (1996) Investigations on the use of the mobile bag technique in ponies. Animal Science 62, 647 (abs).

Hyslop, J J (1998) Modelling digestion in the horse. Proceedings of an Equine Nutrition Workshop, Horserace Betting Levy Board, 7th September, 1998, London.

Hyslop, J J, McLean, B M L, Moore-Colyer, M J S, Longland, A C, Cuddeford D and Hollands, T (1999b) Measurement of cecal outflow rate in ponies using chromium-mordanted feeds. Proceedings of the British Society of Animal Science (In Press).

Hyslop, J J, Stefansdottir, G J, McLean, B M L, Longland, A C and Cuddeford, D (1999a) *In situ* incubation sequence and its effect in degradation of feed components when measured in the cecum of ponies. Animal Science (In Press).

Hyslop, J J, Tomlinson, A L, Bayley A and Cuddeford D (1998) Development of the mobile bag technique to study the degradation dynamics of forage feed constituents in the whole digestive tract of equids. Proceedings of the British Society of Animal Science p129.

Kienzle, E, Radicke, S, Wilke, S, Landes, E and Meyer, H (1992) Preileal starch digestion in relation to source and preparation of starch. Pferdeheilkunde, Sonderheft, 103 – 106.

Macheboeuf, D, Poncet, C, Jestin, M and Martin-Rosset, W (1996) Use of a mobile nylon bag technique with cecum-fistulated horses as an alternative method for estimating pre-cecal and total tract nitrogen digestibilities of feedstuffs. Proceedings of the European Association of Animal Production, 47th Annual Meeting, Lillehammer, Norway, p.296 (abs).

McLean, B M L, Hyslop, J J, Longland, A C, Cuddeford, D and Hollands, T (1999) Development of the mobile bag technique to determine the degradation kinetics of purified starch sources in the pre-cecal segment of the equine digestive tract. Proceedings of the British Society of Animal Science (In Press).

Mehrez, A Z and Orskov, E R (1977) A study of the artificial fiber bag technique for determining the digestibility of feeds in the rumen. Journal of Agricultural Science (Cambridge) 88, 654-650.

Miraglia, N, Martin-Rosset, W and Tisserand, J L (1988) Mesure de la digestibilité des fourrages destinés aux chevaux par la technique des sacs de nylon. Annales de Zootechnie 37, 13-20.

Moore-Colyer, M J S, Hyslop, J J, Longland, A C and Cuddeford D, (1997) Degradation of four dietary fiber sources by ponies as measured by the mobile bag technique. Proceedings of the 15th Equine Nutrition and Physiology Symposium, Fort Worth, Texas. p.118-119.

Orskov, E R and McDonald, I (1979) The estimation of protein degradability in the rumen from incubation measurements weighted according to rate of passage. Journal of Agricultural Science (Cambridge) 92, 499-503.

Owens, F N and Hanson, C F (1992) External and internal markers for appraising site and extent of digestion in ruminants. Journal of Dairy Science 75, 2605-2617.

Paine, C A, Crawshaw, R and Barber, W P (1982) A complete exchange method for *in sacco* estimation of rumen degradability on a routine basis. In: B S A P Occasional Publication No 6, p 177-178. British Society of Animal Science, P O Box 3, Penicuik, Midlothian, EH26 ORZ, UK.

Petry, H and Handlos, B M (1978) Untersuchungen zur bestimmung der verdaulichkeit von nahrstoffen und futter energie mit hilfe der nylon bentel technik beim Schwein. Archives Tierena, 28, 531-543.

Pond, K R, Ellis, W C, Matis, J H, Ferriero, H M and Sutton, J D (1988). Compartment models for estimating attributes of digesta flow in cattle. British Journal of Nutrition 60, 571-595.

Potter, G D, Arnold, F F, Householder, D D, Hansen, D H and Brown, K M (1992) Digestion of starch in the small or large intestine of the equine. Pferdeheilkunde, Sonderheft, 107-111.

Sauer, W C, Jorgensen, H and Berzins, R (1983). A modified nylon bag technique for determining apparent digestibilities of protein in feedstuffs for pigs. Canadian Journal of Animal Science 63, 233-237.

Thrusfield, M. (1997) Veterinary Epidemiology 2nd Edition, Blackwell Science Ltd, Oxford, UK.

Uden, P, Colucci, P E and Van Soest, P J (1980) Investigation of chromium, cerium and cobalt as markers in digesta. Rate of passage studies. Journal of Science, Food and Agriculture 31, 625-632.

NEW METHODS FOR ASSESSING SUBSTRATE UTILIZATION IN HORSES DURING EXERCISE

RAYMOND J. GEOR
The Ohio State University, Columbus, Ohio

There are two major goals in designing diets and feeding regimens for athletic horses: 1) to meet the caloric requirements for maintenance and the work performed; and 2) to optimize substrate availability and utilization during exercise. Although there is a substantial body of knowledge pertaining to the digestible energy requirements for working horses, the effects of different diets and pre-exercise feeding strategies on substrate utilization during exercise are less well understood. Several important questions arise when considering the effects of a dietary manipulation on substrate metabolism during exercise. First, do such manipulations alter substrate availability during exercise? (e.g. greater availability of muscle glycogen or circulating free fatty acids). Second, does the manipulation alter the mix of substrates utilized during exercise? (e.g. greater use of fat when fed a fat-supplemented diet). Finally, is there a change in the relative contributions by muscle (glycogen, triacylglycerols) vs. non-muscle (blood glucose and fatty acids) fuel sources to energy production? Particularly during prolonged athletic activities, such as endurance rides and the speed and endurance test of a three-day event, both the availability of fuel substrates and the efficiency with which these substrates are utilized can affect exercise performance. Therefore, recommendations for feeding these animals must be based on a sound knowledge of the effects of diet and feeding on substrate availability and utilization during exercise.

A number of equine studies have examined the effects of both acute (pre-exercise feeding) and chronic (diet adaptation) dietary manipulations on the metabolic response to exercise in horses. Unquestionably, the results of these studies have provided important information on the horse's response to different feeding practices. However, the majority of these investigations have used measurement of plasma and muscle substrate concentrations as the primary tool for assessment of substrate metabolism. While these measures provide a qualitative assessment of the effects of diet manipulations on substrate metabolism, they do not allow for quantitative assessment of whole body substrate utilization. Rather, techniques such as indirect calorimetry and isotopic tracer methods are necessary for quantification of rates of substrate use.

The objective of this paper is to discuss the various techniques available for assessment of substrate metabolism in the horse during exercise, with a particular emphasis on indirect calorimetry and the stable isotope tracer method.

Fuels for Energy Production During Exercise

Carbohydrate and lipid are the major fuels used by working muscles during exercise. Although there are alterations in protein metabolism during exercise, data from several species indicate that protein is used minimally for energy production. The main *endogenous* fuel reserves are present in skeletal muscle, liver, and adipose tissue. Glycogen present in liver and skeletal muscle represents the storage form of carbohydrate; during exercise, additional glucose is provided by hepatic gluconeogenesis. Two major sources of fat are oxidized during exercise: non-esterified fatty acids (NEFA) released from triacylglycerols stored in adipose tissue and transported by the bloodstream to skeletal muscle, and NEFA derived from triacylglycerol deposits located within skeletal muscle fibers. Several authors have provided estimates of the quantity of carbohydrate and fat stored in the horse. Although these estimates will vary with the breed, age, size, and training status of the horse, a 450-kg horse has approximately 3000-4000 g muscle glycogen, 100-200 g liver glycogen, 1400-2800 g muscle triacylglycerol, and 35,000-40,000 g as adipose tissue triacylglycerol (Harris, 1997). Whereas the endogenous supply of fat is virtually inexhaustible, carbohydrate stores are more limited. In human athletes, decreases in muscle glycogen content and plasma glucose (liver glucose supply) contribute to the onset of fatigue during exercise. Similarly, recent studies in horses have demonstrated that glucose availability is a limiting factor for both moderate- and high-intensity exercise performance (Farris et al., 1995; Lacombe et al., 1999). These observations highlight the need for a greater understanding of factors that govern the supply and utilization of fuel substrates in the horse during exercise.

The relative contribution of different substrates to fuel metabolism during exercise is determined by a number of factors, including the intensity and duration of exercise, the fitness of the horse, and the availability of substrates in plasma and muscle. In other species, including humans, there is unequivocal evidence that the mix of substrates oxidized during exercise is also influenced by the diet consumed. Furthermore, these diet-induced changes in substrate oxidation can have profound effects on exercise performance. A prime example is the combination of exercise training and a low carbohydrate diet that results in decreased muscle glycogen concentrations. Such depletion of the carbohydrate stores severely limits exercise capacity.

Methods for Study of Energy Metabolism in Exercise

Several methods have been used to study substrate metabolism in horses during exercise. These include the analysis of changes in plasma and tissue concentrations of substrates and respiratory gas exchange (indirect calorimetry). More recently, our laboratory has adapted stable isotopic tracer techniques for the study of glucose metabolism in horses during exercise. In this section, I will review each of these methods, with an emphasis on the advantages and limitations of the technique and application of the method for studies that examine the effects of diet and feeding practices on substrate metabolism in horses during exercise.

1. Plasma substrate concentrations

Measurement of the plasma concentrations of glucose, NEFA, glycerol, and various amino acids have been widely employed in equine studies. There are several advantages of this method. In laboratory studies involving treadmill exercise, blood samples can be readily obtained at frequent intervals. Furthermore, laboratory analysis of these samples is straightforward and relatively inexpensive. However, it is important to recognize that plasma concentrations of a substrate result from the difference between the rate of release of a substrate into plasma and its rate of removal from the plasma compartment (tissue uptake). Using glucose as an example, the plasma concentration is the net result of two simultaneous processes: (1) hepatic glucose production, with release of glucose into circulation; and (2) irreversible uptake of glucose by the tissues. Importantly, plasma concentrations provide no quantitative information on this relationship and, as a result, they provide limited insight into the dynamics of glucose production and utilization. Similarly, interpretation of changes in plasma NEFA concentrations during exercise is difficult without measuring rates of NEFA release from adipose tissue and rates of tissue uptake and oxidation.

It should be noted that, in some circumstances, examination of the changes in plasma substrate concentrations provides a *qualitative* assessment of alterations in substrate availability and use. Studies by Lawrence et al. (1995), Stull and Rodiek (1995), and Pagan and Harris (1999) have clearly demonstrated that the nature and timing of a pre-exercise meal markedly alter plasma glucose concentrations during exercise. For example, in the study by Lawrence et al. (1995), consumption of corn grain 2.5 to 3 hours before exercise resulted in a large glycemic response (Figure 1). At the start of a standardized exercise test, plasma glucose concentrations in the fed horses were 1-2 mmol/l higher when compared to trials in which no grain was fed. Conversely, during

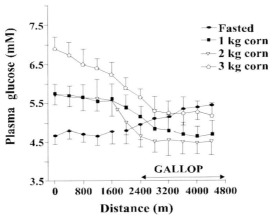

Figure 1. Plasma glucose concentration in exercising horses receiving 0, 1, 2 or 3 kg of corn grain 2.5 to 3 hours before exercise. The first 2400 m of exercise consisted of trotting and walking, while the final 2400 m was undertaken at high intensity (11 m/s). Note 1) the higher glucose concentration at the start of exercise in the horses fed corn grain and 2) that exercise resulted in a decrease in plasma glucose when horses consumed corn but not when they received the controlled treatment (Lawrence et al., 1993)

exercise there were marked decreases in plasma glucose concentrations in trials preceded by corn feeding, whereas plasma glucose was unchanged in the control (no pre-exercise meal) trial. These data provide some evidence for an increase in utilization of plasma glucose under circumstances of increased substrate availability (pre-exercise feeding of carbohydrate). However, estimation of the magnitude of this effect is not possible without measurement of glucose turnover (see below).

2. Tissue samples

Samples of skeletal muscle and, less commonly, liver have been obtained from horses before and after exercise in an attempt to quantify various aspects of metabolism in these tissues. In horses, use of the percutaneous muscle biopsy technique (most commonly, sampling of the middle gluteal muscle) has allowed delineation of some aspects of glycogen metabolism during exercise, in particular the effects of the intensity and duration of exercise on rates of muscle glycogen utilization. Given the importance of glycogen for energy production, measurement of glycogen content in skeletal muscle is an important tool in the assessment of substrate stores. Accordingly, several investigators have used this technique to determine the effects of different feeding strategies on storage and mobilization of muscle glycogen in horses during exercise (see Potter, 1998). For example, Essén-Gustavsson et al. (1991) demonstrated a 12% increase in the resting muscle glycogen concentration of horses fed a carbohydrate-rich diet when compared to control or fat-supplemented diets.

The principal advantage of the sampling technique is the ability to quantify changes in tissue substrate storage and use resulting from dietary and exercise interventions. However, interpretation of biopsy data in relation to whole body metabolism is limited by the static and local nature of the sample. As samples can only be obtained at fixed time points (e.g. before and after a period of exercise), it is not possible to determine the dynamics of muscle glycogen use. Rather, the *net rate* of glycogen utilization can be estimated in these circumstances. Furthermore, the change in glycogen content measured in one muscle (e.g. middle gluteal muscle) may not be representative of glycogen utilization in other working skeletal muscle. Even within a single muscle, differences in the fiber-type composition (percentage of type I vs. type II fibers) of samples collected before and after exercise may confound interpretation of the data. Given that the rate of glycogen utilization differs between fiber types, variance in the percentage of type I vs. type II fibers will lead to erroneous estimates of glycogen utilization rates unless single fiber analysis is performed. Finally, measurement of substrate concentrations in muscle samples does not allow for differentiation of the contributions to energy production by intra- vs. extra-muscularly derived substrates. Clearly, this is a major limitation for studies of substrate metabolism during sustained exercise when a substantial proportion of fuel substrate will be derived from non-muscle sources (i.e. liver, adipose tissue, gut).

Compared to use of the muscle biopsy technique, relatively few equine exercise studies have measured changes in liver substrate content. Collection of liver biopsy samples is an invasive procedure and it can be difficult to obtain an adequate sample, particularly after exercise when the horse's respiratory rate is high. As the liver is the primary source of blood glucose, changes in glycogen content will provide an index of the contribution by liver glycogen to blood glucose supply during exercise. However, during prolonged exercise gluconeogenesis (primarily from lactate, alanine, and glycerol) becomes the predominant mechanism for maintenance of hepatic glucose production. Of course, analysis of liver samples provides no information on the extent to which gluconeogenesis contributes to blood glucose supply during exercise. This disadvantage, together with the invasive nature of the biopsy procedure, limits the usefulness of the liver biopsy technique for equine metabolic studies.

3. Indirect calorimetry

Indirect calorimetry can be used to calculate whole body oxidation rates of carbohydrate and fat. Although the indirect calorimetry technique has been in use since the beginning of the century, there have been surprisingly few equine studies in which this method has been used for estimation of substrate utilization.

Indirect calorimeters are either closed or open. For closed systems, the horse is required to wear a tight-fitting face mask, such that all of the expired air can be collected. Although this method is the most appropriate for resting respiratory gas exchange measurements, closed mask systems severely restrict breathing patterns during exercise and are unsuitable for metabolic studies in this circumstance. On the other hand, open-circuit calorimeters are a valuable tool for equine exercise studies. With open-circuit calorimeters, the horse wears a loose-fitting face mask that does not interfere with normal breathing patterns and, unlike closed systems, does not result in the rebreathing of expired carbon dioxide. This problem is avoided by use of a large fan that draws air through the system at a flow rate several fold higher than the horse's expiratory flow. A sample of the expired air is drawn from the main stream of air flowing away from the face mask and this sample is analyzed for oxygen and carbon dioxide contents. These measurements, together with knowledge of the air flow rate through the system, allow calculation of the horse's oxygen consumption (VO_2), carbon dioxide production (VCO_2), and respiratory exchange ratio (RER). During exercise, these measurements are typically made at frequent intervals (e.g. every 30 s).

From the VO_2, VCO_2, and RER values, calculations of rates of carbohydrate and fat utilization can be made. Several simple equations have been developed for these computations. These calculations are based on the recognition that the three major classes of substrates (i.e., carbohydrates, lipid, and protein), when oxidized, consume O_2 and release CO_2 in a ratio specific for each substrate. This ratio of metabolic gas exchange in the combustion of food is termed the respiratory quotient, or RQ. The RQ is 1.0 for carbohydrates, 0.71 for tripalmitatoyl-glycerol, and 0.80 for protein (Livesey and Elia, 1988). The application of the RQ is based on the assumption that the exchange of oxygen and carbon dioxide measured at the lungs reflects the actual gas exchange from nutrient catabolism in the cell. Therefore, the respiratory exchange ratio (VCO_2 divided by VO_2)

provides a measure equivalent to the RQ. As further discussed below, this assumption is valid under steady-state exercise conditions (constant load, low to moderate intensity exercise), but less valid during higher intensity work. When the RER is at or near 1.0, carbohydrate is the sole source of substrate for energy production. Conversely, RER values approaching 0.71 imply that almost all energy is being derived from lipid sources. Quantitatively, rates (grams per minute) of carbohydrate and fat oxidation can be calculated according to the following formulas:

Carbohydrate oxidation = 4.58 VCO_2 - 3.23 VO_2 (*equation 1*)
Fat oxidation = 1.70 VO_2 - 1.69 VCO_2 (2)

The percentage of energy derived from total carbohydrate and fat oxidation and total energy expenditure (TEE), as kJ/min, can be calculated from:

%CHO = [(RER - 0.71)/0.29] x 100   (3)
%Fat = 100 - %CHO   (4)
TEE   = [(%CHO/100) x VO_2 x 21.1 kJ/l] + [(%Fat/100) x VO_2 x 19.7 kJ/l] (5)

(the values 21.1 and 19.7 represent the number of kilojoules burned per liter of oxygen consumed for 1 gram of carbohydrates and fat, respectively).

The following example is provided to demonstrate the utility of indirect calorimetry for calculation of rates of substrate utilization. In this example, we will assume that a 500 kg horse completes 60 min of treadmill exercise at a workload equivalent to 50% of its maximum rate of oxygen consumption, or VO_{2max} (a fast trot that is maintained throughout the 60 min of work). Values for O_2, VCO_2 (liters of oxygen consumed or carbon dioxide produced per min of exercise), and RER are depicted in Figure 2. During this type of exercise, there will be a steady decline in VCO_2 and RER throughout the exercise bout, indicating greater use of

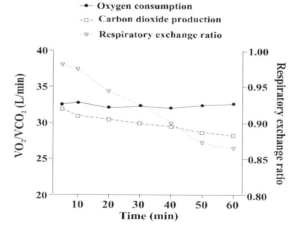

Figure 2. Oxygen consumption (VO_2), carbon dioxide production (VCO_2), and respiratory exchange ration (RER) in a horse during 60 min of exercise at a workload equivalent to 50% of VO_{2max}. Note the steady decreases in VCO_2 and RER during exercise, indicating a progressive increase in energy production from lipid sources.

lipids for energy production as exercise progresses. I have used the respiratory gas exchange values at 5, 15, 30, 45, and 60 min of exercise to quantitate this change in the mix of substrates oxidized with increasing duration of exercise. The calculations below are for the 5 and 60 min time points. Data for all time points are presented in Table 1.

Table 1. Respiratory gas exchange and substrate utilization data in a horse during 60 minutes of exercise at a workload equivalent to 60% of maximum oxygen consumption (VO_{2max}).

Time (min)	5	15	30	45	60
VO_2 (l/min)	32.6	32.2	32.4	32.6	32.7
VCO_2 (l/min)	31.9	30.8	30	29.2	28.3
RER	0.981	0.956	0.926	0.895	0.865
TEE (KJ/min)	684	668	660	641	628
CHOox (g/min)	40.3	37	32.7	28.4	24
FATox (g/min)	1.7	2.75	4.4	6.1	7.8
%CHO	93.5	84.8	74.5	63.8	53.4
%FAT	6.5	15.2	25.5	36.2	46.6

VO_2 = oxygen consumption; VCO_2 = carbon dioxide production; RER = respiratory exchange ratio; TEE = total energy expenditure; CHOox = rate of carbohydrate oxidation; FATox = rate of lipid oxidation; %CHO = percentage of TEE derived from carbohydrate sources; %FAT = percentage of TEE derived from lipid sources.

e.g.   **5 min:** CHO oxidation = (4.58)(31.8) - (3.23)(32.6)
= 40.3 g/min
Fat oxidation   = (1.70)(32.6) - (1.69)(31.8)
= 1.7 g/min
%energy from CHO = [(0.981-0.71)/0.29] x 100 = 93.4%
%energy from FAT = 100 - 93.4 = 6.6%
TEE = [(0.934)(32.6)(21.1)] + [(0.066)(31.8)(19.7)] = 684 kJ/min

60 min: CHO oxidation = (4.58)(28.3) - (3.23)(32.7)
= 24.0 g/min
Fat oxidation   = (1.70)(32.7) - (1.69)(28.3)
= 7.8 g/min
%energy from CHO = [(0.865-0.71)/0.29] x 100 = 53.4%
%energy from FAT = 100 - 53.4 = 46.6%
TEE = [(0.5345)(32.7)(21.1)] + [(0.466)(28.3)(19.7)] = 629 kJ/min

These data highlight two important concepts concerning energy metabolism during low-to-moderate intensity exercise. First, in the early part of exercise carbohydrates are the predominant substrate for energy production, with minimal contribution from lipid sources. Conversely, after one hour of work the contributions by carbohydrates and fats to energy production are almost equal (Figure 3). During longer but lower intensity work, such as that expected of an endurance horse, the proportion of energy derived from fats would be even higher. Second, the greater efficiency of energy production from fats is illustrated by the progressive decrease in the rate of energy expenditure over the course of the exercise bout (Table 1). That is, with increasing exercise duration, less energy (and, therefore, lower heat production) is required to sustain the same amount of work.

Figure 3.

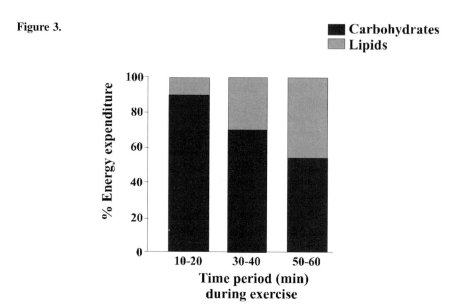

As mentioned, few equine studies have employed the indirect calorimetry method for estimation of substrate utilization during exercise. However, the power of this method for detection of the effects of different diets on substrate metabolism was well demonstrated in a study by Pagan and co-workers (1987). In that study, a 3 x 3 Latin square design was used to determine the effects of feeding diets containing different levels of carbohydrate, fat, and protein. The diets were a 12% crude protein (CP)(as fed basis) commercial horse feed (control diet), a 20% CP feed (high protein diet), or an 11% CP feed containing 15% added soybean oil (high fat diet). Each diet was fed for 4 weeks and exercise tests were performed in the 3rd and 4th weeks of each period. During a long, slow exercise test (105 min at a workload of approximately 35% VO_{2max}), values for RER were significantly lower in the high protein and fat diet trials than in the control diet between 30 min and 90 min (the last measurement) of exercise (Figure 4). These differences in values for RER indicate a remarkable increase in the use of fat for the high protein and fat diets when compared to the control diet. For example, at the 90 min time point, there was about a 50:50 contribution by carbohydrates and fats to energy production when horses were fed the control diet. In contrast, for both the high protein and high fat diets, approximately 85% of energy was derived from lipid sources during the corresponding period.

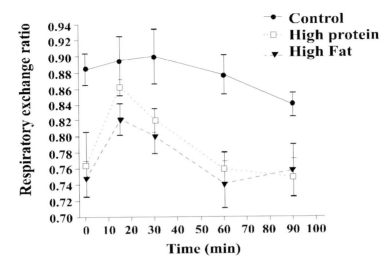

Figure 4. Respiratory exchange ration (mean ± SE) in 3 horses at rest and at 3 time points during a submaximal exercise test (trot at 5 m/s) after 4 weeks on each of three diets (control, high protein, and high fat). The lower RER values for the high protein and fat diets indicated a marked increase in use of lipid for energy production when compared to the control diet. (Pagan et al., 1987)

It should be noted that the above formulas and calculations do not include the contributions made by protein to total energy expenditure. Estimates of protein oxidation require measurement of urinary nitrogen excretion. Because the nitrogen content of mixed proteins is approximately 16%, it is assumed that each gram of urinary nitrogen represents the oxidation of 6.25 grams of mixed proteins. For equine exercise studies (and most human studies), the contribution by protein to energy expenditure is considered to be very small and is ignored for the purposes of calculating substrate oxidation rates. Therefore, these estimates are based on the *nonprotein* RER. The validity of these assumptions has not been tested in horses. However, given the technical difficulties associated with complete urine collection during exercise and the likelihood that protein oxidation makes minimal contribution to energy expenditure, use of the nonprotein RER is the most practical approach.

One of the key assumptions with the indirect calorimetry method is that pulmonary gas exchange truly reflects oxygen consumption and carbon dioxide production at the tissue level. Although this assumption is likely true for oxygen (there are negligible oxygen stores in the body), VCO_2 as measured by analysis of expired air is a reliable estimate of tissue CO_2 production only when there is minimal lactate production, i.e. low-to-moderate intensity, predominantly aerobic exercise. During high intensity exercise, when there is significant lactate production in working muscle, a portion of the CO_2 production measured in expired air will be derived from the body's bicarbonate pool. CO_2 production from this source occurs because bicarbonate is used to buffer lactic acid, with the subsequent formation of carbonic acid and release of CO_2. In this circumstance, estimates of carbohydrate oxidation based on indirect calorimetry measurements will be erroneously high. Therefore, the indirect calorimetry method is *unsuitable* for calculation of substrate oxidation during brief, high-intensity exercise.

4. Isotopic tracer methods

The isotopic tracer technique has been widely employed in metabolic studies involving a variety of species. Until the last decade, the majority of these studies involved use of radioactive tracer substances. However, given the obvious health and safety concerns associated with use of radiochemicals, there has been a sharp decline in the use of radioactive tracers for studies of energy metabolism. Rather, use of tracers labeled with *stable* isotopes (i.e. nonradioactive) has emerged as an important technique for studying substrate metabolism (Wolfe, 1992). A wide variety of metabolic processes involving carbohydrate, lipid, or protein metabolism can be studied using stable isotopes of carbon (^{13}C), hydrogen (^{2}H, deuterium), nitrogen (^{15}N), or oxygen (^{18}O). By definition, the term *isotope* refers to all forms of a given element containing different numbers of neutrons (e.g. ^{12}C and ^{13}C are both stable isotopes of carbon). For the purposes of tracer studies, the term *stable isotope* is defined as the nonradioactive isotope that is less abundant than the most abundant naturally occurring isotope. The stable isotopes most commonly used in the synthesis of metabolic tracers are ^{2}H, ^{13}C, ^{15}N, and ^{18}O (Table 2) (Patterson, 1997). For tracer studies, one or more atoms in the structure of interest is replaced with a stable isotope such as deuterium, e.g. [6,6-^{2}H] glucose is frequently used for the study of glucose kinetics. Here, two ^{1}H atoms at the 6-carbon position on the glucose ring have been replaced by two deuterium (^{2}H) atoms.

Table 2. The relative abundances of the stable isotopes of elements commonly used in metabolic studies. The isotopes of low natural abundance are used in the synthesis of tracer substances.

Element	Stable Isotope	Atom% Natural Abundance
Hydrogen	1	99.985
	2	0.015
Carbon	12	98.89
	13	1.11
Nitrogen	14	99.63
	15	0.37
Oxygen	16	99.76
	18	0.24

Before further explaining this technique, a few terms must be defined. First, when studying the kinetics of a metabolic substrate (e.g. glucose), we refer to the endogenous unlabeled substrate as the *tracee*. The *tracer* is the form of the tracee substance containing one or more stable isotopes (e.g. [6,6-^{2}H]glucose), and the *isotopic enrichment* is the ratio of tracer to tracee in blood samples. Using glucose as the example, the isotopic enrichment is the ratio of labeled to unlabeled glucose in the sample. One of the underlying principles of this method is that the tracer should not significantly change the tracee pool size (the total amount of tracee substance in the body). Therefore, a typical infusion protocol will result in about a 2-3% enrichment of the blood (i.e. in the case of glucose studies, the tracer is 2-3% of the total glucose pool). Nonetheless, highly sensitive analytical methods (gas chromatography-mass spectrometry) allow detection of very small changes in isotopic enrichment, thus permitting study of the effects of various interventions on the kinetics or turnover of the substrate of interest.

In practical terms, the stable isotopically-labeled tracer substance is administered to "trace" the kinetics of production and utilization of that substance. Two main kinetic parameters are calculated. The *rate of appearance, or R_a*, is the total rate of appearance of the substrate into the sampling pool. Physiologically, the R_a is the production rate of a given substrate. For example, for the glucose system, the R_a reflects hepatic glucose production; for studies of fatty acids, the R_a represents release of NEFA from adipose tissue and lipoproteins. Depending on the feeding state of the animal, a portion of the total R_a may also reflect uptake of the nutrient from the gut. The *rate of disappearance, or R_d*, is the rate of loss of substrate from the sampling pool. Physiologically, the R_d is the rate of irreversible tissue uptake, with perhaps a small component from excretion in urine. During exercise, it is well recognized that more than 90% of the R_d for substrates such as glucose and NEFA reflects uptake and utilization by working muscle (Wolfe, 1992). Taken together, the R_a and R_d provide estimates of the rates of production and utilization of a substrate. Thus, when compared to a static measure such as the blood glucose concentration, a significant advantage of the isotopic tracer method is the ability to monitor the dynamics of substrate metabolism during exercise. Furthermore, it becomes possible to obtain estimates of the sources of carbohydrate and lipid when measurements of plasma kinetics are combined with rates of substrate oxidation (from indirect calorimetry).

The utility of the stable isotope method for measurements of glucose turnover in the horse during exercise is illustrated in Figures 5 and 6. In this example, a primed constant-rate infusion of [6,6-^{2}H] glucose was given over a 90 min period of rest, followed by 90 min of exercise at 40% of VO_{2max}. At rest, the rate of tracer infusion was 0.22 µmol/kg body weight per min. During the transition from rest to exercise, the tracer infusion rate was doubled so as to maintain a relatively steady state in isotopic enrichment. At rest, isotopic enrichment was steady at approximately 3% (Figure 5).

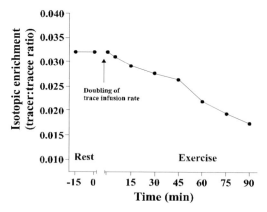

Figure 5. Plasma isotopic enrichment of [6,6-^{2}H] glucose in a horse at rest and during 90 min of exercise at a workload equivalent to 40% of VO_{2max}. Enrichment is expressed as a fraction; therefore, a value of 0.03 is a 3% enrichment. Note the steady decline in enrichment despite the doubling of the tracer infusion rate at the onset of exercise, indicating a large increase in glucose turnover.

Despite the increase in tracer infusion rate, there was a progressive decline in isotopic enrichment during exercise, providing indication of a marked increase in glucose turnover. The magnitude of this change in glucose metabolism is depicted in Figure 6; there were progressive increases in the R_a and R_d of glucose, with values at the end of exercise 6-fold higher than at rest. Importantly, note that plasma glucose concentration was unchanged throughout the trial. Therefore, the tracer method permits detection of large increases in hepatic glucose production (R_a) and tissue glucose utilization (R_d), changes not evident from examination of plasma glucose concentrations. This method offers great potential for future equine studies that examine the effects of different diets on carbohydrate, fat, and protein metabolism.

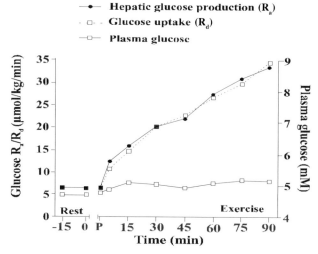

Figure 6. Hepatic glucose production (R_a), glucose uptake (R_d), and plasma glucose concentration in a horse at rest and during 90 min of exercise at a workload equivalent to 40% of VO_{2max}. Note the more than 5-fold increase in glucose production and utilization during exercise, a finding not evident on the basis of plasma glucose concentrations. [P=pre-exercise sample]

Similar to the indirect calorimetry method, the tracer technique is not suitable for assessment of substrate metabolism during high-intensity exercise when the brevity of the exercise task will limit the number of samples that can be obtained and make kinetic analysis difficult. Therefore, the isotopic tracer method is also best suited for studies involving prolonged, low-to-moderate intensity exercise. A further disadvantage of this method is the expense of the instrumentation (gas chromatograph/mass spectrometer) required for measurement of isotopic enrichment. Finally, the cost of the isotope itself can limit applications in the horse. Whereas deuterated tracers are reasonably priced, ^{13}C-labeled tracers (which permit direct measurements of rates of oxidation of that substrate) are currently cost prohibitive for use in horses.

Summary

Several different techniques can be used to study substrate metabolism in the horse during exercise. Although each of these methods has limitations, a combination of indirect calorimetry and stable isotopic tracer methods offers the greatest potential for quantitative analysis of the effects of diet and feeding regimen on substrate metabolism in horses.

References

Essén-Gustavsson B, Blomstrand E, Karlstrom K, et al. Influence of diet on substrate metabolism during exercise. In: 3rd International Conference for Equine Exercise Physiology, ICEEP Publications, Davis, CA, 1991; pp 289-298.

Farris JW, Hinchcliff KW, McKeever KH, et al. Glucose infusion increases maximal duration of treadmill exercise in Standardbred horses. *Eq Vet J* Supplement 18: 357-361, 1995.

Frayn KN. Calculation of substrate oxidation rates in vivo from gaseous exchange. *J Appl Physiol* 55: 628-634, 1983.

Harris P. Energy sources and requirements of the exercising horse. *Annu Rev Nutr* 17: 185-210, 1997.

Lacombe V, Hinchcliff KW, Geor RJ, et al. Exercise that induces muscle glycogen depletion impairs anaerobic but not aerobic capacity of horses. In: 5th International Conference on Equine Exercise Physiology, 1998, p 84 (abstract).

Lawrence L, Soderholm LV, Roberts A, et al. Feeding status affects glucose metabolism in exercising horses. *J Nutr* 123: 2152-2157, 1993.

Livesey G and Elia M. Estimation of energy expenditure, net carbohydrate utilization, and net fat oxidation and synthesis by indirect calorimetry: evaluation of errors with special reference to the detailed composition of fuels. *Am J Clin Nutr* 47: 608-628, 1988.

Pagan JD, Essén-Gustavsson B, Lindholm A, et al. The effect of dietary energy source on exercise performance in Standardbred horses. In: 2nd International Conference for Equine Exercise Physiology, ICEEP Publications, Davis, CA, 1987; pp 686-700.

Pagan JD and Harris PA. Timing and amount of forage and grain affects exercise response in Thoroughbred horses. In: 5th International Conference on Equine Exercise Physiology, 1998, p 113 (abstract).

Patterson BW. Use of stable isotopically labeled tracers for studies of metabolic kinetics: An overview. *Metabolism* 46: 322-329, 1997.

Potter GD. Manipulating energy sources for aerobic and anaerobic muscle metabolism and athletic performance of horses. *Vet Clin Nutr* 5: 10-13, 1998.

Stull C and Rodiek A. Effects of postprandial interval and feed type on substrate availability during exercise. *Eq Vet J* Supplement 18: 362-366, 1995.

Wolfe RR. Radioactive and Stable Isotope Tracers in Biomedicine. Principles and Practice of Kinetic Analysis. New York, Wiley-Liss, 1992.

THE MANY PHASES OF SELENIUM

HAROLD F. HINTZ
Cornell University, Ithaca, NY

Selenium is one of the most interesting of all nutrients. It was first known for its toxicity and was then discovered to be an essential nutrient. The power of its antioxidant activity has been widely acclaimed. The importance of selenium in immune response and as an anticarcinogen has recently received a great amount of attention. In fact, selenium has been claimed to have a protective role in at least 50 diseases of humans (Reilly, 1996). However, Reilly pointed out the evidence is not yet convincing for many of the claims. Casey (1988) rightfully stressed that the claims of many selenophiles might be exaggerated. Selenium is not a panacea. However, there is evidence that selenium can have enormous health benefits for animals and humans.

Selenium as a Toxin

Selenium was first identified as a toxin in the 1930s. Alkali disease of horses and cattle was shown to be caused by selenium. Signs in the horse included hair loss from mane and tail, sloughing of hooves, joint erosion and lameness. Blind staggers characterized by ataxia, blindness, head pressing and respiratory failure, was also thought to be caused by selenium. The mechanism of selenium toxicity is still not clear but the blocking of the function of SH groups involved in an oxidative metabolism within the cells is a likely prospect (Reilly, 1996). According to NRC (1980), the minimum toxic level for horses may vary from 5 to 40 ppm. Such a wide range perhaps relates to differences in availability among sources and other interfering factors in the environment. For example, methylselenocysteine which is found in toxic plants such as *Astragalus bisulcatus* is much more dangerous than sodium salts of selenium (Reilly, 1996).

The identification of selenium as a toxin and the knowledge of the location of high selenium soils has greatly decreased the incidence of selenium toxicity in horses and cattle. For example, Raisbeck et al. (1993) found that prior to the 1940s, reports attributed thousands of animal deaths to selenium each year in Wyoming. However, no substantiated cases were found in reports from 1947 to 1987. Raisbeck et al. contacted veterinarians in Wyoming directly in 1988 and found 4 cases of selenosis (alkali disease) in horses caused by native range and grass hay containing high levels of selenium during the previous three years. They also suggested that blind staggers in ruminants originally attributed to selenium toxicosis was in fact polioencephalomalacia. Selenosis (alkali disease) was reported in horses in western Iowa by Witte et al. (1993). The source was alfalfa hay (19-58 ppm Se) and it was hypothesized that the alfalfa extracted selenium from deeper subsoils after drought-stimulated extensive tap root development. Selenosis has also been reported occasionally because of over-zealous use of selenium supplements and water with high selenium content.

Selenium Deficiency

The essential nature of selenium was demonstrated in the 1950s. Klaus Schwarz in Germany produced liver necrosis in rats fed a brewer's yeast-based diet. Schwarz moved to the United States and continued his experiments. However, he could only produce liver necrosis when brewer's yeast from the United States was replaced with torula yeast. Selenium was identified as the factor in U.S. brewer's yeast that prevented liver necrosis (Schwarz and Foltz, 1957). Selenium was soon shown to be important in the prevention of white muscle disease in cattle, sheep, swine, and horses, and of exudative diathesis in poultry and therefore has had a major impact on animal production.

Tying-up in horses was associated with selenium deficiency in the past, but selenium does not seem to a major factor in cases of tying-up today, probably because of the widespread use of commercial feeds containing selenium. Foreman et al. (1986) reported generalized steatitis in a foal with low blood selenium values but normal vitamin E levels. Reduced reproductive performance and sudden death in horses have also been attributed to selenium deficiency (Reilly, 1996).

Other diseases such as pancreatic atrophy in poultry are caused by selenium deficiency. More recently selenium deficiency has been shown to have an involvement in Keshan disease, a juvenile cardiomyopathy, and Kaschin-Beck disease, characterized by chondrodystrophy found in China in regions where foods contain low levels of selenium.

Selenium and the Immune Response

Selenium is necessary for the development of the acquired immune system. Reilly (1996) pointed out, however, that not all classes of antibodies are affected to the same extent by selenium deficiency and that differences in animal species, age, sex and antigens affect the degree to which antibody production responds to selenium supplementation. Ponies fed 0.22 ppm selenium had a greater immune response than those fed 0.02 ppm (Knight and Tyznik, 1990). The requirement for selenium for optimal immune response in the horse is not known. However, studies in other species show that selenium supplementation at a supranutritional level may not be needed to improve immune response (Reilly, 1996).

Selenium Requirement

NRC (1989) concluded that the selenium requirement for horses was 0.1 ppm on a dry matter basis. However, I prefer to use the suggestion of 0.1 to 0.3 ppm by NRC (1980). It has been shown in studies by Stowe (1967) and Roneus and Lindholm (1983) that 0.1 ppm Se can be adequate to prevent white muscle disease and maintain glutathione peroxidase levels, but as discussed below many factors can influence selenium utilization. However, I feel that situations such as decreased selenium availability could exist under practical situations where 0.1 ppm Se may not be adequate for the health of the horse.

Selenium Availability

Factors such as form of selenium and interrelationships with other dietary ingredients can influence selenium utilization. For example, copper, sulfur, mercury and arsenic can influence selenium. The effect of the increased supplementation of commercial feeds with copper on selenium has not been studied. Stowe (1980), however, found that an oral dose of 20 or 40 mg of copper/kg of body weight prior to giving ponies 6 or 8 mg of selenium/kg of body weight prevented toxicity by selenium in ponies. The copper did not appear to prevent selenium absorption but enhanced excretion.

Arsenic could influence selenium metabolism in horses by increasing biliary excretion of selenium (Traub-Dargatz et al., 1986). Mercury and selenium form a mercury-selenium complex that decreases the activity of both minerals (Goyer, 1997).

Sulfur has been shown to decrease selenium activity in sheep (Hintz and Hogue, 1964) and presumably in horses, but I am not aware of any studies with horses which titrate the effect. Silver and cadmium are examples of other metals that have been shown to influence selenium metabolism in other species but have not been tested in the horse (Jamall and Roque, 1990).

A low vitamin E content of the diet could influence the need for dietary selenium. Sometimes the interaction could be indirect. For example, a diet containing a high content of raw, cull kidney beans can decrease the effectiveness of vitamin E in the prevention of nutritional muscular dystrophy in lambs. Cooking the beans or increasing the selenium supplementation can alleviate the problem (Hogue et al., 1962).

As mentioned earlier Kaschin-Beck disease is associated with selenium deficiency. It has also been suggested that the mold *Fusarium* might also be involved. A toxin extracted from *Fusarium tricenatum* can cause a decrease in collagen microfibroids in chicken embryo chondrocytes. The addition of selenium prevented the decrease (Reilly, 1996). It has been suggested that selenium might help prevent Kaschin-Beck disease by its effect on *Fusarium* in the food (Reilly, 1996). Tibial chondroplasia in growing chicks can be caused by *Fusarium*. Tibial chondroplasia in chicks is similar to osteochondrosis in foals. Would there be a benefit from increasing selenium intake if foals were consuming moldy feed? Were the raw, cull beans fed to sheep to produce nutritional muscular dystrophy contaminated with mold?

The form of selenium can influence activity. As mentioned earlier, methylselenocysteine is more available than sodium salts of selenium. Schwarz and Foltz (1957) demonstrated that the selenium in swine kidney powder was much more effective than other selenium compounds in the prevention of liver necrosis in rats. Selenium in milk was more effective than Na_2SEO_3 in the prevention of exudative diathesis in chicks (Mathias et al., 1965). Many other factors affecting selenium availability are discussed by Combs and Combs (1986).

Few studies have compared the effectiveness of selenium compounds in horses. Podoll et al. (1992) reported no difference between the utilization of dietary selenate and selenite by horses. Pagan et al. (1999) measured selenium utilization in two diets containing 0.41 ppm selenium. In one diet, about 3/4 of the selenium was from sodium selenite and in the second diet it was from selenium enriched yeast (Sel-Plex, Alltech Inc.). The apparent digestibility of yeast selenium was greater than for selenite (57% vs 51%). Selenium retention was 25% greater for yeast selenium than for selenite. Exercise increased urinary excretion of selenium more in the selenite group than in the yeast fed group and plasma selenium remained higher in the selenium yeast group than in the selenite group. The authors concluded that more research is needed to quantify the selenium requirements of horses at various intensities of exercise and to determine the effect of form of selenium on antioxidant status.

Feed and Food Sources of Selenium

The selenium content of grains and forage is determined largely by the selenium content of the soil on which the crop is grown. For example, alfalfa has been found to contain values of .03 to .69 ppm selenium and corn may contain .03 to 1.0 ppm. As mentioned earlier, brewer's yeast in Europe contained low levels of selenium whereas brewer's yeast in the United States contained enough to prevent liver necrosis in rats. Wheat products in the United States are likely to contain higher levels of selenium because much of the wheat is raised in areas of the country that contain selenium. Linseed products also are likely to be raised in higher selenium soils and therefore may contain significant amounts of selenium.

Values for various feedstuffs which can supply a significant amount of selenium are shown in Table 1.

Table 1. Selenium contents of feedstuffs.

Feed	Selenium (ppm)
Brewer's dried grains	0.7-1.0
Dried brewer's yeast	1.0-1.2
Linseed meal	1.0-1.5
Wheat bran	0.6-1.0

Selenized salt is often recommended as a source of selenium. It could be a very useful source for animals grazing forages growing on low selenium soils. Such products containing 30 to 90 ppm selenium have been marketed. The voluntary salt intake of horses will vary depending on environmental temperature, work and on anything that can influence sweat production. One study reported a range of voluntary intakes of 19 to 143 g of salt per day with an average of 53 g (Schryver et al., 1987). If the salt contained 90 ppm selenium, a horse would need to eat 11 g of salt per day to obtain about 1 mg of selenium. If the horse consumed 53 g of selenized salt, it would obtain 4.8 mg. If the horse ate 143 g of selenized salt containing 90 ppm per day, it would obtain 12.9 mg of selenium per day.

The maximum dietary tolerable level of selenium suggested by NRC (1989) is 2 ppm. A 500 kg horse eating 10 kg of feed should be able to tolerate at least 20 mg of selenium daily. Thus, even at the intake of 143 g of salt per day, the tolerable level of selenium would not be exceeded. A horse with a pyschogenic intake of salt could exceed the tolerable level. For non-range conditions, I prefer that the concentrate be the source of selenium rather than free choice salt because the intake of selenium can be more closely controlled.

The selenium content of human food is shown in Table 2. The high selenium contrast of Brazil nuts has not been explained.

Table 2. Selenium in humans foods[a].

Food	Selenium (micrograms)
Brazil nuts (solid unshelled) ½ ounce (4 medium nuts)*	436
Tuna, light 3 ½ ounces	80+
Flounder, 3 ½ ounces	58+
Pork, sirloin, 3 ½ ounces	52
Clams, canned, 3 ½ ounces	49+
Turkey, dark meat, 3 ½ ounces	41
Turkey, white meat, 3 ½ ounces	31
Pasta, cooked, 1 cup	30

[a]Adapted from Environmental Nutrition 22(5):6, 1999.
*Brazil nuts sold in the shell have significantly more selenium than Brazil nuts sold shelled.
+Only 20% is thought to be usable by the body.

As mentioned in the opening paragraph, many health benefits have been claimed for selenium. Reilly (1996) concluded there is strong evidence for the connection of selenium and prevention of heart disease, certain forms of cancer and problems of the immune system but less convincing for the other claims. However, he also pointed out that there is interest in selenium for the prevention of cystic fibrosis, intrahepatic cholestasis of pregnancy and age-related macular degeneration.

One of the most convincing studies showing a relationship between cancer and selenium was conducted by Clark et al. (1996). They reported significant decreases in lung, prostate and colorectal cancer in subjects given 200 μg of selenium per day in a high selenium brewer's yeast tablet supplied by Nutrition 21, LaJolla, CA.

Combs (1998) concluded that the antioxidant protection provided by selenium is directly related to the prevention of overt selenium deficiency and that it corresponds to the expression of SeCys enzymes. He also concluded that other responses, including those involved in anti-carcinogenesis, are related to higher (supranutritional) levels of selenium and those responses may relate to the production of one or more specific intermediary metabolites of selenium. Will higher intakes of selenium by horses have health and performance benefits not previously recognized? Not even "The Shadow" knows, but that possibility exists and should be explored.

References

Casey, C.E. 1988. Selenophilia. Proc. Nutr. Soc. 47:55-62.

Clark, L.C., Combs, G.F., Turnbull, B.W. et al. 1996. The nutritional prevention of cancer with selenium: A randomized clinical trial. J. Am. Med. Assoc. 276:1957-.

Combs, G.F. 1998. Keeping up with selenium. Proc. Cornell Nutrition Conf. P.169-177.

Combs, G.F. and Combs, S.F. 1986. The role of selenium in nutrition. Academic Press. Orlando.

Foreman, J. Potter, K.A. et al. 1986. Generalized steatitis associated with selenium deficiency and normal vitamin E status in a foal. JAVMA 189:83-86.

Goyer, R.A. 1997. Toxic and essential metal interactions. Ann. Rev. Nutr. 17:37-50.

Hintz, H.F. and Hogue, D.E. 1964. Effect of selenium, sulfur and sulfur amino acids on nutritional muscular dystrophy in the lamb. J. Nutr. 82:495-498.

Hogue, D.E., Proctor, J.F., Warner, R.G., Loosli, J.K. 1962. Relation of selenium, vitamin E and an unidentified factor to muscular dystrophy (stiff-lamb or white muscle disease) in the lamb. J. Anim. Sci. 21:25-.

Jamall, I.S. and Roque, H. 1989/1990. Cadmium-induced alterations in ocular trace minerals. Biol. Trace Element Research 23:55-75

Knight, D.A. and Tyznik, W.J. 1990. The effect of dietary selenium on humoral immuno-competence of ponies. J. Anim. Sci. 68:1311-1317.

Mathias, M.M., Allaway, W.H., Hogue, D.E. et al. 1965. Value of selenium in alfalfa for the prevention of selenium deficiencies in chicks and rats. J. Nutr. 86:213-219.

NRC. 1980. Mineral Tolerance of Domestic Animals. National Academy Press, Washington, D.C. P.392-401.

NRC. 1989. Nutrient Requirements of Horses. National Academy Press, Washington, D.C.

Pagan, J.D., P. Karnezos, M.A.P. Kennedy, T. Currier and K.E. Hoekstra. 1999. Effect of selenium source on selenium digestibility and retention in exercised Thoroughbreds. In: Proc 16th Equine Nutrition and Physiology Society.

Podoll, K.L., Bernard, J.B., Ullrey, D.E. et al. 1992. Dietary selenate versus selenite for cattle, sheep and horses. J. Anim. Sci. 70:1965-1970.

Raisbeck, M.F., Dahl, E.R., Sanchez, A. et al. 1993. Naturally occurring selenosis in Wyoming. J. Vet. Diagn. Invest. 5:84-87.

Reilly, C. 1996. Selenium in Food and Health. Blackie Academic. London.

Roneus, B. and Lindholm, A. 1983. Glutathione peroxidase activity in the blood of healthy horses given different selenium supplementation. Nord. Veterinaer Med. 35:337-345.

Schryver, H.F., Parker, M.T., Daniluk, P.D. et al. 1987. Salt consumption and the effect of salt on mineral metabolism in horses. Cornell Vet. 77:122-131.

Schwarz, K. and Foltz, C.M. 1957. Selenium as an intregral part of Factor 3 against dietary necrotic degeneration. J. Amer. Chem. Soc. 79:3292-3295.

Stowe, H.D. 1967. Serum selenium and related parameters of naturally and experimentally fed horses. J. Nutr. 93:60-67.

Stowe, H.D. 1980. Effects of copper pretreatment upon the toxicity of selenium of ponies. Amer. J. Vet. Res. 41:1924-1928.

Truab-Dargatz, J.L., Knight, A. P. and Hamar. D.W. 1986. Selenium toxicity in horses. Comp. Cont. Vet. Ed. 8:771-776.

Witte, S.T., Will, L.A., Olsen, C.R. et al. 1993. Chronic selenosis in horses fed locally produced alfalfa hay. J. Amer. Vet. Med. Assoc. 202:406-409.

STARCH DIGESTION IN THE HORSE

DEREK CUDDEFORD
Edinburgh University, Scotland

Introduction

The rate and extent of starch digestion is determined by:
1. properties of the starch granule;
2. the effect of processing;
3. associated food structures (plant cell walls);
4. transit time through the small intestine;
5. the availability and concentration of enzymes.

These factors will affect the glycemic response of the horse to feeding and the subsequent production of insulin. Resistant starch, together with undigested starch, will pass into the large intestine of the horse where it may be fermented to short chain fatty acids. Resistant starch may escape digestion in the small intestine of the horse because:
1. of physical entrapment within a food, such as in partly-milled grains and seeds (RS_1 starch);
2. starch granules have a B or C crystalline structure which is highly resistant to digestion (RS_2 starch);
3. retrogradation through food processing, mainly in the form of retro-graded amylose (RS_3 starch).

RS_3 starch is entirely resistant to digestion by pancreatic amylase in man (Englyst and Macfarlane, 1986) and this starch is commonly found in peas. RS_1 and RS_2 are quantitatively the most important forms of resistant starch found in horse feeds.

Starches and the Horse

Irrespective of starch source, total tract apparent digestibility is usually very high. For example, Arnold et al. (1981) measured digestion coefficients for corn, oats and sorghum starch as 97.0, 96.7 and 97.0% respectively. Differences are apparent between starch sources when measuring small intestinal or pre-cecal apparent digestibility of starch. Arnold et al. (1981) reported values of 78.2, 91.1 and 94.3% respectively for corn, oats and sorghum. These values were much higher than those reported by Householder (1978) but were explained on the basis of a lower dry matter intake and thus, presumably, a slower rate of passage. Several authors (e.g. Kienzle et al., 1992; Meyer et al., 1993) have confirmed that oat starch is the most digestible of the cereal starches fed to horses, followed by sorghum. Corn starch was much less digestible and, in the report by Meyer et al. (1993), barley starch was the least digestible. These differences are largely explainable on the basis of the differences between the starch granules contained in the different plant materials; oat starch granules are small and easily digested.

Processing can affect starch digestibility, although the magnitude of the effect depends on the nature of the process used. Crude physical treatments such as rolling, crushing or grinding do not significantly improve oat starch digestibility (Kienzle et al., 1992) or that of corn (Meyer et al., 1993). In contrast, fine grinding (<2mm) can improve small intestinal starch digestibility of corn (Kienzle et al., 1992; Meyer et al., 1993). Cooking, either by micronizing or popping (Householder, 1978; Meyer et al., 1993) can lead to significant improvements in pre-cecal starch digestibility, particularly with respect to corn (Meyer et al., 1993). Thus, while some starch sources are poorly degraded when raw, appropriate cooking procedures can significantly improve pre-cecal starch digestion, the best example of this being that of corn (Meyer et al., 1993).

In a review of starch digestion, Potter et al. (1992) suggested that small intestinal starch digestibility declined as starch intake increased; this relationship was confirmed by Kienzle (1994). Potter et al. (1992) further suggested that in meal-fed horses, the upper limit of starch feeding should be 3.5 to 4 g starch/kg body weight/meal. Kienzle et al. (1992) suggested that, depending on the starch source, if the intake exceeded 2 g/kg body weight/meal, there was a risk that the capacity of the small intestine to degrade starch could be exceeded. Thus, there are widely differing estimates of "safety" in terms of meal-feeding starch. A further complication is that of feed interactions; it is naïve to think that starch sources will be degraded in a constant way, irrespective of the basal diet. It has already been pointed out that transit time through the small intestine and availability of enzymes will significantly affect the process of degradation. For example, Meyer et al. (1993) demonstrated that the pre-ileal digestibility of ground corn was reduced by substituting green (*sic*) meal for hay in a mixed ration. Kleffken (1993) showed that this substitution altered the rate of passage of starch in the small intestine and this presumably accounted for the differences measured by Meyer et al. (1993). The unusual results obtained by Hinkle et al. (1983), whereby starch digestibility remained the same or increased with increasing starch intake, may be explained on the basis that roughage feeding declined and thus digesta passage rate would have declined. Animal to animal variability will mean that some horses are better able to digest starch than others are. A basic difference between horses is their effectiveness at comminuting foodstuffs and, since particle size has been shown to correlate with pre-cecal starch digestibility (Meyer et al., 1993), those animals that eat rapidly and/or have poor occlusion will have reduced small intestinal starch digestion. This has been demonstrated by Professor Meyer's group at Hannover (Landes, 1992; Wilke, 1992) in horses fed whole corn. Other differences may exist between horses in terms of enzyme production (Radicke et al., 1992). This view is supported by the fact that small intestinal degradation of ground corn was assisted (from 47.3% to 57.5%) by the addition of amylase (Meyer et al., 1993). Unquestionably, small intestinal passage rate will vary between individual horses, even when fed in the same way. Thus, the notion that there is a "safe" upper level for meal-feeding horses or ponies with starch is questionable in view of the aforementioned factors that significantly impact on the extent of starch digestion in the small intestine.

Cereal Processing- Some Recent Work

Potter et al. (1992) noted the effects of processing on cereals could be masked by differences between sources of starch. In view of this, some experiments were planned to examine the effects of processing on barley utilization by the horse, barley having been previously shown to have a low pre-cecal digestibility (Meyer et al., 1993). Barley of one variety from one field was harvested, dried and then subsequently rolled, micronized or extruded for use in a series of experiments. Animals used were mature Welsh-cross pony geldings (body weight – 280 kg) and each was fitted with a cecal cannula.

In Vivo Studies

This work has been recently reported (McLean et al., 1999a) and involved offering ponies 4 kg DM per day (in two equal meals at 09.00 and 17.00 h) of either 100% hay cubes or one of three diets (different forms of barley) consisting of a 50:50 barley: hay cube mix. The results from this incomplete Latin square changeover design experiment were in accordance with expectation in that different physical processing methods do not alter total tract apparent digestibilities *in vivo* (Potter et al., 1992). Furthermore, the apparent digestibility of the energy (DE) and crude protein of the barley were unaffected by processing; the DE contents of the rolled, micronized and extruded barley were 14.5, 14.8 and 15.0 MJ/kg DM respectively, similar to the NRC (1989) value of 15 MJ/kg DM. Apparent organic matter and starch digestibilities were respectively 841 and 967 for the rolled barley, 846 and 969 for the micronized barley and 837 and 966 g/kg for the extruded barley. *In vivo* total tract apparent digestibility studies give no indication of whether the starch was degraded by mammalian enzymes in the small intestine or by microbial enzymes in the large intestine. These ponies were fed approximately 2.4 g starch/kg body weight/meal and the apparent digestibility of the starch was uniformly high.

In Situ Studies

The three different forms of barley were incubated in the ceca of fistulated ponies to provide time-based degradation data utilizing microbial enzymes (McLean et al., 1998a). Degradation profiles were fitted to the dry matter and starch disappearance data according to Orskov and McDonald (1979). The degradation coefficient 'a' was greatest for the micronized barley, and significantly so ($p < 0.05$) for dry matter loss. As expected, no differences existed between the different forms of barley by 40 h of incubation, although extrusion reduced the effective degradability of dry matter to a small but significant ($p < 0.05$) extent when compared to the rolled barley. This negative effect may have been as a result of complexation during the extrusion process producing indigestible Maillard products. More importantly, at outflow rates (k= 0.50, 0.33 and 0.20) which reflected mean retention times (MRT) of 2, 3 or 5 h in the pre-cecal segment of the digestive tract, the DM degradability was greatest for the micronized cereal, followed by the extruded barley; the rolled barley had the lowest effective degradabilities (B.M.L. McLean-unpublished data). However, the only significant ($p<0.05$) difference was at k=0.50 when the effective DM degradability of micronized barley was 690.7 compared to that of 631.9 g/kg for the rolled barley. When considering starch degradation, there were no significant differences between starch degradation parameters although again, the coefficient 'a' for micronized barley was the highest. In contrast, comparison of the effective degradabilities of starch at k=0.50, 0.33 and 0.20 produced significant differences between the differently processed barleys. At k=0.50, micronized barley was degraded more than the other barleys and significantly so ($p< 0.05$) in respect to extruded barley. This was also the case at k=0.33 and at k = 0.20; the starch of micronized barley was significantly ($p< 0.05$) more degraded than that of the other two barleys, 929.3 compared to 857.1 for extruded and 857.8 g/kg for rolled barley. These important difference between barley sources in terms of starch degradability will have an impact on the nature of the feed residues from a meal that enter the cecum. If we assume a transit time from the esophagus to the ileo-cecal valve of 3 to 4 h, then the provision of micronized barley should result in less undigested starch entering the large intestine compared to when rolled barley is fed. If undigested starch were made available to the microflora within the large intestine of the horse, then changes in the cecal environment would be expected that would be analogous to those measured in the rumen of cows or sheep overfed cereal. There would be a reduction in pH together with a change in the molar proportions of the volatile fatty acids, increased propionate and decreased acetate.

In Vivo Studies

The effects of physical processing of barley on intra-cecal pH and volatile fatty acid parameters have recently been reported (McLean et al., 1998b). The provision of a meal supplying 2.4 g starch/kg body weight in the form of rolled, micronized or extruded barley, together with hay cubes, significantly reduced ($p< 0.05$) cecal pH and acetate and increased propionate, compared to a 100% hay cube diet, when measured 5 h post-feeding. Feeding rolled barley with the hay cubes caused lower ($p< 0.05$) acetate and higher ($p< 0.05$) propionate compared to when micronized barley was fed with the hay; these changes were apparent through most of the day following the 0900 h meal. The micronized barley increased ($p< 0.95$) acetate and lowered ($p< 0.05$) propionate compared to extruded barley at most hourly samplings throughout the day. At 5 h post-feeding, cecal pH values were 6.24, 6.33, 6.38 and 6.48 respectively for rations containing rolled, micronized and extruded barleys with hay cubes and hay cubes alone. Acetate values (mmol/mol) in the cecum were 627, 716, 685 and 764, whereas propionate levels were 302, 221, 250 and 174 respectively. As expected, the hay cubes ration resulted in the highest acetate (764) and the lowest propionate (174) values; the ration containing the micronized barley had the closest values, 716 and 221 respectively.

Mobile Nylon Bag Studies

The mobile nylon bag technique has been used to study the degradation dynamics of forages over the whole length of the digestive tract of equids (Hyslop et al., 1998) and in the pre-cecal segment (Moore-Colyer et al., 1997). Recently McLean et al. (1999b) have used the technique to investigate the degradation of purified wheat starch, chemically modified wheat starches and a purified pea starch. Bag transit times (following introduction by nasogastric tube into the stomach) to the cecum varied between 1 and 7.5 h. Pea starch was the least well degraded (probably due to the presence of resistant starch, RS_3) in the small intestine whereas wheat starch required extensive chemical modification (cross-linking) before pre-cecal degradation was significantly ($p< 0.5$) reduced. It was concluded that the mobile nylon bag technique could be used to successfully model feed degradation dynamics over time in the pre-cecal segment of the equine digestive tract. Thus, the technique was used to measure pre-cecal losses of dry matter and starch from the differently processed barleys. Initial results for the effective degradabilities of the dry matter of rolled, micronized and extruded barleys at k=0.50, 0.33 and 0.25 showed that extruded barley had significantly ($p< 0.05$) lower effective degradability at all assumed MRT; there were no differences between rolled or micronized barleys (B.M.L. McLean – unpublished data). The starch data have yet to be modelled.

Discussion

Willard et al. (1977) showed that cecal pH was affected by diet; pH was significantly ($p<0.05$) lower at 4, 5 and 6 h post-feeding 6 kg sweet feed to horses compared to when hay was fed. Furthermore, the molar proportions of acetate and propionate were altered; there was more propionate *pro-rata* when concentrate was fed and lactic acid levels were higher. An interesting observation by these authors was that horses fed the concentrate-only diet chewed wood and practiced coprophagy. The lowest mean cecal pH recorded by Willard et al. (1977) was 6.12, 6 h post-feeding concentrate.

Essentially, the results obtained by Willard et al. (1977) showed that feeding large amounts of starch-rich feed affects the cecal environment in terms of pH and in terms of the proportions of volatile fatty acids formed therein. The latter effect had already been recorded by Hintz et al. (1971) when they examined the effects of feeding different forage: concentrate ratios.

Garner et al. (1975) developed an experimental model for the induction of laminitis in horses. A gruel of 85% corn starch and 15% wood cellulose was introduced via a cecal fistula at the rate of 17.6 g/kg body weight in order to create a "grain overload." Assuming a dry matter (values not given) for the gruel of 200 g/kg, then the starch supplied would have been about 3 g starch/kg body weight. Grain overload, together with gastrointestinal disease, are the most common predisposing factors to laminitis in horses (Slater et al., 1995). Prior to grain overload, mean cecal pH was 7.18 and this fell to 5.72, 8 h after the overload. At this pH and by this time, *Lactobacillus spp* had significantly ($p<0.05$)

increased in number (Garner et al., 1978). These authors concluded that low cecal pH causes the death of favorable organisms, releasing endotoxins, and that lactic acid accumulates and causes a generalized lactic acidosis. Radicke et al. (1991) measured the effect of feeding 1 to 2, 2 to 3, or 3 to 4 g of starch/kg body weight/meal using either oats or corn as a source of the starch. As expected, at all levels of intake, cecal pH was lower when corn was fed, and the differential between the two cereals increased in proportion to starch intake. Increasing oat starch intakes to the levels asserted as "safe" by Potter et al. (1992) did not significantly reduce cecal pH. However, feeding the same level of corn starch caused a marked reduction in cecal pH to values close to 6. Radicke et al. (1991) considered that a cecal pH of 6 represented sub-clinical acidosis, and below 6 there was considerable risk in terms of the development of clinical conditions and in terms of imbalancing cecal fermentation. This was in accord with the results obtained earlier by Garner et al. (1978). Recently, Johnson et al. (1998) fed a changing ration over a 4-week period to 400/500 kg horses. Initially, they were fed 8 kg hay, then 6 kg hay and 2 kg concentrate, then 4 kg hay and 4 kg concentrate and in the fourth week, 2 kg hay and 6 kg concentrate. The daily ration was fed in two equal parts so ultimately the horses received a meal of 3 kg concentrate and 1kg of hay. If the concentrate contained 550 g starch/kg, then the starch supplied would have been between 3.3 and 4.1 g/kg body weight, within the Potter et al. (1992) recommendations. However, the authors measured a declining fecal pH from about 6.6 to below 6 and noted abnormal behaviors similar to those recorded by Willard et al. (1977). Unfortunately, Johnson et al. (1998) failed to confirm the relevance of fecal pH to cecal pH at post mortem. A recent survey (L Paul-unpublished data) of feces from horses fed different diets confirmed that the nature of the diet can affect the fecal pH value. Horses at grass produced feces with a pH of 6.75 (n=34), racehorses fed 70% oats and 30% forage had an average fecal pH of 6.38 (n=20) and forage-fed ponies had a value of 6.49 (n=6); there was considerable between-animal and circadian variation. Johnson et al. (1998) showed that the daily feeding of 225 mg of virginiamycin prevented the extreme fall in fecal pH associated with the increased level of concentrate feeding although at the highest level, fecal pH was still significantly (p< 0.05) reduced compared to horses only fed hay. The starch was mostly cereal in origin, supplied by wheat and barley, so that there would be no expectation of low pre-cecal starch digestion; unfortunately, the authors did not describe the form of the concentrate. The fact that virginiamycin appeared to have some "protective" effect, possibly through suppressing lactic acid production (Rowe et al., 1994), suggests that, even with relatively low starch intakes, horses may suffer cecal dysfunction without enduring very low cecal pHs. This is in accordance with the views of Radicke et al. (1991).

Finally, another situation in which cecal function may be compromised is when horses and ponies consume grass. Fructan is a fermentable polysaccharide stored in grass which can comprise 5 to 50% of the grass DM. Fructan cannot be digested by mammalian enzymes so that if it escapes hydrolysis in the stomach, it will enter the cecum undegraded. If there is a 30% loss of non-fructan DM in the small intestine (extrapolated from data of McLean et al., 1998b) then it is possible to calculate the likely amount of fermentable fructan entering the cecum. Thus, if 25% of grass DM is fructan, then it will comprise about 32% of the DM entering the cecum. McLean et al. (1998b) measured significant (p< 0.05) reductions in cecal pH when only small amounts of starch entered the cecum (approximately 20% of the DM) which suggests that the ecosystem is quite sensitive to small amounts of rapidly fermentable substrate. Thus, the consumption of fructan-rich grass by horses may cause changes analogous to those measured when starch escapes small intestinal digestion; this may, perhaps, partially explain the occurrence of laminitis at grass. Unfortunately, grass fructan levels are immensely variable, unlike the level of starch in a cereal, and thus it is impossible as yet to predict likely fructan intake. However, we do know that, unlike starch, fructan cannot be degraded in the small intestine of the horse although it may be susceptible to hydrolysis in the stomach.

It is apparent that horse feeders must minimize the flow of fermentable polysaccharide to the large intestine of the horse to maintain health and to maximize substrate (glucose) availability to the performance horse. While this goal is achievable with starch by simply feeding highly degradable micronized cereals little and often, the regulation of fructan intake remains an irresolvable problem.

Acknowledgements

Recent work at Edinburgh, in collaboration with the Welsh Institute of Rural Studies and the Institute of Grassland and Environmental Research, was funded by the Horserace Betting Levy Board (Project 612) and Dodson and Horrell Ltd.

References

Arnold, FF, Potter, G D, Kreider, J L, Schelling, G.T. and Jenkins, W L (1981). Carbohydrate digestion in the small and large intestine of the equine. Proceedings of the 7th Equine Nutrition and Physiology Symposium, Warrenton, Virginia, USA p 19-22.

Englyst, H N and Macfarlane, G T (1986) Breakdown of resistant and readily digestible starch by human gut bacteria. Journal of Science, Food and Agriculture 37, 699-706.

Garner, H E, Coffmann, J R, Hahn, A W, Hutcheson, D and Tumbleson, M E (1975) Equine laminitis of alimentary origin: an experimental model. American Journal of Veterinary Research 36, 441-449.

Garner, HE, Moore, J N, Johnson, J H, Clark, L, Amend, J F, Tritschler, L G, Coffmann, J R, Sprouse, R F, Hutcheson, D P, and Salem, C A (1978) Changes in the cecal flora associated with the onset of laminitis. Equine Veterinary Journal 10, 249-252

Hinkle, D K, Potter, G D and Kreider, J L (1983) Starch digestion in different segments of the digestive tract of ponies fed varying levels of corn. Proceedings of the 8th Equine Nutrition and Physiology Symposium, Kentucky. p 227-230.

Hintz, H F, Argenzio, R A and Schryver, H F (1971) Digestion coefficients, blood glucose levels and molar percentages of volatile fatty acids in intestinal fluid of ponies fed varying forage: grain ratios. Journal of Animal Science 33, pp 992- 995.

Householder, D D (1978) Prececal, postileal and total tract digestion and growth performance in horses fed concentrate rations containing oats or sorghum grain processed by crimping or micronizing. PhD Dissertation , Texas A & M University, College Station, USA.

Hyslop, J J, Tomlinson, A L, Bayley A and Cuddeford D (1998) Development of the mobile bag technique to study the degradation dynamics of forage feed constituents in the whole digestive tract of equids. Proceedings of the British Society of Animal Scienc p.129 .

Johnson, K G, Tyrrell, J, Rowe, J B and Pethick, D W (1998). Behavioral changes in stabled horses given non-therapeutic levels of virginiamycin. Equine Veterinary Journal 30, 139-143.

Kienzle, E (1994) Small intestinal digestion of starch in the horse. Revue Medicine Veterinaire 145, 199-204.

Kienzle, E, Radicke, S, Wilke, S, Landes, E and Meyer, H (1992) Preileal starch digestion in relation to source and preparation of starch. Pferdeheilkunde, Sonderheft, 103 – 106.

Kleffken, D (1993) Praeileale Verdauung von Getreidestärke (Gerste/Mais) in Abhangigkeit von Zubereitung, Rauhfutterangebot und Amylasezusatz beim Pferd. Tierärztlische Hochschule, Hannover, Dissertation.

Landes, E (1992) Amylaseaktivität sowie Konzentration organischer Säuren im Jejunum-und Cecumchymus des Pferdes nach Hafer-und Maisfütterung. Tierärztlische Hochschule Hannover, Dissertation.

McLean, B M L, Hyslop, J J, Longland, A C and Cuddeford, D (1998b) Physical processing of barley and its effects on intra-cecal pH and volatile fatty acid parameters in ponies offered barley-based diets. Proceedings of the 2nd Annual Conference of the European Society of Veterinary and Comparative Nutrition. Vienna, Austria, p 61.

McLean, B M L, Hyslop, J J, Longland, A C, and Cuddeford, D (1998a) Effect of physical processing on *in situ* degradation of barley in the cecum of ponies. Proceedings of the British Society of Animal Science p 127.

McLean, B M L, Hyslop, J J, Longland, A C, Cuddeford, D and Hollands, T (1999a) *In vivo* apparent digestibility in ponies given rolled, micronized or extruded barley. Proceedings of the British Society for Animal Science (In Press).

McLean, BM L, Hyslop, J J, Longland, A C, Cuddeford D and Hollands, T (1999b) Development of the mobile bag technique to determine the degradation kinetics of purified starch sources in the pre-cecal segment of the equine digestive tract. Proceedings of the British Society of Animal Science (In Press).

Meyer, H, Radicke, S, Kienzle, E, Wilke, S and Kleffken, D (1993) Investigations on preileal digestion of oats, corn and barley starch in relation to grain processing. Proceedings of the 13th Equine Nutrition and Physiology Symposium, Florida, USA, p 92-97.

Moore-Colyer, M J S, Hyslop, J J, Longland, A C and Cuddeford, D (1997) The degradation of organic matter and crude protein of four botanically diverse feedstuffs in the foregut of ponies as measured by the mobile bag technique. Proceedings of the British Society of Animal Science, p 120.

NRC (1989) Nutrient requirement of horses, 5th edition. National Academy Press, Washington, D C.

Orskov, E R and McDonald, I (1979) The estimation of protein degradability in the rumen from incubation measurements weighted according to rate of passage. Journal of Agricultural Science (Cambridge) 92, 499-503.

Potter, G D, Arnold, F F, Householder, D D, Hansen, D H and Brown, K M (1992) Digestion of starch in the small or large intestine of the equine. Pferdeheilkunde, Sonderheft, 107-111.

Radicke, S, Kienzle E, and Meyer, H (1991) Preileal apparent digestibility of oats and corn starch and consequences for cecal metabolism. Proceedings of the 12th Equine Nutrition and Physiology Symposium, Calgary, Canada. p 43-48.

Radicke, S,Landes, E, Kienzle, E, and Meyer, H (1992) Activity of amylase in the digestive tract of horses in relation to diet type. Pferdeheilkunde, Sonderheft, 99-102.

Rowe, J B, Lees, M J and Pethick, D W (1994) Prevention of acidosis and laminitis associated with grain feeding in horses. Journal of Nutrition 124, 2742S-2744S.

Slater, M R, Hood, D M and Carter, G K (1995) Descriptive epidemiological study of equine laminitis. Equine Veterinary Journal 27, 364-367.

Wilke, S (1992) Zur praeilealen Verdaulichkeit von Hafer und Mais verschiedener Zubereitungen beim Pferd. Tierärztliche Hochschule Hannover, Dissertation.

Willard, J G, Willard, J C, Wolfram, S A and Baker, J P (1977) Effect of diet on cecal pH and feeding behaviors of horses. Journal of Animal Science 45, 87-93.

EFFECT OF CORN PROCESSING ON GLYCEMIC RESPONSE IN HORSES

K.E. HOEKSTRA[1], K. NEWMAN[2], M.A.P. KENNEDY[1] AND J.D. PAGAN[1]
[1]Kentucky Equine Research, Inc., Versailles, KY
[2]Alltech, Inc., Nicholasville, KY

Summary

An experiment was conducted to evaluate how cracking, grinding or steam processing affects starch digestibility of corn, using glycemic response as an indirect measure of prececal starch digestibility. In a replicated 3 x 3 Latin square design, six mature horses were fed either cracked, ground, or steam-flaked corn (2 g/kg of BW in a single meal) and 1% BW/d hay over six periods, each lasting 8 to 10 d. At the end of each period, horses were fed their respective grain meals and blood samples were taken at 30-min intervals for 8 h. Area under the curve, mean glucose and lactate, peak glucose and lactate, and time to peak glucose and lactate were determined. Steam-flaked corn produced a greater glycemic response than cracked or ground corn. Peak glucose was also greater for steam-flaked corn. Results of this study indicate that steam flaking alters glycemic response (and presumably starch digestibility) to a much greater extent than grinding or cracking.

Introduction

Previous research has demonstrated that prececal starch digestibility of corn is improved by grinding or popping (Meyer et al., 1993). In this study, digestibility increased 17 and 61% following grinding and popping of the whole kernel. Crushing or rolling the corn kernel did not improve the digestibility of starch in the small intestine. It was concluded that starch granules are not significantly altered from their original form in the whole kernel when crushed or rolled. However, when further processing occurs, such as grinding, the structure of the starch granule is altered, allowing the granule to be more available to digestive enzymes and thus improving nutrient availability (Fahrenholz, 1994). Steam rolling, extruding, or micronizing corn (all of which involve heating the kernel) may result in even greater digestibility than grinding or cracking. This experiment was conducted to determine if steam processing alters glycemic response, an indirect measure of prececal starch digestibility.

Materials and Methods

Management of Horses
Six mature geldings, four Arabians and two Thoroughbreds, with a mean age of 8.4 y (range of 6 to 10) and mean BW of 493 kg (range of 405 to 645), were used

in the experiment. Horses were maintained on pasture during the day (0800 to 1600) and housed in 3.0 m x 3.0 m box stalls overnight (1600 to 0800). Horses were not subjected to forced exercise, but were given free access to exercise while on pasture.

Treatments
Horses were used in a replicated 3 x 3 Latin square design to evaluate how processing affects starch digestibility. Glycemic response was used as an indirect measure of prececal starch digestibility. Horses were fed either cracked, ground, or steam-flaked corn (2 g/kg of BW in a single meal) and 1% BW/d alfalfa-grass hay over six periods. Each period lasted 8 to 10 d. Amounts fed were based on BW established prior to the first period. Grain rations were fed once per day (0700) and hay was divided into two equal feedings per day (1600 and 2200). Horses were offered free choice salt in 2-kg salt blocks when in box stalls and had free access to water at all times. Horses were allowed free access to grazing while maintained on pasture. Three-day transition periods occurred at the beginning of each new period to gradually introduce new diets.

Sample Collection
At the end of each period, horses were housed in box stalls and an 8-h blood collection was conducted. Horses were catheterized in the jugular vein prior to feeding and a fasting blood sample was taken to determine baseline glucose (mg/dl) and lactate (mmol/L) values. Horses were then fed their respective treatment diets and blood samples were taken at 30-min intervals for 8 h. Plasma concentrations of glucose and lactate were determined using a YSI Model 2300 STAT glucose and L-lactate analyzer (Yellow Springs Instrument Co., Inc., Yellow Springs, OH).

Statistical Analysis
Area under the curve, mean glucose (mg/dl) and lactate (mmol/L), peak glucose (mg/dl) and lactate (mmol/L), and time to peak glucose and lactate (min) were determined. Plasma glucose and lactate concentrations were statistically analyzed by the general linear model procedure for analysis of variance. Period, horse, and diet were included in the model.

Results

The glycemic response of each grain was compared using a glycemic index where each feed's glucose area under the curve was expressed relative to cracked corn (Figure 1). Steam-flaked corn produced a greater glycemic response than cracked or ground corn (P < .05; Table 1). Peak glucose was also greater for steam-flaked corn (P < .01). Horses on the cracked corn diet demonstrated greater peak glucose than those on ground corn (P < .01). Plasma glucose concentrations were consistently lower for cracked and ground corn treatments, when compared with steam-flaked corn, from 90 to 180 min post-feeding during sample collection (P < .05; Figure 2). Time to peak glucose was unaffected by processing. Area under the curve for lactate was greater for steam-flaked corn when compared with ground corn (P < .05; Table 2). Plasma lactate concentrations were greater for steam-flaked corn than cracked or ground corn between 90 and 150 min post-feeding (P < .05; Figure 3). Mean lactate, peak lactate, and time to peak lactate were unaffected by processing.

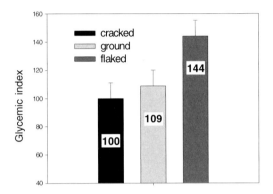

Figure 1. Glycemic index for all diets.

Table 1. Area under the curve, mean glucose, peak glucose, and time to peak glucose for all diets.

	Area under curve	Mean (mg/dl)	Peak (mg/dl)	Time to glucose peak (min)
Dietary Treatment				
Cracked Corn	1,734[a]	95.6[a]	114.8[b]	62.5
Steam-Flaked Corn	2,500[b]	98.4[b]	125.1[c]	72.5
Ground Corn	1,887[a]	95.4[a]	109.1[a]	62.5
SEM	191	42	1.77	5.2
Statistical significance	.05	.01	.01	NS[d]

[abc] Treatments lacking a common superscript differ (P < .05)
[d] Not significant

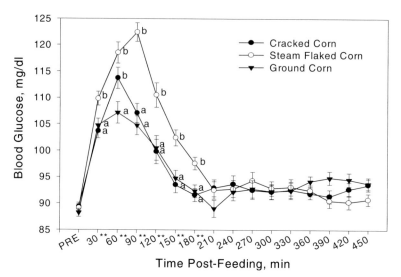

Figure 2. Plasma glucose concentrations post-feeding.
    [ab] Treatments lacking a common superscript differ (P < .05)
    [**] Treatments different at given sample time (P < .01)

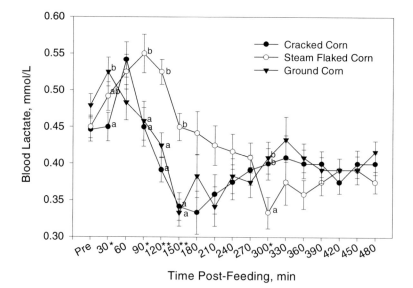

Figure 3. Plasma lactate concentrations post-feeding.
    [ab] Treatments lacking a common superscript differ (P < .05)
    [*] Treatments different at given sample time (P < .05)
    [**] Treatments different at given sample time (P < .01)

Table 2. Area under the curve, mean lactate, peak lactate, and time to peak lactate for all diets.

	Area under curve	Mean lactate (mmol/L)	Peak lactate (mmol/L)	Time to peak (min)
Dietary Treatment				
Cracked Corn	5.25[ab]	.40	0.57	97.5
Steam Flaked Corn	8.81[b]	.43	0.64	140.0
Ground Corn	3.56[a]	.41	0.59	102.5
SEM	1.41	.08	0.03	37.3
Statistical significance	.05	NS [c]	NS	NS

[ab]Treatments lacking a common superscript differ (P < .05)
[c]Not significant

Discussion

Results of this study indicate that steam-flaked corn alters glycemic response (and presumably prececal starch digestibility) to a much greater extent than grinding or cracking. Similar improvements in starch digestibility have been found in cattle fed steam-flaked corn, when compared to cattle fed whole, ground, or dry-rolled corn (Theurer, 1986).

Higher peak concentrations of glucose in the plasma of horses fed steam-flaked corn indicates that steam flaking results primarily in small intestinal digestion, avoiding excessive microbial fermentation in the large intestine and greater lactate production by the hindgut. Processing of starch, starch intake levels, and source and timing of forage feeding have a large effect on prececal starch digestibility, and the misapplication of these factors has the potential to result in digestive disorder such as colic or laminitis (Meyer et al., 1993). High lactate and glucose concentrations demonstrated in the present experiment occurred at similar collection times (between 90 and 150 min post-feeding), suggesting that lactate produced and absorbed into the bloodstream may have resulted from bacterial fermentation of soluble carbohydrate in the nonglandular region of the stomach (Meyer, 1983).

Conclusions

Steam flaking alters glycemic response to a much greater extent than grinding or cracking. Further research is necessary to determine if other processing techniques that involve heating the corn kernel, such as extruding or micronizing, will result in similar improvements in glycemic response. Additional research is also required to determine if glycemic improvements have a beneficial effect on performance of the horse.

References

Fahrenholz, C. 1994. Cereal grains and byproducts: What's in them and how are they processed? In: J.D. Pagan (Ed.) Advances in Equine Nutrition. p 57-70. Nottingham University Press, Nottingham.

Meyer, H. 1983. The pathogenesis of disturbances in the alimentary tract of the horse in the light of newer knowledge of digestive physiology. In: Proc. Horse Nutr. Symp., Uppsala, Sweden. p 95-108.

Meyer, H., S. Radicke, E. Kienzle, S. Wilke, and D. Kleffken. 1993. Investigations on preileal digestion of oats, corn, and barley starch in relation to grain processing. In: Proc. Equine Nutr. and Phys. Soc. Symp. p 92-97.

Theurer, C.B. 1986. Grain processing effects on starch utilization by ruminants. J. Anim. Sci. 63(5):1649-1662.

ASSESSMENT OF SELENIUM STATUS IN HORSES

I. VERVUERT, M. COENEN, M. HÖLTERSHINKEN, M. VENNER AND P. RUST
Institut für Tierernährung, Tierärztliche Hochschule, Hannover, Germany

The essentiality of selenium (Se) for membrane integrity, growth, reproduction and immune response in horses is well established. The dietary Se requirement in the horse is estimated to be 0.15-0.2 (Gesellschaft für Ernährungsphysiologie, 1994) or 0.15-0.3 mg/kg, respectively (NRC, 1989) and plasma Se levels between 100-200 µg/l are considered adequate. The objective of this study was to assess Se status in horses under various management and feeding conditions. Se in plasma and in feed samples as well as GSH-Px activity in whole blood as an indicator of Se status were measured in 304 horses of different ages, breeds and feeding regimes. Blood was collected by jugular venipuncture into vacuum tubes containing lithium heparin. Se concentration in plasma and in feed samples was determined by atomic absorption spectroscopy. GSH-Px activity in whole blood was analyzed by photometry and related to hemoglobin in blood. Results are presented as mean ± s.d. Regression analysis was used to evaluate the relationship between GSH-Px activity in whole blood and Se concentration in plasma. Plasma Se concentration (n=304) ranged from 16 to 291 µg/l with a mean of 116 µg/l, and GSH-Px activity extended from 2 to 190 U/g hemoglobin with a mean of 92.6 U/g hemoglobin. The relationship between plasma Se concentration and GSH-Px activity is best described by the linear regression equation: $y = 0.78x + 43.1$, $r = 0.49$. There was wide variation in measures of Se status, even if horses received the same feed supplement. 26% of the horses had plasma Se levels below 70 µg/l (dietary Se <0.15 mg/kg DM), but there was no evidence of myopathy in performance horses or reproductive difficulties in mares associated with low plasma Se levels. Horses that were given a mineral supplement containing selenium (255 mg/kg Se DM) had critically high plasma Se levels (about 300 µg/l). The results of this study raise some questions with regard to the consequences of a marginal selenium supply. In spite of low plasma Se levels, no medical problems were observed. Beyond that there is a high variability in Se status among horses even under the same feeding and management conditions.

EFFECT OF DIFFERENT CEREAL GRAIN ON EQUINE DIET DIGESTIBILITY

S. SARKIJARVI, M. SAASTAMOINEN AND L. KAJANTO
Agricultural Research Centre of Finland, Animal Production Research, Finland

Diets were based on hay silage alone or hay silage with 33% oats, naked oats, barley, wheat or corn on a dry matter basis. Dietary treatments were assigned to horses in a 6 x 6 balanced Latin square design. Six mature Finnhorses were fed three times a day to meet Finnish nutrient requirements for light work. A preliminary feeding period of 23 days was followed by a 5-day collection period. Representative feces samples were taken twice a day during the collection period. Lignin was used as an indirect marker for the estimation of apparent digestibility. The mean dry matter (DM) and organic matter (OM) digestibilities of hay silage were 0.580 g/kg^{-1} and 0.594 g/kg^{-1}, respectively. Including cereals in the equine diet increased the apparent digestibilities of nutrients, but the effect was different for various grains. Inclusion of naked oats, wheat or corn to the diet had the greatest improvement in digestibility with organic matter digestibilities being 0.728 g/kg^{-1}, 0.728 g/kg^{-1} and 0.726 g/kg^{-1}, respectively. Inclusion of barley in a hay silage diet improved OM digestibility eleven percent and that of oats six percent. Differences in protein digestibilities were somewhat smaller, but the order of the grains was similar. Barley had better crude protein digestibility than corn. The diet containing naked oats had the highest digestibility of crude protein, 0.812 g/kg^{-1}. Naked oats seemed to be comparable to wheat based on diet digestibilities. Therefore, hulling of oats might be a recommended way to improve nutritive value of oats. Furthermore, according to recent studies the digestibility of starch in oats is better than in corn because of the physical structure of oat starch. More research is being conducted to compare digestibilities of different varieties of oats.

CHEMICAL CHARACTERIZATION AND IN VITRO FERMENTATION OF HIGH TEMPERATURE DRIED ALFALFA AND GRASS HAY

J. MURRAY[1], A.C. LONGLAND[1] AND M. MOORE-COLYER[2]
[1] IGER, Aberystwyth, U.K.
[2]Aberystwyth University, U.K.

Grass hay is the traditional conserved forage for equines in the UK. However, much UK grass hay is of low nutritive value and despite increased intakes, many equines fed a hay-only diet are in negative energy balance. Thus, there is increasing interest in alternative conserved forages. High temperature dried (HT) alfalfa provides a source of high quality protein and energy from digestible fiber. It also contains valuable micronutrients such as calcium, vitamins A and E, and the B vitamins thiamin, riboflavin, biotin and folic acid. As gas is produced during the fermentation of feedstuffs, an automated gas production system developed for evaluating feeds for ruminants was employed to follow the fermentation kinetics of ground grass hay (H), high temperature dried chopped alfalfa (A), molassed chopped high temperature dried alfalfa (MA) and ground high temperature dried alfalfa (GA). These four feeds were incubated with an equine hindgut microbial inoculum, and the evolution of gas was monitored over 96 h. End point dry matter (DM) disappearance of feeds was related to total gas pool. The total volume of gas evolved was greatest from A > MA > GA >H. Initial rates of gas production were greatest from GA > A > H. In conclusion there is greater potential for maintaining equines on diets containing alfalfa than on hay alone. Furthermore, this technique appears to have considerable potential for evaluating forages for equines in vitro.

RELATIONSHIP BETWEEN NDF AND HAY INTAKE IN HORSES: A REVIEW OF PUBLISHED STUDIES

A. ST. LAWRENCE, R.J. COLEMAN AND L.M. LAWRENCE
University of Kentucky, Lexington, KY

Forage is an important part of any horse feeding program, and accordingly it is important to be able to accurately predict the amount of forage a horse will consume. A review of the literature was undertaken to determine if a relationship exists between forage chemical composition and voluntary dry matter intake (VDMI) in mature horses. Six studies representing VDMI of 21 different forages (4 alfalfa and 17 grass hays) were used. Grass hays included both cool and warm season varieties. The only forage chemical components that were consistently reported across all studies were crude protein (CP) and neutral detergent fiber (NDF). Because of this, the relationships of CP to VDMI and NDF to VDMI were examined. The CP content of the grass hays ranged from 8.0 to 17.5% while the legume hays ranged from 14.8 to 20.9%. The NDF content of the grass hays ranged from 59.5 to 74.9% while the legume hays ranged from 40.9 to 56.3%. All studies used mature horses that were allowed ad libitum access to long hay. Only studies that reported animal weights specific to each trial and measured intake over a minimum of 4 d were used. All reported VDMI values were converted to g/kg bodyweight/d. Crude protein was not strongly related to VDMI (r^2 = .14, grass hays; r^2 = .34, grass and legume hays combined). The relationship of NDF to VDMI for grass hays alone is expressed by the equation $y = 124.55 + .0155x^2 - 2.5742x$ (r^2 = .67) where x = % NDF and y = g/kg bodyweight/d VDMI. The relationship for grass and legume hays combined is expressed by the equation $y = 18.377 - .0051x^2 + .3895 x$ (r^2 = .53). No equation was calculated for alfalfa hays alone because only four VDMI values were available. The lower correlation between NDF and VDMI when both grass and legume hays were analyzed may be related to differences in the structure and digestibility of grasses and legumes. The data reviewed from these papers suggest that, when compared to CP, NDF values will provide a better prediction of the dry matter intake of mature horses fed long stem hay.

EFFECT OF AN ALUMINUM SUPPLEMENT ON NUTRIENT DIGESTIBILITY AND MINERAL METABOLISM IN THOROUGHBRED HORSES

K. A. ROOSE, K. E. HOEKSTRA, J. D. PAGAN, R. J. GEOR
Kentucky Equine Research, Inc., Versailles, KY

Summary

The effect of aluminum supplementation on nutrient digestibility and macro- and micromineral balance was studied in balance trials in mature Thoroughbred horses (n=4) in a replicated 2 X 2 Latin square experiment, with each period lasting four weeks. A 5-day complete digestion trial was performed at the end of each period. The treatments were: 1) a basal diet that consisted of 2 kg/d unfortified sweet feed, 6.8 kg/d mixed hay and 1 oz sodium chloride, and containing 159.90 ppm of aluminum; and 2) the basal diet plus 224 g of an aluminum-containing supplement (30301 ppm aluminum in the form of dihydroxy-aluminum sodium carbonate and aluminum phosphate), and providing 931 ppm aluminum. There was no effect of aluminum supplementation on nutrient digestibility or the metabolism of calcium, phosphorus, magnesium, zinc, copper and boron. However, urinary iron excretion was higher ($P < 0.05$) for the aluminum-supplemented diet compared to the basal diet. It is concluded that short-term consumption of a diet containing 930 ppm aluminum has negligible effect on nutrient digestibility and mineral metabolism in horses.

Introduction

In ruminants and horses, it has been reported that high levels of aluminum in the diet adversely affect the metabolism of other minerals, particularly calcium and phosphorus (Allen, 1984; Schryver et al., 1986). Allen (1984) reported that diets containing greater than 1500 ppm aluminum (as $AlCl_3$) reduced phosphorus absorption and increased the dietary phosphorus requirement in sheep. Allen (1984) also found that a single large dose of aluminum in ruminants resulted in a significant decrease in phosphorus incorporation into tissue, including blood, liver, kidney, brain, muscle and bone. In ponies, Schryver et al.(1986) demonstrated a 30% decrease in phosphorus absorption in ponies fed a diet containing 4500 ppm aluminum. Although calcium absorption was unaffected by this level of aluminum intake, the ponies were in negative calcium balance because of greater urinary excretion of calcium compared to the basal diet. These authors speculated that the increase in urinary loss of calcium was due to reduced use of calcium in the formation of hydroxyapatite, presumably because of the lack of phosphate. On the other hand, aluminum intake of 1500 ppm or less does not appear to adversely affect mineral metabolism in horses. In the study by Schryver et al. (1986), a diet containing 1370 ppm aluminum had no effect on macro- or micromineral balance. Similarly, in ruminants, there was little effect of aluminum on mineral metabolism at levels of aluminum intake below 1000 ppm.

119

Gastric ulceration is common in horses, particularly those in race training. Antacids, basic compounds that neutralize acid in the gastric lumen, are frequently administered to horses for the treatment and prevention of gastric ulcers (MacAllister, 1999). An antacid specifically formulated for horses has recently become available (Neigh-Lox™). As the active ingredient in this product contains aluminum, one concern is the effect of high aluminum intake on mineral metabolism. Therefore, the current study was undertaken to determine the effects of ingestion of this aluminum-containing product on nutrient digestibility and mineral balance in Thoroughbred horses.

Materials and Methods

Four mature Thoroughbred geldings (mean age of 12 years, mean body weight 568 kg), were used in a replicated 2 x 2 Latin square design. Each dietary period consisted of a 21-day adaptation period and a five day complete collection digestion trial. Horses were turned out into paddocks during the day with muzzles to prevent grazing (0830 to 1600) and were housed in 3.0 m x 3.0 m stalls overnight (1600 to 0830). The basal diet (CON) consisted of 6.8 kg mixed hay, 2 kg unfortified sweet feed and 1 oz sodium chloride (39.34% sodium and 60.66% chloride) (Table 1). The grain ration was divided into two equal feedings (0700 and 1600 h), and forage was provided at 0700, 1600 and 2200 h. Fresh water was available at all times. The treatment (AL) diet was the basal diet plus 8 oz per day (4 oz in each grain meal) of an aluminum-containing supplement (Neigh-Lox™), KPT Technology Inc., Midway, KY) (Table 1). The basal and aluminum-supplemented diets were 159 and 931 ppm aluminum per kg of feed, respectively, and provided approximately 1 and 12 mg of aluminum/kg body weight/day.

After a 3-week period on each diet, the horses were fitted with collection harnesses (Equisan PTY Ltd., Melbourne, Australia) that allowed complete and separate collection of urine and feces. Intake of feed and water and excretion of feces and urine were measured daily for 5 days (during the last week of each dietary period). Composite samples of feed, feces and urine representing the 5 days of collection were prepared. Blood samples for determination of serum biochemistry were taken before and after each dietary period. Feed and feces were analyzed for dry matter, crude protein, acid detergent fiber, neutral detergent fiber, hemicellulose, fat, nonstructural carbohydrates, calcium, phosphorus,

TABLE 1. Nutrient composition of the diet.

Nutrient	Grain[2]	Forage	Supplement*
Dry matter, %	85.88	90.77	91.65
Crude protein (CP), %[1]	0.98	19.50	8.76
Acid detergent fiber (ADF), %[1]	8.92	26.67	2.20
Neutral detergent fiber (NDF), %[1]	22.20	42.83	8.11
Hemicellulose, %[1,3]	13.28	16.16	5.91
Fat, %[1]	4.53	1.92	6.88
Ash, %[1]	3.00	8.82	18.73
Nonstructural carbohydrates (NSC),%[1,4]	9.29	6.93	57.52
Calcium, %[1]	0.14	1.19	1.04
Phosphorus, %[1]	0.33	0.31	2.53
Magnesium, %[1]	0.17	0.16	0.11
Potassium, %[1]	0.87	2.41	0.64
Sodium, %[1]	0.08	0.02	1.67
Chloride, %[1]	0.00	0.00	0.00
Iron, ppm[1]	119.00	141	421
Zinc, ppm[1]	29.00	18.7	6.20
Copper, ppm[1]	10.10	8.90	10.40
Manganese, ppm[1]	24.50	53.40	31.40
Aluminum, ppm[1]	79.60	80.30	30301.40
Boron, ppm[1]	6.125	27.24	4.412

[1] 100% dry matter basis
[2] 45% cracked corn, 45% whole oats, 10% molasses
[3] Hemicellulose = NDF - ADF
[4] NSC = 100 - (CP + NDF + Fat + Ash)
* Neigh-Lox™

magnesium, potassium, sodium, chloride, iron, zinc, copper, manganese, aluminum and boron. Urine samples were analyzed for calcium, phosphorus, magnesium, potassium, sodium, chloride, iron, zinc, copper, manganese, aluminum and boron (Rock River, Watertown, WI, USA). The apparent digestibility of each nutrient was calculated. Serum chemistry measurements were made using an automated analyzer (Cobas Mira Analyzer, Rood and Riddle, Lexington, KY, USA).

Data were analyzed by analysis of variance with general linear model procedures. The model included horse, period and treatment as main effects. Treatment comparisons were made by a paired student t-test.

TABLE 2. Apparent digestibility of horses receiving the control diet or the diet supplemented with aluminum.

% Digestibility	Treatment		
	CONTROL (n = 4)	ALUMINUM[1] (n = 4)	SEM
Dry matter	58.65	58.09	0.78
Crude protein (CP)	67.61*	65.97	0.24
Acid detergent fiber (ADF)	21.11	19.28	1.55
Neutral detergent fiber (NDF)	35.72	34.45	1.71
Hemicellulose	57.18	56.55	2.02
Fat	28.91	32.00	4.40
Ash	35.23	30.61	1.83
Nonstructural carbohydrates (NSC)	86.83	87.52	0.25
Calcium	35.55	40.90	1.00
Phosphorus	-35.75	-28.75	2.42
Magnesium	-6.52	-5.50	2.47
Potassium	62.54	58.75	1.35
Sodium	18.20	24.86	4.28
Chloride	77.77	75.98	1.90
Iron	-155.34	-131.89	33.55
Zinc	-262.53	-187.78	15.62
Copper	-189.86	-155.78	26.00
Manganese	-147.65	-151.44	6.94
Aluminum	-303.15*	563.64	37.25
Boron	-58.85	-39.83	10.79

[1]Neigh-Lox™

*Significant difference between Control and Aluminum-supplemented diet, P < 0.05 (SEM, standard error of the mean)

TABLE 3. Urinary mineral excretion in horses receiving the control diet or the diet supplemented with aluminum.

	Treatment		
	CONTROL (n = 4)	ALUMINUM[1] (n = 4)	SEM
Calcium (g)	19.07	20.57	1.94
Phosphorus (g)	0.30	0.33	0.02
Magnesium (g)	3.83	4.21	0.32
Potassium (g)	102.00	98.29	2.05
Sodium (g)	5.89	6.33	0.70
Chloride (g)	71.16	68.39	2.09
Iron (mg)	201.40	255.70*	6.00
Zinc (mg)	41.73	43.97	3.90
Copper (mg)	37.18	38.17	4.63
Manganese (mg)	4.81	5.87	0.53
Aluminum (mg)	205.75	198.10	18.94
Boron (mg)	114.45	127.56	10.08

[1]Neigh-Lox™

*Significantly greater (P < 0.05) compared with Control treatment

TABLE 4. Effect of basal and high aluminum intakes on mineral metabolism in horses fed the two diets.

	Intake	Urine	Feces	Retained
Aluminum Treatment				
		Calcium		
Basal	135.2	34.0	85.8	15.4
High	138.0	36.3	82.6	19.1
		Phosphorus		
Basal	44.2	0.50	60.2	-16.5
High	53.3	0.60	68.6	-16.0
		Magnesium		
Basal	22.8	6.8	24.1	-8.2
High	23.1	7.4	24.5	-8.9
		Potassium		
Basal	291.7	180.9	108.4	2.3
High	292.0	173.2	123.1	-4.3
		Sodium		
Basal	24.5	10.6	19.3	-5.4
High	30.6	11.2	23.5	-4.1
		Chloride		
Basal	140.1	126.4	31.4	-17.7
High	140.2	121.2	34.1	-15.1
		Iron		
Basal	1.92	0.36	4.95	-3.40
High	2.06	0.46	5.15	-3.55
		Zinc		
Basal	295.8	71.6	1020.7	-796.5
High	314.9	79.9	945.2	-710.2
		Copper		
Basal	128.7	67.9	245.9	-185.1
High	131.8	66.7	219.2	-154.0
		Manganese		
Basal	662.2	8.7	1215.6	-562.1
High	669.3	10.6	1276.9	-618.1
		Aluminum		
Basal	1.12	0.36	4.10	-3.34
High	12.32	0.36	16.1	-4.10
		Boron		
Basal	318.4	201.9	240.1	-123.6
High	317.7	229.9	185.1	97.2

Calcium, phosphorus, magnesium, potassium, sodium, chloride, iron and aluminum values are mg/kg of body weight/day. Values for zinc, copper, manganese and boron are µg/kg body weight/day.

TABLE 5. Serum biochemistry values in horses receiving the control diet or the diet supplemented with aluminum.

	Treatment		
	CONTROL	ALUMINUM[1]	
	(n = 4)	(n = 4)	SEM
Blood urea nitrogen (mg/dl)	17.75	19.00	1.60
Creatinine (mg/dl)	1.42	1.45	0.10
Aspartate aminotransferase (iu/l)	259.50	425.75	189.09
Total bilirubin (mg/dl)	2.20*	1.62	0.09
Direct bilirubin (mg/dl)	0.40	0.28	0.07
Alkaline phosphatase (iu/l)	77.25	65.25	4.14
Lactate dehydrogenase (iu/l)	190.25	253.00	85.14
Creatine phosphokinase (iu/l)	149.25	938.75	803.13
Sorbitol dehydrogenase (iu/l)	142.50	160.50	60.7
g-glutamyl transferase (iu/l)	13.75	11.75	0.71
Albumin (g/dl)	3.50	3.43	0.08
Calcium (mg/dl)	3.65	13.4	0.24
Phosphorus (mg/dl)	3.73	3.50	0.22
Glucose (mg/dl)	88.25	88.00	5.80
Sodium (mmol/l)	137.00*	133.75	0.25
Potassium (mmol/l)	3.50	3.63	0.05
Chloride (mmol/l)	100.00	98.25	0.63
HCO$_3$ (mmol/l)	25.00	24.00	0.41

[1]Neigh-Lox™
*Significantly greater (P < 0.05) compared with Aluminum treatment

Results

Consumption of a diet with a moderately high aluminum content (930 ppm) had minimal effect on nutrient digestibility and mineral balance. High aluminum intake resulted in a significant (P<0.05) reduction in the apparent digestibility of crude protein, but this difference was very small (67.61% in CON vs. 65.97% in AL) (Table 2). In general, external balance of the macro- and microminerals also was unaffected by the high aluminum intake. However, urinary excretion of iron was significantly (P<0.05) greater when the horses consumed the high aluminum diet (Table 3). There was no effect of high aluminum intake on the absorption, excretion or retention of calcium, phosphorus, sodium, potassium, chloride, magnesium, zinc, copper, manganese and boron (Table 4). Retention of aluminum was not affected by the level of aluminum in the diet. For both dietary periods, serum biochemical variables were within reference limits for the analytical laboratory (Table 5). Serum sodium and total bilirubin were significantly (P<0.05) higher for the control diet compared to the aluminum-supplemented diet.

Discussion

The results of this study indicate that short-term consumption (one month) of a diet containing moderately high levels of aluminum (930 ppm) has negligible effect on nutrient digestibility and external mineral balance in horses. These findings are consistent with the results of a previous study in ponies, wherein an aluminum intake of 1370 ppm had no effect on mineral metabolism (Schryver et al., 1986). Furthermore, aluminum intakes of less than 1000 ppm in ruminants do not adversely affect mineral metabolism (Allen, 1984). The primary concern with high aluminum intake is reduced phosphorus absorption (Allen 1984; Schryver et al., 1986). Although the mechanism for this inhibition of phosphorus absorption has not been determined, it has been hypothesized that insoluble, nonabsorbable aluminum-phosphate complexes form in the intestinal tract (Allen, 1984). Importantly, 50% of the aluminum in the supplement fed in the present study was in the form of aluminum phosphate. This form of aluminum would be expected to have minimal effect on the digestion and absorption of dietary phosphorus. It should also be emphasized that, in equids, adverse effects of dietary aluminum on calcium and phosphorus metabolism have only been documented at extremely high levels of intake (4500 ppm or more than four times greater than the aluminum intake in the present study).

In summary, the results of this study indicate that short-term consumption of an aluminum-containing nutritional supplement, when fed at the manufacturer's recommended level of intake, has negligible effect on nutrient digestibility and external mineral balance in mature horses.

References

Allen, V. G. (1984) Influence of dietary aluminum on nutrient utilization in ruminants. Journal of Animal Science, 59(3): 836-843.

MacAllister, C. G. (1999) A review of medical treatment for peptic ulcer disease. Equine Veterinary Journal, Supplement 29: 45-49.

Schryver, H. F., Millis, D. L., Soderholm, L. V., Williams, J. and Hintz, H. F. (1986) Metabolism of some essential minerals in ponies fed high levels of aluminum. Cornell Veterinary, 76: 354-360.

FORAGE ANALYSIS: THREE POINTS TO CONSIDER

PAUL K. SIROIS
Dairy One Forage Lab, Ithaca, NY

Introduction

Great strides have been made over the last few decades in our understanding of nutrient requirements for livestock. As we strive to do a better job formulating rations and meeting those nutrient requirements, better knowledge of feed composition becomes essential. Simply using average or tabular values is no longer sufficient to describe forages. Commercial feed analysis is now a routine part of much ration development. As testament to that fact, the Dairy One Forage Lab began operations in 1974 (as the NY DHIC Forage Lab) and processed 5000 samples that first year. Today, the lab processes in excess of 115,000 samples per year.

This paper will address three relevant aspects of the analysis process:

1. Representative sampling.
2. How not to submit a hay sample for analysis.
3. Near infrared reflectance spectroscopy (NIR).

Representative Sampling

Obtaining a representative sample is the first and most critical step of the analysis process. Unfortunately, it is often the most overlooked. Laboratories are staffed and equipped to do the best job possible of analysis. Quality assurance (QA) programs are employed to monitor the integrity of results. Internal QA programs usually involve analyzing daily check samples of known value and are used to insure consistency of results. Many labs also participate in external sample check programs. Sponsoring organizations submit periodic samples of unknown value to participating labs. Results are sent back to the sponsoring organization and compared to the results of other participating labs. The function of the external program is to insure that results are consistent with other labs in the industry. Thus, internal check programs serve to maintain consistency from day to day and external programs insure consistency from lab to lab.

The labs, however, have no control over the first and most important step, that of obtaining a representative sample. Labs can only analyze what they receive. They will do as good a job of analyzing a well taken sample as they will a poorly taken sample. In the latter case, you wind up with a good analysis of a poor sample. Thus, it is the responsibility of the "sample taker" to obtain a representative sample.

A paper presented by Martin et al. (1988) demonstrated the importance of multiple subsampling to form a composite sample. Twenty bales of alfalfa hay from the same lot were individually sampled. The core samples from each bale were individually analyzed by NIR. The results in Table 1 illustrate the variation in nutrient composition from bale to bale. For example, protein ranged from 18.2 – 22.4% and NDF from 33.7 – 54.1%. Relative feed value (RFV) is a forage score based on ADF and NDF reflecting the digestibility and intake potential of haycrop forages for ruminants. The RFV ranged from 103 – 184.

The individual samples were then combined to form a composite. As can be seen in Table 1, the composite analysis was equivalent to the arithmetic mean of the 20 individual samples. This clearly illustrates the importance of gathering multiple subsamples to truly reflect the quality of a lot of hay.

Table 1. Quality tests of single bales of alfalfa hay from the same lot.*

Bale No.	DM%	CP%	ADF%	NDF%	RFV
1	87.9	18.2	35.3	44.6	128
2	86.7	18.4	35.8	48.7	117
3	86.6	18.4	36.1	44.3	128
4	87.3	18.9	32.5	39.0	152
5	88.4	19.8	31.4	38.3	156
6	87.1	19.8	32.7	41.5	142
7	85.9	20.3	32.7	40.0	148
8	88.0	20.3	31.5	38.5	156
9	85.6	20.3	36.9	54.1	103
10	85.5	20.4	32.1	40.6	146
11	87.4	20.5	32.0	39.2	152
12	86.9	20.5	32.5	39.1	151
13	86.4	20.8	31.5	41.2	145
14	86.2	20.8	33.4	42.0	139
15	88.0	21.2	30.3	35.7	170
16	84.7	21.3	31.4	38.5	156
17	86.8	21.4	29.3	33.9	181
18	89.9	21.5	28.6	33.7	184
19	85.2	21.9	32.1	40.3	148
20	87.8	22.4	29.4	37.0	166
Minimum	84.7	18.2	28.6	33.7	103
Maximum	89.9	22.4	36.9	54.1	184
Average	86.9	20.4	32.4	40.5	148
Composite A	88.1	20.7	31.5	40.7	147
Composite B	88.0	20.3	31.7	41.0	146

*All results DM basis.
Adapted from Martin et al. (1988). The data were sorted by crude protein.

In practice, it is not uncommon for people to sample one to three bales. If, by chance, the poorest bale is selected, ration recommendations would result in overfeeding and vice versa. Both situations will have nutritional and economic consequences as feed cost is typically one of the largest costs on any livestock operation.

Thus, if you are willing to invest the time, effort and dollars in forage analysis to better formulate rations, it is in everyone's best interest to do the best job possible obtaining a representative sample.

In practice, the greater the number of subsamples the better. In reality, sampling 10 – 12 bales should provide a good representative sample.

Sampling Techniques

HAY

Now that you know it is important to take and composite multiple subsamples to submit for analysis, what is the proper way to collect those subsamples? The following are ***unacceptable*** when submitted for analysis:

1. A flake or slab of hay.
2. A handful of hay pulled from a bale.
3. A handful of hay grabbed from the manger.
4. A handful of hay grabbed and cut up with scissors.

It is very difficult for the lab to obtain a representative sample from these submissions. Any sampling of dry forages that involves grabbing a handful of material usually results in a subsample that is poorer in quality than the actual nutrient content. This is particularly true with alfalfa, because grab sampling usually results in a fistful of stems with the finer and more fragile leaves shaking off. Leaves contain most of the nutrients, being higher in both protein and digestibility than the stems. Any procedure that results in leaf loss will have a negative impact on the analyzed value. The opposite also holds true; any sampling that results in concentrating the leaves will make the sample look better than the forage actually is.

The only way to obtain a proper hay sample is by using a bale probe or corer. This is typically a metal tube from 38 – 48 cm (15 – 18 in) long and sharpened at one end. Depending upon the type of probe, it is either hand operated or may be coupled to an electric drill. Bales should be probed in the center of the small, square end. The probe takes a representative cross section as it spins and cuts its way through the bale. The resulting core sample will proportionately reflect the leaf and stem material in the bale. Typically, obtaining and combining 10 – 20 core samples will form a good composite sample.

PASTURE

The key to sampling pasture is sampling multiple sites. Randomly select 12 – 20 sites where the animals have been grazing and clip a handful of forage at grazing height. Grazing height is the level to which the animals are consuming. For example, if the grass is 25 cm (10 in) high and the animals are consuming the top 15 cm (6 in), it is only the top 15 cm that should be submitted for analysis. All subsamples should be combined and thoroughly mixed in a clean plastic bucket to form a composite (further cutting the forage into 5 – 8 cm (2 – 3 in) pieces aids in blending). Take a 0.5 kg (1 lb) sample, pack tightly in a plastic bag, seal and freeze for 12 hours prior to submission. Freezing will help prevent marked chemical changes due to respiration or fermentation.

It is also important to remember that pasture is wet. This may seem obvious, but not many people realize its significance. For example, pasture is typically about 20% dry matter. A 500 gram sample submitted for analysis will be split and half will be used for analysis. Upon drying the 250 gram split sample, 50 grams will be left. After grinding, 45 grams will be left for analysis or less than 10% of the original weight.

In several instances, we have received pasture samples of 6 – 20 blades of grass. This is not enough sample to even begin attempting an analysis.

Table 2. Weight losses from drying and grinding prior to analysis.*

Wet Weight,g	Dry Weight,g	Post Grind Weight	Post Grind Loss,g	Post Grind Loss,%	Dry Matter,%
10	1.8	1.1	0.7	38.9	18.0
20	3.7	3.0	0.7	18.9	18.5
30	5.7	4.7	1.0	17.5	19.0
40	7.4	6.6	0.8	10.8	18.5
50	9.6	8.5	1.1	11.5	19.2
60	11.1	10.0	1.1	9.9	18.5
70	12.4	11.2	1.2	9.7	17.7
80	14.5	13.6	0.9	6.2	18.1
90	17.2	15.5	1.7	9.9	19.1
100	18.9	17.2	1.7	9.0	18.9
200	34.7	33.1	1.6	4.6	17.4
300	56.5	51.5	5.0	8.8	18.8
400**	**76.5**	**59.5**	**17.0**	**22.2**	**19.1**
500	94.2	87.2	7.0	7.4	18.8

* The same grass sample was used for all original weights.
**Unexplained large sample loss post-grinding.

Table 2 illustrates starting and ending sample weights after drying and grinding. With the exception of the 400 gram samples, the average sample loss during initial preparation is 12.5%. Thus, it is important to take into account the moisture level of the sample prior to submission to insure that the lab has an adequate amount of material for analysis. The wetter the sample, the greater the amount of sample required.

As mentioned above, it is advisable to freeze or dry pasture samples prior to submission. Samples shipped internationally should be predried to avoid spoilage and marked chemical changes during shipment. Samples can easily be dried in a microwave oven. Dried samples are also less expensive to ship.

A lot of money is invested in the shipping and analysis process. Greater economic potential rides on the results. The highest return per dollar invested is realized if time is taken at the outset to follow simple collection procedures.

NIR Technology

Near infrared reflectance (NIR) spectroscopy is a sophisticated analytical technique used for determining the chemical characteristics of agricultural and food products, pharmaceuticals and beverages. It is based on the fact that each of the major chemical components of a sample has near infrared absorption properties that can be used to differentiate one component from another. Once a sample has been dried and ground, a NIR analysis can be completed in about 60 seconds, yielding up to 18 nutrients.

Advantages:
1. Accuracy – advancements in computer hard and software have provided the tools to take full advantage of the technology.
2. Speed – customers demand fast turnaround time. NIR analyses can be completed in under 24 hours.
3. Cost – analyses are typically half the price of wet chemistry.
4. Labor efficiency – more analyses can be completed in a shorter period of time with less labor.
5. Safety – eliminates the use of hazardous chemicals.
6. Environmentally friendly.

Disadvantages:
1. Expense and time required to build new calibrations.

More on Accuracy

Several factors influence the accuracy of NIR measurements:
1. Does the component have NIR reflectance properties? Each of the major organic feed components has absorption characteristics (due to vibrations arising from the stretching and bending of H bonds associated with C, O, N) in the near

infrared region that are specific to that component (Marten et al., 1985). NIR is most sensitive to organic compounds.Compounds lacking the above properties will not calibrate as well.

2. Robustness of the calibration set. There must be adequate variation in the population of samples used for developing the calibration. The variation must be inherently reflective of the sample population as a whole. The more closely the calibration set resembles the sample population, the better the performance of the calibration.

3. Accuracy of the reference method. A NIR measurement can only be as accurate as the reference method used to develop the calibration. For example, crude protein is accurately measured by Kjeldahl and can be calibrated quite well. The detergent fiber methods of analysis are not as precise, and therefore are less well determined by NIR. For example, the standard error of calibration (SEC) of the Dairy One hay calibration for CP is 0.63 while for ADF is 1.52 (Table 3).

4. Calibration updates. New varieties and hybrids are introduced every year. In order to keep calibrations current, they must be continually expanded to include new genetics. Software routines ease the process of identifying new samples for calibration expansion. Each sample has its own spectral fingerprint (spectra). Software comparisons of new spectra to existing spectra in the calibration database identify samples to add for expansion.

In an ideal world, calibrations would exist for each different forage species. For example, alfalfa, timothy and tall fescue would have their own individual calibrations. In the real world, most samples are mixtures of different species and are often not well identified. Thus, typical commercial calibrations are developed to cover a broad range of samples. For example, a typical hay calibration starts with a poor grass hay (5% CP) and ends with a high quality alfalfa (28% CP). The goal is to cover a wide variety of qualities, species and mixtures.

When all of the above are taken into account, excellent calibrations can be developed. Table 3 compares the proficiency of NIR to wet chemistry. Following the evolution of software, calibration refinements will lead to further enhanced accuracy. NIR is recognized by the Association of Official Analytical Chemists (AOAC) as an official method.

Table 3. Dairy One NIR & wet chemistry standard errors of analysis for major components.

Component	RSQ	NIR SEC	Wet SE
CP%	0.99	0.63	0.26
ADF%	0.95	1.52	1.26
NDF%	0.97	2.36	1.40
Ash%	0.90	0.66	0.40
Fat%	0.81	0.31	0.15

RSQ = r squared statistic for Dairy One hay calibration.
NIR SEC = standard error of calibration.
Wet SE = standard error for repeated measures from Dairy One 1/00 QA report averaged over several feed types.

Minerals

NIR does not measure minerals directly. Minerals are indirect measurements based on relationships with other components. Predicted mineral values will be better than average tabular values, but it must be understood that they may not be the absolute values. Given these restrictions, the results are quite good. Table 4 lists the statistics for the Dairy One hay calibration.

Table 4. Dairy One NIR & wet chemistry standard errors of analysis for minerals.

Component	RSQ	NIR SEC	Wet SE
Ca%	0.89	0.169	0.068
P%	0.70	0.039	0.029
Mg%	0.73	0.041	0.010
K%	0.85	0.276	0.469

RSQ = r squared statistic for Dairy One hay calibration.
NIR SEC = standard error of calibration.
Wet SE = standard error for repeated measures from Dairy One 1/00 QA report averaged over several feed types.

NIR mineral results are routinely used by the feed industry for ration formulation. Wet chemistry minerals should be substituted when precise formulation is required for exceptional circumstances. This could be for rations where the animals are not performing as expected or for high performance rations where fine tuning to the "nth" degree is desired.

The bottom line is: if mineral concentrations and their balance with other elements is of paramount concern, wet chemistry minerals should be used for ration balancing.

The Future

The next generation of NIR software is available and will soon be in commercial use. In the past, *global* calibrations were developed for specific feed types. For example, thousands of hay samples would be collected from which 500 - 2000 would be used for calibration development. A single hay global calibration would then be used to analyze all future samples. In the next generation, databases will be built for a particular feed. When a sample is scanned, its spectra will be compared to all of the spectra in the database. The software will then select 100 samples that most nearly resemble the sample being scanned and develop a calibration specifically for that sample. Thus, a *local* calibration will be developed for each sample as it is analyzed. This will eliminate the need for broad based calibration development. The focus will shift to database expansion and increasing diversity. Thus, calibrating will no longer be an issue as it will be handled on an individual sample basis. The end result is enhanced accuracy.

Summary

Routine feed analyses are an essential component of ration development. For meaningful results, time must be invested in the sample collection process. Multiple subsamples must be taken to form a representative composite. The greatest return per dollar invested will be realized by following a few simple procedures to insure that a good sample is submitted for analysis.

NIR forms the basis for most commercial feed analyses. At Dairy One, NIR accounts for 69% of all analyses performed. Enhancements in computer hard and software have enabled the industry to make full use of this powerful technology. The next great leap will be the use of local rather than global calibrations. Individual calibrations will be developed for each sample as it is analyzed. This should result in enhanced accuracy. NIR expansion in the market will continue as new calibrations are introduced.

References

Marten, G.C., J.S. Schenk and F.E. Barton II, editors. 1985. Near Infrared Reflectance Spectroscopy (NIRS): Analysis of Forage Quality. USDA, Agricultural Handbook No. 643.

Martin, N.P., R.L. Ellingboe and J.G. Linn. Proceedings 18th National Alfalfa Symposium, p. 92-102, March 2 & 3, 1988, St. Joseph, Mo. Missouri Cooperative Extension and Certified Alfalfa Seed Council.

HARD TO FIND NUTRIENTS FOR RATION EVALUATIONS: FILLING IN THE HOLES

MIKE LENNOX

Kentucky Equine Research, Inc., Versailles, KY

Meaningful ration evaluations are as dependent on the values for the nutrient concentrations of the feed ingredients used as they are on the estimation of the nutrient requirements of the horse. Accurate and complete sets of nutrient values for the feed ingredients used in horse rations are seldom published. If ration evaluations are to stay current as recommendations for horse requirements are updated, more emphasis must be placed on creating and maintaining a database of ingredients used in horse rations.

Information that is currently available has been borrowed from information used in other species. Ingredients that horse rations share with the monogastric diets are well documented for their vitamin, mineral and amino acid content. The fiber contents of ingredients common in ruminant diets are also well documented, although the fiber component most often used in ruminant calculations, acid detergent fiber (ADF), falls short of being a complete analysis for equine diets. This leaves some holes in information that is available to include in equine diets. Many of the forage ingredients that are common in equine diets are not common in ruminant diets and have not been routinely analyzed for the fiber components. These same ingredients are even less common in monogastric diets and are therefore seldom analyzed for amino acid, mineral or vitamin content. It will be the responsibility of those using these ingredients in horse rations to accumulate the information to fill these holes.

Values for nutrient concentrations to be used in a ration evaluation can come from one of four sources.

Laboratory analysis of the ingredient

Chemical analysis of a feed product or ingredient can supply values for many of the nutrients that are used directly for ration evaluation. Most of the mineral components as well as the protein levels can be measured directly and the results used as reported for formulation. This is the method of choice for ingredients that are unique to a specific ration, such as when a client uses a particular local hay. Nutrients that are usually measured directly and used for the evaluation are dry matter, protein, calcium, phosphorus, magnesium, sodium, potassium, copper, zinc, manganese and iron.

Laboratory analysis of the ingredient with a calculation

Some nutrients are not measured directly by lab analysis but can be calculated from the results of related measurements. Energy, although not a single nutrient by definition, is used as one in most ration evaluations. Energy cannot be measured directly in the laboratory, so it is calculated from the values of nutrients that can be measured directly and contribute to the energy-producing characteristics of that feed. Amino acids can be measured directly for a feed sample but for single samples are more often calculated using the crude protein measurement and a calculation of the normal amino acid concentration in the particular feed.

Feed composition tables

Values used in evaluations can be taken from feed composition tables for the ingredient that most closely resembles the one being used. The values can be from printed tables like those published in Nutrient Requirements of Horses (NRC, 1989) or the default values supplied in formulation programs. Tables produced for use with other species can be used if care is taken so that the units used to quantify the nutrients are appropriate for horses.

The guaranteed analysis of a product

Any commercial product must have a guaranteed analysis for some nutrients. The nutrients that are guaranteed are dependent on the regulation that is applicable in the state where they are to be sold. These guarantees are usually very limited in number and by no means a complete list of nutrients in the product. Published promotional literature of commercial products often has a more complete list of nutrients than that of the guaranteed analysis.

Calculations to be Used in Conjunction with Laboratory Analysis

Energy

The variation in energy content of the feed ingredients causes greater adjustment to the formulation of equine diets than any other nutrient. The lack of a direct measurement of energy necessitates the estimation of the energy content from the nutrient fractions that can contribute to the total energy-creating potential of the feed ingredient. The more detailed the partitioning of these fractions by laboratory analysis, the more accurate the estimate of the energy content of that ingredient. Some discussion continues about the best system of estimating the energy content of ingredients from chemical analysis. It is important to note that whatever system is used to estimate the energy content of the ingredients must match the system used to estimate the requirements of the horse. A direct measurement of the ability of the horse to derive this calculated energy must be in the loop somewhere. The principle for each of these systems is similar. An estimation of each nutrient fraction that will contribute to energy production by the horse can be made by chemical analysis. The energy-generating capability of each fraction is estimated. The sum of these fractions represents the total energy content of that ingredient. The equations used to estimate the energy content from the chemical components are numerous and vary to match the nutrient fractions that are available from laboratory analysis for entry in the equation.

Equations used to estimate the digestible energy content of feeds for horses.

Parameters needed	Equation to estimate energy DE
ADF	DE = 3.57 - .0401 ADF
NDF	DE = 4.88 - .0769 NDF + .0489 NDF^2
ADF, CP	DE = 4.22 - .111 ADF + .0332 CP + .00112 ADF^2
NDF, CP	DE = 5.18 - .0783 NDF + .0491 NDF^2 - .0199 CP
ADF, NDF, CP, EE, NSC, Ash	DE = 2260 + 14.17 CP - 11.48 ADF - 4.88 (Hemicellulose) + 57.2 EE + 24.38 CHO - 31.77(Ash)
TDN	DE = .255 + .0366 TDN
DDM	DE = .273 + .0351 DDM

ADF= acid detergent fiber, NDF= neutral detergent fiber, CP = crude protein, TDN= total digestible nutrients, CHO = soluble carbohydrate, EE = ether extract, DDM= digestible dry matter, hemicellulose = NDF-ADF

Amino acids from protein

The protein concentration of most feedstuffs is well documented. Analysis of the protein content of forages and grains is more common than any other laboratory test. However, analysis of the amino acid profile of the feeds for equine diets is rare. The amino acid profiles of the ingredients used in monogastric diets, mostly the grains, are well documented. Values for these ingredients can be taken from feed composition tables and used directly in the evaluation. The values for forage products are more difficult to find. As forage can be a large proportion of the equine diet, values for these ingredients must be estimated if a meaningful comparison is to be made with the daily requirement of the horse. The only amino acid with established and widely accepted requirement levels for horses is lysine. In the absence of absolute values for lysine, an estimate calculated from the crude protein level of the ingredient and usual lysine content of the protein found in that ingredient is preferable to using a value of zero or leaving the entry blank. Additional tables and calculations will be needed if other amino acids are to be used in equine ration evaluations. The following are equations that can be used in the absence of known or tabular values for lysine when the crude protein level is known.

Feed type	Equation to estimate lysine from crude protein
Alfalfa hay	Crude protein x .042
Clover hay	Crude protein x .050
Timothy hay	Crude protein x .030
Bermuda grass hay	Crude protein x .035
Bluegrass pasture	Crude protein x .027
Oat hay	Crude protein x .031
Corn silage	Crude protein x .051

Acid base balance from macrominerals

The concentrations of the minerals sodium and potassium are well documented in most ingredients used in horse rations. Even in the absence of tabular values for these minerals they are routinely measured directly by laboratory analysis. The horses' requirements for these minerals are stated as absolute amounts per day or as concentrations within the total diet. For these requirements, the calculations can be made. There is increasing interest in the acid base balance of horse diets. For this calculation the values of the other strong anions, chloride and sulfur, must also be included. To date no specific requirement for chloride has been used so the chloride values of many ingredients used in equine rations have remained unreported. The chloride content of ingredients can be analyzed in the laboratory but is seldom included in the analysis packages. The tabular values for these minerals will become more available for forage ingredients as the dairy industry uses the acid base calculations in ration formulation. The chloride content of some common feeds in horse rations is as follows:

Feed	Chloride concentration %
Alfalfa hay	0.31
Grass hay	0.46
Oats	0.13
Corn	0.04

Trace Minerals not Determined by Lab Analysis and Seldom Listed in Tables

Selenium

Selenium is seldom included in the mineral analysis package of most forage laboratories. The specialized procedures needed to accurately test for selenium make this test cost prohibitive for most individual samples. The selenium content of plant materials is directly proportional to the selenium level of the soil in which the plant was grown. For this reason tabular values for selenium are less precise than other minerals unless the feed composition table is specific to a given region. Localized information available from agronomy extension agents can be a valuable resource for estimating selenium content of feed ingredients grown within a specific region.

Iodine

Iodine levels of feed ingredients are rarely complete in feed composition tables. Laboratory procedures to analyze iodine in feed ingredients for a single sample or ration evaluation are usually cost prohibitive. Testing for iodine is seldom included in forage analysis packages. When iodine levels are reported for feed ingredients, the unit of measure is often to only one decimal place of parts per million. If the requirement for iodine is reported at 0.1 ppm, the reporting of any iodine in the forage portion of the diet would fully meet the requirement. For iodine to be meaningful in ration evaluations, the accuracy of both the reporting of the nutrient levels and the requirements may need to be reassessed.

Vitamin A

Vitamin A levels reported for feed ingredients are not a direct measurement of vitamin A content but a calculation of the carotene content multiplied by a conversion factor. In the NRC requirements for horses the conversion factor used is 400 IU of vitamin A for each g of carotene. Vitamin manufacturers suggest using different conversion factors for each ingredient and for animals of different species and ages. This single conversion factor would suggest that diets containing a majority of the ration as alfalfa would need no additional supplementation of vitamin A. Research has shown that such a diet would not equal the performance of a diet with added vitamin A. For ration evaluations containing a large percentage of alfalfa, the use of this conversion factor will always show the vitamin A supplied far in excess of the requirement. This is not a concern if this excess is from the inclusion of carotene in the diet. However, if these apparent excesses from carotene calculations mask an oversupply of formed vitamin A or cause complacency towards oversupplementation of vitamin A, the validity of the ration evaluation may be compromised.

Vitamin E

Vitamin E contents of forages used in horse diets are seldom analyzed but are often reported in feed ingredient tables. The forage ingredients of the diet often meet the vitamin E requirement for normal health and well-being very easily. The increased demand for vitamin E as an antioxidant in performance diets is most often not dependent on the forage portions of the diet; the higher levels of vitamin E intake will usually come from the supplemented part of the diet. Vitamin E is one of the more expensive nutrients added to equine rations so oversupplementation of diets because of inaccurate or missing values during ration evaluation can be critical.

Feedstuff	Vitamin E (IU/kg)
Alfalfa hay	40
Clover hay	40
Timothy hay	30
Oats	15
Soybean meal	3
Wheat middlings	20
Grass hay	10

Nutrients that May be Used in the Near Future for Ration Evaluations

Fatty acid components of rations

At present there is no requirement set for the amount of fat in a horse's diet and no requirement for any specific fatty acid as there is for linoleic acid in other species. The pet food industry now pays very close attention to specific fatty acid components of the diet. The commercial poultry industry is also concerned both with feeding a minimum amount of a given fatty acid and the ratio between fatty acids in the diet. To make adjustments like this in horse rations, more complete values for the nutrients in ingredients commonly used in equine diets will be needed. One ratio that is currently being used in other species is the ratio of omega 3 and omega 6 fatty acids within the diet. Common equine rations would be proportionally high in omega 6 fatty acids and almost devoid of omega 3. Most of the ingredients that would be used to manipulate this ratio would have known levels of all fatty acid components. Quantifying the amount of omega 6 in the basal ration would be necessary to calculate the ratio. Values for fatty acid composition are published for most ingredients that are used in human and monogastric diets. However, there are few published figures for the fatty acid content of the forages which will always comprise a significant portion of horses' diets. As these are ingredients that are not commonly used in diets for any other species for which fatty acids requirements are established, tabular values will not be soon coming. The responsibility of analysis for these nutrients will fall to those who use them in feed formulation.

Chromium, boron, aluminum and silicon

Chromium, boron, aluminum and silicon are all minerals that will have an increasing visibility in equine diets. All of these minerals are required in amounts less than the microminerals that are currently added to equine diets. The technology is available to analyze for these minerals in feed ingredients and values are sometimes listed in feed composition tables. In spite of this, the concentration of these minerals in the feed ingredients will have little effect on the formulation of feeds. The measurement of the elemental levels of these ultra trace minerals is of minor importance compared to their chemical form. If these minerals are to be added to the diets, they will be in specific chemical form. The absolute levels of these minerals are a concern only if they are suspected of being in excess and potentially toxic. Although they may be added to equine diets as complexes in the future, the level in feed ingredients should have little effect on feed formulation.

As progress is made in updating the horse's requirements for other nutrients, information on the nutrient composition of feed ingredients used in the diets may lag behind. This may prevent the widespread use of equine nutrition advances by those formulating equine rations. These less common nutrient values may be accessible for the ingredients that horse diets share with monogastric diets. Some of the carbohydrate fraction partitioning of forages to estimate energy can be found in information from the dairy industry. Documenting the new nutrients of interest in ingredients that are exclusive to horse diets will be the responsibility of the same people formulating and evaluating horse rations in the future.

References

Association of Official Analytical Chemists (AOAC International) - Official Methods of Analysis, 15th edition. 1990.
BASF. Animal Nutrition 6th Edition. 2000.
Dale, Nick. Ingredient analysis table: 1999 Edition. Feedstuffs, July 30, 1999; 24-31.
Degussa-Huls Corporation Annual Corn Report. 1998.
Fonnesbeck, P.V. Estimating digestible energy and TDN for horses with chemical analysis of feeds. 1981. J. Anim. Sci. 53 (Suppl. 1): 241.
Griewe-Crandell, K.M. 1993. Vitamin A depletion in Thoroughbreds: A comparison of pasture and non-pasture feeding regimes. Proc. 13th Equine Nutr. Physiol. Soc. Symp. 1-2.
NRC. Nutrient Requirements of Horses. Fifth Revised Edition. 1989.
Pagan, J.D. Horse NRC centers around energy, vitamin E biggest change. Feedstuffs, May 29, 1989. Vol. 60, No. 22; 10-11.
United States - Canadian Tables of Feed Consumption. Third Revision. 1982.
Van Amburgh, M.E., J.E. Voorhees, and J. B. Robertson. 1999. Total dietary and soluble fiber content of selected ruminant feeds. Proc. Cornell Nutr. Conf. for Feed Manufacturers. 196.

GLUCOSAMINE INHIBITS NITRIC OXIDE PRODUCTION IN EQUINE ARTICULAR CARTILAGE EXPLANTS

J.I. FENTON, K. CHLEBEK-BROWN, J.P. CARON, M.W. ORTH
Michigan State University, East Lansing, MI

Osteoarthritis (OA) is a progressive degradation of articular cartilage that is a common cause of lameness for athletic horses. Oral supplementation of compounds that prevent cartilage degradation and/or joint injury is an attractive solution for lameness. Glucosamine is a potential antiarthritic compound currently being marketed. It is a naturally occurring, non-toxic molecule that decreased pain and improved mobility in osteoarthritic joints in a number of human studies. In vitro data suggests that glucosamine may increase the synthetic activity of chondrocytes. However, the biochemical basis to support its potential as an antiarthritic agent is not well documented. One molecule thought to accelerate the progression of OA is nitric oxide. Equine chondrocytes have recently been shown to produce nitric oxide in response to two arthritogenic molecules, lipopolysaccharide (LPS) or recombinant interleukin-1ß (rhIL-1ß). In addition, increased concentrations of nitric oxide and interleukin-1 are observed in synovial fluid from diseased joints. In vitro data suggest that nitric oxide activates cartilage degrading enzymes, specifically metalloproteinases. Therefore, the objective of this study was to determine whether glucosamine could inhibit nitric oxide production in equine articular cartilage explants. Articular cartilage was obtained from the carpal joint of horses (2-8 years old) sacrificed for reasons other than joint problems. Three 3.5 mm cartilage discs were biopsied from the weight bearing region of the proximal articular surface (weighing 40-60 mg total) and were placed in each well with 1 ml of Dulbecco's modified Eagle's medium: F12 (1:1) + 10% fetal bovine serum. Media was exchanged daily. The recovered media was stored at 4EC until analyzed. Explants were maintained in basal media two days prior to the start of treatments. Varying concentrations of LPS (10 or 50 µg/ml), rhIL-1 (50 ng/ml), and glucosamine (0.25, 2.5, or 25 mg/ml) were then added to wells. Treatments were done in triplicate. Control wells without LPS, rhIL-1ß, and glucosamine were run with each experiment. Nitric oxide released into media peaked the first day after adding LPS or rhIL-1ß and returned to baseline concentrations by the third day. In three separate experiments, the addition of 25 mg/ml of glucosamine to the explant culture in the presence of LPS or rhIL-1ß prevented the increase in nitric oxide production. Additionally, at 10 µg/ml LPS, 2.5 mg/ml glucosamine partially inhibited nitric oxide production. Our results suggest that glucosamine could prevent cartilage degradation by inhibiting nitric oxide production. Future experiments will focus on how glucosamine inhibits nitric oxide and whether these results are replicated in *in vivo* experiments.

THE INFLUENCE OF FRUIT FLAVORS ON FEED PREFERENCE IN THOROUGHBRED HORSES

M.A.P. KENNEDY, T. CURRIER, J. GLOWAKY AND J.D. PAGAN
Kentucky Equine Research, Inc., Versailles, KY

Introduction

Many feed flavors are added to equine rations in an attempt to improve palatability. Burton et al. (1983) studied the effect of adding three different feed flavors (apple, caramel and anise) to a pelleted horse diet. They found no improvement in feed consumption with these particular flavors. Fruit flavors are often added to horse feeds and supplements. There has been little work to substantiate their effectiveness in improving palatability. Therefore, the following preference test was conducted to evaluate if rate of intake of oats could be influenced by the addition of fruit flavors.

Methods and Materials

Eight mature Thoroughbreds (2 mares and 6 geldings) were used in a replicated 4 x 4 Latin square two choice preference test. The four flavors tested were apple, cherry, teaberry and citrus on a wheat midds carrier. Two horses per period were offered each treatment along with a control in buckets hanging side by side in a 10' x 10' box stall. Each bucket contained 2 kg of whole oats. The control had 100 ml of water added just prior to feeding. Two grams of each flavor were dissolved in 100 ml of water and added to the treatment bucket just prior to feeding. This level of flavor inclusion is the amount typically added to horse feeds. The horses were allowed to eat from both buckets for 5 minutes. The amount of oats remaining in each bucket at the end of this time was recorded. Each treatment was offered for three consecutive days, in the morning and in the afternoon. The control and treatment buckets were switched at each feeding.

Results and Discussion

There was a trend towards higher consumption of the flavored oats as compared to the control. The flavored oats were consumed at an average rate of 147 g/minute while the rate of intake for the plain oats was 130.5 g/minute. The cherry flavored oats were consumed at a significantly higher rate of intake compared to the control (133 g/min versus 153 g/min) ($p<.05$).

The order of preference of the plain oats and flavored oats can be described as:

$$\text{Control} < \text{teaberry} = \text{citrus} = \text{apple} < \text{cherry}$$

The results of this study suggest that the palatability of oats can be slightly improved by the addition of fruit flavors. Cherry appears to have the greatest influence on palatability. More research is needed to more clearly define the role that flavors play in enhancing palatability of horse rations.

VOLUNTARY INTAKE OF LOOSE VERSUS BLOCK SALT AND ITS EFFECT ON WATER INTAKE IN MATURE IDLE THOROUGHBREDS

M.A.P. KENNEDY[1], P. ENTREKIN[1], P. HARRIS[2] AND J.D. PAGAN[1]
[1]Kentucky Equine Research, Inc., Versailles, KY
[2]Waltham Centre for Pet Nutrition, Leicestershire, UK

Introduction

Little is known about voluntary salt intake in horses. Schryver et al. (1987) measured voluntary salt intake in unexercised horses. In this study, salt intake from salt blocks averaged 53 g/day. Voluntary intake of loose versus block salt has not been determined. Therefore, the purpose of the present study was to measure the voluntary intake of loose versus block salt over time and evaluate how salt intake affects water consumption.

Methods and Materials

Four mature Thoroughbred geldings were used in this 8-week switchback experiment. During week 1 of each 4-week period the horses were not offered salt but were allowed free choice access to water. Throughout weeks 2-4 the horses were offered either free choice loose or block salt and free choice water. Salt intake was measured weekly by weighing the salt (~1.8 kg) at the beginning and end of each week. The salt was oven dried prior to weighing. At the end of the first 4-week period, the treatments (loose vs block) were reversed and a second 4-week trial was conducted.

Each horse was housed in a 10' x 10' box stall with salt holders attached to the back of each stall. Daily, the horses were allowed 2 hours of free choice exercise in paddocks without access to water, and they were also walked on a horse walker for 1 hour. While in the paddocks, the horses wore muzzles that prevented grazing. Throughout the entire study, the horses were fed alfalfa/ timothy hay cubes at 2% body weight. Two water buckets were continuously available in each stall. Daily water consumption was measured.

Results and Discussion

Salt consumption for the loose salt treatment was considerably higher than for the block salt for weeks 2 & 3 of the experiment (54 versus 22 g/day and 51 versus 20 g/day, respectively)(p<.05). With both salt types, there was a small drop in intake between weeks 2 and 3. During week 4, loose salt intake dropped significantly (from 51 to 30 g/day)(p<.05) while block salt intake increased (from 20 to 27 g/day).

147

There was a fairly large drop in daily water consumption in both treatment groups during the second week of salt availability (loose salt - 2.13 l/day: block salt - 2.74 l/day). During the third week of salt availability, there was a further drop of 0.92 l/day for loose salt and essentially no change in water consumption for the horses with block salt.

CONCLUSIONS

Horses with free choice access to loose salt drank significantly more water than when offered block salt ($p < .05$). Salt intake was more consistent from week to week when offered in a block form. This level of salt consumption equaled around 20-27 grams/day which is very close to the level recommended for a mature idle horse by the 1989 NRC.

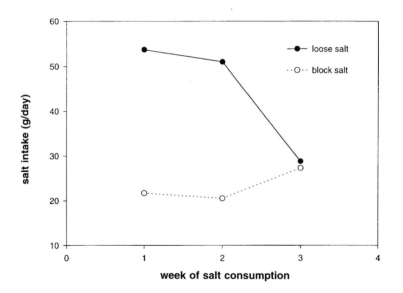

Water Intake
(loose vs block salt)

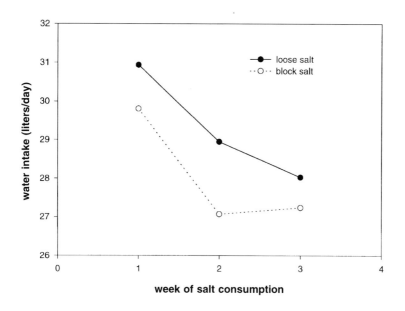

GROWTH AND DEVELOPMENT

ENERGY, PROTEIN AND AMINO ACID REQUIREMENTS FOR GROWTH OF YOUNG HORSES

EDGAR A. OTT
University of Florida, Gainesville, FL

Energy, protein and amino acid intake have the greatest impact on the growth and development of growing horses. Somehow we all know that this statement is true but we seem to push it to the back of our minds and focus on minerals and vitamins, the catalysts for quality development. In fact, if you read the popular magazines one gets the impression that energy and protein are the causes of all of the developmental problems in our young animals. Let's see if we can put energy and protein in their proper place in our feeding programs.

Energy and protein, and consequently amino acid, requirements for the young horse are a function of the size of the animal (maintenance needs) and the rate at which the animal is growing. Conversely, the rate at which an animal grows is a function of how well we meet his energy and protein needs. If you want to maximize the growth rate of the animal you must meet the animal's energy and protein needs. If you want to restrict the growth rate of the animal you restrict the energy or the protein intake.

Energy Requirements

Energy is required for all body functions. But energy is not a chemical that one can analyze for but the result of the metabolic processing of substrates including glucose, fatty acids (both long chain and short chain) and the carbon chains of proteins not required for other functions. There are two procedures for determining the energy requirements of a growing animal, the factorial approach and the feeding trial. The factorial approach involves the determination of the energy needs of the animal for maintenance and adding the energy required for growth. The NRC (1989) committee found that the data in the literature revealed that the daily digestible energy requirement for maintenance of mature horses was: DE (Mcal) $= 1.4 + .03$ Bw_{kg}. Since there were no direct measurements of the maintenance needs of growing horses it was assumed that the relationship was constant regardless of age. Thus a foal weighing 300 kg has a DE requirement of: DE (Mcal) $= 1.4 + .03(300) = 10.4$ Mcal/day. His growth requirement, which is added to the maintenance requirement, is based on tissue energy deposition and efficiency of the conversion of dietary energy to tissue energy. Tissue energy deposition is represented by the equation: DE (Mcal) $= (4.81 + 1.17X - 0.023X^2)(ADG)$ where X = age in months and ADG is average daily gain in kg. Thus the equation for daily digestible energy requirement for the growing horse is: DE (Mcal) $= (1.4 + 0.03$ $Bw_{kg}) + (4.81 + 1.17X - 0.023X^2)(ADG_{kg})$. The proof, of course, is in how the animals respond to the calculated energy requirement intake. We will look at that a little later in the presentation.

Protein Requirements

Protein requirements of the growing foal are a function of the amino acid requirements of the foal, the amino acid content of the feed ingredients and the digestibility of those amino acids. The protein concentration required in the diet is a function of how well the available amino acid content of the diet matches the amino acid needs of the animal. Since the requirement of the animal can be divided into both a maintenance and a growth component, and since both energy and protein restriction will reduce growth of the animal (Ott et al., 1986), a constant relationship exists between energy and protein needs. This relationship is represented by the following: CP (g/day) for weanlings = 50 g/Mcal DE, CP (g/day) for yearlings = 45 g/Mcal DE. These relationships exist only if the protein meets the amino acid needs of the animal and may be in excess of the animal's needs if the diet meets the amino acid needs with less protein.

Amino Acid Requirements

Most of the data on the amino acid requirements of the growing horse have been feeding trials that compare the growth of young horses on diets providing various concentrations of amino acids (Breuer and Golden, 1971; Hintz et al., 1971; Potter and Huchton, 1975; Ott et al., 1981). This work started back in the late 1960s and has continued into the 1990s. Lysine was documented to be the first limiting amino acid in most natural horse feeding programs, especially grass-based programs. This is because grasses and cereal grains tend to be quite low in lysine, requiring that the diet be supplemented with a more concentrated source of this amino acid. Soybean meal is the most likely source of higher lysine protein for horse feeds but animal protein products and a combination of other plant sources and lysine may also be used. There is general agreement that lysine is the first limiting amino acid in most equine diets; however, there is less evidence regarding the second limiting amino acid. Methionine, tryptophan and threonine are all candidates as the second limiting amino acid in growing horses because they are limiting in other species. The data on methionine supplementation for growing horses are mixed. Some experiments show a growth response to methionine supplementation, some show a negative effect and some no response at all (Borton et al., 1973; Meakim, 1979). Tryptophan is generally higher in horse feeds than in diets for other nonruminant animals so it is not likely to be limiting. Threonine may very well be the second limiting amino acid in grass-based feeding programs for horses.

The addition of 0.1% threonine to a 10.5% protein concentrate supplemented with 0.2% lysine gave a small but consistent response over the lysine response when fed with coastal bermudagrass hay (Table 1, Graham et al., 1994). This provides evidence that at least for grass-based programs, threonine is the second limiting amino acid for the growing horse.

Table 1. Influence of lysine and threonine supplementation on growth of yearling horses.

	Concentrate[1]		
	10.5% CP	10.5% CP + 0.2% lys	10.5% CP +0.2% lys +0.1% thr
Av. daily gain, kg	0.57	0.64	0.67
Heart girth gain, cm	9.7	10.1	11.3
Withers height gain, cm	5.0	5.1	4.8
Hip height gain, cm	4.6	5.0	4.8

[1]Fed with coastal bermudagrass hay

Growth Rates of Young Horses

The growth rate of the foal at any point in time is the function of the age of the animal, his genetic potential for growth, his previous growth history and the nutrients available at that point in time. For example, a Thoroughbred foal with the genetic potential to weigh 1000 lb at 14 months of age may get there by several different routes. The most efficient program, and the one least likely to cause development problems, is the program that results in the nice smooth growth curve that is typical of most texts and would represent the result of plotting the growth of a large number of foals. However, in real life this nice smooth curve does not represent the way most foals grow, although it certainly is an excellent goal for which to strive. A number of factors influence the growth rates of foals. Availability of adequate nutrients is important to maximizing, or at least optimizing, growth. The required nutrients change as the animal grows and the growth rate changes. The transition from suckling to weanling is a problem for some foals and they will sometimes only maintain themselves or will actually lose weight during this transition. Others will experience various health problems that will slow their development at some point during their growth phase. The key to minimizing the variability in the animals is to provide an optimal quantity of nutrients for each phase of the growth cycle. This requires that each animal be individually fed a concentrate balanced to be fed with the selected forage.

Meeting the Energy Requirements of the Growing Foal

The daily energy requirements of the foal can only be satisfied by providing the foal with a diet of suitable energy density that will allow the animal to consume enough energy to meet his needs. Yearlings restricted to forage will usually be able to consume enough energy to provide for the maintenance needs of the animal and support some growth. However, seldom will the animal be able to consume enough energy to support a growth rate compatible with the animal's genetic potential (Roquette et al., 1985; Hansen et al., 1987; Webb et al., 1989). The concentrate (grain ration) therefore becomes a key component in providing adequate energy. With a typical grain concentrate providing 1.4 Mcal/lb (3.08 Mcal/kg) and a forage providing 0.82 Mcal/lb (1.80 Mcal/kg), as fed, the ratio of concentrate to forage can be used to optimize energy intake while controlling the energy density of the diet to minimize digestive problems. The NRC (1989) recommended energy intake of growing foals can usually be met with a concentrate to forage ratio of about 70:30 for weanlings and 60:40 for yearlings. Our data varied from these recommendations somewhat in that our weanlings averaged 62:38 (Table 2) and the yearlings averaged 64:36 (Table 3). Since horses of this age will consume about 2.4 to 2.6 lb/100 lb BW of as fed feed daily, this would mean that a 600 lb weanling would consume 9.0 to 9.6 lb of concentrate (60 - 64%) and 5.4 to 6.0 lb of hay (36 - 40%) daily. This is compatible with the generally accepted principle that horses should consume at least 1.0% of their BW in forage daily to maintain good digestive function. Some variation around this ideal ratio is acceptable and may even be expected but care should be exercised not to let it vary too heavily toward the concentrate.

Table 2. Daily dry matter intake by weanling foals[1].

	Mean ± SE	Range
Concentrate, %[2]	61.9 ± 0.3	60 - 63
Hay, %[3]	38.1 ± 0.3	37 - 40
Concentrate DM, %BW	1.45 ± 0.01	1.38 - 1.52
Hay DM, %BW	0.89 ± 0.01	0.84 - 0.97
Total DM, % BW	2.34 + 0.02	2.24 - 2.44

[1] 86 weanlings, ave. wt. 252 ± 4 kg
[2] Concentrate fed to appetite for 1.5 hr twice daily
[3] Hay was fed at 1.0% BW

Table 3. Daily dry matter intake of yearlings[1].

	Mean ± SE	Range
Concentrate, %[2]	63.6 ± 0.5	58 - 69
Hay, %[3]	36.4 ± 0.5	31 - 42
Concentrate DM, % BW	1.52 ± 0.02	1.33 - 1.82
Hay DM, % BW	0.85 ± 0.02	0.75 - 1.00
Total DM, % BW	2.37 ± 0.03	2.16 - 2.65

[1] 230 head, ave. wt. 355 ± 6 kg
[2] Concentrate was fed to appetite for 1.5 hr twice daily
[3] Hay was fed at 1.0% BW/day

Based on our data (Table 4), the energy intake of a 252 kg weanling gaining 0.80 kg/d and consuming 1.45% BW of DM from the concentrate and 0.89% BW DM from forage would be 16.5 Mcal/d. This is less than the 17.80 Mcal DE the NRC (1989) indicates this size foal would require when gaining 0.80 kg/d. Based on our data (Table 5), the energy intake of a 355 kg yearling gaining 0.57 kg/d and consuming 1.52% BW of DM from the concentrate and 0.85% BW of DM from forage would be 23.79 Mcal/d. This is greater than the NRC (1989) indicates this size yearling gaining 0.57 kg/d requires. These differences may be well within the expected variation that exists between animals and feeding programs or they may indicate that we need to revisit the energy requirements of growing horses to refine the recommendations.

Table 4. Daily energy intake by 252 kg weanling (6 mo. of age) gaining 0.80 kg/day.

252 kg x 0.0145 = 3.65 kg DM intake x 3.30 Mcal/kg =	12.06 Mcal
252 kg x 0.0089 = 2.24 kg DM intake x 1.98 Mcal/kg =	4.44 Mcal
	16.50 Mcal
NRC (1989)	
252 kg weanling gaining 0.80 kg/day requires	17.80 Mcal/d

Table 5. Daily energy intake by 355 kg yearling (13 mo. of age) gaining 0.57 kg/day.

355 kg x 0.0152 = 5.40 kg DM intake x 3.30 Mcal/kg =   17.82 Mcal

355 kg x 0.0085 = 3.02 kg DM intake x 1.98 Mcal/kg =   5.97 Mcal

                                                        23.79 Mcal

NRC (1989)
355 kg yearling gaining 0.57 kg/day requires   21.20 Mcal/d

Another consideration is the addition of fat to the concentrate to increase the energy density of the concentrate. The addition of 5% fat to the concentrate will increase the energy density of the concentrate by about 10%. This means that the animal can meet his energy needs with less concentrate. Our experience with this type of program (Table 6) is that it lowers the concentrate and subsequently the feed intake of the animal and decreases the blood insulin concentrations which may be advantageous in helping minimize bone development problems.

Table 6. Influence of fat addition to the concentrate on feed intake and growth of yearling horses.

	Concentrate	
	Basal	Basal + 5% fat
Concentrate intake, kg	5.74	5.38
Hay, kg	3.25	3.20
Total, kg	8.99	8.58
ADG, kg	0.61	0.59
Withers height gain, cm	2.25	2.76
Heart girth gain, cm	7.05	8.23
Body length gain, cm	5.95	6.31
Hip height gain, cm	4.20	3.88

Meeting the Protein and Amino Acid Needs of the Growing Foal

Based on our data (Table 7), the protein content of a concentrate can be reduced by 2 to 3% if 0.2% lysine is added to the concentrate and still support maximum growth response from the growing horses. Therefore, if you need a 15% concentrate to meet the needs of a yearling, a 12% protein concentrate with 0.2% added lysine will be comparable. It is important to provide the mineral and vitamin concentrations you would include in the 15% protein product because the foals will grow at the rate expected from the higher protein product. Although the data are much less convincing, it may also be possible to lower the protein even more, perhaps to 10.5 or 11.0%, by adding 0.1% supplemental threonine. The economics of that program probably favors the higher protein concentrate at this time. Lower threonine prices in the future may change that recommendation.

Table 7. Influence of lysine supplementation on growth of yearling horses[1].

	Concentrate[2]		
	14.5% CP	12.0% CP	12.0% CP + 0.2 % lysine
Initial weight, kg	340	330	330
Final weight, kg	441	417	431
Av. daily gain, kg	0.72	0.62	0.72
Withers height gain, cm	5.7	6.0	6.4
Heart girth gain, cm	14.8	12.2	16.3
Body length gain, cm	10.9	10.9	11.0
Feed/gain	16.5	17.9	15.0

[1] Ott et al., 1981
[2] Fed with coastal bermudagrass hay (1% BW)

Conclusions

Our current recommendations are shown in Table 8. At typical intake ratios between 70:30 and 60:40 concentrate to forage, these energy, protein and amino acid concentrations will provide appropriate nutrient intakes to support maximum growth in most weanlings and yearlings. When specified lysine concentrations can be provided using lower protein concentrations, similar growth responses can be expected. If the energy content of the concentrate is increased by the addition of fat, it is probably appropriate to increase the protein and/or lysine concentrations to compensate for the lower feed intake.

Table 8. Protein and lysine recommendations for concentrates for growing horses (as fed).

	Forage			
	Grass		Legume	
	Weanling	Yearling	Weanling	Yearling
Dig. Energy, Mcal/kg	3.00	3.00	3.00	3.00
C. Protein, %	18.0	15.0	15.0	12.0
Lysine, %	0.85	0.65	0.65	0.45

References

Borton, A., D. L. Anderson and S. Lyford. 1973. Studies of protein quality and quantity in the early weaned foal. 3rd ENPS, Gainesville, FL p. 19-22

Davison, K. E., G. D. Potter, J. W. Evans, L. W. Green,P. S. Hargis, C. D. Corn and S. P. Webb. 1989. Growth and nutrient utilization in weanling horses fed added dietary fat. 11th ENPS, Stillwater, OK p. 95-100

Graham, P. M., E. A. Ott, J. H. Brendemuhl and S. H. TenBroeck. 1994. The effect of supplemental lysine and threonine on growth and development of yearling horses. J. Anim. Sci. 72:380-386

Hansen, D. K., F. M. Rouquette, Jr., G. W. Webb, G. D. Potter and M. J. Florence. 1987. Performance of yearling horses on pasture and supplemental feed. 10th ENPS, Ft. Collins, CO p.25

Meakim, D. W. 1979. Bone mineral content determination of the equine third metacarpal via radiographic photometry and the effect of dietary lysine and methionine supplementation on growth and bone development in the weanling foal. MS Thesis, Univ. FL

NRC, 1989. Nutrient Requirements of Horses, 5th Revised Ed. National Academy Press, Washington, DC.

Ott, E. A., R. L. Asquith, J. P. Feaster, and F. G. Martin. 1979. Influence of protein level and quality on growth and development of yearling foals. J. Anim. Sci. 49:620-628.

Ott, E. A., R. L. Asquith, and J. P. Feaster. 1981. Lysine supplementation of diets for yearling horses. J. Anim. Sci. 53:1496-1503

Ott, E. A. and R. L. Asquith. 1986. Influence of level of feeding and the nutrient content of the concentrate on growth and development of yearling horses. J. Anim. Sci. 62:290-299

Ott, E. A. and J. Kivipelto. 1998 Influence of dietary fat and time of hay feeding on growth and development of yearling horses. J. Equine Vet. Sci. (In press)

Rouquette, Jr., F. H., G. W. Webb, and G. D. Potter. 1985. Influence of pasture and feed on growth and development of yearling Quarter Horses. 9th ENPS, E. Lansing, MI p. 14

Webb, G. W. W. A. Hussey, B. E. Conrad, and G. D. Potter. 1989. Growth of yearling horses grazing kleingrass or bermudagrass pastures. 11th ENPS, Stillwater, OK p. 267

INFLUENCE OF HOUSING ON BONE GROWTH AND CARTILAGE METABOLISM IN WEANLING HORSES

R. A. BELL, B. D. NIELSEN, M. ORTH, J. SHELLE, J. CARON, AND C. HELESKI
Michigan State University, East Lansing, MI

Young horses are often housed in stalls as opposed to being out on pasture allowing free access to exercise. The objective of this study is to determine whether housing weanling horses in stalls is detrimental to bone growth and development compared to bone growth in weanling horses maintained on pasture. Additionally, the effect of stalling versus pasture-rearing horses on cartilage metabolism will be determined. Stalling may be detrimental to bone growth due to a lack of exercise. Exercise allows the bone to respond to changes ensuring bone mass capable of withstanding pressures (loading). Without the necessary exercise, bone will begin degrading until the bone is adapted to minimal loading. The degradation of cartilage in response to stalling may be due to cartilage needing a mechanical signal (exercise) to stimulate new proteoglycan synthesis to provide a greater compressive resistance and without this exercise/mechanical signal, cartilage will begin degrading. Numerous studies have demonstrated that depriving animals of free exercise is detrimental to bone and muscle strength. Laying hens housed in battery cages have weaker bones than those housed in perches, demonstrating that bone strength is related to the amount of movement allowed. Stalled horses have also shown a decrease in the rate of bone formation, an increase in bone resorption, and a decrease in bone mineral content. Horses allowed free access to exercise showed a decrease in incidence of osteochondrosis and an increase of newly synthesized proteoglycans. Our hypothesis is that stalling retards bone growth and development and is detrimental to cartilage growth. To test this hypothesis, eighteen Arabian weanlings (from four to five months of age), from Michigan State University Horse Teaching and Research Center, will be age matched and randomly assigned to two treatment groups, pastured horses (P, n=9) and stalled horses (S, n=9). The same design will be used with 18 Quarter horse weanlings from the Michigan State University Merillat Equine Center. Group S will be confined to stall housing with no access to free exercise while Group P will be maintained on pasture. Horses will be maintained on the study for three months and will be individually fed concentrate twice daily and allowed free choice roughage meeting NRC recommendations for feeding weanlings. Blood samples will be taken every two weeks via jugular venipuncture and serum will be collected. The serum will be analyzed for deoxypyridinoline (marker for bone resorption), osteocalcin (marker for bone development), and keratan sulfate (marker for cartilage degradation). Dorsal-palmar radiographs of the left front leg will be taken every 28 days to determine the radiographic bone aluminum equivalence (RBAE) in order to estimate the mineral content, and hence bone strength of the third metacarpal.

INFLUENCE OF HOUSING ON BEHAVIOR IN WEANLING HORSES

C.R. HELESKI, B.D. NIELSEN, C.M. CORN, A.J. ZANELLA
Michigan State University, East Lansing, MI

Opinions abound regarding the contentedness of horses in various housing situations. However, supporting evidence is scarce. Housing effects on behavior have been examined in hamsters (Arnold and Estep, 1990), pigs (Blackshaw and McVeigh, 1984; Ekesba, 1981; Hemsworth and Beilharz, 1979; Zanella et al., 1991), calves (Brown and Leaver, 1978; Dantzer et al., 1983; Friend et al., 1985), hens (Dawkins, 1976), rats (Steplewski et al., 1987) and dogs (Hubrecht et al., 1992). Two hypotheses will be examined. The first is that pastured weanlings will show effects of better welfare than will stalled weanlings. This will be evaluated by analyzing the percentage of time each individual spends performing sterotypies, such as cribbing or stall walking; evaluating their response times and number of mistakes in a maze test, both baseline and at the end of the experiment; looking at biological indicators of stress via saliva samples (catacholamines and glucocorticoids); comparing growth rates; evaluating cell-mediated immune function in response to a weak antigen; analyzing differences in group interactions; and evaluating differences in daily ethograms. The maze test and evaluation of group interactions will be tested at day zero of the project and day ninety of the project. Salivary samples will be taken hourly for one twenty-four hour period each month. Percentage of time spent engaged in sterotypies and daily ethogram will be evaluated via video collection for a twenty-four hour time period once each week. Height, weight and cannon circumference will be measured monthly to chart growth rates. Immune function will be compared from day zero to day ninety. The second hypothesis is that pastured weanlings will be easier to handle than stalled weanlings. The weanlings in each group will receive the same number of minutes handling each day. Their response will be scored by blind evaluators who will score them once each month during routine grooming, farrier work, clipping, leading and standing with a handler. Eighteen Arabian weanlings (four to five months of age) at Michigan State University's Horse Teaching and Research Center will be age matched and randomly assigned to two treatment groups beginning in August, 1998. Eighteen Quarter Horse weanlings (four to five months of age) at Michigan State University Merillat Farm will also be age matched and randomly assigned to two treatment groups beginning in May, 1998. Weanlings will be on the study for three months and will be fed concentrate twice daily and fed ad libitum roughage. The diet will meet or exceed NRC recommendations (1989). Stalled weanlings will be handled minimally each day for the purpose of moving to a different stall during stall cleaning. Pastured weanlings will also be handled minimally each day for the purpose of placing them in outside, individual feeding areas while they eat their concentrate.

STALLING YOUNG HORSES ALTERS NORMAL BONE GROWTH

K.E. HOEKSTRA, B.D. NIELSEN, M.W. ORTH, D.S. ROSENSTEIN,
H.C. SCHOTT, J.E. SHELLE
Michigan State University, East Lansing, MI

A major concern with young performance horses is the high incidence of skeletal injury. Young, growing horses transferred from pasture to stalls prior to yearling sales or commencement of training may be predisposed to injury. A decrease in bone density of the third metacarpal has been demonstrated in young racehorses soon after the onset of training. However, it is unknown whether the decrease in bone density was the result of bone remodeling caused by increased strain rates on the bone associated with training or bone modeling resulting from decreased strain rates associated with a change in housing from pasture to stalls at the start of training. Transferring young horses from pasture to stalls has been shown to result in decreased osteocalcin concentrations (indicative of osteoblastic activity), indicating a slowdown in the rate of bone formation due to decreased physical activity. Studies of other species have demonstrated similar decreases in bone strength in response to confinement rearing. Though horses on pasture may not run excessively, only a few fast strides per day may be necessary to prevent bone loss associated with limited physical activity available to horses housed in stalls. Consequently, pastured horses may have a skeletal structure that is better prepared for training and competition. If so, the common practice of housing yearling horses in stalls prior to yearling sales or commencement of training causes concern about the effects of stalling on bone growth. This study was designed to determine if bone development is negatively affected when yearlings are taken from pasture to be housed in stalls and allowed limited exercise. In addition, the consequential effects of the change in housing on bone modeling/remodeling at the onset of training were determined. Sixteen Arabian yearlings, with an average age of 18.6 mo, were pair-matched by age and randomly placed into two groups. One group was housed in box stalls while the second group was kept on pasture. Radiographs of each horse's left front leg were taken every 28 d to measure mineral content of the third metacarpal, as determined by radiographic bone aluminum equivalencies (RBAE). Blood samples were taken every 14 d to determine serum osteocalcin concentrations and 24-hour urine collections were taken every 28 d to measure urinary deoxypyridinoline. After an 84-d pre-training period, six horses from each group were randomly selected to complete a 56-d training period. Analysis of the radiographs showed that stall-housed horses had a decrease in lateral RBAE from d 0 to d 28 ($P < .05$). Pasture-reared horses had greater lateral RBAEs at d 28, 56, and 140 ($P < .05$), and a tendency to be greater at d 112 ($P = .07$). The change in RBAEs of the medial cortex from d 0 tended to be greater in the pastured horses ($P < .1$). Serum osteocalcin concentrations were lower in the stalled horses at d 14 ($P < .05$). Following d 14, the osteocalcin concentrations in the stalled horses returned to baseline. Urinary deoxypyridinoline (indicative of bone resorption) was greater at d 28 in horses housed in stalls than horses maintained on pasture (P < .01). Following d 28, deoxypyridinoline in the stalled horses returned to baseline. It appears that housing yearlings in stalls may negatively affect normal bone growth experienced by yearlings allowed to remain on pasture. Although it was not tested in this study, free access to exercise may have provided sufficient loading on the legs of pastured horses to promote normal bone growth. Results suggest that housing yearling and two-year-old horses in stalls without access to forced or free exercise impairs normal bone growth, compared with horses maintained on pasture. Initial training did not appear to alleviate the negative effects of stalling on bone formation.

THE EFFECT OF A NUTRITIONAL SUPPLEMENT ON THE INCIDENCE OF OCD LESIONS, FUTURE PERFORMANCE AND BONE-RELATED LAMENESS OF STANDARDBRED WEANLINGS

K.L. WAITE, B.D. NIELSEN, M.W. ORTH, D.S. ROSENSTEIN AND B.A.L. LEVENE
Michigan State University, East Lansing, MI

Osteochondritis dissecans (OCD) is a developmental orthopedic disease in young animals that occurs due to a disturbance of growth cartilage such that cartilage fails to develop into healthy, weight-bearing bone. While epidemiological data suggests that OCD may be present in as many as 25% of all young horses, little information is available regarding the long term effects of OCD on future performance of the equine athlete. Research has suggested a number of factors which may play a role in the development of OCD. However, a single definitive cause has yet to be determined. A number of researchers have shown that low Cu concentration in the diet of young horses may increase the incidence of OCD lesions, epiphysitis and intermittent lameness. Additional work has demonstrated an increase in bone density, hence bone strength, as determined by radiographic bone aluminum equivalence (RBAE) in horses fed increased levels of calcium over those recommended by the NRC (1989). Sodium zeolite A (SZA) as a source of dietary silicon has been shown to decrease the incidence of tibial dyschondroplasia in poultry and to decrease the incidence of athletic injury to racehorses. Thus, we hypothesize that feeding a supplement to young horses containing increased levels of Cu, Ca and silicon will decrease the incidence of OCD, improve bone metabolism as estimated by serum markers and decrease future incidence of bone-related lameness. To test our hypothesis, 100 Standardbred weanlings will be randomly assigned to dietary treatment groups: Control (C), Low (L) and High (H) (Table 1).

Table 1. Content of dietary supplement (Consolidated Nutrition, L.C.) by treatment.

Treatment	C	L	H
Ca	.40%	.50%	.70%
Cu	10 ppm	20 ppm	40 ppm
SZA	.00	.92%	2.80%

Each farm will house horses on all three treatments, and horses will remain on their respective treatment for 12 months. Feed will be analyzed for mineral concentration, energy density and protein content. At day 0 and at the conclusion of the project, horses will have radiographs taken of the hock, stifle and fetlocks to assess the presence of OCD lesions. Radiographs will be analyzed by an individual blinded to treatment. Serum samples will be collected at day 0, 100, 200 and at the conclusion of the project for the analysis of osteocalcin, a biochemical marker of bone synthesis, as well as C-terminal cross-linked telopeptides of type I collagen (ICTP), a biochemical marker of bone resorption. Differences between treatment groups will be statistically analyzed for repeated measures using PROC MIXED (SAS). The feeding trial will conclude before horses are sold in fall yearling sales and enter training. One year after the completion of the feeding trial, race records of all study horses will be obtained and differences by treatment in official starts, average race times, money earned and injuries will be analyzed.

THE ROLE OF VITAMINS IN GROWTH OF HORSES

STEPHEN E. DUREN AND KATHLEEN CRANDELL
Kentucky Equine Research Inc., Versailles, Kentucky

Equine nutritionists have long concerned themselves with both the absolute amounts and the balance of minerals in the diets for young, growing horses. The role of vitamins in growth is often ignored. A primary reason for the lack of interest may be that all but vitamins A and E are produced in the body, so it is assumed that even young growing horses are producing everything they need. Vitamins C and D as well as the B vitamin niacin are produced by the horse; the rest of the B vitamins and vitamin K are produced by microbes in the horse's cecum and large intestine (Lewis, 1995). Many of these vitamins have key roles in the regulation of calcium and phosphorus and in proper skeletal development. The following paper will discuss several vitamins with known influences on skeletal growth and indirect relationships with healthy growth. In addition, this paper will evaluate the potential need for vitamin supplementation in growing horse diets.

Vitamins Required in the Diet

The two important vitamins which the horse cannot produce itself are the fat-soluble vitamins, A and E. Because they are fat-soluble, they require at least a small amount of fat in the diet to be absorbed properly. The amount of fat typically found in green grass is usually enough to aid in the absorption of these nutrients, under normal conditions.

VITAMIN A

Vitamin A is best known for its role in vision but also has functions in reproduction, gene expression, differentiation of epithelial cells, embryogenesis and growth. Vitamin A is found in abundant quantities in fresh green forages in the form of carotenes, which are converted to vitamin A by enzymes in the intestinal mucosal cell of the horse. The importance of fresh green forage in maintaining good health in the horse was well understood even before the early days of vitamin A research (Howell et al., 1941). Once forage is cut, there is rapid oxidation of carotenes (up to 85% within the first 24 hours and then about 7% per month during storage) which results in hay being practically devoid of carotenes after extended storage (McDowell, 1989). Horses on hay-only diets demonstrated depletion of vitamin A liver stores over a relatively short period of time (Fonnesbeck and Symons, 1967; Greiwe-Crandell et al., 1995). The inability of hay to supply adequate vitamin A to the horse is the rationale behind the inclusion of vitamin A in commercial horse feed mixes and supplements. Manufacturers cannot know whether a client is going to give a horse access to pasture or not.

169

Vitamin A has a distinct role in growth of the horse. Both deficiency and toxicity of vitamin A adversely affected growth, body weight, rate of gain and heart girth within young growing ponies (Donoghue et al., 1981). This retardation of growth may have reflected impaired cell proliferation and differentiation. Bone remodeling is modulated by vitamin A in the growing animal. Vitamin A's role in bone remodeling is in the proper functioning of osteoclasts, the bone cells responsible for resorption of bone. Without sufficient vitamin A, excessive deposition of periosteal bone occurs. The appearance of bones in vitamin A deficiency is actually shorter and thicker than normal (Fell and Mellanby, 1950). This is in part caused by the dysfunction of osteoclasts, but also by a reduction in the degradation of glycosaminoglycans and the synthesis of proteoglycans also caused by deficiency (Dingle et al., 1972). In addition, effects of overall bone changes may result in mechanical pressure on certain nerves, such as the optic or auditory nerves, which can result in blindness and/or deafness.

It is possible that some of the systemic effects of vitamin A on growth, as well as the poor growth usually associated with vitamin A deficiency, are related to its effects on growth hormone secretion. Vitamin A takes different functional forms once it is working in the body, one of which is retinoic acid. Retinoic acid has been found to affect growth hormone regulation (Sporn et al., 1994). Retinoic acid can synergize with either thyroid hormone or glucocorticoids to enhance the transcriptional activity of the growth hormone gene, and subsequently of growth hormone secretion from cells (Bedo et al., 1989). Retinoic acid is also essential for embroyonic development. Retinoic acid has been implicated in the expression of hox genes, which determine the sequential development of various parts of a developing fetus and in the morphogenesis of the vertebrate limb (DeLuca, 1991; Thaller and Eichle, 1987).

Intakes estimated to prevent signs of severe vitamin A deficiency (9.7 to 10.6 IU/kg body weight/day) in horses by Stowe (1968) are suggested to be far too low for maximal growth in young horses by Donoghue and coworkers (1981), who recommended 63 IU/kg body weight per day. This is above what the NRC is currently recommending, which is 45 IU/kg body weight per day. For horses grazing sufficient quantities of green pastures, their vitamin A requirement can be met entirely by the carotenes in the forage (Greiwe-Crandell et al., 1997). In northern states and countries, vitamin A supplementation is particularly important because of the short growing season of grasses. Weaning foals supplemented with 40,000 IU of vitamin A per day (~160 IU/kg body weight per day) along with hay and oats had improved serum levels during winter and spring, but supplementation had no effect during the summer when the horses were on pasture (Mäenpää et al., 1988). Toxicity levels for vitamin A are estimated to be around 1000 IU/kg body weight per day (NRC, 1989). Carotenes have not been found to be toxic at any level in the horse because it appears that the enzyme which converts carotenes to vitamin A is found in limited amounts in the body and therefore excessive conversion is not possible.

VITAMIN E

Vitamin E may not be directly linked to growth, but its roles in immune response, nerve and muscle function, and its powerful antioxidant properties make it vital to the health of the young growing horse. Together with selenium, vitamin E acts to maintain normal muscle function, aid in the prevention of muscular disease, and provide antioxidant protection to body tissue, particularly cell membranes, enzymes and other intracellular substances, from oxidation induced damage (McDowell, 1989). A deficiency of vitamin E may cause a variety of different symptoms and pathological changes, which may include nutritional muscular dystrophy (weak and poorly oxygenated muscles) and poor immunity to diseases (e.g. recurrent cold and cough) (Moore and Kohn, 1991; Bendich, 1993). Growth rates were found to decrease during periods of illness in Thoroughbred weanling foals (Greiwe-Crandell, unpublished research). Both vitamin E and selenium may help leukocytes and macrophages survive the toxic products that are produced during phagocytosis of invading bacteria (McDowell, 1989). The importance of vitamin E in the diet of the young growing horse deserves more attention in research.

Vitamin E, like vitamin A, is found in abundant quantities in fresh green forage, and horses consuming adequate quantity of green forage have not been found to have vitamin E deficiency. However, vitamin E rapidly disappears during harvesting for hay, with 30 to 85% being lost initially and further loss occurring during storage. The amount of vitamin E in hay is quite variable, depending on the type of forage and the harvesting procedures. Because of the large number of horses that have access only to hay, manufacturers routinely add vitamin E to commercial feed mixes. Vegetable oils are relatively high in vitamin E but are generally not fed in sufficient quantities to make a big impact on the supply of this vitamin in the diet. The increasing popularity of high fat feeds may have an impact on the fortification of vitamin E in feeds since these feeds require extra vitamin E to prevent oxidation of fat in the feed. Horses are not very efficient in storing vitamin E, although body stores may contain sufficient vitamin E to cover four months of inadequate intake in a well-repleted adult. Foals born to mares depleted in vitamin E may have little or no reserves, which would make them more susceptible to infectious diseases (McDowell, 1989). The NRC has estimated adequate intakes of vitamin E for young growing foals at 80 IU/kg of diet. Mäenpää and coworkers (1988) were able to improve serum vitamin E levels in weanling horses through the long, grassless winters in Finland with 400 IU/day of vitamin E. No signs of vitamin E toxicity in the horse have been reported (NRC, 1989). However, extrapolating from other animals, the NRC has set an upper safe limit of 20 IU/kg of body weight per day.

Vitamins Produced in the Body

Vitamins C, D and all of the B vitamins are produced in the body of the horse by either enzymatic conversion or microbial synthesis of substrates normally found in the diet. Since the production of these vitamins is hard to quantitate, there are still questions on precisely what external or internal factors influence production of these vitamins and whether synthesis is adequate to meet the demands of a growing horse.

VITAMIN C

The well-publicized role of vitamin C as a powerful antioxidant overshadows the many other functions it has in the body. In the forms of ascorbate (ascorbic acid) and dehydroascorbic acid, vitamin C acts as a cofactor or co-substrate for eight different enzymes in the body (Ziegler and Filer, 1996). Perhaps the most important role of vitamin C for the growing horse is its function in collagen synthesis and repair. Collagen is the tough, fibrous, intercellular material (protein) that is the principal component of skin and connective tissue, the organic substance of bones and teeth, and ground substance between cells. Three enzymes require ascorbate for proline or lysine hydroxylation in collagen synthesis. Without lysine there is poor formation of cross-links in collagen fibers. Without hydroxyproline (from proline) non-fibrous collagen is formed instead of fibrous collagen, which results in unstable extracellular matrix. Without proper collagen synthesis, proper growth is not possible.

Vitamin C is also important in energy production and hormone and amino acid synthesis. Two enzymes require ascorbate for biosyntheses of carnitine, which is used by mitochondria for transmembrane electron transfer in ATP synthesis. Two more enzymes require ascorbate for corticosteroid hormone biosynthesis. Vitamin C is also needed for production of the amino acid tyrosine.

The antioxidant action of vitamin C is that of an electron donor (or reducing agent) for intra- and extracellular chemical reactions (Ziegler and Filer, 1996). The ability of vitamin C to work both inside and outside the cell is due to its water-soluble nature and because it is distributed throughout the body in the body water. Because oxidants (normally reduced by ascorbate) may affect DNA transcription or could damage DNA, protein or membrane structures, ascorbate may have a central role in cellular oxidant defense. Uncontrolled oxidation in the body could affect growth and development. Also ascorbate can transfer electrons to tocopherol radicals (reduced vitamin E) in lipid particles or membranes. Essentially, vitamin C can help recycle vitamin E.

Vitamin C deficiency is not normally observed in horses because they can synthesize ascorbic acid from glucose in the liver. Animals that cannot synthesize vitamin C (such as humans and primates) are lacking in the enzyme L-gluconolactone (Lewis, 1995). Supplemental vitamin C for the horse has not been found to affect blood levels unless it is given in very large amounts (Stillions et al., 1971). The absorption of vitamin C in horses appears to occur in the ileum by the process of passive diffusion (Lewis, 1995). Production of vitamin C in the liver can be limited, and in some circumstances, the supply may not be adequate to meet the requirement of the horse. Supplementation of vitamin C may be beneficial during dietary deficiencies of energy (in particular glucose and glucose substrates), protein, vitamin E, selenium, and iron; rapid growth; stressful situations like performance for competition; transportation; a change of environment; disease (bacterial and viral infections); and parasitic infection (McDowell, 1989).

Since supplemental vitamin C is poorly absorbed in the horse, research has been done to find the amounts that will show increases in blood levels and which forms are more effective. Daily supplementation had an effect on plasma levels while single feeding did not (Snow et al., 1987). Daily supplementation of 20 g of vitamin C did make a difference in plasma ascorbate in racehorses in training (Snow and Frigg, 1987b). Feeding more than 4.5 g of ascorbic acid did not increase plasma concentrations any more than 20 g (Snow et al., 1987). Ascorbyl palmitate appears to be better absorbed in the horse than ascorbic acid or ascorbyl stearate (Snow and Frigg, 1987a). No toxic levels of vitamin C intake have been observed in the horse (NRC, 1989).

VITAMIN D

Vitamin D is known as the sunshine vitamin since it is made on the skin from 7-dehydrocholesterol by a reaction catalyzed by ultraviolet (UV) light. Studies conducted near the turn of the 19th century documented severe bone-deforming disease in children with limited exposure to sunlight, compared to healthy children with normal exposure to the sun. This established a link between the sun, the sunshine vitamin (vitamin D) and skeletal health.

The function of vitamin D is maintenance of calcium homeostasis in the blood (McDowell, 1989). Circulating calcium is used for normal mineralization of bone as well as for a host of other body functions. Parathyroid hormone (PTH) and calcitonin function with vitamin D to control blood calcium and phosphorus concentrations. When blood calcium is low, the parathyroid is stimulated to release PTH. PTH travels to the kidney and stimulates 1-hydroxylation of 25-OH vitamin D to form the active vitamin (1,25 OH vitamin D). Active vitamin D then stimulates intestinal calcium uptake, stimulates bone mineral release and stimulates resorption of calcium by the kidney, all in an effort to restore blood calcium levels (Linder, 1991). Calcitonin regulates high serum calcium by depressing gut absorption, halting bone demineralization and slowing reabsorption in the kidney.

Since vitamin D is readily synthesized and absorbed from the skin, is it necessary to supplement vitamin D in the diet? El Shorafa and coworkers (1979) determined young Shetland ponies kept outdoors in Florida do not need vitamin D supplementation. However, ponies deprived of sunlight with no vitamin D supplementation lost their appetite and had difficulty standing. Oral supplementation with vitamin D (approximately 400 IU/kg diet) prevented the signs of vitamin deficiency (El Shorafa et al., 1979).   In modern horse production systems young show horses are often kept out of the sunlight to prevent dulling of the hair coat. For horses not exposed to sunlight or artificial light with an emission spectrum of 280-315 nm, the National Research Council (NRC, 1989) has established a requirement for dietary vitamin D. Growing horses require 800 IU of vitamin D per kilogram of diet dry matter according to the NRC (1989). If an 800 lb year-ling eats 2% of its body weight per day in dry feed, this would equate to a vitamin D requirement of 5800 IU/day. Eating a diet consisting of 50% fresh, sun-cured alfalfa hay (1800 IU/kg) would easily satisfy the requirement established by the NRC, 1989. In addition, feeding a grain fortified with vitamin D at a concentration of 1100 IU/kg would supply the yearling with a daily vitamin D intake (hay + grain) of 10,500 IU/day. The actual vitamin D intake would likely be less than calculated since vitamin D is lost at a rate of 7.5% per month with hay storage (Lewis, 1995). The NRC (1989) establishes the maximum tolerance level 2200 IU/kg of diet dry matter, a value of 16,000 IU/day in the above example.

Conservative supplementation of vitamin D in the grain concentrate portion of the diet seems to be warranted due to the important role of vitamin D in bone growth and the variability young show horses may have in the opportunity for exposure to sunlight. In addition, losses in vitamin content of forage with prolonged storage make prediction of dietary vitamin D content difficult. Vitamin D should not be given in an effort to treat developmental orthopedic disease (DOD) by increasing calcium and phosphorus absorption and bone mineralization. DOD has not been shown to be caused by vitamin D deficiency and supplementation with vitamin D will not make up for diets that are not properly fortified with calcium and phosphorus. Oversupplementation of vitamin D to horses is toxic and results in extensive mineralization of cardiovas-cular and other soft tissues (Harrington and Page, 1983). Care should be taken to remain well below the maximum tolerance level (2200 IU/kg diet) established by the NRC, 1989.

VITAMIN K

Vitamin K was the last fat-soluble vitamin to be discovered (McDowell, 1989). For many years, vitamin K has been known for its blood-clotting function. Vitamin K is essential to activate several blood clotting factors including prothrombin (factor II) as well as factors VII, IX and X. These four blood-clotting proteins are synthesized in the liver in inactive precursor forms and are converted to biologically active proteins by the action of vitamin K (Suttie and Jackson, as cited by McDowell, 1989). The method of this activation is through carboxylation of specific glutamic acid residues associated with the inactive proteins (McDowell, 1989).

Recently, the carboxyglutamyl residues have been found in other proteins associated with a variety of tissues. Most notable is osteocalcin, a protein involved in bone metabolism. Osteocalcin is responsible for binding to hydroxyapatite and facilitating bone mineralization. Undercarboxylated osteocalcin does not bind hydroxyapatite with the same affinity as carboxylated osteocalcin (Knapen et al., 1989). If vitamin K is in short supply, one would expect to find irregularities in blood clotting along with undercarboxylated osteocalcin. However, it is suspected that osteocalcin is more sensitive to low vitamin K activity than are the blood clotting proteins. Vermeer et al. (1996) concluded the liver is capable of efficiently extracting the required amount of vitamin K from the bloodstream, even at low circulating vitamin K concentrations. This is probably less so for extrahepatic tissues, notably bone. Therefore, it seems possible that bone tissue may be vitamin K deficient, while liver, and thus the blood clotting mechanism, is vitamin K adequate. Knapen et al. (1991) have speculated that osteocalcin is a far more sensitive marker for vitamin K status than are the blood coagulation factors.

If vitamin K has a positive effect on net bone formation, it might be expected that vitamin K antagonists (coumarin) have an opposite effect. Pastoreau et al. (1993) reported that lambs treated with vitamin K antagonists (warfarin) had strongly decreased bone formation indicated by a 30% lower bone mass in three months compared to controls. A deficiency in vitamin K would be expected also to have negative consequences for bone health. In humans, vitamin K intake is reported to decrease with age (Jie et al., 1995), and subjects with increased concentrations of undercarboxylated osteocalcin had a sixfold increased risk of hip fracture (Szulc et al., 1993). Supplementation of vitamin K increases the serum markers for bone formation, including osteocalcin and bone alkaline phosphatase, and may reduce urinary calcium and hydroxyproline excretion, a well-known marker for bone resorption (Knapen et al., 1989).

The NRC (1989) has not established requirements for vitamin K fortification of equine diets. Natural sources of vitamin K are phylloquinone (K-1), found in green leafy plants, and menaquinone (K-2) which is produced by bacteria in the digestive system. Both phylloquinone and menaquinone are converted to the active vitamin (hydroquinone) in the liver (Lewis, 1995). The NRC (1989) states that if the intake or intestinal synthesis of vitamin K are inadequate, horses will have an increased susceptibility to hemorrhage. With new functions of vitamin K being explored, the previous statement may no longer be true. Research being conducted at Colorado State University with exercising horses is studying the concentration of carboxylated and uncarboxylated osteocalcin as it relates to microfractures in bone (Siciliano, personal communication). Preliminary results of this research have encouraged the research group to take a closer look at vitamin K in equine diets. Another area that requires study is vitamin K status of young, growing horses. Foals undergo significant bone growth prior to having a completely functional hindgut and prior to eating a significant amount of green forage, the two principal sources of vitamin K. With current research interest, look for nutrient requirements for vitamin K in horse diets in the near future. At the present time, the NRC (1989) indicates that oral intake of phylloquinone (K1) and menadione (K3) appear to be essentially innocuous in horses (NRC, 1989).

References

Bedo, G., P. Santisteban, and A. Aranda. 1989. Retinoic acid regulates growth hormone gene expression. Nature 339:231.

Bendich, A. 1993. Physiological role of antioxidants in the immune system. J. Dairy Sci. 76:2789.

De Luca, L.M. 1991. Retinoids and their receptors in differentiation, embryogenesis, and neoplasia. FASEB J. 5:2924.

Dingle, J.T., H.B. Fell, and D.S. Goodman. 1972. The effect of retinol and retinol binding protein on embryonic skeletal tissue in organ culture. J. Cell Sci. 11:393.

Donoghue, S., D.S. Kronfeld, , S.J. Berkowitz, and R.L Copp. 1981. Vitamin A nutrition in the equine: growth, serum biochemistry, and hematology. J. Nutr. 111:365.

El Shorafa, W.M., J.P. Feaster, E.A. Ott and R.L. Asquith. 1979. Effect of vitamin D and sunlight on growth and bone development of young ponies. J. Anim. Sci. 48:882.

Fell, H.B., and E. Mellanby. 1950. Effect of hypervitaminosis on fetal mouse bone cultivated in vitro. Br. J. Med. 2:535.

Fonnesbeck, R.V., and L.D. Symons. 1967. Utilization of the carotene of hay by horses. J. Anim. Sci. 26:1030.

Greiwe-Crandell, K.M., D.S. Kronfeld, L.A. Gay, and D. Sklan. 1995. Seasonal vitamin A depletion in grazing horses is assessed better by the relative dose response test than by serum retinol concentration. J. Nutr. 125:2711.

Harrington, D.D. and E.H. Page. 1983. Acute vitamin D3 toxicosis in horses: Case reports and experimental studies of the comparative toxicity of vitamins D2 and D3. JAVMA 182:1358.

Howell, C.E., G.H. Hart, and N.R. Ittner. 1941. Vitamin A deficiency in horses. Am. J. Vet. Res. 2:60.

Jie, K.S.G., M.L. Bots, C. Vermeer, J.C.M. Witteman and D.E. Grobbee. 1995. Vitamin K intake and ostocalcin levels in women with and without aortic atherosclerosis: a population-based study. Atherosclerosis 116:117.

Knapen, M.H.J., K. Hamulyak and C. Vermeer. 1989. The effect of vitamin K supplementation on circulating osteocalcin (bone Gla-protein) and urinary calcium excretion. Ann. Intern. Med. 111:1001.

Knapen, M.H.J., K.S. Jie, K. Hamulyak and C. Vermeer. 1991. Vitamin K deficiency redefined. Thromb. Haemost. 65:671.

Lewis, L.D. 1995. Equine Clinical Nutrition: Feeding and Care. Williams and Wilkins, Media, PA, USA.

Linder, M.C. 1991. Nutritional Biochemistry and Metabolism with Clinical Applications. Appleton and Lance, Norwalk, CT, USA.

Mäenpää, P.H., A. Pirhonen, and E. Koskiene. 1988b. Vitamin A, E, and D nutrition in mares and foals during the winter seasons: Effect of feeding two different vitamin-mineral concentrates. J. Anim. Sci. 66:1424.

McDowell, L.R. 1989. Vitamins In Animal Nutrition: Comparative Aspects to Human Nutrition. Academic Press, San Diego, CA, USA.

Moore, R.M. and C.W. Kohn. 1991. Nutritional muscular dystrophy in foals. Compendium for Continuing Education 13/3:476.

NRC, 1989. Nutrient Requirements of Horses. 5th Revised Edition. National Academy Press, Washington D.C.

Pastoureau, P., P. Vergnaud, P. Meunier, P.D. Delmas. 1993. Osteopenia and bone-remodeling abnormalities in warfarin-treated lambs. J. Bone Miner. Res. 8:1417.

Snow, D.H. and M. Frigg. 1987. Plasma concentrations at monthly intervals of ascorbic acid, retinol, ß-carotene and tocopherol in two Thoroughbred racing stables and the effects of supplementation. Proc. 10th Equine Nutr. and Physiol. Symp., Colorado State University, p.55.

Snow, D.H. and M. Frigg. 1987. Oral administration of different formulations of ascorbic acid to the horse. Proc. 10th Equine Nutr. and Physiol. Symp., Colorado State University, p.617.

Snow, D.H., S.P. Gash and J. Cornelius. 1997. Oral administration of ascorbic acid to horses. Equine Vet. J. 19:520.

Sporn, M.B., A.B. Roberts, and D.S. Goodman. 1994. The Retinoids: Biology, Chemistry, and Medicine (2nd Ed.) Raven Press, Ldt., New York.

Stowe, H.D. 1967. Reproductive performance of barren mares following vitamins A and E supplementation. Am. Assoc. Equine Pract. p. 81.

Stillion, M.C., S.M. Teeter and W.E. Nelson. 1971. Ascorbic acid requirement of mature horses. J. Anim. Sci. 32:249.

Szulc, P., M.C. Chapuy, P.J. Meunier and P.D. Delmas. 1993. Serum under carboxylated osteocalcin is a marker of the risk of hip fracture in elderly women. J. Clin. Invest. 91:1769.

Thaller, C., and G. Eichele. 1987. Identification and spatial distribution of retinoids in the developing chick limb bud. Nature 327:625.

Vermeer, C., B.L.M.G. Gijsbers, A.M. Craciun, M.C.L. Groenen-VanDooren and M.H.J. Knapen. 1996. Effects of vitamin K on bone mass and bone metabolism. J. Nutr. 126:1187S.

Ziegler, E.E. and L.J. Filer, Jr. 1996. Present Knowledge in Nutrition, Seventh Edition. ILSI Press, Washington D.C.

THE EFFECT OF DIETARY SELENIUM SOURCE AND LEVEL ON BROODMARES AND THEIR FOALS

K.M. JANICKI, L.M. LAWRENCE, T. BARNES AND C.I. O'CONNOR

University of Kentucky, Lexington, KY

Fifteen pregnant mares were blocked by foaling date and randomly assigned to one of three selenium (Se) supplements: 1 mg Se/d (I1) or 3 mg Se/d (I3) as sodium selenite, or 3 mg Se/d (O3) as Se-enriched yeast (Alltech, Inc., Nicholasville, KY). Mares received their treatments daily for approximately 55 d pre-foaling and 56 d post-foaling. Blood samples were taken from each mare prior to supplementation and at 2 wk intervals throughout the study, including at time of foaling. A single colostrum sample was taken from each mare prior to the foal suckling. Blood samples were obtained from foals at 12 h, 2, 4, 6, and 8 wk post-foaling. Serum and colostrum samples were assayed for IgG concentration. Mares were weighed approximately 1 wk prior to foaling. Mares and foals were weighed 12 h post-foaling, and at 2 wk intervals for 56 d. Se amount or form did not affect colostrum IgG concentration or foal serum IgG concentration at 12 h. To adjust for pre-treatment differences in IgG concentration among groups, mare IgG data were analyzed for treatment differences using the initial IgG concentration as a covariate. Mare IgG concentrations were not affected by treatment at 4 or 2 wk pre-foaling, or at foaling. Failure of passive transfer occurred in two foals in treatment I3, and a third foal in this group did not complete the study. Consequently, serum IgG data from the remaining foals in group I3 and all foals in group O3 were combined and compared to foals in group I1. Foals from mares receiving 3 mg Se/d (either I3 or O3) had higher concentrations of IgG at 2 wk (P<.05), and at 4 and 8 wk (P<.1) compared to foals from mares receiving I1. Average daily gain of foals (1.5 kg/d for O3, 1.4 kg/d for I3, and 1.4 kg/d for I1) was not affected by treatment (P>.1). Placental weight and time to placental expulsion were not affected by Se amount or form. Supplementing mares with 3 mg Se/d may be beneficial by increasing foal IgG concentrations during the first few months of life when foals are most vulnerable to disease.

NUTRITIONAL STATUS OF BROODMARES AND FOALS ON COMMERCIAL FARMS IN ONTARIO

L.S. HUBER[1,3], J.C. PLAIZIER[1], B.W. MCBRIDE[1], G.L. ECKER[2] AND
A.F. CLARKE[3]
[1]University of Guelph, Guelph, Ontario
[2]Equine Performance Group, Puslinch, Ontario
[3]Equine Research Centre (Guelph) Inc., Guelph, Ontario

The trace minerals copper and zinc are required for normal growth, development and immune function. The objective of this study was to monitor the trace mineral status of broodmares in gestation and to determine influences on colostrum and foal serum. Eighty-one Thoroughbred and Standardbred broodmares on nine commercial breeding farms in Ontario were monitored monthly during the last trimester and during parturition. Blood samples for copper and zinc analyses were obtained at each visit and from foals at birth. Colostrum samples were analyzed for protein, immunoglobulin G (IgG), copper and zinc. The digestible energy, crude protein, copper and zinc daily intakes of the mares were estimated. Thoroughbreds and Standardbreds had a mean colostrum copper content and standard error of 0.71 ± 0.04 and 0.61 ± 0.03 mg/l, respectively. Mean broodmare serum copper levels showed a decrease during gestation. These values were in the low range of adequate serum levels for adults. Foaling month significantly ($P \leq 0.005$) affected mare serum copper levels. Mare serum copper levels and colostrum copper concentrations were positively correlated ($P \leq 0.05$). Both were positively correlated to estimated dietary protein ($P \leq 0.05$, $P \leq 0.05$) and energy intake ($P \leq 0.05$, $P \leq 0.05$), respectively. Foal serum copper levels at birth were positively correlated to colostrum IgG ($P \leq 0.05$) and protein ($P \leq 0.05$). There was an increase in mare serum zinc levels during gestation but these levels were dependent upon farm. Differences in mare colostrum copper levels could be due to farm management or genetic differences between breeds. Dietary and (or) farm management practices could influence the serum copper and zinc status of broodmares.

MINERAL INTAKE FROM MILK AND PASTURE IN NURSING FOALS

A. MATSUI, Y. INOUE AND Y. ASAI
Japan Racing Association, Tokyo, Japan

A nursing foal should consume sufficient minerals to avoid incidence of developmental orthopedic disease (DOD). The quantity of minerals obtained from milk and pasture, and whether they are sufficient for a foal's mineral requirement, are not well known. For this reason it is difficult to decide when to provide foals with creep feed and its appropriate composition. The objective of this study was to measure the amount of mineral intake from milk and pasture and to examine the appropriate nutritional management of a nursing foal. Eight Thoroughbred mares and foals were used for the study of milk consumption and mineral intake from milk. The amount of daily milk consumption was measured by the weight-suck-weight method at 3 days, 1 week, 3, 7, 10 and 17 weeks of age. Milk samples were taken by hand at the same time and the amount of Ca, P, Mg, Zn and Cu intake from milk was measured. Milk consumption of foals at 3 days, 1 week, 3, 7, 10 and 17 weeks of age was 19.0, 19.4, 16.5, 16.2, 13.0 and 11.4 kg/day, respectively. This shows that the amount of daily milk consumption decreased with age after the foal reached 3 weeks of age. From 3 days to 17 weeks of age, Ca, P, Mg, Zn and Cu intake from milk decreased 20.3 to 6.9 mg, 16.3 to 5.0 mg, 1.9 to 0.3 g, 64.1 to 25.9 mg and 12.3 to 3.6 mg, respectively. Daily pasture consumption of five foals was estimated by use of an indicator (lignin) method. All feces excreted in a day were collected and lignin in feces and pasture was analyzed. Daily Ca, P, Mg, Zn and Cu intake from pasture increased from 3 days to 17 weeks of age. The mineral intakes at the age of 17 weeks were 10.9 mg, 8.42 mg, 3.6 g, 52.8 mg and 15.4 mg, respectively. In relation to NRC (1989) recommendations for Ca, P, Mg, Zn and Cu, mineral intake from milk and pasture was inadequate between 1 and 3 weeks of age. These results indicate that it would be desirable to begin creep feeding a nursing foal as soon as it becomes old enough to consume feed.

ESTABLISHMENT OF NORMAL GROWTH RATES OF THOROUGHBREDS IN JAPAN

Y. INOUE, A. MATSUI AND Y. ASAI

Japan Racing Association, Tokyo, Japan

Nutrient requirements for growing horses are calculated based on age, body weight, mature body size, and average daily gain. It is important to study normal growth in horses and to obtain detailed information. A previous report (Hintz et al., 1979) has been accepted worldwide, but these data cannot be applied adequately in Japan. For a few decades, fast growth of foals and yearlings has been promoted and has been advantageous to sellers in Japan, but simultaneously the incidence of bone formation problems in the growing process has increased. The purpose of this study was to obtain the normal growth rates of Thoroughbreds in Japan. Body weight, withers height, heart girth and cannon bone circumference of approximately 270 foals of the Hidaka district in Japan were measured every 2 months from 1 to 15 months of age. The incidence of epiphysitis was monitored and the relationship to the dietary contents at weaning was studied. The body weight curve up to approximately 6 months of age was similar to data for horses with a 600 kg mature weight while between 6 and 15 months of age, the body weight curve was intermediate to 500 kg and 600 kg mature body weight (NRC, 1989). The body weight and withers height curves in this study were similar to those of horses in Kentucky (Pagan et al., 1996). In this study, epiphysitis of the fetlock joint was observed in some foals from 3 to 7 months of age. These foals weighed heavier at 1 month of age and their growth rates were more rapid than unaffected foals. Epiphysitis of the knee joint was observed in some foals at approximately 15 months of age. These foals weighed lighter while nursing, but were inclined to grow rapidly after weaning compared to unaffected foals. There were high correlations between the incidence of epiphysitis in fetlock joints after weaning and poor intakes of dietary copper, zinc, calcium and protein at this stage of growth. These results revealed that normal growth rate and criteria for nutritional supplementation of Thoroughbreds in Japan were necessary.

NUTRIENT REQUIREMENTS

NUTRIENT REQUIREMENTS OF THE YOUNG EQUINE ATHLETE

BRIAN D. NIELSEN
Michigan State University, East Lansing, MI

In many segments of the horse industry, horses enter training in their yearling or two-year-old year. This creates a situation for the owner or trainer that is quite different than what is experienced by many other stockmen or horsemen. Not only does the individual who is feeding this young animal have to worry about providing the correct ration for optimal growth, but he also has to be concerned about providing the correct balance of nutrients for exercise as well. In the past, much research has focused on feeding the growing horse. Additionally, there have been plenty of nutritional studies that have centered on the adult athlete. Unfortunately, the amount of research that has centered directly on the young athlete is quite limited. And when one searches for studies focused on young horses at the time they first enter training, the amount of information is even more limited. The reason for this is twofold. First, the process of training young horses is very labor intensive. Hence, to put enough horses in training for a controlled study proves to be quite challenging. Second, because of limited resources, researchers are forced to use animals that have not been very competitive in their respective events. It is rather uncommon for researchers to be able to put a horse into training before it is tested to determine how competitive it will be. Therefore, in order to examine the nutrient requirements of the young equine athlete, we are often forced to draw conclusions from adult horses in training and combine them with growth studies conducted with young horses.

Energy

The energy requirements of young horses entering training, and the relationship of the energy requirements to other nutrients, is one of the first items that needs to be addressed. A horse can meet its energy requirements through the consumption of both carbohydrates and lipids. Whereas carbohydrates have typically been the main energy source for horses, the addition of fat to the equine diet is now quite commonplace. Fat can be used to increase the energy of the diet without greatly increasing total intake. Studies have shown that fat can be as much as 20% of the diet with no harmful effects on the animal (Duren et al., 1987). This can provide a very useful means of meeting the increased energy demands of exercising animals. However, as energy is the most obvious metabolic factor that differs between the resting and exercising state, we need to ask whether the requirements for other nutrients vary in direct proportion to energy demand (Frape, 1988). In essence, as we increase the energy requirements of the horse as it begins training, it is yet to be fully determined as to whether the requirements of other nutrients (i.e. protein, Ca, P) should be based on a percentage of the total diet, grams of nutrient per day or on the basis of a nutrient to calorie ratio. To explore this issue, research in the young exercising horse needs to focus on two questions. First, does a difference in physiological response to exercise or growth exist as the nutrient to calorie ratio changes, and second, does voluntary food intake increase in proportion to an increased energy demand, thus increasing intake of all other dietary nutrients (Frape, 1988). With little research having been done in the young horse in training, we can only speculate as to whether there is a difference in physiological response when the nutrient to calorie ratio increases or decreases. At this point, if one is increasing the caloric density of the diet, it may be advisable to increase the other nutrients in the diet so that the ratio remains the same. Above average weight gains have been associated with the onset of bone abnormalities (Thompson et al., 1987). In addition, it has been reported that inadequate concentrations of protein and minerals relative to the energy concentration of the diet can result in skeletal problems (Potter, 1982). However, when growing horses are fed high energy diets, no adverse effects have been reported as long as diets provided suitable nutrient to calorie ratios (Hansen et al., 1987; Scott et al., 1987). Of course, this becomes a moot point if a horse is able to consume sufficient quantities of the feedstuff to meet its energy requirements without having to change the caloric density of the feed.

It is when lipids are used to increase energy density that care needs to be taken to balance the ration for the rest of the nutrients unless one is using a manufactured feed that already contains added fat and has been balanced to take this into account. By top-dressing a commercially prepared feed with fat, one is diluting the total nutrient balance. Hence, it becomes necessary to rebalance the ration. Does this mean one should avoid using fat in the diet of young performance horses? Definitely not. While carbohydrates serve as the main source of energy for high-intensity, short-duration activity, lipids serve as the predominant fuel for low-intensity, long-duration activity (Snow, 1992). As a result, there are obvious benefits in feeding fat to horses involved in endurance or aerobic activities, but there are also many benefits to horses competing anaerobically. This includes a possible glycogen-sparing effect (Pagan et al., 1987; Meyers et al., 1987). This effect would allow fats to be used preferentially, instead of glycogen, during non-anaerobic activity so that more glycogen is available for anaerobic activity. Other beneficial effects include an increase in muscle glycogen content (Oldham et al., 1989; Hughes et al., 1995), a decreased thermal load (Scott et al., 1993) and a reduction in performance times (Oldham et al., 1989). Because of its numerous beneficial effects, it is fortunate that horses can adapt to a fat-supplemented diet rather quickly. Thoroughbred horses in training have demonstrated a digestive adaptation to fat within one week of supplementation and increasing muscle glycogen concentration was seen within 21 days of receiving an added-fat diet (Hughes et al., 1995). Therefore, though the benefits are seen rapidly, it is important that the athletic horse receives the added fat early enough to be of full benefit in competition.

Protein

Protein requirements of young horses always generates a lot of interest due to the perceived notion that protein requirements for the athlete are much higher than for the idle animal. However, this is not necessarily true (Patterson et al., 1985). Additionally, there are negative effects associated with feeding an excess of protein. The first is expense. Protein tends to be an expensive nutrient, so the feeding of excess protein is usually a waste of money. Many feeds formulated for athletic horses tend to have large amounts of both energy and protein. While the extra energy is usually needed, one should determine whether the extra protein is beneficial. One reason for the misconception that feeds containing higher concentrations of protein are better for the athletic horse is that they tend to be more expensive. Most people naturally assume that items that are more expensive are better and since most people would like to feed the best feed available to their performance horses, they tend to feed the high protein feeds, regardless of whether there is an actual need for it. One study that looked specifically at feeding two concentrations of protein (10% and 20%) to two-year-old horses in training found no consistent benefits or detrimental effects in feeding the higher concentration of protein (Frank et al., 1987). The researchers reported that the extra cost of the high protein ration was not justifiable based on the performance of the horses. Another problem with feeding too much protein is that in order to get rid of the extra nitrogen, there will be an increase in urine production to help remove extra ammonia that is produced. The increased urine production leads to increased water intake (Hintz et al., 1980; Ralston, 1988), wetter stalls and an increased cost for bedding materials. Additionally, the increase in ammonia has potentially negative effects for horses housed in stalls with poor ventilation. It is believed that the decreased air quality resulting from the additional ammonia may increase respiratory problems for the athletic horse.

Despite the potentially negative consequences of feeding extra protein, the total protein requirement of the young horse entering training is slightly increased from maintenance. Freeman et al. (1988) found an increase in nitrogen retention, presumably resulting from increased muscle hypertrophy, when mature horses entered a training program and became conditioned. However, it is felt that if a DE:CP ratio is maintained, the additional DE intake needed for work will result in consumption of enough additional protein to meet the exercising horse's protein requirement (Freeman et al., 1988; Hinkle et al., 1981).

Minerals

Much work has been done on Ca balance in mature, idle horses and growing horses (Hintz et al., 1986). However, there appears to be little work conducted on 18- to 24-month old horses entering training to determine changes in mineral balance. It has been presumed that any increase in Ca requirements due to exercise would be met by increased Ca intake associated with increased dry matter intake (NRC, 1989). However, this has not been documented in young horses entering training. A study with 53 yearling Quarter Horses placed into race training demonstrated a substantial decrease in optical density of the third metacarpal during the first two months of training (Nielsen et al., 1997). This was followed by an increase that continued through the duration of the study. In a subsequent project, 10 untrained Quarter Horse geldings were put into race training and were fed a diet balanced to meet NRC recommendations to further investigate the influence of early training on bone metabolism (Nielsen et al., 1998a). A similar decrease in the mineral content of the third metacarpal was observed during the first two months of the project. The study also indicated a potential deficiency in dietary Ca in the young horses as they entered training. A follow-up study was conducted to investigate feeding different concentrations of Ca and P to two different groups of two-year-olds placed into training for the first time (Nielsen et al., 1998b). While the NRC (1989) recommends 0.32% Ca in the total diet for long yearlings in training, and 0.31% Ca for two-year-old horses in training on an as-fed basis, after completing this study (Nielsen et al., 1998b), it appears that the recommendation is too low when formulating diets for young horses entering intense training. It is probable that the ideal concentration of Ca is 0.40% as Schryver et al. (1978) found no additional benefit from feeding 0.60% Ca compared to 0.40% Ca to yearling Standardbreds in training. Though no additional benefit was seen by feeding the higher concentration of Ca (Schryver et al., 1978), the exercise being performed by horses in that study may not have been loading the skeletal system sufficiently to initiate a strong remodeling response (Nunamaker et al., 1991). However, there was definitely a benefit in feeding 0.38% Ca compared to 0.31% Ca as demonstrated by an increase in mineral content of the third metacarpal and an increased Ca retention in horses on the high mineral diet (Nielsen et al., 1998b). However, this experiment used NRC guidelines for Ca concentration in the diet, rather than the total intake of Ca in grams per day. When total Ca intake (g/d) is considered, the control diet provided only 80 to 85% of the requirement and the high mineral diet provided about 100%. When the adequacy of the NRC recommendation is considered from this perspective, it is arguable as to whether the data can be used to support a contention that the current recommendation is too low. Hence, one needs to decide whether it is more appropriate to base mineral requirements on a given quantity per day or whether it is more appropriate to feed as a percent of total intake.

Interestingly, a study was conducted examining factors associated with shin soreness in human athletes (Myburgh et al., 1988). The study contained 25 athletes that developed shin soreness and 25 control athletes who matched the injured athletes in age, sex and sport. Only three athletes who developed shin soreness consumed the recommended dietary allowance (RDA) of Ca, whereas 15 control athletes met the requirement. Additionally, only two of the control athletes consumed under half the RDA of Ca compared to ten of the injured athletes. As a result, it was reported that low Ca intake was significantly ($P < .005$) related to shin soreness. This seems to indicate that inadequate Ca intake may play a role in skeletal injuries.

The recommended concentration of P in the diet as suggested by the NRC appears to be adequate. No differences between treatments or days were seen in P retention in the study by Nielsen et al. (1998b). Hence, there appeared to be no benefit in feeding additional P above what the NRC recommends, other than to maintain a constant Ca:P ratio. Based on the Ca and P concentrations investigated in that study, it appears that the equine is quite capable of regulating P retention.

It is difficult to draw definite conclusions from the Mg balance results in the study by Nielsen et al. (1998b), as Mg concentrations were greater than what the NRC recommends. Additionally, once the feed was analyzed, it was discovered that the Mg concentration was not uniform throughout the project. Furthermore, there were no major differences in the amount of Mg fed to the control horses and the high mineral treatment group. However, additional supplementation of Mg may be greater than NRC (1989) recommendations. There was an increase in Mg retention in the latter part of the study in the high mineral group. This was the result of decreased urinary Mg, despite increased Mg intake. Since this is when the biggest increase in mineral content of the third metacarpal occurred in the high treatment group, there exists a potential benefit of feeding additional Mg during periods of high bone formation. This, however, is speculative at best. Additionally, the fact that the horses were in a negative Mg balance when fed Mg concentrations averaging 0.15% of the diet indicates that the currently recommended concentrations of Mg are inadequate.

Besides a potential increase in dietary mineral need for changes in bone metabolism, the additional sweating that accompanies exercise can cause an increased need for some minerals. Schryver et al. (1978) found that the total excretion of Ca and P in the sweat of mature polo horses ranged from 80 to 145 mg of Ca and 11 to 17 mg of P in a 20-min exercise period. Hoyt et al. (1995a) estimated sweat loss of Ca to be 8.5 mg per Mcal DE consumed above maintenance for work and sweat loss of P to be 10.7 mg per Mcal DE consumed above maintenance for work. Thus, it does not appear that sweating greatly increases the need for Ca and P. In contrast, an increased need for Na, Cl and K to replace that lost in sweat has been reported (Hoyt et al., 1995a; Meyer, 1987). For horses that are exercising hard and producing a lot of sweat, electrolytes should be provided.

Copper is another mineral that plays an important role in proper bone and cartilage development of young horses. A low Cu concentration in the diet appears to have links to osteochondrosis. A negative correlation between Cu concentration and perceived affliction of metabolic bone disease has been reported by Knight et al. (1985). Knight et al. (1990) also reported a reduction in prevalence and severity of osteochondrosis and other developmental cartilage lesions in foals fed elevated amounts of Cu. Gunson et al. (1982) reported several cases of foals with severe generalized osteochondrosis. The foals were suspected of having chronic zinc/cadmium toxicosis as lesions of their joint cartilage were similar to those seen experimentally in animals fed diets high in zinc. This may be partially due to the high levels of zinc interfering with normal copper absorption as similar cartilaginous lesions were also seen in foals fed a copper deficient diet (Bridges and Harris, 1988).

In addition to osteochondrosis, low Cu concentrations have been linked to epiphysitis in the fetlock joint of weanlings (Asai et al., 1993). In a study conducted by Hurtig et al. (1990), foals fed a diet containing 7 ppm copper had greater incidence and severity of angular limb deformities, mild flexural deformities, epiphysitis, intermittent lameness and osteochondrosis lesions than weanlings fed 30 ppm of copper. Hence, it appears that the NRC (1989) recommendation of 10 ppm is minimal and that by increasing the concentration, a reduction may be seen in the incidence of developmental orthopedic disorders.

One other mineral that may need special attention is selenium. Depending on the part of the country in which feedstuffs are grown, there may be a need for selenium supplementation. Forages and grains grown in the northeastern United States are often low in Se (Kubota et al., 1967). Hence, if one lives in this part of the country, it is advisable to evaluate the Se concentrations in the rations of their horses to ensure that they are adequate.

Vitamins

Research has not been able to demonstrate any drastic need for changes in vitamin requirements for the young exercising horse, with one possible exception. Horses undergoing heavy training may develop a lethargic, depressed condition described as "track sour" in the racing industry. Often these horses lose their appetite. As they are often receiving large quantities of energy in their diets, there is a chance that these horses are deficient in B-vitamins, particularly thiamin, which are necessary for utilizing extra dietary energy. Thiamin is needed for the metabolism of pyruvate. If the absorption of microbially synthesized thiamin in the intestine is not adequate, there may be a dietary requirement (Carroll et al., 1949). Topliff et al. (1981) reported that supplementation of thiamin is necessary in hard working horses receiving a traditional diet. Hence, one should be aware of the signs of a horse becoming "track sour" and supplement the diet if needed.

Water

The importance of a good, clean water supply to young horses in training is obvious. As horses increase their workload, there is also an increase in sweating rates. This is especially true during hot weather or when horses are performing endurance-type activities. While being ridden 9.6 km daily, Thoroughbreds have been reported to have sweat losses of 15.6 g/kg of body weight (Hoyt et al., 1995b), while almost 40 liters of water have been reported to have been lost during an 80-km endurance ride (Snow et al., 1982). Hence, there is an additional need in exercising horses for water to replace that lost as sweat. If electrolytes are added to the water, it is recommended that additional water also be available to the animal. Some horses will refuse to drink enough water if it is supplemented with electrolytes and this could leave the horse prone to dehydration.

Other Tips

As previously stated, the energy needs of a horse can increase greatly upon entering training. Though part of this increased demand can be made up through the addition of fat to the diet, it will usually become necessary to increase the carbohydrate portion of the diet as well. When increasing carbohydrates in the diet, it is recommended that the total daily intake be divided into as many feedings as possible. Increasing the intake without increasing the number of times fed will increase the likelihood of digestive problems such as colic.

Despite all of the recommendations being set forth on how to achieve optimum performance, one last item should be mentioned. Though there is an increased nutritional demand placed upon the young horse as it enters training, one management practice used by some in the industry is to choose to ignore that increase in demand, and for good reason. Often when a young horse is being "broken" to ride, the trainer will not increase the intake of the animal until after four to six weeks of training. This is to make the animal easier to control. One saying used by some horse trainers is "work the mental through the physical." A horse that is "full of energy" will be more difficult to control and train during the early stages of training. As a result, it is not uncommon to wait until the horse has made sufficient progress in its training before making large increases in its feed. This would tend to make the nutritional requirements of that young horse even more complicated. Despite that problem, this practice will continue and is one that many recommend.

Literature Cited

Asai, Y., Y. Mizuno, O. Yamamoto and H. Fujikawa. 1993. Requirements of copper and zinc for foals in connection with the incidence of epiphysitis. Anim. Sci. Tech. (Jpn.) 64(12):1193-1200.

Bridges, C.H., and E.D. Harris. 1988. Experimentally induced cartilaginous fractures (osteochondritis dissecans) in foals fed low-copper diets. J. Am. Vet. Med. Assoc. 193:215.

Carroll, F.D., H. Goss and C.E. Howell. 1949. The synthesis of B vitamins in the horse. J. Anim. Sci. 8:290.

Duren, S.E., S.G. Jackson, J.P. Baker and D.K. Aaron. 1987. Effect of dietary fat on blood parameters in exercised Thoroughbred horses. In: J.R. Gillespie and NE Robinson. eds. Equine Ex. Phys. 2. Davis, CA: ICEEP Publications:674-685.

Frank, N.B., T.N. Meacham and J.P. Fontenot. 1987. Effect of feeding two levels of protein on performance and nutrition of exercising horses. Proc. 10th ENPS. p. 579.

Frape, D.L. 1988. Dietary requirements and athletic performance of horses. Equine Vet. J. 20(3):163.

Freeman, D.W., G.D. Potter, G.T. Schelling and J.L. Kreider. 1988. Nitrogen metabolism in mature horses at varying levels of work. J. Anim. Sci. 66:407.

Gunson, D.E., D.F. Kowalczyk, C.R. Shoop, and C.F. Ramberg, Jr. 1982. Environmental zinc and cadmium pollution associated with generalized osteochondrosis, osteoporosis, and nephrocalcinosis in horses. J. Am. Vet. Med. Assoc. 180:295.

Hansen, D.K., F.M. Rouquette, G.W. Webb, G.D. Potter and M.J. Florence. 1987. Performance of yearling horses on pasture and supplemental feed. Proc. 10th ENPS. p. 25.

Hinkle, D.K., G.D. Potter, J.L. Kreider, G.T. Schelling and J.G. Anderson. 1981. Nitrogen balance in exercising mature horses fed varying levels of protein. Proc. 7th ENPS. p. 91.

Hintz, H.F., H.F. Schryver, and J.E. Lowe. 1986. Calcium for pregnant mares and growing horses. Equine Prac. 8(9):5.

Hintz, H.F., K.K. White, C.E. Short, J.E. Lowe and M. Ross. 1980. Effects of protein levels on endurance horses. Proc. 72nd Amer. Soc. Anim. Sci. (Abstract):202.

Hoyt, J.K., G.D. Potter, L.W. Greene and J.G. Anderson. 1995a. Mineral balance in resting and exercised miniature horses. J. Equine Vet. Sci. 15(7):310-314.

Hoyt, J.K., G.D. Potter, L.W. Greene, M.M. Vogelsang and J.G. Anderson. 1995b. Electrolyte balance in exercising horses fed a control and a fat-supplemented diet. J. Equine Vet. Sci. 15(10):429-435.

Hurtig, M.B., S.L. Green, H. Dobson and J. Burton. 1990. Defective bone and cartilage in foals fed a low-copper diet. Proc. Amer. Assoc. Equine Pract. p. 637-643.

Hughes, S.L., G.D. Potter, L.W. Greene, T.W. Odom and M. Murray-Gerzik. 1995. Adaptation of Thoroughbred horses in training to a fat supplemented diet. Equine Vet. J. 18:349-352.

Knight, D.A., A.A. Gabel, S.M. Reed, L.R. Bramlage, W.J. Tyznik and R.M. Embertson. 1985. Correlation of dietary mineral to incidence and severity of metabolic bone disease in Ohio and Kentucky. Proc. 31st Amer. Assoc. Equine Pract. p. 445.

Knight, D.A., S.E. Weisbrode, L.M. Schmall, S.M. Reed, A.A. Gabel, L.R. Bramlage and W.J. Tyznik. 1990. The effects of copper supplementation on the prevalence of cartilage lesions in foals. Equine Vet. J. 22(6):426-432.

Kubota, J., W.H. Allaway, D.L. Carter, E.E. Cary and V.A. Lazar. 1967. Selenium in crops in the United States in relation to selenium-responsive diseases of animals. J. Agr. Food Chem. 15:448-453.

Meyer, H. 1987. Nutrition of the Equine Athlete. In: J.R. Gillespie and NE Robinson. eds. Equine Ex. Phys. 2. Davis, CA: ICEEP Publications:644-673.

Meyers, M.C., G.D. Potter, L.W. Green, S.F. Crouse and J.W. Evans. 1987. Physiological and metabolic response of exercising horses to added dietary fat. Proc. 10th ENPS. p. 107-113.

Myburgh, K. H., N. Grobler and T. D. Noakes. 1988. Factors associated with shin soreness in athletes. Phys. Sports Med. 16(4):129.

National Research Council. 1989. Nutrient Requirements of Horses (5th Ed.). National Academy Press, Washington, DC.

Nielsen, B.D., G.D. Potter, L.W. Greene, E.L. Morris, M. Murray-Gerzik, W.B. Smith and M.T. Martin. 1998a. Characterization of changes related to mineral balance and bone metabolism in the young racing Quarter Horse. Accepted for publication in J. Equine Vet. Sci.

Nielsen, B.D., G.D. Potter, L.W. Greene, E.L. Morris, M. Murray-Gerzik, W.B. Smith and M.T. Martin. 1998b. Response of young horses in training to varying concentrations of dietary calcium and phosphorus. Submitted for publication in J. Equine Vet. Sci.

Nielsen, B.D., G.D. Potter, E.L. Morris, T.W. Odom, D.M. Senor, J.A. Reynolds, W.B. Smith and M.T. Martin. 1997. Changes in the third metacarpal bone and frequency of bone injuries in young Quarter Horses during race training — observations and theoretical considerations. J. Equine Vet. Sci. 17(10):541-549.

Nunamaker, D. M., D. M. Butterweck and J. Black. 1991. In vitro comparison of Thoroughbred and Standardbred racehorses with regard to local fatigue failure of the third metacarpal bone. Am. J. Vet. Res. 52:97.

Oldham, S.L., G.D. Potter, J.W. Evans, S.B. Smith, T.S. Taylor and S.W. Barnes. 1989. Storage and mobilization of muscle glycogen in racehorses fed a control and high-fat diet. Proc. 11th ENPS. p. 57-62.

Pagan, J.D., B. Essen-Gustavsson, A. Lindholm ad J. Thornton. 1987. The effect of dietary energy source on exercise performance in Standardbred horses. In: J.R. Gillespie and NE Robinson. eds. Equine Ex. Phys. 2. Davis, CA: ICEEP Publications:686-700.

Patterson, P.H., C.N. Coon, and I.M. Hughes. 1985. Protein requirements of mature working horses. J. Anim. Sci. 61:187.

Potter, G.D. 1982. Feeding young horses for sound development. Proc. Texas A&M Horse Short Course. p. 35-42.

Ralston, S.L. 1988. Nutritional management of horses competing in 160 km races. Cornell Bet. 78:53.

Schryver, H.F., H.F. Hintz and J.E. Lowe. 1978. Calcium metabolism, body composition, and sweat losses of exercised horses. Amer. J. Vet. Res. 39(2):245.

Scott, B.D., G.D. Potter, J.W. Evans, J.C. Reagor, G.W. Webb and S.P. Webb. 1987. Growth and feed utilization by yearling horses fed added dietary fat. Proc. 10th ENPS. p. 101.

Scott, B.D., G.D. Potter, L.W. Greene, M.M. Vogelsang and J.G. Anderson. 1993. Efficacy of a fat-supplemented diet to reduce thermal stress in exercising Thoroughbred horses. Proc. 13th ENPS. p. 66-71.

Snow, D.H. 1992. A review of nutritional aids to energy production for athletic performance. Equine Athlete. 5(5):1-9.

Snow, D.H., M.G. Kerr, M.A. Nimmo and E.M. Abbott. 1982. Alterations in blood, sweat, urine and muscle composition during prolonged exercise in the horse. Vet. Rec. 110:337.

Thompson, K.N., S.G. Jackson, and J.R. Rooney. 1987. The effect of above average weight gains on the incidence of radiographic bone aberrations and epiphysitis in growing horses. Proc. 10th ENPS. p. 5-9.

Topliff, D.R., G.D. Potter, J.L. Kreider, G.T. Jessup and J.G. Anderson. 1981. Thiamin supplementation for exercising horses. Proc. 7th ENPS. p. 167.

COMPARISON OF THE DIGESTIBLE ENERGY (DE) AND NET ENERGY (NE) SYSTEMS FOR THE HORSE

PAT A. HARRIS

WALTHAM Centre for Pet Nutrition, Leicestershire, U.K.

The horse can be described as a monogastric herbivore or a non-ruminant herbivore which is suited to the digestion and utilization of high fiber diets due to continual microbial fermentation primarily within the hindgut (cecum and colon). Domestication, and an increasing demand for horses to perform under circumstances that require energy intakes above those able to be provided by their more 'natural' diet of fresh forage, has resulted in the inclusion, in particular, of cereal grains and their by-products as well as supplemental fat in many horse diets. Such additions may be made in the form of the raw material or processed raw material or a manufactured compound feed. The upper part of the gastrointestinal tract has a relatively small capacity and the horse has digestive and metabolic limitations to high grain (high soluble carbohydrate) diets. Large grain meals may overwhelm the digestive capacity of the stomach and small intestine leading to the rapid fermentation of the grain carbohydrate in the hindgut. This may result in one of a number of disorders including colic, diarrhea and laminitis. Providing the right amount of energy from the appropriate sources without compromising the digestive system is therefore very important especially to the performance horse. Recently it has been suggested that the net energy system may be a more appropriate way of describing both the energy content of feeds as well as the energy requirements of horses. This paper will explore certain aspects of both the NE and the more traditional DE systems and will look at :

- Energy sources and how the horse obtains its energy from the different feedstuffs.
- Energy content of the different feedstuffs and its availability from such feedstuffs.
- Energy requirements.

Definition

A typical dictionary definition of energy would be 'Physics a. The capacity of a body or system to do work. b. A measure of this capacity measured in joules (SI) units.' Energy *per se* is therefore not a nutrient and the precise definition of energy, especially in a nutritional sense, may be complex.

Energy Sources and Efficiency of Utilization

Practically, certain nutrients in a horse's diet provide the energy intake for that individual following conversion of their chemical energy to other forms of chemical energy, mechanical energy and heat. Dietary energy is provided to the horse by four principal dietary energy sources :

1) Hydrolyzable carbohydrates e.g. starch.
2) Cellulose, pectins, hemicelluloses, etc. (i.e. non-starchpolysaccha-rides: a component of dietary fiber).
3) Fats *(normally less than 3% total feed intake but most horses able to digest and utilize fat efficiently - not usually recommended to feed at more than 10% of the diet and any fat should be introduced gradually).*
4) Proteins *(not a nutritionally preferred option as an energy source: ergogenically inefficient; nitrogen must be removed, as excess protein is not stored, resulting in increased water requirements and potentially higher ammonia levels in the stable).*

In general, a high proportion of the available starch ingested is degraded to glucose before absorption in the small intestine (S.I.) (unless the digestive capacity of the S.I. is overwhelmed). However, a proportion of the starch and a varying proportion of the dietary fiber (depending on the extent of lignification) will be subjected to microbial fermentation. This occurs primarily in the large intestine, producing predominantly short chain or volatile fatty acids, some of which may be used directly as an energy fuel by the gut cells themselves while the majority is absorbed and converted to either glucose or fat .

Figure 1.

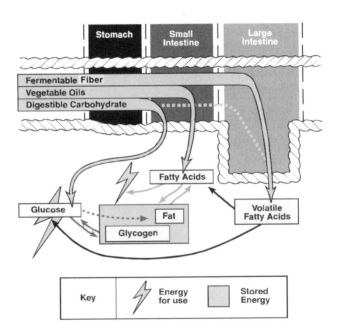

The fermentation process is ultimately less efficient than obtaining energy from carbohydrate sources directly via glucose. This helps explain why feeds with a high fermentable fiber content provide less useable energy than those feeds with a high digestible carbohydrate content. The extent to which cereal starch provides glucose or volatile fatty acids, as the end result of digestion, will depend on its prececal digestibility which, in turn, will vary according to the feedstuff under consideration and the extent and nature of the processing it has been subjected to. If excessive starch reaches the hindgut it will be rapidly fermented resulting in high levels of lactic acid and a number of potential adverse sequellae.

The main fuel sources available for energy production by the horse both at rest and during exercise are considered to be *carbohydrate* in the form of muscle glycogen or blood glucose; *fat* in the form of muscle or plasma triglyceride, plasma free fatty acids and muscle stores of *adenosine triphosphate* (ATP) and *phosphocreatine* (PCr) :

1. Anaerobic metabolism of glucose occurs more quickly but less efficiently (i.e. less ATP produced) than in the case of aerobic metabolism.
2. Fat can only be metabolized aerobically. This is comparatively slow but very efficient, with less of the gross chemical energy being lost as heat.
3. Energy is very quickly available from the very limited stores of ATP and PCr.

There are a large number of factors that affect the proportion of energy that can be derived from each potential energy source, during different energy bouts, including:

1. The intensity and duration of exercise;
2. The muscle fiber composition of the horse;
3. The diet and the fitness (coupled with the training regimen) of the horse.

It has been suggested that plasma free fatty acids are quantitatively the most important energy source (especially during submaximal exercise) in the horse; but as the intensity of exercise increases, the relative contribution of fat to total energy production decreases so that during very high intensity exercise, the catabolism of carbohydrate accounts for the majority of energy.

Energy Content of Feedstuffs

At present there are three main ways to describe the energy potential of a horse feed: total digestible nutrients (TDN), digestible energy (DE) and net energy (NE). Each of these has been determined in a number of ways over the years with TDN becoming less popular recently. Further confusion results from the fact that two units of energy are in common use in the horse industry: the joule (J) predominantly in Europe and the calorie in the United States (4.184 J is taken to be equivalent to 1 calorie).

Figure 2.

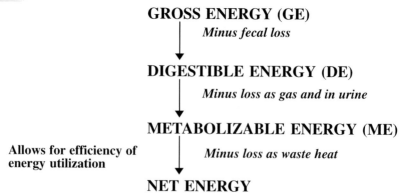

GROSS ENERGY (GE)

Minus fecal loss

DIGESTIBLE ENERGY (DE)

Minus loss as gas and in urine

METABOLIZABLE ENERGY (ME)

Allows for efficiency of
energy utilization

Minus loss as waste heat

NET ENERGY

= energy available for life and movement

The energy value of any feedstuff, as well as the total diet, for the horse will depend upon the relative amounts of hydrolyzable and fermentable substrates that it contains. Determination of a feed's energy value using *in vivo* methods, especially in the horse, tends to be time consuming, labor intensive, costly and often highly impractical. Therefore, as with many other species, effort has concentrated on finding methods for assessing the energy values of feeds using prediction equations. At the moment these tend to be based on the chemical composition of the feed, which may not truly reflect its functional aspects.

TDN

The energy content of rations has been calculated as the percent total digestible nutrients in a number of ways as illustrated in Table 1. Conversion factors have been used to convert TDN values to today's more commonly used DE values. These may not be appropriate. The most frequently used factor is based on work in ruminants which resulted in an average conversion factor of 2000 kcal of DE being equivalent to 1 lb TDN or 4.41 Mcal DE/kg TDN. However, subsequent work in ruminants suggested that this conversion factor was strongly influenced by the level of digestible protein. In the horse limited work in this area has been carried out, but in one study of five pony stallions the DE to TDN relationship was found to be 4.648 Mcal/kg for a hay diet and 4.624 Mcal/kg for a hay and concentrate diet with similar crude protein levels (Barth et al., 1977), suggesting that the 4.41 conversion factor could result in substantial errors.

DE

Most commonly the energy content of horse feed is referred to by its DE or digestible energy content, i.e. fundamentally the gross chemical energy in the feed minus the energy lost in the feces. Using standard digestibility balance studies,

Table 1.

Examples of equations that have been used to determine TDN and DE from
the chemical composition of feed include:
a) TDN = % DCP + % DNFE + (%DEE x 2.25)
DE = 2 x TDN (%) in Mcal/kg
b) TDN = 78.1 - (1.01 x ADF %) + (0.823 x CP %)
DE = 0.255 $\pm$ (0.0366 x TDN %)
2.2
c) DE (kcal/kg DM) = 2118 + 12.18 (CP) - 9.37(ADF) -3.83(NDF- ADF)
+ 47.18(fat) + 20.35(NSc) - 26.3(Ash)
d) DE (MJ/kgDM) = DCP x 0.023 + DEE x 0.0381 + (DCF+DNFE) x 0.0172
e) DE (MJ/kg DM) = 11.1 + 0.0034 CP + 0.0158 CF - 0.00016 CF^2

DCP = Digestible crude protein	ADF = Acid detergent fiber
DNFE = Digestible nitrogen free extract	NDF = Neutral detergent fiber
DEE = Digestible ether extract	DCF = Digestible crude fiber
NSc = 100 - CP - EE - NDF - ash	CP = Crude protein
	CF = Crude fiber

the digestible energy content of a ration can be estimated *in vivo*. This does not
provide a truly accurate measurement because fecal energy includes energy origi-
nating from endogenous sources as well as undigested feed and bacteria. It is
nevertheless a very useful practical guide. However, this is an expensive and
time consuming way to determine DE and so a number of equations to estimate
DE content of feedstuffs have been quoted in the literature (see Table 1 for some
examples) and continual modifications are being proposed to improve the predictive nature of these equations.

Net Energy

The DE system tends to overestimate the actual mechanical energy potential of a
high fiber feed compared with a high soluble carbohydrate feed, as fiber
predominantly produces VFAs which are not used as efficiently as glucose. The
French Net Energy system (primarily developed by the Institut National de la
Recherche Agronomique [INRA]) was developed to allow for the differences in
utilization of the metabolizable energy available from different feeds,
depending on the proportion of the end products of digestion produced and the
biochemical pathways used by these end products to produce mechanical
energy (as outlined above).

It uses the horse feed unit (HFU) or in French, l'unite fouragire cheval (UFC).
The UFC corresponds to the net energy value (2250 kcal) of one kg standard
barley (87% DM) in a horse at maintenance. The UFC value of a particular feed
is calculated by dividing its NE content in kcal by that of barley, i.e. 2250.

The NE UFC/kg DM for corn, barley, oats, maize silage, hay and straw are 1.35, 1.16, 1.01, 0.88, 0.53 and 0.28, respectively.

Comparing this system with the DE/ME system, the energy value of straw for example has been reported to be 41% that of barley using DE values, 37% that of barley using ME values and 29% that of barley using NE. Using this as an extreme example, if a 500 kg horse was being fed 3 kg barley and 7 kg hay and one wanted to replace the barley, *in energy terms*, with straw :

· Using DE values one would calculate having to feed about 7.5 kgs of straw.

· Using NE values (to get the same useable energy) one would calculate having to feed <u>about 10.5 kgs of straw.</u>

The discrepancy between the actual energy value to the horse for different feeds as calculated by these various methods increases as the cell wall contents increase. So if you fed a diet which contained certain high fiber feedstuffs according to its DE value you would tend to overestimate the actual useable energy available from that diet and the horse might either lose weight or lack sufficient energy.

For the purpose of this paper the NE value for oats (as fed) according to the French Net Energy system is taken to be ~2.0 MCal/kg and for hay ~1.0 – 1.25 Mcal/kg.

Figure 3.

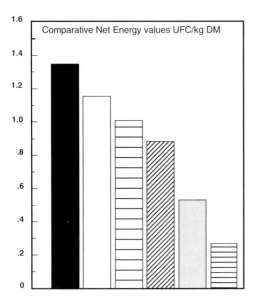
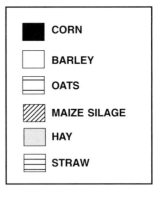

Net Energy Available to the Horse from the Diet

A number of methods or models have been used to estimate the energy partitioning of a diet and therefore the energy available. Most of these rely on a mixture of calculated, determined and assumed efficiency factors and have not been validated fully under field conditions. However, they provide at the moment the most reliable way of determining the true or net energy available from a diet.

The UFC of forages or concentrates (raw materials), it is suggested, can be predicted directly from their chemical composition, although these predictive values are more accurate if the DE is known (Martin-Rosset et al., 1994.):

1. Forages
   UFC = 0.0557 + 0.0006 CC + 0.2489 DE (r^2 = 0.996)
   UFC = 0.825 - 0.0011 CF + 0.0006 CP (r^2 = 0.69)
2. *Concentrates* - raw materials
   UFC = -0.134 + 0.0003 CF -0.0004 CP + 0.0003 CC + 0.3160 DE
   (r = 0.99)(where CC = cytoplasmic carbohydrate, CP = crude protein,
   CF = crude fiber).

It should be noted, however, that an additional correction has been recommended for high fat diets. For compound feeds it has been suggested that the UFC value can also be predicted (Martin-Rosset et al., 1994) from chemical composition using a number of equations. The difference between the values obtained using a composite equation and the addition of the individual feed component UFC values is said to be between 0.2-1.3 UFC/100 kg organic matter.

The French Net Energy system (an empirical system) for calculation is illustrated in Figure 4 and can be compared with a partitioning model (a more physiological system), as illustrated in Figure 5, for determining NE, ME, and DE values (Kronfeld, 1996). *The net energy values can be directly compared only for maintenance, but not for work, as explained below.* Using one of the suggested composite equations appropriate for forages (Martin-Rosset et al., 1994) or the system outlined in Figure 4, the net energy available for maintenance from a 100% 'average' timothy hay works out to be around 5.2 MJ/kg which is very close to that predicted, using the partitioning system (Kronfeld, 1996) for a similar timothy hay diet of around 5.4 MJ/kg. Similarly a compound feed with fat supplementation gives a French Net Energy value of around 9.5 MJ/kg for maintenance compared with 9.3 MJ/kg derived using partitioning. Further comparisons need to be made both theoretically and in the field before conclusions can be made as to which system provides an overall advantage.

Unfortunately accurate digestibility data for the various different energy sources within different feeds are not currently available, and the effect of different feed types and processing techniques on energy availability are not fully understood. Therefore, the amount of energy each nutrient type within a particular ration will really provide to an individual animal cannot be accurately determined. Differing analytical methods and terminology used also add to the confusion. Both systems, as illustrated in Figures 4 and 5, therefore can only provide a guide to the energy content of a diet.

Energy Requirements for Maintenance

Definitions include 'the daily food intake that maintains constant body weight and body composition of a healthy adult horse with zero energy retention at a defined level of activity in comfortable surroundings.' Several equations for estimating maintenance energy requirements have been derived over the years and will not be discussed in detail here. A number of factors including the individual, body composition, gender, environmental temperature, whether in work or not, age, lifestage, breed, temperament and season may affect actual maintenance energy requirements.

Energy Requirements for Exercise

Many factors also influence the additional requirements for exercise including:
· Weight of the rider;
· Weight of the tack;
· Ability of the rider;
· Degree of fatigue;
· Condition and training of the animal;
· Diet composition and
· Environmental conditions.

In addition, the nature of the exercise itself will vary according to its intensity, duration and contour (terrain).

The efficiency of the conversion of chemical energy (derived ultimately from the diet) to mechanical work (work efficiency or k_w) is only about 20-25%. Most of the energy released appears as heat.

Figure 4 (page 207): System for determining the net energy of a diet using the French Net Energy system devised by INRA (based on Martin-Rosset *et al.*, 1994). Examples only of the various equations available are given.

Figure 5 (page 208): System for determining the net energy available from a certain diet (in MJ) *for maintenance* based on a physiological partitioning system (Kronfeld, 1996), on an as-fed basis.

Figure 4.

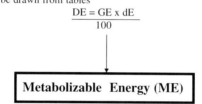

Forages

$$\boxed{\textbf{Gross Energy (GE)}}$$

- Predicted from the CP content
- Conversion factor will vary according to type of hay

Concentrates
GE = 5.72 CP + 9.5 EE + 4.79 CF + 4.17 NFE + 'A'
The 'A' values depend on the kind of feed cereals

$$\boxed{\textbf{Digestible Energy (DE)}}$$

- Predicted from GE and the digestibility of energy (dE) efficiency factor
dE = 0.0340 + 'B' + 0.9477 x digestible organic matter
B = +1.1 for concentrates: -1.1 for forages.
dOM can be predicted from crude fiber content for forages to some degree
dOM for concentrates need to be drawn from tables

$$DE = \frac{GE \times dE}{100}$$

$$\boxed{\textbf{Metabolizable Energy (ME)}}$$

For all feeds
100 (ME/DE) = 84.07 + 0.165 CF - 0.276 CP + 0.184 CC

For protein rich feeds $\geq$ 30% CP on DM basis
100 (ME/DE) = 94.36 + 0.110 CF - 0.275 CP
ME/DE $\approx$ 0.78 - 0.8 for oil meals
0.91 wheat straw
0.84 - 0.88 hays
0.90 - 0.95 for cereals → \ for feedstuff ME = $\frac{ME}{DE}$ ratio x DE

value for that feedstuff

$$\boxed{\textbf{Net Energy (NE)}}$$

NE   =   (ME x efficiency of ME utilization [k_m]   = X
NE (UFC) = $\frac{X}{2250}$

k_m varies with feed composition and is predicted from the chemical composition, e.g.
Forages
100k_m = 57.56 - 0.0110 CF + 0.0105 CP + 0.0270 CC + 0.0150 DOM
100k_m = 71.64 - 0.0289 CF + 0.0148 CP
Cereal by-products
100k_m = 94.41 - 0.0237 OM - 0.0022 CP + 0.0121 CC
100k_m = 82.27 - 0.0248 CF - 0.0160 CP

continued

Example for Barley
>   GE = 3854 Kcal/kg
>   dOM = 0.83 (from Tables) \ dE = 0.8 DE = 3076 Kcal/kg
>   ME/DE = 0.93 \ ME = 2864 Kcal/kg
>   km = 0.785 \ NE = 2250 Kcal/kg
>   1 UFC = 2250 Kcal NE

CP = Crude protein, EE = Ether extract, CF = Crude fiber, NFE = Nitrogen free extract
CC= Cytoplasmic carbohydrates
dOM=Digestible organic matter
OM = Organic matter

Figure 5.

Gross Energy/Intake Energy (IE)

- Calculated from the relative proportion of fat, carbohydrate and protein using conversion factors (heats of combustion) of 38.9, 17.5 and 23.7 KJ/g respectively

$$\downarrow$$

Digestible Energy (DE)

- Using estimated digestibility factors $(K_d = DE/IE)$ - various estimates from other experiments
- Factors will vary according to nature of diet
- Factors of 0.74 and 0.65 x IE for protein and fat respectively for a diet without supplemental fat
- Factors of 0.61 and 0.82 x IE for protein and fat respectively for diet with supplemental fat
- Factors of 0.8 x IE for NSC
- Factors of 1.0 x IE for CHO-H ie where hydrolyzable carbohydrate CHO-H = nonstructural carbohydrate (NSC) x 0.8
- Factors of 0.25 - 0.5 x IE for fermentable carbohydrate where fermentable carbohydrate = NDF + 0.2 x NSC)

$$\downarrow$$

Metabolizable Energy (ME)

- Takes into account energy loss in gas produced during fermentation as well as via urea in the urine
- 0.78 x DE for protein (loss of energy as urea)
- 0.95 x DE for fermentable carbohydrate (loss of energy as methane)

$$\downarrow$$

Net Energy Available for Maintenance (NE)

- Uses estimated efficiency factors (K_m)
- K_m 0.7; 0.89; 0.85 and 0.63 x ME for amino acids, long chain fatty acids, glucose, short chain fatty acids respectively

NB. Amount of energy available for the work of maintenance or exercise will depend on the exercise undertaken and the heat produced ie K_w may not be the same as K_m.

$$\downarrow$$

Net energy available for work (NE_w)
K_w of 0.228 for glucose oxidation : 0.245 for long chain fatty acid oxidation

Digestible Energy Requirements

DE requirements can be worked back from the energy expenditures. Alternative guidelines have been given for directly determining the DE requirements for different types of physical activity on a minute-by-minute basis or even more general empirical equations can be used. The following equation, for example, was produced for horses whose work load (kg x km) was not greater than 3560 (i.e. not applicable to endurance horses) and whose body weight was maintained (Anderson et al., 1983). This therefore assumes that this body weight was desirable:

DE (Mcal/day) = 5.97 + 0.021W + 5.03X - 0.48X^2
(where W = body weight in kg : X = Z x distance travelled in km x 10^{-3} :
Z = weight of horse, rider and tack in kg.)

Such an equation makes no allowance for speed which may be important in the horse.

The NRC (1989) suggests even more general equations as a guide:
· Light work (e.g. pleasure riding, bridle path riding, etc.) 1.25 x Maintenance DE
· Moderate work (e.g. ranch work, jumping, etc.) 1.5 x Maintenance DE
· Intense work (e.g. racing, polo, etc.) 2 x Maintenance DE

Net Energy Requirements

Estimated net energy requirements, in addition to the daily maintenance net energy requirements, for a number of types of work (based on work by INRA) are shown in Table 2. It is not possible to directly compare these net energy requirements with those derived based on the partitioning system described above (Kronfeld, 1996) because the French Net Energy system assumes the same efficiency for work (K_w) as for maintenance (K_m), i.e. $K_w \sim K_m$, whereas in the partitioning system, K_w (~ 0.2 - 0.25) is taken to be much lower than K_m (~ 0.7 - 0.8). i.c. the French NE system states that the NE value for hay is around 1.0 – 1.25 Mcal/kg whether the hay is used for work or maintenance whereas in the partitioning system hay has a value of around 1.29 Mcal (5.4 MJ) for maintenance and only 0.39 MCal (1.63 MJ) for work. Because the requirements are estimated using the same assumptions, those made under the French Net Energy system appear far higher than those estimated using the partitioning system (see examples page 210).

Table 2.

Type of Work	Guide to Mcal Net energy requirements/hr
Very light (@ 50% walk 50% trot)	0.45 - 1.125
Light lesson (@ 50% walk 40% trot 10% canter)	2.25 - 3.375
Moderate lesson (@ 20% walk 10% canter 10% jumping 60% trot)	3.375 - 4.5
Hard lesson	5.625 - 6.75
Light hack >3hrs/day (@ 90% walk 10% trot)	1.125
Light short ride 1-2hrs/day (@ 50% walk 45% trot 5% canter)	3.375
Moderate outdoor training (@20% walk 10% jumping 15% cantering 45% trot)	4.5 - 5.625
Intense training/competition (@ 10% walk 15% cantering 15% jumping 40% trot)	5.625 - 7.895

Estimated net energy requirements needed to be added to the daily maintenance requirement based on a 560 kg body weight horse carrying a 100 kg load (INRA 1984).
NE requirements for maintenance according to (Martin-Rosset et al., 1994)

$$^{ab}NE = 84 \ W^{0.75} \quad (Kcal/day)$$
$$^{ab}NE = 0.038 \ W^{0.75} \quad or \ (UFC/day)$$

[a] + 10-20% for stallions [b] + 5-15% for working horses to take into account the rise in overall energy metabolism, where W = Body weight (kg).

Practical Application

The three-day event, and in particular the cross-country day, is one of the most demanding equestrian sports and provides an indication of the energetic load of horses working at different speeds. Using a number of assumptions and theoretical calculations, the metabolic energy (ME) requirements of the various stages of the cross-country day have been estimated for a 575 kg (total weight + rider) horse: Phase A ~7.9 MJ; Phase B ~ 6.2 MJ; Phase C ~ 17.25 MJ & Phase D (the cross-country phase) ~ 13.45 MJ. Total 44.8 MJ ME. Overall, it was estimated that the total energy expenditure (ME) was 44,850 kJ if no allowance for jumps was made and 46,151 kJ if such an allowance was made (Jones & Carlson, 1995).

Using this example as above, it has been suggested that if the 44.85 MJ of ME was used with 20% efficiency (i.e. $K_w = 0.2$) it would mean that 8.97 MJ of energy had been available for work (i.e. net energy), the remainder being lost in heat production. If higher K_w values are used, then obviously a greater percentage would have been used for work (Kronfeld, 1996), e.g. if the 44.8 MJ of ME was used with 24% efficiency it would mean that ~11 MJ of energy had been available for work (i.e. net energy), the remaining *nearly three quarters being lost in heat production.*

In reality (Jones & Carlson, 1995) the expenditure in the field is likely to be higher than that predicted because additional energy will be required to overcome wind resistance, raise and lower the center of mass with the terrain, allow for periods of deceleration and acceleration, allow for horizontal translation over a jump and so on. Such allowances would need to be made if more accurate estimates are to be established, but these estimations provide very good baseline data.

A partition model based on the NE requirements of a 500 kg horse undergoing the cross-country phase of the three-day event as described above is illustrated in Figure 6. The values for ME, DE, and GE demonstrated (Kronfeld, 1996) were calculated based on a net energy requirement of around 11 MJ for the competition work and efficiency factors based on a diet comprised of 45% timothy hay, 45% oat grain and 10% vegetable oil (taking CP = 8.9%; fat 13.1%; hydrolyzable carbohydrate 22.3% and fermentable carbohydrate 42%). Several assumptions were specified for the model and efficiency factors were selected that applied to this diet. However, it gives a good illustration of how energy partitioning in a competition horse fed a likely competition diet may occur. In practice many such three-day event horses would be fed proportionally more hydrolyzable carbohydrate and less fermentable carbohydrate and not all the energy used would be expected to have been provided that day.

When similar calculations as used in Figure 6 were, however, applied to a 100% timothy hay (CP = 8.6%; fat 2.3%; hydrolyzable carbohydrate 12.8%; fermentable carbohydrate 59.8%) diet they showed that in order to provide the same net energy (required for maintenance, other work as well as the actual competition), an intake energy of 348 MJ would have been required from the hay diet. This corresponds to an intake of around 22 kg of hay or ~ 4.5% body weight. Horses would not be able to ingest as much hay as this daily, which helps to explain why hay only diets are unsuitable for intensively exercising animals.

The relative value of replacing hay with cereal grains and fats to meet the energy requirements for strenuous exercise appears to be explained, in principle, better by the NE system than by the more commonly used DE system.

This example provides support for using the net energy system which enables different feedstuffs to be compared directly according to their ability to provide mechanical energy. As stated above, this approach relies on knowing the NE requirements for the various exercise patterns and these are not currently readily available. In addition it relies on both the requirements and the availabilities from the feed having been determined using the same criteria,

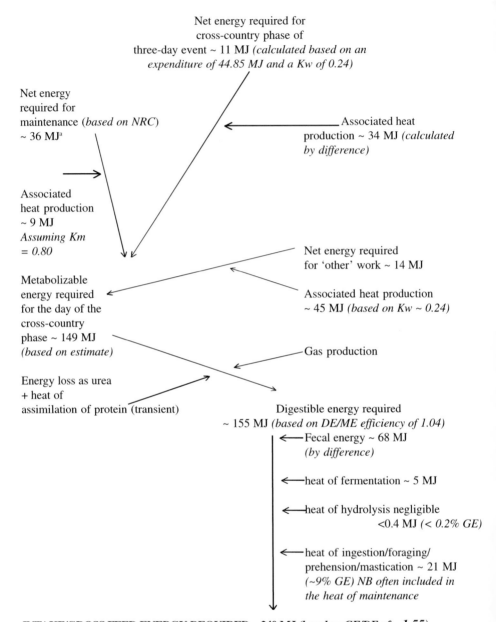

Figure 6. Schematic picture to illustrate how the amount of gross energy required from a particular diet can be determined if the net energy requirements for work and maintenance are known or can be calculated. Based (Kronfeld, 1996) on a 500 kg horse competing in the cross-country phase of a three-day event being fed a diet of 45% timothy hay, 45% oats and 10% vegetable oil. [a]Probably underestimated as the maintenance requirements of exercising horses may be higher than those of sedentary animals.

i.e. both determined via the French Net Energy system or the partitioning system. Mixing systems for determining the net energy available from food or required from food for exercise at present is likely to cause more confusion as outlined below

Practical Examples of Differences Between Partitioning and French Net Energy System

The 575 kg (total weight + rider) horse in the three-day event example above requires, based on the partitioning system, about 6 Mcal (25 MJ) of net energy for all activity (above maintenance) compared with a similar-sized horse undertaking 90 min work at a moderately intense level based on the French Net Energy system having additional (above maintenance) net energy requirements of around 9 Mcal (38 MJ) (assuming that this includes the allowance for the warm-up period, recovery period, etc.).

Daily net energy requirements, using the French Net Energy system, of around 6.9 UFC or 20.25 MCal or 84.7 MJ/day for a 500kg horse undertaking 2 - 4 hours outside exercise have been reported (INRA, 1990). Such net energy intakes would be difficult to achieve unless the net energy values of feeds for work are considered to be equivalent to those for maintenance, which is the case for the French Net Energy system, but not the partitioning system described by Kronfeld. Each system therefore at present must be used in isolation, which may cause confusion as illustrated in the following example:

1. French system:
According to Table 2 the NE requirements for a 560 kg horse plus rider
 undergoing 2 hours worth of moderate exercise are:
-        Maintenance = (84 x 115) + 10% = 10.6 Mcal (44.35 MJ)
-        Work = 2 x 4.6 = 9.2 Mcal (38.5 MJ)
TOTAL = 19.8 Mcal (82.84 MJ)

According to the French system:
-        Good hay provides about 1.25 Mcal NE/kg
-        Oats provide about 2.0 Mcal NE/kg
In a 50:50 mix – total intake 12 kg
-        Hay provides 6 x 1.25   = 7.5 Mcal
-        Oats provide 6 x 2.0   = 12 Mcal
TOTAL = 19.5 Mcal = very close to requirements.

2. Partitioning system
IF USE SAME REQUIREMENTS (i.e. French NE requirements) and 50:50
diet hay and oats
Maint = 10.6 Mcal
-        Hay provides 1.29 x 3.2   = 4.13 Mcal
-        Oats provide 2.0 x 3.2   = 6.4
TOTAL very close to requirements
Work = 9.2 Mcal

continued

·   Hay provides 0.39 x 2.8 = 1.1
·   Oats provide 0.62 x 2.8 =1.9
TOTAL = 3.0 Mcal **ONE THIRD OF ESTIMATED REQUIREMENTS**

An owner would need to feed more than <u>11 kg of hay and 11 kg</u> of oats to meet the calculated French requirements if using the partitioning estimates of feed NE content. In the partitioning system less energy would have been calculated as being required for work (ie. 20-25% of the ME rather than 70-75%).

Present Disadvantages of the NE over the DE System

The net energy system has a number of potential advantages, but at the moment an English version of all the tables and factors used is not readily available; most countries use the DE system and DE values for horse feedstuffs are more routinely available. The NE system relies on the fact that maintenance requirements for energy account for the largest part of the total energy requirement which may not be true for certain performance animals. It also assumes that the UFC value for a particular feed is the same for maintenance as for work, which due to the increased heat production associated with work, may not be a valid assumption as discussed. The use of a NE system also implies amongst others that the NE requirements of the horse have been accurately described. Certain of the equations used to predict ME and therefore the NE values of feeds, etc. appear to have very low correlation values ($r^2 = 0.45$), whereas others as illustrated above are very high and at present full justification of their use does not appear to be available. However, it is possible, as more information becomes available, that this system will become more generally applicable and widely used.

Conclusion

Although much more work is needed before we fully understand the energy requirements of the exercising horse and how we can optimally manipulate the diet to provide this level of energy, there have been a number of useful developments in the last few years. The relative value of replacing hay with cereal grains and fats to meet the energy requirements for strenuous exercise appears to be explained, in principle, better by the French NE system or the energy partition model than by the more commonly used DE system. Work is needed to enable the NE system to be more widely applicable. In a number of other areas a unified approach would help to minimize the confusion currently present and help maximize progress in this area.

References

Anderson CE, Potter GD, Kreider JL, Courtney CC. 1983. Digestible energy requirements for exercising horses. *J. Anim. Sci.* 56(1): 91-95.

Barth KM, Williams JW, Brown DG. 1977. Digestible energy requirements of working and non-working ponies. *J. Anim. Sci.* 44: 585-589.

INRA 1984 . Tables de la valeur nutritive des aliments pour le cheval. In: *Le Cheval. Reproduction-Selection - Alimentation- Exploitation.* Eds R Jarrige, W Martin-Rosset. pp661 - 689.

INRA 1990. In *Alimentation des chevaux.* Ed W Martin-Rosset p 232. Versailles. INRA Publications.

Jones JH, Carlson G. 1995. Estimation of metabolic energy cost and heat production during a three-day event. *Equine Vet. J.* Suppl. 20: 23-31.

Kronfeld DS. 1996. Dietary fat affects heat production and other variables of equine performance under hot and humid conditions. *Equine Veterinary Journal Suppl 22 (Vol 11)* pp 24 - 35.

Martin-Rosset W, Vermorel M, Doreau M, Tisserand JL, Andrieu. 1994. The French horse feed evaluation systems and recommended allowances for energy and protein. *Livestock Production Sci.* 40: 37-56.

NRC 1989. *Nutrient requirements of horses* 5th edition. Washington DC. National Academy press.

ESTABLISHING NUTRIENT REQUIREMENTS OF HORSES AND OTHER ANIMALS IN AN ERA OF MODELING

GARY L. CROMWELL[1] AND CHARLOTTE KIRK BAER[2]
[1] *University of Kentucky, Lexington, KY*
[2] *National Academy of Sciences, National Research Council*

Energy and essential nutrients such as amino acids, minerals, and vitamins are required by horses and other animals for the various processes of life, including maintenance, growth, reproduction, lactation, and work. Having accurate estimates of dietary nutrient requirements of animals is important. Deficiencies of even a single nutrient in a diet will limit an animal's performance and well-being, and diets with excessive nutrients are expensive and contribute to environmental pollution.

The National Research Council (NRC) plays a major role in establishing the nutrient requirements of horses and other animals. The NRC is a private, nonprofit organization that was established in 1916 to provide advice to the U.S. government on issues of science and technology. The NRC is the "working arm" of The National Academies, which includes the National Academy of Sciences (NAS), an honorary society instituted 137 years ago (in 1863) by President Abraham Lincoln through an act of Congress. Members are elected to the Academy based on their contributions to science. The NRC has ten major units, one of which is the Board on Agriculture and Natural Resources (BANR). The Committee on Animal Nutrition (CAN) is the oldest standing committee of the BANR and the NRC. Dating back to the establishment of the NRC itself, CAN has addressed issues of animal feeding since 1917. Its series of publications on nutrient requirements, written by subcommittees and overseen by CAN, covers nearly 30 species of food, companion, and laboratory animals. These reports have been translated into seven languages and are used worldwide.

The first of the nutrient requirement publications, Recommended Nutrient Allowances for Swine and Recommended Nutrient Allowances for Poultry, were published in 1944. These were concise documents (the swine publication was 10 pages) that identified the nutrients known at that time to be essential for pigs and poultry and that listed dietary requirements for some of these nutrients. The following year, similar publications were released for beef cattle, dairy cattle, and sheep. The first publication of this type for horses, Recommended Nutrient Allowances for Horses, was published in 1949. In 1961, there was a second printing of this publication with a new title, Nutrient Requirements of Horses. Since the initial release in 1949, this publication has been revised four times, in 1966, 1973, 1978, and 1989.

It has now been 11 years since the last edition of Nutrient Requirements of Horses was published. New research findings have emerged over the past 10-15 years and a better understanding of the finer points of equine nutrition has occurred. As a result, the previous edition is in serious need of revision. The

CAN recognizes this fact and has recommended that the publication be revised. The BANR and the NRC have approved CAN's recommendation to convene a subcommittee to produce a new revision.

NRC's Process of Producing Reports

The process by which nutrient requirement publications are prepared is relatively simple, but it is thorough, rigorous, and somewhat time-consuming. Once the study proposal has been approved, and an understanding exists between the sponsor and BANR, the study may commence. Studies may have one or several sponsors from government or the private sector, or both. Because the NRC offers a one-of-a-kind service, not duplicated by other organizations, it does not compete for federal contracts. The NRC provides a public service, which is supported by the users of its products. In the case of nutrient requirement publications, the reports are produced by the nonprofit NRC and are published and disseminated throughout the world by the nonprofit National Academy Press. It is through the dedicated work of volunteer experts and the financial support of end-users that the reports are made widely available for use in the industry, government, research, and teaching communities.

The search for candidates for subcommittee membership is initiated by staff with input and oversight from the relevant boards. In defining the areas of expertise that should be represented on a subcommittee and identifying individuals qualified to serve, the staff reviews scholarly literature and consults widely with members of the National Academies, CAN, knowledgeable authorities, and professional associations. Sponsors may offer suggestions but are not responsible for selecting subcommittee members. Subcommittee members are chosen on the basis of their experience in the various areas of nutrition, and after careful review by CAN and BANR, they are appointed by the chair of the NRC, who also is president of the NAS. Subcommittee members serve as individuals, not as representatives of organizations or interest groups. Members are sought with background and experience in academia, government, and industry. Each person is selected on the basis of his or her expertise and good judgment, and is expected to contribute accordingly to the study. Potential sources of bias and conflict of interest are significant issues that are taken into consideration in the selection of subcommittee members.

A successful report is the result of a dynamic group process, requiring that subcommittee members be open to new ideas and innovative solutions, and be willing to learn from one another and from other individuals who provide input. Subcommittees are expected to be evenhanded and to examine all evidence dispassionately. The CAN subcommittees review the world's literature, particularly research published since the last edition. Although all interested parties should be heard and their views given serious consideration, one of the subcommittee's primary roles is to separate fact from opinion, and analysis from advocacy. Scientific standards are essential in evaluating all arguments and alternatives. Experience suggests that completing the consensus process and writing a report that clearly represents the subcommittee's findings is the most difficult, frustrating, yet rewarding aspect of serving on a study subcommittee. The report is used by many audiences--regulatory, research, and industry, to name a few--and for this reason, the report must be of the highest quality. Although each subcommittee may approach the drafting of its report differently, every report is the collective product of a group process.

Like all good science, reports should be based on fact and rigorous analysis. All CAN reports must undergo an independent review by anonymous expert panels of reviewers. Review is a multitiered process, which ensures the level of highest quality that sets the NRC apart from any other organization. While the report is reviewed by expert scientists, a review coordinator and monitor, as well as the NRC's report review subcommittee, oversee the entire process.

Upon completion of the report, it is published by the National Academy Press. The Press prints and sells the report at cost, and strives to make the report widely available to users on a global scale.

The Swine Model - A New Approach to Estimating Requirements

Several subcommittees have used a modeling approach to more precisely estimate nutrient requirements. The beef and swine subcommittees utilized models in their 1996 and 1998 publications and the dairy subcommittee is currently utilizing models to estimate requirements of dairy cattle. To illustrate the value of models in estimating nutrient requirements, let me review some of the background that went into the development of the models used in the NRC's Nutrient Requirements of Swine publication.

Prior to the last edition of the swine publication, requirement estimates for growth were based mainly on weight classes of pigs and single estimates were made for pregnant and lactating sows without consideration for genetic differences among pigs. Environmental factors that are now known to have a profound influence on the pig's nutrient requirements also were not considered. Simply put, this "one size fits all" approach to defining nutrient requirements is no longer acceptable in the current highly technical arena of swine nutrition.

A new approach was taken by the NRC subcommittee to produce more accurate estimates of nutrient requirements that take into consideration not only the pig's body weight, but also its accretion rate of lean (protein) tissue, gender, health status, and various environmental factors. Similarly, to accurately estimate nutrient needs of gestating and lactating sows, one needs to account for body weight, weight gain during gestation, weight loss during lactation, number of pigs in the litter, weight gain of the litter (a reflection of milk yield), and certain environmental factors.

A series of integrated mathematical equations was needed for the new edition to account for the many factors that are now known to influence nutrient requirements. These integrated equations provide the framework for "modeling" the biological basis of predicting requirements. Unfortunately, no models were available that met the needs for the task at hand. While there were a number of commercial models used in the swine industry, the equations are tightly guarded for proprietary reasons. Also, most simulation models predict animal performance as an output when certain levels of nutrients and various environmental factors were given as inputs. The NRC models are just the opposite; they predict the levels of nutrients (model output) needed to achieve a certain level of production under a given set of environmental conditions (model inputs). Therefore, it was necessary for the subcommittee to develop their own models.

Five principles guided the NRC swine subcommittee as it developed its models. The models were: (1) made for easy use by people with varying levels of nutritional expertise and with limited information; (2) developed for continued relevance for several years to come (i.e., until the next edition will be published); (3) intended to be structurally simple, so they can be understood easily by users; (4) developed to be transparent so that all of the equations are available to the user; and (5) firmly anchored to empirical data at the whole-animal level rather than being simply based on theoretical values.

The tedious task of building the models from scratch took the subcommittee over two years. As the models were developed, they were tested with previously established requirements and with data sets from recently conducted research. The models were modified and refined to ensure that the requirement estimates produced were supported by empirical data. Finally, the models were validated with independent data from experiments that had not been used in the construction of the models. To allow nutritionists to understand the inner workings of the models, all of the equations are presented in the publication.

Three independent models were developed: a growth model, a gestation model, and a lactation model. The growth model estimates amino acid requirements of pigs from weaning to market weight, and the gestation and lactation models estimate energy and amino acid requirements of gestating and lactating sows. Along with energy and amino acid requirements, the software also allows the user to estimate mineral and vitamin requirements, which are based on mathematical equations. The models are included in commercially produced, user-friendly software on a compact disk (CD-ROM) that accompanies the NRC publication.

Swine Growth Model

This model estimates the amino acid requirements of pigs over the range of 20 to 120 kg of body weight. The model estimates requirements on an available amino acid basis, which is based on true absorption of amino acids at the terminal ileum. The amino acid requirements are estimated on a daily (grams/day) basis. The estimates are the sum of the pig's daily requirements for (1) maintenance and (2) deposition of whole-body protein.

Lysine is recognized as the first limiting amino acid in essentially all practical swine diets, so it is estimated first. The daily lysine requirement for maintenance is related to metabolic body weight (kilograms of weight raised to the 3/4 power) and is considered as 0.036 g of lysine per kg of $BW^{0.75}$. The daily lysine requirement for protein deposition is considered as 0.12 g of true ileal digestible lysine per gram of whole-body protein accreted. The value of 0.12 entails two components: the lysine content of whole-body protein (~7.0%), and the partial efficiency of incorporation of digestible lysine into whole-body protein (~58%) (i.e., 7.0/0.58 = 0.12).

Naturally, to use this information, the daily rate at which pigs deposit protein in the body tissues must be known. This can be calculated if one knows the rate at which pigs deposit lean tissue in the carcass, which can be determined from equations utilizing data from packing plant kill sheets. In the model, whole-body protein is calculated from carcass fat-free lean, using a conversion factor of 2.55. In other words, carcass lean is divided by 2.55 to obtain whole-body protein. Thus, a pig having 325 g/day of carcass lean gain is predicted to have a whole-body protein gain of 127 g/day (325/2.55=127).

The user inputs mean daily lean gain (lean gain per day averaged over the period from 20 to 120 kg of body weight), and the model calculates the daily protein accretion at any given weight. This calculation is based on the overall mean lean gain per day along with an assumed shape to the lean gain (protein accretion) curve. The model uses a default curve (Figure 1) to make this calculation. The shape assumes that daily protein accretion rate accelerates during early growth, reaches a plateau, then the rate declines during the finishing period. Pigs with different lean growth rates have the same general pattern of protein accretion but the heights of the curves will differ (Figure 2). The user can input different shaped curves, if desired.

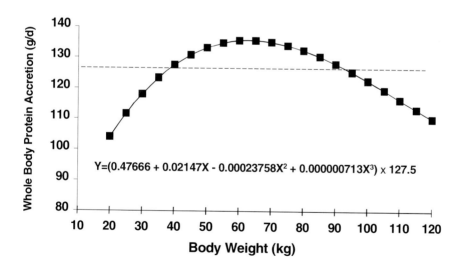

$$Y=(0.47666 + 0.02147X - 0.00023758X^2 + 0.000000713X^3) \times 127.5$$

Figure 1. Potential whole body protein accretion rate of pigs of high-medium lean growth rate with a carcass fat-free lean gain averaging 325 g/day from 20 to 120 kg of body weight using the NRC growth model. The lean growth rate of 325 g/day is converted to a mean whole-body protein accretion rate of 127.5 g/day (325/2.55 = 127.5).

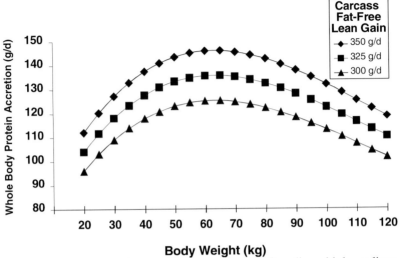

Figure 2. Whole body protein accretion rates of pigs of medium, high-medium, and high lean growth rates with carcass fat-free lean gains averaging 300, 325, and 350 g/day from 20 to 120 kg of body weight as estimated

After the daily lysine requirement is determined, the requirements for the other amino acids are estimated using blends of amino acids patterns. The "ideal protein" concept is used, which assumes that there is an ideal pattern of amino acids for maintenance and an ideal pattern for body protein accretion. These two patterns are blended depending on the proportion of lysine being used for maintenance and protein accretion in a given situation. Requirements are then converted to a percentage basis by dividing the daily amino acid requirements by the daily feed intake. Daily feed intake is predicted from body weight using a digestible energy (DE) intake equation in the model (Figure 3). Alternatively, feed intake can be entered by the user.

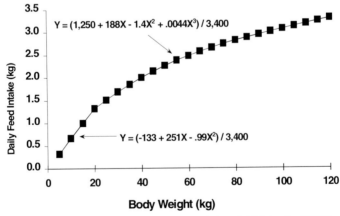

Figure 3. Estimated feed intake of a diet containing 3,400 kcal of DE/kg of pigs from 3 to 120 kg of body weight as estimated by the NRC growth model.

The dietary percentage requirements of amino acids differ depending upon the gender of the pig (females and males require higher percentages of amino acids than castrates), environmental temperature, space allowance, and energy density of the diet. These factors, which are user inputs, affect energy intake, which, in turn, affects the requirements on a percentage basis. From the true ileal digestible amino acid requirements, apparent ileal digestible and total amino acid requirements are then calculated from equations. All three sets of requirements are presented to the user.

The growth model effectively estimates amino acid requirements of pigs based on their genetic ability to deposit lean tissue. Figure 4 illustrates the dietary lysine requirements, expressed as a percentage of the diet, from 20 to 120 kg body weight in pigs (1:1 mix of females and castrates) with medium, high-medium, and high lean growth rates (300, 325, and 350 g/day of carcass fat-free lean) under standard conditions.

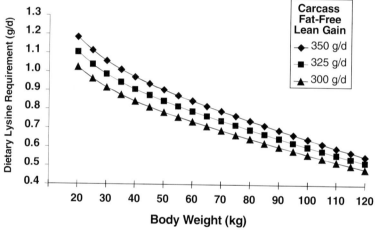

Figure 4. Dietary lysine requirements (%) of pigs of medium, high-medium, and high lean growth rates with carcass fat-free lean gains averaging 300, 325, and 350 g/day from 20 to 120 kg of body weight as estimated by the NRC growth model. The requirements are for total lysine, assuming a corn-soybean meal mixture.

The model adjusts the lysine requirements if changes are made in the gender ratio, energy density of the diet, environmental temperature, animal space, or other factors.

The amino acid requirements for the young pig (3 to 20 kg) are estimated strictly from empirical data, due to a lack of information needed to model the requirements in the same manner as for growing-finishing pigs. The model uses an equation that fits a curvilinear regression line across estimated lysine requirements for several weight categories. The other amino acids are then handled by blending ideal patterns for maintenance and protein accretion.

Mineral and vitamin requirements are not modeled but are estimated from an exponential equation that fits the requirements, on a percentage basis, according to the mean body weights of several weight categories. An example is shown in Figure 5. Daily requirements are then generated by multiplying dietary percentage requirements by daily feed intake.

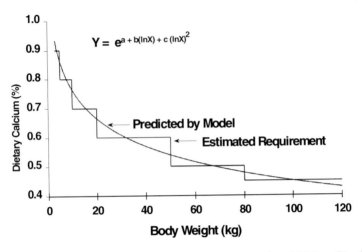

Figure 5. Dietary calcium requirement (%) of pigs from 3 to 120 kg of body weight using the generalized exponential equation in the NRC growth model.

Swine Gestation and Lactation Models

These two models operate much like the growth model in terms of the amino acid requirements. Daily amino acid requirements of gestating sows are based on maintenance and the estimated protein accretion in the sow's body and in the products of conception (uterus, fetuses, placental tissues and fluids, etc.), which are calculated based on user inputs of breeding weight, gestational weight gain, and assumed litter size. Similarly, daily amino acid requirements of lactating sows are based on maintenance, body protein gain or loss during lactation, and milk production of the sow. These are calculated by the model from user inputs of the sow's postpartum weight and projected lactation weight change, lactation length, number of pigs nursed, and daily weight gain of the nursing pigs.

As with the growing pig, lysine is estimated first, then blends of ideal ratios of amino acids for maintenance, protein accretion, milk synthesis, and body protein breakdown are used to estimate requirements for the other amino acids. Daily energy requirements and feed intake are estimated by the model and the percentage requirements are then calculated. The user also has the option of entering daily energy intake; in that case, gestation weight gain or lactation weight loss (or gain) is then calculated by the model. Mineral and vitamin requirements are not calculated by the sow models. Instead, they were estimated by the committee on a dietary concentration basis, then the daily amounts are calculated by multiplying the dietary concentration by the daily feed intake.

Model Supporting Information

A comprehensive set of appendix tables in the tenth edition of Nutrient Requirements of Swine gives added information relative to the models. All of the equations in the models are given such that one could produce the models in spreadsheet form if one so desires. This would allow the user to interface the model with other programs, such as feed formulation programs. A second appendix gives procedures for determining the mean lean gain of pigs based on initial and final estimates of carcass fat-free lean content. A third appendix gives procedures for developing alternative carcass lean gain curves or whole-body protein accretion curves. Finally, two additional appendices include a user's guide and help screens to aid in the use and application of the models.

Software containing the model programs is included on a CD-ROM that accompanies the publication. The software can be easily loaded to a personal computer and it produces both screen and printed copies of requirements. The software also contains detailed feed composition tables that are available on screen or as hard copy.

Obtaining the Models

The NRC publication Nutrient Requirements of Swine, Tenth Revised Edition and the compact disk containing the model programs can be purchased from National Academy Press, 2101 Constitution Avenue N.W., Lockbox 285, Washington, DC 20055. A discount is provided if the order is placed online from the web bookstore at: www.nap.edu. The publication and model as well as other NRC publications can be downloaded from the National Academy Press web site at www.nap.edu, then click on "Reading Room." For more information, contact Charlotte Kirk Baer, Committee on Animal Nutrition, Board on Agriculture and Natural Resources, National Research Council, 2101 Constitution Avenue, N.W., Washington, D.C. 20418.

Summary

The NRC plays an important role is establishing nutrient requirements of horses and other animals used for food, service, recreation, companionship, and other purposes. Nutrient Requirements of Horses, last published in 1989, is becoming outdated and needs to be revised. Approval has been granted by the NRC to pursue a revision of the horse publication and it is CAN's desire that the public support for this revision will lead us to a new era of estimating nutrient requirements for horses. A modeling approach taken by several NRC subcommittees (beef cattle, swine, and dairy cattle) represents a major step forward in estimating nutrient requirements. Models allow nutritionists to formulate diets and develop feeding programs that will more precisely meet the requirements of animals under widely varying conditions.

ADVANCES AND GAPS IN ENERGY NUTRITION

LAURIE M. LAWRENCE
University of Kentucky, Lexington, KY

Although energy is not a nutrient per se, it is one of the most important dietary essentials. The current Nutrient Requirements of Horses (NRC, 1989) provides recommendations for daily digestible energy (DE) intake of horses of various body weights during maintenance, gestation, lactation, growth and work. The text of the document provides an explanation of the derivation of the actual values, as well as discussion regarding energy source and feed intake. The nutrient requirements given in the 1989 publication represented what the subcommittee considered to be minimum amounts needed for normal health, production and performance. When possible, the committee based recommended daily energy intakes on published research reports. Because extensive research was not available for every class of horse, the subcommittee also used "calculations designed to extrapolate information over the total population" and/or "the subcommittee's experience in applying information to field situations" (NRC, 1989). Since 1989, enough new research in equine nutrition has been published to refine some (but not all) existing recommendations for daily energy intakes. There are also several topics that could be included in the text of the document or included in the tables. In the future, it may be possible to develop a multilevel model to estimate energy requirements in horses. A multilevel model would provide information in the same format currently available but could also provide flexibility to users who want to customize recommendations for specific situations.

Energy Requirements for Growth

Using the results of several studies with growing horses, the 1989 NRC determined DE requirements to be related to body weight (BW), average daily gain (ADG) and age. Body weight was used to determine the requirement for maintenance; ADG and age were used to estimate the amount of energy necessary for growth. The age component was necessary to account for differences in the efficiency of DE use for gain by horses of different ages. When DE recommendations in the 1989 NRC are compared to the recommendations in the previous edition (1978), it is obvious that the method adopted in 1989 resulted in significant changes to some of the values. For example, in 1978, the DE requirement for yearlings (500 kg mature weight; .55 kg ADG) was 16.8 Mcal/d. The 1989 value is about 17% higher (19.7 Mcal/d). One method of evaluating the accuracy of nutrient recommendations is to compare predicted values to values observed in actual feeding trials. Figure 1 depicts information from several growth trials reported in the last decade. The amount of DE necessary for the horses in each trial as predicted by the 1989 NRC (based on age, body weight and ADG) was calculated and compared to the actual amount of DE fed. Except in a few cases, the predicted value exceeded the actual requirement. Differences between predicted and actual DE intakes could be attributed to an incorrect prediction equation for

DE requirements. However, in most of the studies, the amount of DE intake fed was calculated from feed composition and not actually measured; therefore, some of the difference could be attributed to inaccuracy in estimating DE intake. In addition, the values in Figure 1 represent mean values for gender groups or various treatments within specific studies. To perform an accurate evaluation of the 1989 prediction equation for DE in growing horses, it will be necessary to have individual horse data.

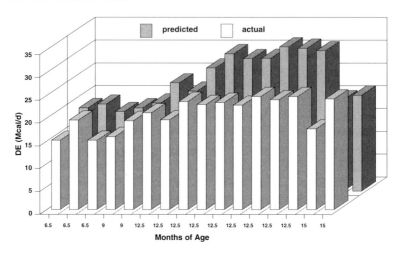

Figure 1. Predicted and observed digestible energy requirements of growing horses. (Calculated from data of Cymbaluk et al., 1990; Graham et al., 1994; Ott and Asquith, 1995; Ott and Kivipelto, 1999; Savage et al., 1993).

When the last edition of Nutrient Requirements of Horses was compiled, there were no studies on the DE requirements of long yearlings and two-year-olds in training. Clearly, this is an area where major gaps in knowledge still exist. The DE requirements in 1989 were formulated using the DE recommendations for horses in moderate work plus an allowance for gain. Since that time a number of studies have examined the nutrition of young horses in training (Hiney et al., 1999; Nielsen et al., 1993; Pagan et al., 1995; Pendergraft and Arns, 1993), but none have focused on DE requirements and these studies did not report enough infor-mation on DE intake or ADG to allow comparisons to the 1989 values.

In the 1978 edition of Nutrient Requirements of Horses, nutrient recommendations were included for nursing foals at three months of age. These recommendations were deleted in the 1989 edition but could be reinstated in the next version. Ideally, nutrient recommendations should be included for foals from birth. Information generated in the last decade on average growth curves in nursing foals (Breuer et al., 1996; Jelan et al., 1996; Pagan et al., 1996) could easily be combined with information on milk composition to estimate the nutrient requirements of foals from birth through two months of age. In addition, several papers have described daily gains under various milk replacement programs for early weaned foals (King and Nequin, 1989; Knight and Tyznik, 1985; Lawrence et al., 1991; Pagan et al., 1993). Recommendations for young foals should also consider the need for supplementation programs when different management programs are used (Breuer et al., 1996).

Energy Requirements for Reproduction and Lactation

The 1989 NRC recommends that pregnant non-lactating mares receive maintenance levels of DE intake during the first 8 months of gestation and that DE should be increased 11%, 13% and 20% above maintenance in the 9th, 10th and 11th months, respectively. These recommendations were based on the research of Meyer and Ahlswede (1978) that found the majority of fetal growth to occur in the last trimester of gestation. Two studies have reported weight changes in pregnant mares during various stages of gestation (Kowalski et al., 1990; Lawrence et al., 1992a). These studies found that pregnant mares entering the last trimester of gestation gain relatively little weight in that period. In one study, the majority of weight gain occurred in the second trimester of gestation. Neither study measured DE intake; however, the results of the studies could be used to provide guidelines for expected body weight gains in pregnant mares.

Digestible energy requirements for lactation are based on an expected milk production of 3% of BW in early lactation and 2% of BW in late lactation (NRC, 1989). Doreau and coworkers (1992) reported the mean milk production of French draft mares (725 kg) to increase from 2.6% BW in the first week of lactation to 3.9% BW in the eighth week of lactation. The mares in their study were given ad libitum access to high forage or high concentrate diets. Mares on the high forage diet consumed 50.8 Mcal DE/d in week four of lactation and were in positive energy balance based on an average daily gain of .18 kg. For the mares fed the high concentrate diet, mean daily DE intake at week four was 66 Mcal/d, and weight gain was 1.18 kg/d. Based on the responses to these two diets, the DE intake of these mares for lactation with zero body weight gain would be about 48 Mcal/d. Based on BW and actual milk production, the expected daily DE requirement (NRC, 1989) would be about 42 Mcal/d. Griewe-Crandell et al. (1997) fed 7 kg of concentrate and grass hay (ad libitum) to 575 kg lactating mares and reported that BW was maintained. If total feed intake is estimated at 2.5% of BW, the estimated energy intake of the lactating mares would have been about 40 Mcal/d, compared to the NRC (1989) estimate of 32 Mcal/d. For both studies, the actual intakes seem to be somewhat higher than the NRC (1989) recommended intakes.

In 1989, the DE requirement for breeding stallions was estimated to be similar to that for horses in light work (1.25 x maintenance). In 1993, Siciliano and coworkers reported results from a field study at a large central Kentucky stallion operation. The amount of feed offered per day was recorded by farm employees on a daily basis. Stallions were weighed once a month. Because the authors reported that data collected early in the study were more consistent than data collected later, only data from February and March will be used for comparison here. In February and March, when stallions were covering 5 to 25 mares per month, the amount of DE offered was just slightly higher than the 1989 recommendation for 600 kg stallions. The results suggest that the current recommendations are realistic, although the authors did not account for any pasture intake by the horses, which might have increased the difference between actual DE intakes and recommended DE intakes.

Energy Requirements for Work

In the fourth edition of Nutrient Requirements of Horses (1978), the only specific recommendations for energy intakes of horses performing different levels of work were found in a table in the text of the document (Table 1).

Table 1. DE use during various activities as listed in Nutrient Requirements of Horses (1978).

Activity	DE/hour (kcal) per kg BW (above maintenance)
Walking	0.5
Slow trotting, some cantering	5.0
Fast trotting, cantering, some jumping	12.5
Cantering, galloping, jumping	23.0
Strenuous effort (polo, racing)	39.0

To arrive at a daily DE intake for exercising horses using the 1978 NRC, users had to calculate the amount of energy used in work and add that to the maintenance requirement. In 1989, the subcommittee on horse nutrition adopted a more user-friendly approach, categorizing work as light, moderate, or intense, and estimating the daily DE requirements as 25%, 50%, and 100% above maintenance for these categories, respectively. The 1989 recommendations reflected a blending of information from controlled research studies, feeding surveys and practical experience. A number of studies published since 1989 can be used to evaluate the existing recommendations. Gallagher and coworkers (1992a;1992b) surveyed the diets of Thoroughbreds and Standardbreds in race training. Estimated DE intakes were 28 to 31 Mcal/d for Standardbreds (mean BW of 449 kg) and 31 to 36 Mcal/d for Thoroughbreds (mean BW of 505 kg). The NRC (1989) recommendations for intense work would be 28.8 Mcal/d for Standardbreds and 33 Mcal/d for Thoroughbreds. Southwood and coworkers (1993) in Australia reported daily DE intakes of 30.8 Mcal/d for Thoroughbreds and 31.5 Mcal/d for Standardbreds. In regard to moderate exercise, data from several controlled laboratory experiments can be compared to the NRC (1989) requirements. Taylor et al. (1995) reported that 420 kg horses undergoing moderate work three to four times per week maintained body weight when they received 19 to 22 Mcal DE/d. Using similar horses and a similar exercise program, Graham-Thiers et al. (1999) reported an ADG of .3 kg/d when horses received approximately 28 Mcal DE/d. The NRC (1989) recommendation for 420 kg horses in moderate work is about 21 Mcal/d. Powell (1999) reported that 540 kg horses working five days per week at a moderate intensity maintained BW when fed 27 Mcal/d. The

NRC (1989) recommendation is about 26.5 Mcal/d. Using the information from these studies, it appears that the NRC (1989) recommendations for moderate and intense work are relatively realistic. Because energy intake impacts body condition, several studies have surveyed body condition scores in horses involved in different athletic events. The results of these studies suggest that there is a broad range of acceptable condition scores for performance horses (Gallagher et al., 1992a and 1992b; Garlinghouse et al., 1999; Lawrence et al., 1992b).

Environmental Effects

The 1989 NRC states that all recommendations should be applied with consideration to digestive and metabolic differences among horses, health status, variation in nutrient availability in feed ingredients, previous nutritional status and climatic and environmental conditions. However, few data were available prior to 1989 to allow quantification of some of these considerations. One area that can now be quantified to some extent is the effect of environment on DE requirements. Cymbaluk and Christison (1990) reviewed studies with mature horses and growing horses. They suggested that the maintenance requirement of growing horses is increased by about 33% in cold housing conditions and by more than 50% in severely cold conditions. They also suggested that the daily maintenance requirement for mature, 500 kg horses is increased .408 Mcal of DE for every degree below the lower critical temperature (LCT). One of the difficulties in estimating environmental effects on requirements is determining the LCT. The LCT may vary for horses of different ages or body sizes and can change as horses become cold adapted. In addition, other factors such as presence of precipitation or wind will contribute to heat loss and result in increased maintenance energy requirements even though air temperature may be above the LCT. Data from other species such as dairy or beef cattle could be used to estimate the combined effects of wind, precipitation and cold on heat loss and maintenance energy requirements in horses.

Weight Gain in Adult Horses

The condition scoring system first introduced by Henneke and coworkers in 1983 has been widely accepted and used by nutritionists and managers within the horse industry. The 1989 NRC suggested that broodmares be maintained at a condition score of at least 5. For practical application of this recommendation, it is important to know how much weight gain is necessary to increase the body condition score of a mare from 4 to 5, and how much DE is required to accomplish this gain. This question is important for other groups of horses, including performance horses where weight lost during a long competition season may need to be replenished in a relatively short lay-up period. The 1989 publication indicated that an increase in condition score could be achieved if DE intake was increased 10 to 15% above maintenance. However, recent information suggests that this recommendation may not be realistic for many situations. The amount of DE required per kilogram of gain will depend on several factors including composition of gain and composition of diet. The amount of DE required per kilogram of gain typically increases with maturity. The NRC (1989) suggests that 19.8 Mcal of DE (above maintenance) are required per kg of gain by two-year-old horses. In a study by Heusner (1993) mature horses required approximately 24 Mcal of DE above maintenance per kilogram of gain. In the study by Graham-Thiers et al. (1999), a daily DE intake of 7 Mcal above maintenance and exercise resulted in an ADG of .3 kg, which suggests a value of 23 Mcal DE/kg gain. From these studies, it seems reasonable to estimate that at least 20 Mcal of DE are required per kilogram of gain for mature horses. Heusner (1993) found that weight gains of 33 to 45 kg were associated with an increase in condition score of approximately two units (from 4 to 6) in mature horses (approximately 480 to 580 kg). Consequently, it appears that each unit of condition score increase requires about 16 to 20 kg of weight gain. Using these assumptions, Table 2 shows the estimated amount of energy above maintenance required to increase the condition score of a horse from 4 to 5 over different periods of time.

Table 2. Estimated increase in DE intake necessary to change the condition score of a 500 kg horse from 4 to 5[A].

Time Period to Accomplish Gain	Daily DE above Maintenance(Mcal/d)	% Increase in DE above Maintenance
60 d	5.3 to 6.7 Mcal	32 to 41%
90 d	3.6 to 4.4 Mcal	22 to 27%
120 d	2.7 to 3.3 Mcal	16 to 21%
150 d	2.1 to 2.7 Mcal	13 to 16%
180 d	1.8 to 2.2 Mcal	11 to 14%

[A] Assumptions: 1 unit of change of condition score requires 16 to 20 kg of gain and 1 kg gain requires 20 Mcal DE above maintenance.

Dietary Energy Sources

The principal sources of energy in typical equine diets are structural carbohydrates (SC) and nonstructural carbohydrates (NSC). High intakes of grain concentrates and other feeds high in NSC have been associated with increased risk of colic and laminitis. A potential addition to the next version of Nutrient Requirements of Horses would be a guideline on maximum amounts of nonstructural carbohydrates and/or minimum daily amounts of structural carbohydrates. Using ileal fistulated ponies and horses, researchers have now estimated the upper limit of starch digestion for the equine small intestine (Cuddeford, 1999; Potter et al., 1992). This information could be incorporated into the dietary energy recommendations to ensure feeding practices that promote the health and well-being of the horse. However, to make a maximum starch or NSC recommendation usable, it would be necessary to adjust the feed composition tables to include information on starch or NSC content. Describing minimum SC intakes could be approached by recommending daily minimum neutral detergent fiber (NDF) or acid detergent fiber (ADF) intakes. Information regarding the effect of ADF and NDF content of forages on voluntary intake could help refine the concentrate to forage ratios recommended for various classes of horses.

Considerable research has been conducted in the last 20 years on the effect of dietary energy form on metabolic and endocrine regulation in growing and working horses. Doreau et al. (1992) have demonstrated that dietary energy form can alter milk composition such that a high concentrate diet favors higher lactose production and a high roughage diet favors higher fat production. Despite the relatively large number of studies that have evaluated effects of different energy forms on horses, this area is still not completely understood. Further research is required to understand the relationships between energy source, energy amount and hormonal regulation of metabolism during growth, lactation and exercise.

Estimating Energy Requirements

Although many other species express energy requirements in units of metabolizable or net energy, the DE system still appears to be the most practical for horses. In 1989, the DE system was used because little information regarding the ME or NE value of common horse feeds was available. This situation still exists. However, because the amount of information regarding energy content of feeds may expand in the next decade, adding flexibility to the means of expressing and calculating energy requirements might be considered. One way to do this would be to design a multilevel model in which Level 1 is a straightforward listing of DE requirements using the same format as the 1989 Nutrient Requirements of Horses, and incorporating any necessary adjustments in current recommendations. In general, Level 1 would represent requirements for horses fed average diets in average environments. As in the current system, users would be able to specify some characteristics of the horse, such as mature body weight or rate of gain. In Level 1, users would not be able to change any of the basic assumptions used to determine the requirements.

To increase flexibility and to allow for the incorporation of new information, a higher level model could be added. In Level 2, energy requirements would still be expressed in terms of DE, but the user would have some flexibility to adjust some factors used in the determination of individual requirements. In the 1989 publication, DE requirements for early lactation are based on assumptions about the amount of milk mares produce (3% of BW) and the amount of DE needed to produce a kilogram of milk (.792). In a Level 2 model, a user could alter these two assumptions. For example, if the user had information suggesting that milk production was higher than 3% of body weight, then the assumptions could be adjusted to the applicable higher value. In a Level 2 model, the user could also make adjustments to the amount of DE required per kilogram of milk. For instance, it is expected that the digestible energy in a high fat, high starch diet is more efficiently converted to net energy than the digestible energy in a high fiber diet. As a result, the amount of DE required when horses consume a high fat, high starch diet would be actually less than when a high fiber diet is consumed.

The concept of a multilevel model is not novel. Nutrient Requirements of Beef Cattle (NRC, 1996) utilizes a two-level model. Level 1 is suggested for use when information on feed composition is limited and when the user is not experienced in using, interpreting, and applying the inputs and results from Level 2. The second level requires a greater understanding of feed ingredients and a greater knowledge of specific animals or management conditions. In the beef cattle scenario, Level 2 was envisioned as having several purposes including being a tool for evaluating feeding programs while accounting for more of the variation in animal performance in specific production settings.

Although any models constructed for horses will probably be much less complex than models constructed for other species, a two-level model will still allow users to fine-tune some requirements if they have the knowledge to do so. Obviously there is a potential downside to allowing user input. Users without sufficient expertise, or with inaccurate information, could make changes that could result in large errors in estimating requirements. These errors could have profound effects on animal well-being. Consequently, it would be essential to design Level 1 as a completely protected model, and Level 2 as a partially protected model. Partial protection could include an alert to the user whenever requirements calculated in Level 2 deviate from the values determined in Level 1 by more than a predetermined amount, perhaps 10 to 20%.

References

Breuer, L., R. Zimmerman, and J. Pagan. 1996. Effect of supplemental energy intake on growth rate of suckling Quarter Horse foals. Pferdeheilkunde 12 (3):249.

Cuddeford, D. 1999. Starch digestion in the horse. Proc. 1999 Equine Nutr. Conference for Feed Manufacturers, KER, Inc. Lexington KY. 129.

Cymbaluk, N. and G. Christison. 1990. Environmental effect on thermoregulation and nutrition of horses. In, Clinics of North America: Equine Practice. 6(2):355.

Cymbaluk, N., G. Christison and D. Leach. 1990. Longitudinal growth analysis of horses following limited and ad libitum feeding. Equine Vet J. 22:198.

Doreau, M., S. Boulot, D. Bauchart, J.P. Barlet, and W. Martin-Rosset. 1992. Voluntary intake, milk production and plasma metabolites in nursing mares fed two different diets. J. Nutr. 122:992.

Gallagher, K., J. Leech and H. Stowe. 1992a. Protein, energy and dry matter consumption by racing Thoroughbreds: A field survey. J. Equine Vet Sci. 12:43.

Gallagher, K., J. Leech and H. Stowe. 1992b. Protein, energy and dry matter consumption by racing Standardbreds: A field survey. J. Equine Vet Sci. 12:382.

Garlinghouse, S., R. Bray, E. Cogger and S. Wickler. 1999. The influence of body measurements and condition score on performance results during the 1998 Tevis Cup. Proc. 16th Equine Nutr. Physiol. Symp. Raleigh, NC. 403.

Graham, P., E. Ott, J. Brendemuhl, and S. TenBroeck. 1994. The effect of supplemental lysine and threonine on growth and development of yearling horses. J. Anim Sci. 72:380.

Graham-Thiers, P., D.S. Kronfeld and K.A. Kline. 1999. Dietary protein influences acid-base responses to repeated sprints. Equine Vet J. Suppl. 30:463.

Griewe-Crandell, K., D. Kronfeld, L. Gay, D. Sklan, W. Tiegs and P. Harris. 1997.

Vitamin A repletion in Thoroughbred mares with retinyl palmitate or B-carotene. J. Anim. Sci. 75:2684.

Henneke, D, G. Potter, J. Kreider and B. Yates. 1983. Relationship between condition score, physical measurements and body fat percentage in mares. Equine Vet J. 15:371.

Heusner, G. 1993 Ad libitum feeding of mature horses to achieve rapid weight gain. Proc 13th Equine Nutr. Physiol. Symp. Gainesville, FL. 86.

Hiney, K., G. Potter, S. Bloomfield and P. Gibbs. 1999. Radiographic and biochemical measures of skeletal response to pretraining and race training in two-year-old Quarter Horses. Proc. 16th Equine Nutr. Physiol. Symp. Raleigh, NC. 40.

Jelan, Z., L. Jeffcott, N. Lundeheim and M. Osborne. 1996. Growth rates in Thoroughbred foals. Pferdeheilkunde 12(3):291.

King, S. and L. Nequin. 1989. An artificial rearing method to produce optimum growth in orphan foals. J. Eq. Vet Sci. 9:319.

Knight D. and W. Tyznik. 1985. Effect of artificial rearing on the growth of foals. J. Anim Sci. 60:1-5.

Kowalski, J., J. Williams and H. Hintz. 1990. Weight gains of mares during the last trimester of gestation. Equine Practice 12:6.

Lawrence, L., J. DiPietro, K. Ewert, D. Parrett, L. Moser and D. Powell. 1992a. Changes in body weight and condition in gestating mares. J. Equine Vet Sci 12:355.

Lawrence, L., S. Jackson, K. Kline, L. Moser, D. Powell and M. Biel. 1992b. Observations on body weight and condition of horses in a 150-mile endurance ride. J. Equine Vet Sci. 12:320.

Lawrence, L., M. Murphy, K. Bump, D. Weston and J. Key. 1991. Growth responses in hand-reared and naturally reared Quarter Horse foals. Equine Practice. 13:19.

Meyer, H. and L. Ahlswede. 1978. The intra-uterine growth and body composition of foals and the nutrient requirements of pregnant mares. Anim. Res. Dev. 8:86.

Nielsen, B., G. Potter, E. Morris, T. Odom, D. Senor, J. Reynolds, W. Smith, M. Martin and E. Bird. 1993. Training distance to failure in young racing Quarter Horses fed sodium zeolite A. Proc. 13th Equine Nutr. Physiol. Symp. Gainesville, FL. 5.

NRC. 1996. Nutrient Requirements of Beef Cattle. National Academy Press, Washington DC.

NRC. 1978. Nutrient Requirements of Horses. National Academy Press, Washington DC.

NRC. 1989. Nutrient Requirements of Horses. National Academy Press, Washington DC.

Ott, E. and R. Asquith. 1995. Trace mineral supplementation of yearling horses. J. Anim. Sci. 73:466.

Ott, E. and J. Kivipelto. 1999. Influence of chromium tripicolinate on growth and glucose metabolism in yearling horses. J. Anim. Sci. 77:3022.

Pagan, J., I. Burger and S. Jackson. 1995. The long term effects of feeding fat to 2-year-old Thoroughbreds in training. Equine Vet J. Suppl. 18:343.

Pagan, J. , S. Jackson and S. Caddel. 1996. A summary of growth rates in Thoroughbreds in Kentucky. Pferdeheilkunde 12(3):285.

Pagan, J., S. Jackson and R. DeGregorio. 1993. The effect of early weaning on growth and development in Thoroughbred foals. Proc. 13th Equine Nutr. Physiol. Symp. Gainesville, FL. 76.

Pendergraft J. and M. Arns. 1993. Tall fescue utilization in exercised yearling horses. Proc. 13th Equine Nutr. Physiol. Symp. Gainesville, FL. 106.

Potter, G., F. Arnold, D. Householder, D. Hansen and K. Bowen. 1992. Digestion of starch in the small or large intestine of the equine. Pferdeheilkunde. 3/4:109.

Powell, D. 1999. Effect of feed restriction and dietary calorie source on hormonal and metabolic responses of exercising horses. Ph.D. Thesis. University of Kentucky, Lexington KY.

Savage, C., R. McCarthy and L. Jeffcott. 1993. Effects of dietary energy and protein on induction of dyschondroplasia in foals. Equine Vet. J. Suppl 16:74.

Siciliano, P., C. Wood, L. Lawrence and S. Duren. 1993. Utilization of a field study to evaluate the digestible energy requirements of breeding stallions. Proc. 13th Equine Nutr. Physiol. Symp. Gainesville, FL. 293.

Southwood, L., D. Evans, D. Hodgson, W. Bryden and R. Rose. 1993. Nutrient intake of horses in Thoroughbred and Standardbred stables. Aust. Vet. J. 70:164.

Taylor, L., P. Ferrante, D. Kronfeld, and T. Meacham. 1995. Acid-base variables during incremental exercise in sprint-trained horses fed a high-fat diet. J. Anim. Sci 73:2009.

PROTEIN AND AMINO ACIDS

EDGAR A. OTT
University of Florida, Gainesville, FL

The daily protein requirement of the horse is a function of the endogenous nitrogen loss and the nitrogen deposition and secretion by the animal. Since the horse has only minimal ability to utilize non-protein nitrogen, most of the animal's nitrogen needs must be provided in the form of protein or amino acids. Like other animals, the horse has a need for specific quantities of essential amino acids and a need for a pool of nonessential amino acids. The dietary protein requirement of the animal is a function of the amino acid needs of the animal, the amino acid composition of the dietary protein, the digestibility of the protein and perhaps the restructuring of the amino acid profile of some of the protein in the hindgut of the animal.

NRC (1989) Recommendations

With the exception of the lysine requirements of the growing horse, the protein requirements are expressed as crude protein requirements. In most situations, the crude protein needs of the animal were determined directly from the literature or from calculations from information on the digestible protein needs.

Maintenance - The daily protein requirement for a horse at maintenance is primarily a function of endogenous nitrogen loss. Data in the literature indicate that this is 0.49 to 0.68 g of digestible protein (DP)/kg BW daily. The committee selected 0.60 g DP/kg BW as appropriate for most horses. If we assume that maintenance horses are likely going to sustain themselves on grass pasture or grass hay and that the digestibility of that forage is 46%, the crude protein requirement of the animal is calculated by dividing the 0.60 g DP by 0.46. This gives a value of 1.30 g CP/kg BW or 650 g CP daily for a 500 kg horse. The requirement can also be expressed as a function of the energy needs of the animal and would be 40 g CP/Mcal DE/daily. There was no information available on the amino acid requirements for maintenance although Slade et al. (1970) demonstrated that the nitrogen requirements of horses at maintenance could be satisfied with a lower nitrogen intake from fish meal than from corn gluten meal. He attributed this primarily to the increased digestibility of the fish meal but the response could have also been due to the amino acid composition of the protein.

Reproduction - The protein requirements for maintenance were considered adequate for early gestation. This is because 60 to 65% of the fetal development takes place during the last 90 days of gestation. This means that only 35 to 40% of the development takes place during the first 8 months. This would mean that a 500 kg mare with a requirement of 300 g DP/d would deposit about 5 g DP in the fetus daily during early gestation. It would seem that protein deposition during early gestation is of little significance to the mare, especially when we consider

237

that the typical feeding program of 10 kg grass forage containing 8% CP would provide at least 368 g DP daily without supplementation. During late gestation the protein requirement increases due to the more rapid deposition of fetal tissues. During this time the mare will deposit about 26 g DP daily or about a 9% increase in her needs. This was handled by increasing the CP/Mcal DE from 40 to 44 and is probably quite liberal.

The protein requirement of the lactating mare is maintenance plus the protein required for milk production. This was calculated by dividing the protein content of the milk by the efficiency with which the animal converted dietary protein to milk protein and multiplying that value by the milk production of the animal. The efficiency with which dietary protein is converted to milk protein was estimated from dairy cattle data. Production figures were obtained from the literature and were blocked into early (3% BW/d) and late (2% BW/d) lactation even though the lactation curve is fairly well documented.

Growth - Based on numerous feeding trials where energy and protein intakes could be calculated, the DE requirements of weanlings and yearlings were estimated by the equation:

$$DE\ (Mcal/d) = (1.4 + 0.03\ BW) + (4.81 + 1.17x - 0.023x2)(ADG)$$
$$where\ x = age\ in\ months\ and\ ADG\ is\ in\ kg/d$$

Based on data from at least ten feeding trials with growing horses it was concluded that there was a relationship between protein and energy needs for growth. This relationship was calculated to be 50 and 45 g CP/Mcal DE daily, for the weanlings and yearlings, respectively. Some of these same experiments allowed the calculation of the lysine requirements which were estimated to be 2.1 and 1.9 g lysine/Mcal DE daily, respectively. Based on these recommendations lysine needs to be at least 4.2% of the protein for growing horses for maximum efficiency.

Work - The protein requirement of the working horse is influenced by muscle hypertrophy as the animal trains, increased muscle protein content and perspiration loss. Sweat contains 1.0 to 1.5 g N/kg sweat and horses can lose as much as 5 kg sweat/100 kg BW daily; therefore a 500 kg horse could lose 234 g protein daily in the sweat (25 kg x 1.5 g/kg x 6.25). The recommendation for the working horse is to keep the CP/Mcal DE ratio constant at 40 g CP/Mcal DE. Doubling the energy expenditure would double both the energy requirement and the CP intake. A 500 kg horse would therefore consume 656 g CP at maintenance and 1312 g CP at intense work. This increased intake should more than meet the added protein needs of the working horse.

Research since the NRC (1989) Publication

Maintenance - There is little new information on the protein-amino acid requirements of horses at maintenance. This suggests that there is a general acceptance of the recommendations at the DP level. Some may question whether the 46% digestibility assumed by the NRC (1989) is applicable to all maintenance situations. Additional information is available on the digestibility of forages. Lieb et al. (1993) determined the digestibility of four hays: rhizoma peanut (15.9% CP), alfalfa (19.7% CP), Coastal bermudagrass (14.4% CP), and bahiagrass (7.1% CP). The apparent digestibilities of the four were 70, 79, 63 and 25%, respectively. Dugan et al. (1993) compared Coastal bermudagrass (14.8% CP) and flaccidgrass (8.1% CP) and found digestibilities of 68.4 and 53.9%, respectively. In a study of four bermudagrass varieties containing 10.3 to 11.1% CP (McCann et al., 1995) the protein digestibility was 63.5 to 66.7%. Comparisons of the protein digestibilities of Coastal bermudagrass (8.28% CP), Matua (10.90% CP), and alfalfa (16.44% CP) by Sturgeon et al. (2000) found digestibilities of 60.56, 64.11, and 76.38%, respectively. These studies verify that low protein forages have low digestibility but suggest that moderate protein forages are much better protein sources than NRC (1989) assumed. The above data are plotted in Figure 1 to illustrate the relationship between protein content of the forage and its apparent protein digestibility.

Figure 1. Relationship of hay protein to protein digestibility
Y= 2.01x + 33.15, r=.51

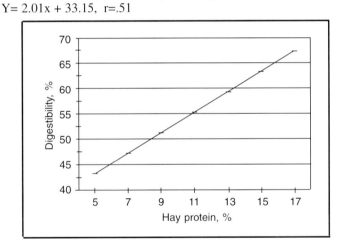

The apparent total tract digestibility does not tell the whole story. The amount of protein that is digested in the foregut greatly influences the amino acid availability to the animal. Thus feeding programs that provide protein in a form that is readily digested in the foregut will have higher value to the animal than those programs where a considerable portion of the protein is digested in the hindgut. This may be especially true for combinations of forage and concentrate. If the concentrate provides sources of highly digestible protein, the digestibility will be considerably higher than shown above. Farley et al. (1995) have demonstrated that for semi-purified diets providing 5, 9.5,14 and 16.5% CP where soybean meal provided all of the supplemental protein, the apparent digestibilities were 62.9, 78.8, 85.8, and 87.4% and the prececal true digestion was 72%.

Gestating and lactating mares - The requirements for protein for the gestating and lactating mare have been calculated by the NRC (1989) based on tissue deposition during gestation and milk composition and production during lactation. Not many feeding trials are available to verify or modify those calculations. Glade and Luba (1990) fed mares one of two programs during the last week of gestation and for seven weeks of lactation. One group of mares was fed a complete pelleted feed containing 17.92% CP at 1.25% BW/d or a mixture of 92% of the complete pellet and 8% soybean meal fed at 1.20% BW/d. The two diets provided similar energy intakes but the soybean meal supplemented diet provided slightly more protein and a different amino acid mixture. Milk from the mares collected on day seven was higher in alanine, aspartic acid, glutamic acid, histidine, isoleucine, leucine, lysine, methionine, phenylalanine, proline, serine, threonine, tyrosine, and valine than milk from unsupplemented mares. Plasma from the foals from mares fed soybean meal was higher for histidine, leucine, lysine, methionine, and valine than the plasma from unsupplemented mares, resulting in increased foal growth based on increased height gain. This work suggests that the lactating mare will respond to amino acid availability by altering milk composition and that the foal will respond to this improved amino acid balance by increasing growth.

Growing horses - Although it is obvious that the protein requirement of the growing horse is influenced by the amino acid content of the protein, we still have very little information on the amino acid requirements. Saastamoinen and Koskinen (1993) demonstrated that weanling foals 7 to 12 months of age would respond to protein quality. They supplemented a basal diet with milk protein or barley protein and measured increased weight gain (P < .01) and heart girth (P < .05) by the milk supplemented foals. Graham et al. (1994) provided evidence that for diets composed of 40% Coastal bermudagrass hay and 60% concentrate, threonine is the second limiting amino acid for yearling horses. This work verified that the yearling needs 1.9 g lysine/Mcal DE and demonstrated that 1.7 g threonine/Mcal gave greater growth than 1.5 g threonine/Mcal. Similar response was reported by Staniar et al. (1999). They supplemented mares and foals with either a 14% CP concentrate or a 9% CP concentrate with 0.6% added lysine and 0.4% added threonine. The foals on both programs grew at the same rate. The NRC (1989) recommendations for protein and lysine were verified in a study by Coleman et al. (1997). They fed weanlings a 60:40 alfalfa hay:concentrate diet providing NRC (1989), 109% NRC and 112% NRC recommendations for protein and lysine. The diets provided 48.5, 52.4 and 57.3 g protein/Mcal and 1.99, 2.24, and 2.58 g lysine/Mcal. They concluded that the NRC (1989) provided adequate protein and lysine to support maximum growth under this type of feeding system. The foals grew at 0.72 kg/d. A similar trial by Wall et al. (1997) also found no difference in growth response of yearlings fed 40:60 ratios of hay to concentrate using bermudagrass hay supplemented with soybean meal vs alfalfa based diets. The rate of gain on these animals was 0.44 kg/d, probably because the animals were older and they had limited access to the feed. In this study the animals consumed only 1.9 % BW of DM daily. The results suggest that even though the animals on the soybean meal supplemented diet had greater nitrogen retention, the two protein sources were similar in their ability to support growth.

Working horses - The effect of exercise on protein requirements has been investigated by several groups. Pagan et al. (1997) demonstrated that exercise would reduce DM digestibility but had no effect on protein digestibility. Graham-Thiers et al. (1999) compared a 7.5% CP diet with added lysine and threonine to a 14.4% CP diet for working horses. The low protein diet seemed to have little or no adverse effect on the physiological factors related to energy or protein metabolism, suggesting that the amino acid supplemented low protein diet was adequate to meet the needs of the animals. High protein intakes appear to have little effect on the metabolism of the working horse. The extra amino acids are apparently deaminated as blood urea-N was higher in the animals receiving the high protein diet but blood ammonia levels were not affected. Blood lactate levels were depressed by the high protein diet suggesting that the high protein diet depressed glycogen availability or that glycogen was spared by the availability of amino acids that were used directly in energy metabolism. Blood alanine was lower in the horses on the high protein diet.

Amino acids are known to play an important role in energy metabolism in working muscle. The branched chain amino acids leucine, isoleucine and valine provide a source of energy that should reduce lactic acid production during strenuous work. Glade (1989) demonstrated that the administration of 25 g of a mixture of leucine, isoleucine, valine, glutamine and carnitine in a ratio of 7:5:5:0.2:0.1 before and after strenuous exercise for 9 weeks resulted in a measurable decrease in plasma lactate (P<0.05) and decreased heart rate for horses doing treadmill work. Horses exercised to fatigue were shown to have a small increase in muscle leucine (P < 0.05) and lysine (P < 0.05), a larger increase in alanine (P < 0.001) and a decrease in glutamate (P < 0.001)(Miller-Graber et al., 1990). Glutamate plays a key role in intermediary metabolism by providing α-ketogluterate for the tricarboxylic acid cycle and it can be transaminated to alanine. Alanine can be transported to the liver where it is converted to pyruvate and ultimately glucose. It is therefore likely that as animals approach fatigue, amino acids may play a key role in meeting substrate needs.

This need may, however, be met by adequate amounts of high quality protein. The administration of supplemental branched chain amino acids, L-alanine, L-leucine, isoleucine, and L-valine one hour before training of Standardbred racehorses had no measurable beneficial effect on energy metabolism in the animals (Stefanon et al., 1999). Likewise, horses fed branched chain amino acids before and after treadmill training, three times each week provided no evidence of benefit when compared to unsupplemented animals (Casini et al., 1999).

Formulating to Meet the Protein and Amino Acid Needs of the Horse

It is evident from the above data that the protein requirement of the horse is influenced by how well the protein provides for the amino acid needs of the animal. It appears that lysine and threonine are the first limiting amino acids in grass-based feeding programs, but we have little or no data on the horse's needs for, or the availability of, the other essential amino acids. This presents a special challenge when formulating diets since the amino acid composition of typical feed ingredients used in horse feed varies considerably. If we examine the amino acid requirements of growing swine (NRC, 1999) we find a recommended relationship between lysine and the other essential amino acids. It should be noted that the amino acid content of the typical swine diet may vary considerably from the ideal diet. A comparison of this relationship with the same information for the growing foal gives us an insight into some of the challenges facing scientists as we seek to refine the protein and amino acid requirements of the horse. In general, the current recommendation for lysine is considerably lower, as a percent of the total protein, for horses than for swine. This leads to considerable differences in the lysine to other essential amino acid ratios in typical horse diets compared to swine recommendations. This variation can either frustrate us or challenge us to gain more information on the amino acid needs of the horse.

It appears from available data that although all horses may have specific requirements for amino acids, the growing foal is most sensitive to this balance. Using the NRC (1989) recommendations, the yearling (12 months) weighing 325 kg and growing at 0.65 kg/d will require 21.3 Mcal DE, 956 g protein and 40 g lysine. If we feed the yearling 40% Coastal bermudagrass hay (1.8 Mcal DE.kg, 8.5% CP, and 0.3% lysine, as fed) and 60% concentrate (3.0 Mcal DE, 14% CP and 0.60% lysine, as fed), we will meet the animal's requirements for all three nutrients if he consumes 8.45 kg/d (2.6% BW). This program exceeds his protein requirement but just provides his lysine needs. However, by adding L-lysine to the concentrate we could provide this animal with a concentrate containing the same energy and lysine content but only 12% CP (Table 1). This would drop his CP intake to 66 g below his requirement (NRC, 1989) and still provide adequate lysine and enough protein and lysine to achieve the same gain as the 14% protein product will support. The advantage of this approach includes 1) more accurate formulation to meet the animal's amino acid needs, 2) lower protein content for those horsemen concerned about high protein diets, 3) possible economic advantage, and 4) lower nitrogen excretion and less nitrogen load on the environment.

Based on the data available on working horses cited above, we may be able to apply this same approach to mature horses.

Table 1. Nutrient recommendations for concentrates for growing horses (as fed).

	Hay:Conc.	DE Mcal/kg	CP %	Lysine %	Threonine %
Weanlings					
Concentrate	33:67	3.0	16.7	0.79	0.60
Alt. Conc. 1[1]	33.67	3.0	14.0	0.79	0.60
Alt. Conc. 2[2]	40:60	3.3	15.5	0.88	0.67
Yearlings					
Concentrate	40:60	3.0	14.3	0.62	0.52
Alt. Conc. 1[1]	40:60	3.0	12.0	0.62	0.52
Alt. Conc. 2[2]	46:54	3.3	13.3	0.69	0.58

[1]Alternate Conc. 1 has 0.2% added lysine and 0.1% added threonine.
[2]Alternate Conc. 2 has 5% added fat and will need to have about 10% higher concentrations of nutrients.

References

Casini, L., D. Gatta, L. Magni, B. Colombani. 2000. Effect of prolonged branched-chain amino acid supplementation on metabolic response to anaerobic exercise in Standardbreds. J. Equine Vet. Sci. 20:120-123.

Coleman, R. J., G. W. Mathison, L. Burwash, and J. D. Milligan.1997. The effect of protein supplementation of alfalfa cube diets on the growth of weanling horses. 15th Equine Nutr. Phy. Sym., Ft. Worth, TX p. 59-64.

Dugan, K. G., K. R. Pond, J. C. Burns, D. S. Fisher, R. A. Mowery, J. A. Moore, and T. G. Goodwin. 1993. 13th Equine Nutr. Phy. Sym. Gainesville, FL p. 11-15.

Farley, E. B., G. D. Potter, P. G. Gibbs, J. Shumacher, and M. Murray-Gerzik. 1995. Digestion of soybean meal protein in the small and large intestine. 14th Equine Nutr. Phy. Sym., Ontario, CA p. 24-29.

Glade, M. J. 1989. Effects of specific amino acid supplementation on lactic acid production by horses exercised on a treadmill. 11th Equine Nutr. Phy. Sym. Oklahoma State U. p. 244-251.

Glade, M. J. and N. K. Luba. 1990. Benefits to foals of feeding soybean meal to lactating broodmares. Equine Vet. Sci. 10:422-428.

Graham, P. M., E. A. Ott, J. H. Brendemuhl, and S. TenBroeck. 1994. Effect of supplemental lysine and threonine on growth and development of yearling horses. J. Anim. Sci. 72:380-386.

Graham-Thiers, P. M., D. S. Kronfeld, K. A. Kline, T. M. McCullough, and P. A. Harris. 1999. Dietary protein level and protein status during exercise, training and stall rest. 16th Equine Nutr. Phy. Sym., Raleigh, NC p. 104-105.

Lieb, S., E. A. Ott, and E. C. French. 1993. Digestible nutrients and voluntary intake of rhizoma peanut, alfalfa, bermudagrass and bahiagrass hays by equine. 13th Equine Nutr. Phy. Sym., Gainesville, FL p. 98-99.

McCann, J. S., G. L. Heusner, and G. Burton. 1995. Digestibility of four bermudagrass cultivars in mature horses. 14th Equine Nutr. Phy. Sym., Ontario, CA p. 84.

Miller-Graber, P. A., L. M. Lawrence, E. Kurcz, R. Kane, K. Bump, M. Fisher, and
J. Smith. 1990. The free amino acid profile in the middle gluteal before and
after fatiguing exercise in the horse. Equine Vet. J. 22:209-210.

NRC. 1989. Nutrient Requirements of Horses. 5th Revised Ed. National
Academy Press, Washington, DC.

Pagan, J.D., P. Harris, T. Brewster-Barnes, S.E. Duren, and S. G. Jackson. 1997. The
effect of exercise on the digestibility of an all forage or mixed diet in
Thoroughbred horses. 15th Equine Nutr. Phy. Sym., Ft. Worth, TX p. 128-132.

Saastamoinen, M. T. and E. Koskinen 1993. Influence of quality of dietary protein
supplement and anabolic steroids on muscular and skeletal growth of
foals. Anim. Prod. 56:135-144.

Slade, L. M., D. W. Robinson and K. E. Casey. 1970. Nitrogen metabolism in
nonruminant herbivores. I. The influence of nonprotein nitrogen and protein
quality on the nitrogen retention of adult horses. J. Anim. Sci. 30:753-760.

Staniar, W. B., J. A. Wilson, L. H. Lawrence, W. L. Cooper, D. S. Kronfeld, and
P. A. Harris. 1999. Growth of Thoroughbreds fed different levels of protein
and supplemented with lysine and threonine. 16th Equine Nutr. Phy. Sym.,
Raleigh, NC p. 88-89.

Stefanon, B., C. Bettini, P. Guggia. 2000. Administration of branched-chain
amino acids to Standardbred horses in training. J. Equine Vet. Sci. 20:115-119.

Wall, L. H. III, G. D. Potter, P. G. Gibbs, and G. W. Brumbaugh. 1997. Growth of
yearling fillies fed alfalfa or soybean meal. 15th Equine Nutr. Phy. Sym.,
Ft. Worth, TX p. 3-7.

FEEDING STANDARDS FOR ENERGY AND PROTEIN FOR HORSES IN FRANCE

WILLIAM MARTIN-ROSSET
Centre of Research of Clermont-Ferrand/Theix, France

Thanks to the increase in knowledge since 1965 in the USA and Europe, INRA created new feeding systems for horses in France in 1984: the UFC[1] system for energy and the MADC[2] system for protein (INRA, 1984). These systems provide sets of tables that give the nutritive value of the feeds and the nutrient requirements of the horses, respectively. Both are expressed according to the same feed evaluation systems, either UFC for energy or MADC for protein. The goals of these systems are to allow:

1) an accurate comparison of the nutritive value of feedstuffs;
2) the formulation of well balanced rations to achieve a production goal;
3) the prediction of the animal performance when amount and quality of rations are known.

The validity of the UFC and MADC systems was tested through many feeding trials using mares, growing horses and working horses to create the new French feeding standards. These systems and the proposed feeding standards were updated in 1990 by the INRA group on the basis of further experiments and feeding trials, which focused on energy metabolism and mare and growing horse nutrition. In the 1990s, research was conducted at INRA on routine methods for predicting the net energy values of feeds and on nitrogen digestion. These systems are official feeding standards in France (INRA, 1990), and they are used in an increasing number of western European countries[3] and in some eastern European countries[3], with or without local adaptations (Miraglia and Oliveri, 1990; Staun, 1990; Smolders, 1990; Austbo, 1996). But there is still an increasing demand for a new scientific basis of equine nutrition and feeding management in the world to improve equine nutrition evaluation as pointed out by Pagan and Jackson in the proceedings of the 1998 KER Equine Nutrition Conference for Feed Manufacturers (KER, 1998).

---

[1] UFC : Horse Feed Unit (Unité Fourragère Cheval in French)
[2] MADC : Horse Digestible Crude Protein (Matières Azotées Digestibles Cheval in French)
[3] INRA 1990 Handbook has been translated and edited in Spanish 1993, in Italian 1994, in Romanian 1996, English and Polish translations are in progress. Portugese translation may occur.

Feed Evaluation Systems

The Horse Net Energy System (UFC)
A new horse net energy system was proposed and introduced in France (INRA, 1984) to evaluate and express the energy value of feeds and to recommend energy allowances for horses. The scientific concepts, the bases and the structure of the UFC system were precisely established and validated through several studies (indirect calorimetry and feeding trial) carried out on horses and ponies (Vermorel and Martin-Rosset, 1997). The mode of utilization of the UFC system for rationing horses was described in detail in a handbook devoted to end-users (INRA, 1990).

The French horse net energy system is based on two concepts:
1) maintenance is the major component of energy expenditure in most horses - 50 to 90% (Martin-Rosset et al., 1994);
2) the net energy (NE) value of nutrients for both maintenance and work (physical activity) depends on the free energy (ATP) produced by oxidative catabolism (Vermorel et al., 1984; Vermorel and Martin-Rosset, 1997).

The NE value of feeds is calculated through a stepwise procedure (Figure 1) from:
1) their gross energy (GE) content;
2) their digestible energy (DE) as measured in horses;
3) the ratio between metabolizable energy (ME) and DE as determined in horses;
4) the efficiency of ME utilization for maintenance (km) :

$$NE = ME \times km \text{ or } NE = DE \times \frac{ME}{DE} \times km.$$

The km is computed from the energy cost of eating, the assumed proportions of absorbed energy supplied by the various nutrients and the efficiencies of nutrient energy utilization:

For concentrate feeds :

$$Km = 0.85 E_{Gl} + 0.80 E_{LCFA} + 0.70 E_{AA} + (0.63 \text{ to } 0.68) E_{VFA}$$

For forages :

$$Km = 0.85 E_{Gl} + 0.80 E_{LCFA} + 0.70 E_{AA} + (0.63 \text{ to } 0.68) E_{VFA} - 0.14 (76.4 - ED)$$

where E is the percentage of absorbed energy supplied by glucose or lactate (Gl), long chain fatty acids (LCFA), amino acid ME (AA) and volatile fatty acids (VFA). ED is energy digestibility (%) of the feed. The last term [- 0.14 (76.4 - ED)] of the prediction equation corresponds to the cost of eating for forages.

As a result, **NE content of a reference feed, such as barley in Europe, accounts for 2.250 Mcal*/kg fresh material for a horse at maintenance** and is stated:

Net energy content of one kg of standard barley (87 % DM) is :
GE = 3.854 Mcal*/kg fresh material
OMD[1] = 0.83   ED = 0.80   DE = 3.076 Mcal*/kg
ME/DE = 0.931   ME = 2.864 Mcal*/kg   Km = 0.785
NE = 2.248 Mcal*/kg, e.g. NE = 2.250 Mcal*/kg

---
* 1 Megajoule = 4.19 Mcal   [1]Organic Matter Digestibility

Figure 1. Stepwise procedure to determine the net energy value of feeds in the UFC system (Vermorel & Martin-Rosset, 1997)

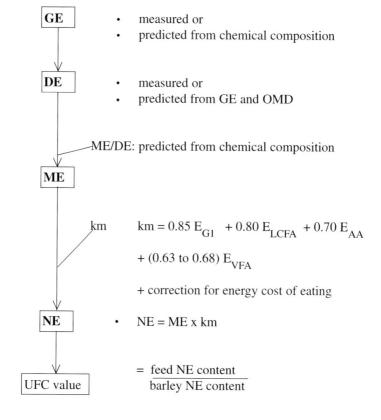

In France and in most countries of western Europe, the NE content of feed is related to that of a reference feed (barley) and expressed in feed units, which allows easy comparison of the feeds and their substitution. For horses, INRA proposed the Horse Feed Unit (Unité Fourragère Cheval or UFC in French):

$$\text{UFC value of a feed} = \frac{\text{NE content of feed}}{\text{NE content of barley}}$$

In 1984 and then in 1990, the NE content was stated by INRA to be 2.200 Mcal/kg and the INRA feed composition and nutritive value tables and nutrient requirement tables were created referring to this value of barley (Vermorel et al., 1984 ; Vermorel and Martin-Rosset, 1987).

In the year 2000, one feed unit corresponds to the NE content (2.250 Mcal) of one kg of standard barley (87% DM) for horses at maintenance (Vermorel and Martin-Rosset, 1997). But the energy values in feed tables and nutrient requirements (INRA, 1990) have not yet been changed; as a result the NE referring value of barley is still 2200 kcal/kg. The feed and requirement tables will be adjusted in 2001 in the next expected revision of INRA.

In this paper NE content of one kg of barley is maintained at 2.200 Mcal/kg (INRA, 1990).

Expressed per kg DM, the NE values range from 2.97 Mcal (maize), 2.55 Mcal (barley), 2.22 Mcal (oats), 1.94 Mcal (maize silage), 0.88 Mcal - 1.47 Mcal (hays) to 0.57 Mcal (straw) and the UFC values of feeds range from 1.35 (maize), 1.16 (barley), 1.01 (oats), 0.88 (maize silage), 0.40- 0.67 (hays) to 0.26 (straw).

Although organic matter digestibility (OMD) is the major factor of feed energy value, the further steps from DE to ME, then from ME to NE, increase the discrepancies between feeds. For instance, the DE, ME and NE contents of poor quality grass hay amount to 49%, 41% and 35% of those of barley, respectively.

The UFC system is an empirical model for predicting the NE value of feeds for horses. It does not portend to give the true energy values of feeds but the values are closer than a DE system would predict. It is now well demonstrated that methane and urinary energy losses and utilization of ME for maintenance (or fattening) vary with diet composition in horses as in other species. In a DE system, the energy of protein rich feeds and forages is overestimated (by about 15% for cereal byproducts, 25-30% for oil meals and 30-35% for hays), whereas that of feeds rich in starch is underestimated (Table 1).

Although feed digestibility is the main variation factor of NE value in all existing energy systems, the main limit of accuracy of the UFC system now consists in the percentage of energy supplied by the main nutrients. For instance, digestion in the small intestine seems to vary with starch origin (oats vs. barley, maize or sorghum), grain processing and starch supply (Potter et al., 1992a and 1992b; Meyer et al., 1993). However, significant errors in these estimates have relatively small effects on km. For instance, in the case of wheat bran, a 20% underestimation of glucose supply, involving an increase in the estimate of the VFA supply, causes a 0.4% unit error in km, a relative error of only 0.5% (Vermorel and Martin-Rosset, 1997).

Using km to predict the energy value of feeds for horses in various physiological situations certainly causes errors in the cases of lactation and growth. However, in ruminants the ratio of km to kl (lactation) is relatively constant whatever the feed (Van Es, 1975). The situation could be similar in horses and the NE values of feeds should be similar for maintenance and lactation. Furthermore, energy requirements for lactation account for about 50% of total energy requirements (Doreau et al., 1988). The differences in efficiency of ME utilization for maintenance (km) and fattening (kf) or growth (kpf) are certainly higher than for lactation, especially in the case of forages, but requirements for growth account for only 10 to 30% of total energy requirements in light breeds and 20 to 40% in heavy breeds (Agabriel et al., 1984).

Table 1. Energy values of feeds related to that of barley in a DE system and in the UFC system. Relative difference between the two systems for each feed (Vermorel and Martin-Rosset, 1997).

As fed	DE System	UFC System	DE % NE
Maize starch	116	131	88
Maize	111	115	96
Wheat	108	109	99
Barley	100	100	100
Oat	90	85	105
Wheat feeds, fine	102	94	108
Wheat bran	88	77	114
Wheat bran, coarse	81	71	115
Maize, gluten feed	96	83	116
Pea (seeds)	107	99	108
Fababean (seeds)	102	90	113
Beet pulp, dehydrated	85	74	116
Linseed meal (36% CP)	98	79	124
Soyabean meal (45% CP)	113	91	124
Peanut meal (48-50% CP)	114	87	130
Lucerne hay	70	54	130
Good quality grass hay	64	49	132
Bad quality grass hay	49	35	137

The NE supplied by the feeds of a ration are additive.

Example: A diet composed of 8 kg of grass hay (1.125 Mcal NE/kg fresh material) and 5 kg of barley (2.200 Mcal NE/kg fresh material) supplies 9.00 + 11.00 (respectively) = 20.00 Mcal NE.

The Horse Digestible Crude Protein System (MADC)
The Horse Digestible Crude Protein System is based upon two concepts (Jarrige and Tisserand, 1984 ; Tisserand and Martin-Rosset, 1996):
1)   nitrogen value of feedstuffs depends on the amount of amino acids (AA) truly provided by the feedstuffs;
2)   the amount of AA provided by feedstuffs depends on the site of digestion in the digestive tract (small intestine vs. large intestine).

Non-protein nitrogen (NPN) may account for 10 to 30% of total nitrogen in forages. In concentrates, NPN is very low. The true digestibility of nitrogen (N) from the small intestine estimated from the studies carried out in the USA and in Europe in the 1970s with markers in slaughtered or fistulated animals (Jarrige and Tisserand, 1984; Martin-Rosset et al., 1994 ; Cuddeford, 1997 ; Potter et al., 1992-1995) or performed recently at INRA with mobile nylon bag technique (MNBT) in fistulated horses (Cordelet, 1990; Macheboeuf et al., 1995-1996) or in Scotland (Moore-Colyer et al., 1998) range from 30-50% for hays, 60-70% for grass, 60% for dehydrated alfalfa to 70-80% for grains and cakes (Tisserand and Martin-Rosset, 1996). For hays, true N digestibility does not seem to be affected by cell wall content (Figure 2). For concentrates, there is a relevant relationship between the true N digestibility in the total tract calculated by INRA (INRA, 1984-1990) and the true N digestibility measured by INRA with the MNBT method in the small intestine of fistulated horses (Figure 3).

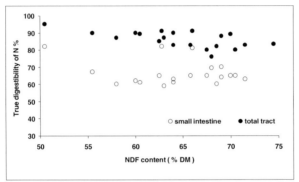

Figure 2. Effect of Neutral Detergent Fiber (NDF) content on the true digestibility of nitrogen of 20 different hays measured in the small intestine by MNBT methods (from Macheboeuf et al., 1995 and Martin-Rosset et al., 2000).

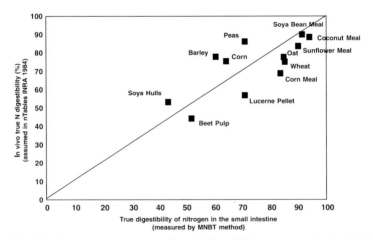

Figure 3. Relation between nitrogen true digestibility of 12 concentrates in the total tract assumed in INRA tables (1984) and true digestibility of nitrogen measured by MNBT method (Macheboeuf et al., 1996) in the small intestine by INRA (1994-1996) (Martin-Rosset et al., 2000).

The true digestion coefficients of feed N in the large intestine measured in slaughtered equines, in ileal-fistulated ponies (Jarrige and Tisserand, 1984; Martin-Rosset et al., 1994; Cuddeford, 1997; Glade, 1983-1984; Potter et al., 1992-1995) and with mobile nylon bag technique at INRA (Cordelet, 1990; Macheboeuf, 1995-1996) range between 75-90% for forages and concentrates (cereals and cakes). Finally, fecal nitrogen is composed of indigestible feed nitrogen mostly bound to fiber (so-called acid detergent insoluble nitrogen or ADIN), endogenous nitrogen (3 g N/kg DMI) and microbial protein nitrogen (50-60% of total N in feces) (Meyer et al., 1993). The amount of soluble N is rather low (Nicoletti et al., 1980). Feed proteins (and endogenous proteins) are degraded partially in the large intestine as amino acids (AA), peptides, and ammonia and resynthesized to microbial protein according to available energy and the type of nitrogen (Robinson and Slade, 1974 ; Meyer, 1983 ; Jarrige and Tisserand, 1984 ; Martin-Rosset et al., 1994). Microbial protein thereafter provides free AA and ammonia. It was assumed by INRA that no more than 10-30% of nitrogen in the large intestine would be absorbed as AA and peptides, which is in accordance with what it is known in pigs and ruminants. These assumptions are drawn from the work conducted in equines on the efficiency of AA absorption either in vivo with isotopic techniques and labelled bacteria with N_{15}, C_{14}, S_{35} (Slade et al., 1971; Goodbee and Slade, 1972; Wysocki and Baker, 1975; McMeniman et al., 1987; Schmitz et al., 1990), the technique of infusion of homoarginine into the cecum or overdosing different AA infused into the colon (Tisserand et al., 1996) and in vitro with cecal (Freeman et al., 1989 and 1991) or with colon mucosa (Bochröder et al., 1994).

From the above considerations, the protein value of feeds for the horse is referred to as the sum of feed and microbial AA absorbed in the small and large intestines, respectively, and expressed in Horse Digestible Crude Protein (Matières Azotées Digestibles Cheval or MADC in French).

The attempt to evaluate amounts of absorbable AA in the whole digestive tract on the basis of true digestibility measured in the small intestine and in the large intestine points out that digestible crude protein (DCP) overestimates the value of forage expressed in MADC by 10 to 30% (Table 2). As a result, MADC content of the main group of forages is calculated by reducing their DCP content by:

1) 10% for green forages;
2) 15% for hay and dehydrated forages;
3) 30% for good quality grass silages;

or multiplying DCP content respectively by a factor K which is 0.90, 0.85, 0.70 for the forages listed previously (Table 4). There is no limitation for concentrate. As a result K= 1.0.

Table 2. Assessment of the amount of absorbable intestinal amino acids (expressed in AIAA) provided by some feedstuffs. Comparison between forages and concentrates (in g/kg DM)[a].

Feedstuffs	Crude protein		Small intestine			Large intestine				Total tract	
	Total g	Non-aminated g	Entry g	true digestibility	AIAA g	Entry g (1)	true digestibility	AIAA % (2)	AIAA g	MADC g	DCP g
Rich concentrates (crude fiber 8%)	180	9	171	0.85	145	26	0.90	10	2	147	148
								30	7	152	
				0.75	128	43		10	4	132	
								30	12	140	
Green grass at early grazing stage	180	18	162	0.70	113	49	0.80	10	4	117	128
								30	12	125	
				0.60	97	65		10	5	102	
								30	16	113	
Barley-corn mixture	110	5	105	0.85	89	16	0.90	10	1	90	90
								30	4	93	
				0.75	79	26		10	2	81	
								30	7	86	
Grass-hay (at heading)	110	11	99	0.50	50	49	0.75	10	4	54	65
								30	11	61	
				0.40	40	59		10	4	44	
								30	13	53	
Grass silage	110	28	82	0.50	41	41	0.75	10	3	44	65
								30	9	50	
				0.40	33	49		10	4	37	
								30	11	44	

[a] Jarrige and Tisserand, 1984
(1) Entry : in grams of aminated nitrogen provided by feeds
(2) Percentage of alimentary proteins absorbed as amino acids and peptides ; AIAA: absorbable intestinal amino acid

This type of expression allows the comparison of the feeds and their substitution on the basis of their AA supply. The MADC content per kg DM of cereal straw is 0 but ranges from 30 to 80 g/kg for grass hays and from 80 to 100 g/kg for alfalfa hays according to stage and harvest conditions. The MADC content averages 79 g/kg DM in maize grain, 92 g/kg DM in barley, 98 g/kg DM in oats and reaches 496 g/kg DM in soyabean meal. The MADC supplied by the feeds of a ration are additive.

Example: A diet containing 8 kg of grass hay (50 g MADC/kg fresh material) and 2 kg of barley (80 g MADC/kg fresh material) supplies 400 g MADC + 160 g MADC (respectively) = 560 g MADC.

In a few situations, the expression has to be refined by essential AA content for specific requirements (for example, lysine for growth) (NRC, 1989); threonine is questioned (Hintz and Cymbaluk, 1994).

Tables of the Nutritive Value of Feeds
The chemical composition, the energy value (NE) and protein value (MADC) of feeds were computed in a new set of tables by INRA in 1984 and 1990. The tables include 150 feeds, including fresh forages, silages, dried forages, roots, tubers, cereals, seeds and their byproducts.

In addition, sets of equations were given in the introduction of the tables to predict energy and protein values of feeds from their chemical composition provided by laboratory analysis.

The NE value of feeds (expressed in Mcal or UFC) can be predicted accurately either from tabulated values stated by INRA (1984-1990) or from laboratory analysis (Tables 3a and 3b). The accuracy can be improved by using the digestible organic matter (DOM) content of feeds which can be estimated by the enzymatic (pepsine cellulase) method (Martin-Rosset et al., 1996b) or the near infrared spectrophotometric method (Andrieu et al., 1995 and 1996). The gas production method might be promising using fecal mocula (Lowman et al., 1997; Macheboeuf and Jestin, 1998), but the method is heavier than enzymatic or NIRS methods and less reliable as pointed out by recent studies (Macheboeuf et al., 1998a and 1998b; Cuddeford, 1999).

The nitrogen value of feeds can be predicted in the MADC system from the chemical composition of feeds (namely % CP) by using the appropriate relationships (Table 4) based on in vivo (total and/or partial) digestibility determination in the horse (Martin-Rosset et al., 1984; Macheboeuf, 1995-1996).

As a result, the nutritive value of feeds can be predicted in two ways :

1) either directly by the tables when the information concerning the feeds is available
2) or indirectly by using the appropriate equations (Tables 3 and 4) when laboratory analyses are provided using the most updated methods.

Table 3a. Prediction equations of the UFC value of feeds for horses from chemical composition and digestible organic matter or digestible energy content (Martin-Rosset et al., 1994 and 1996).

	RSD	R²
2.a - Forages (n = 47)		
N° 2.1 UFC = 0.825 -1.090 CF + 0.555 CP	0.043	0.832
N°2.2 UFC=0.568-0.650CF +0.687CP+1.804CC	0.031	0.922
N° 2.3 UFC = - 0.124 + 0.254 CC + 1.330 DOM	0.012	0.988
N°2.4 UFC=-0.056 +0.562 CC +0.0619 DE	0.007	0.996
2.b - Concentrates; Raw materials (n = 51)		
N ° 2.5 UFC = 0.815 - 0.947 CF + 0.0345 CP + 0.582 CC	0.060	0.931
N° 2.6 UFC = 0.131- 0.628 CF - 0.282 CP + 1.340 DOM	0.041	0.967
N° 2.7 UFC = - 0.730 - 0.722 CP + 0.572 OM + 0.0941 DE	0.033	0.979
N°2.8 UFC=-0.134+0.274CF-0.362CP+0.316CC+0.0755DE	0.017	0.995
2.c - Compound feeds		
N° 2.9 UFCo =1.326 -1.937 CFo - 0.135 Cpo	0.060	0.956
N° 2.10 UFCo =1.333 -1.684 ADFo - 0.096 Cpo	0.060	0.958
N° 2.11 UFCo =1.173 -1.605 CFo + 0.051 CPo + 0.215 STAo	0.043	0.976
N° 2.12 UFCo =1.181 -1.397 ADFo + 0.082 CPo + 0.214 STAo	0.040	0.978
N° 2.13 UFCo =1.219 - 0.852 ADFo - 0.287 NDFo - 0.857 Llo + 0.034 CPo + 0.207 STAo	0.031	0.988

Abbreviations and units
UFC : horse feed unit (per kg DM) ;
UFCo = horse feed unit (per kg OM)
DM: dry matter ;
OM: organic matter;
CF: crude fiber;
CP: crude protein;
CC: cytoplasmic carbohydrates
    (water soluble carbohydrates + starch);   (kg per kg dry matter)
ADF: acid detergent fiber;
NDF: neutral detergent fiber
CFo, Cpo and others with 'o' designation :
    expressed in kg /kg organic matter.
DOM: digestible organic matter (kg/kg DM)
DE: digestible energy (MJ/kg DM)

Table 3b. Average content in water-soluble carbohydrates of forages (% of dry matter) (INRA, 1988).

	WSC	Starch		WSC	Starch
Green forages			**Roots, tubers and**		
Ryegrass			**byproducts**		
Seeding year	3-10		Fodder beet	62	
First growth, leafy	10-15		Sugar beet molasses	62	
stem elongation	10-20		Sugar beet pulp, dried	7	
heading	10-20		Cassava, pellets	3	70
flowering	10-15		Cassava, roots	5	77
Stemmy regrowth	10-15		Potato pulp, dried	1	35
Leafy regrowth	5-10		Sugar beet, molasses		
			Residues	9	
Other grasses			Jerusalem artichoke		63
Seeding year	3-8				
First growth	5-10		Potato		60-65
Regrowth	4-8		Turnip	40	
Lucerne and red clover		Traces	**Cereals**		
First growth, early bud	6-10		Oats	2	44
Late flowering	3-5		Wheat	2	67
2nd and 3rd growth	3-6		Maize	2	72
			Barley	3	59
			Rye	3	62
			Sorghum	1	70
			Rice, paddy	2	70
			Rice, polished	1	85
			Triticale	5	62
White clover	3-4				
			Cereal byproducts		
Maize (whole plant)			Wheat byproducts		
Milk stage (25% DM)	18	16	Low-grade feed flour	5	60
Dough stage (28% DM)	14	24	Feed meal (CF < 6 %)	5	40
Flint stage (33% DM)	10	30	Middlings (CF > 6 %)	5	27
Kale	20-30		Bran	5	22
			Coarse bran	5	17
Conserved forages			Maize byproducts		
Hay, first cut			Gluten feed	3	28
Natural grassland	4-8		Gluten meal	16	
Ryegrass	8-15		Solubles		9
Other grasses	3-8		Bran	6	27
Legumes	2-4		Germ meal	1	25
Hay, regrowth	3-5		Barley byproducts		
Grass silage			Brewers' grain, dried	5	7
Without additives	0-2		Malt sprouts	13	9
With additives			Rice byproducts		
Ryegrass	2-6		Broken rice	2	77
Other grasses	1-2		Low-grade flour	6	23
Maize silage					
Milk stage (25% DM)		20			
Dough stage (30% DM)		28			
Flint stage (35% DM)		34			

(WSC: water-soluble carbohydrates)

Table 4. Relationships between DCP content (g/kg DM) and CP content (g/kg DM) in forages[1].

	n	Relationships	RSD	R
Fresh forages				
Natural grassland, grasses and legumes	14	DCP = - 27.33 + 0.8614 CP	± 7.7	0.967
		DCP = - 74.52 + 0.9568 CP + 0.1167 CF	± 6.3	0.980
Hays				
Natural grassland grasses	47	DCP = - 25.96 + 0.8357 CP	± 7.1	0.968
Legumes	25	DCP = - 29.95 + 0.8673 CP	± 9.2	0.933
All forages	72	DCP = - 27.57 + 0.8441 CP	± 8.6	0.964

[1] Martin-Rosset et al., 1984a and 1994

Calculation of MADC value of feedstuffs with INRA system (INRA, 1984-1990)

MADC = DCP x k

k = 1 for concentrates
k = 0.90 for green forages
k = 0.85 for hays and dehydrated forages
k = 0.80 for straws and their byproducts with high lignin content
k = 0.70 for good grass silages

Requirements and Recommended Allowances

Definitions and Methods for Determination

In France, nutrient requirements and allowances are clearly distinguished. The requirements stand for physiological expenditure of horses for maintenance, pregnancy, lactation, growth and exercise. The requirements are covered by the nutrients of the ration and by the body reserves when the amount of nutrients supplied is inadequate. The nutrient allowances represent the amount of the nutrients provided by the ration. A recommended allowance is the amount of nutrients which should be supplied to horses to achieve a desirable level of performance allowed by their potential. The animals are assumed to be in good health, well managed and housed during the winter period. These should be considered as optimum allowances, which cover at least the requirements. Exceptions are:

- Draft mares (700 kg BW) where a moderate and controlled used of body reserves is assumed during the winter period to reduce feed costs.
- Growing horses bred for school-riding or hacking where limited growth is assumed during the winter period but a compensatory growth period is expected during the subsequent summer period to achieve an optimum body weight at late breaking.
- Exercising horses where a moderate and controlled used of body reserves is assumed at short term (a few days) during the training period to avoid physiopathologic disorders related to large variation in workload and subsequent daily nutrient intake.

The allowance can be estimated by a factorial method from metabolic data and/or by feeding experiments according to the physiological function. With the factorial method, amount of energy or proteins fixed or exported is divided by the metabolic efficiencies of metabolizable energy or the efficiencies of the digestible crude protein which are specific to the physiological function where these efficiencies are known. With the feeding method, the allowances are determined by long-term feeding trials (and energy or nitrogen balances) conducted with a high number of animals. In these experiments, energy and nitrogen intakes are related to the true performances.

Energy Requirements and Recommended Allowances

In the horse, as in other farm animals, a distinction between maintenance and production requirements has been made, although overall metabolism is influenced by variations in animal expenditures.

Maintenance Requirements

These have been assessed from feeding trials conducted at the end of the last century by Wolff et al., (1880-1898); Grandeau et al., (1888-1904), former feeding standards (Olsson and Rudvere, 1955) and from more recent feedings trials (Breuer, 1968; Stillions and Nelson, 1972; Anderson et al., 1983) and indirect calorimetry trials (Hoffmann et al., 1967; Wooden et al., 1970; Knox et al., 1970) to 140 kcal DE/kg (BW 0.75) or 120 kcal ME/kg (BW 0.75), thus 84 kcal NE/kg (BW 0.75) or 0.038 UFC/kg (BW 0.75) (Vermorel, Jarrige and Martin-Rosset, 1984). These requirements were checked with feeding and calorimetry trials in horses of light breeds (Martin-Rosset and Vermorel, 1991). The daily NE requirements reach 8.80 Mcal and 11.44 Mcal for geldings weighing 500 and 700 kg, respectively. The maintenance requirements are increased by 10 to 20% or not according to the breed (Potter et al., 1987; Vermorel, Martin-Rosset, 1997) or for stallions (Axelsson, 1949; Brody, 1945; Nadal'Jack, 1961; Kossila et al., 1972; Anderson et al., 1983); working horses are increased by 5 to 15% to take into account the rise of the overall energy metabolism (Kellner, 1909) and the importance of spontaneous activities related to temper in exercising horses.

Work Requirements

When a horse moves, the energy expenditure exceeds the maintenance requirements. This additional expenditure results from the work done by the skeletal muscles, the increased work done by the respiratory, cardiovascular and other organs, and the increased tone of all the other muscles. The increase in oxygen consumption is the best measurement of increased energy expenditure. It was first measured in laboratory conditions in horses walking on a treadmill (Zuntz and Hagemann, 1898; Hornicke et al., 1974; Person, 1983; Pagan et al., 1987) and later in more natural conditions, with increasingly elaborate mobile equipment, in horses pulling loads (Brody, 1945), harnessed to a sulky (Karlsen and Nadal'Jack, 1964) and finally saddled (Hornicke et al., 1974; Hornicke et al., 1983; Meixner et al., 1981; Pagan and Hintz, 1986b).

The oxygen consumption, which is about 3 ml/min/kg live weight at rest, increases linearly with speeds up to a gallop (550 m/min) for 560 kg saddle horses with riders (Meixner et al., 1981). At the highest speeds studied, 600 to 700 m/min, the oxygen consumption is 100 ml/min/kg live weight. In well-trained horses, the maximum consumption may reach 125 to 140 ml/min/kg live weight. At this peak level, anaerobic metabolism predominates. The oxygen consumption remains high after the effort ceases and returns only gradually to its rest level. This additional requirement, which is a measure of the oxygen debt, is partly used to metabolize the lactate accumulated during the effort (oxidation and synthesis of glycogen).

Table 5. Variations of energy expenditure with velocity in the horse: unit energy cost (INRA, 1984) [1].

Situation	Energy expendinture		
	Velocity (m/mn)	(Kcal/mn)	Times of maintenance (maintenance = 1)
Waiting without rider	0	11.5	1.1
Waiting with rider	0	12	1.2
Walk	110	50	2.5
Slow trot	200	110	10
Normal trot	300	160	15
Fast trot (*)	500	350	35
Normal gallop	350	210	20
Fast gallop (*)	600	420	40
Maximum velocity (*)		600	60

[1] The energy expenditures were calculated from the oxygen consumption (and oxygen debt) by Meixner, Hornicke and Ehrlein (1981) in horses of 560 kg body weight carrying a load of 100 kg (rider + tack + apparatus). For the walk, energy expenditure was calculated from various data (Brody, 1945; Hoffmann et al., 1967; Nadal'Jak, 1961; Zuntz and Hagemann, 1898).

(*) Value calculated from the maximum oxygen consumption of horses estimated by Vermorel, Jarrige and Martin-Rosset (1984) and the oxygen debt.

From all these data, INRA has calculated the **energy costs of locomotion per meter/minute** (m/min) when the oxygen debt is included, at standard velocity for different gaits in a horse weighing 560 kg and riding with a 100 kg load: rider + saddlery (e.g. the unit energy costs) (Table 5). The energy expenditure was calculated from the measurement of oxygen consumption multiplied by the thermal equivalent (kcal/l) corresponding to the RQ (respiratory quotient) calculated at each measurement point (Brody, 1945).

Because the energetic efficiency falls as the velocity increases, energy expenditure rises exponentially (Zuntz and Hageman, 1898; Brody, 1945; Nadal'Jak, 1961; Hoffman et al., 1967; Vogelsang, 1981; Pagan and Hintz, 1986b; Martin-Rosset, 1993). Relative to rest, the energy expenditure of a 560 kg horse (carrying a 100 kg load comprising rider, saddle and measuring equipment) is 11.5 cal/mn. The energy expenditure of the horse during locomotion, calculated using all the available data (Hornicke et al., 1983; Karlsen and Nadal'Jak, 1964; Hoffman et al., 1967; Meixner et al., 1981; Thomas and Fregin, 1981; Pagan and Hintz, 1986b), is increased tenfold at a slow trot (200 m/min), 40-fold at a fast gallop (600 m/min) and about 60-fold at top velocity (Table 5).

When the velocity is a constant walking pace (80 to 100 m/min) the energy expenditure of locomotion expressed in kcal per live weight per horizontal meter traveled is roughly constant, 0.55 to 0.31 kcal/kg/m (Hoffman et al., 1967; Zuntz and Hagemann, 1898). The energy expenditure associated with the effort is proportional to the distance up to 24 km. It represents an increase in the energy expenditure of the horse at rest of 7% per hour walking (Hoffman et al., 1967). It is proportional to the duration of the work. However the energy cost of horizontal locomotion is multiplied by 2.5 on a 10% slope, and by 15 when the horse clears an obstacle measuring 1 m.

But, the **total daily energy expenditure for work,** which is additional to the maintenance expenditure of the animal at rest, results from four variables :

1) The duration of the work, which is easy to measure and indeed is standardized to some extent in stables and riding schools.
2) The intensity of the work. The energy expended doing one hour of work is extremely wide ranging (unlike that required to produce 1 kg of milk or even 1 kg of growth).
3) The ancillary effects of the work, or prolongation of the expenditure incurred reverting to the resting state (e.g. oxygen debt), and demand before start of work (e.g., response to stress), such as a characteristic increase in heart rate when the horse is saddled (Thomas and Fregin, 1981).
4) A general elevation of the level of metabolism, likely during intensive work (e.g. training for competition).

We need to evaluate an order of magnitude for **the energy cost in NE of a standardized hour of work** in the main conditions in which horses are used (Figure 4). Two evaluation methods can be envisaged, one factorial (analytical method), the other the global method (feeding experiments).

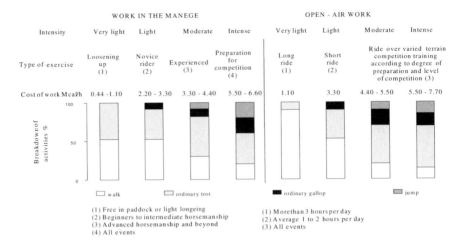

Figure 4. Approximate evaluation of energy cost per hour of work done by a sport horse according to the nature and duration of the work, additional to daily maintenance requirements (from INRA, 1984 and 1990; Martin-Rosset et al., 1994)

The analytical method breaks down the hour of work into periods of different intensities, each assigned a corresponding unit cost, as shown in Table 5. Even if the two components (1) and (2) are correctly evaluated, which is very difficult, uncertainty remains concerning the weighting to be given to the two others (3) and (4), possible interactions between different periods, and metabolizable energy efficiencies for activities of different intensities.

The global method comes down to measuring the quantity of energy required for adult horses in different situations to maintain constant body weight and body condition. This is the most satisfactory method both in practice and from a physiological standpoint, because it takes into account all four variables. However, it is applicable only in certain well-defined situations, and requires determination of nutritive value of feeds, weighing food intake and horses, and measuring variations in body reserves which is seldom done to state an energy balance.

INRA has used the analytical method to calculate **the cost of an hour of work** from its components (Figure 4) and the unit cost of each (Table 5). Also taken into account were the ancillary effects of anticipation and remanence. This loading, which ranges from 10 to 20%, can only be very approximate using current knowledge. To test its reliability, it was applied during long term experiments to horses at the National Riding School at Saumur in France (n = 80 horses used for high level riding instruction and for competition: B and C jumping classes and two and three star three-day events) and at the Animal Technology Teaching Centre at Rambouillet near Paris, France (n = 24 horses used for medium level riding instruction and for competition such as C and D jumping classes and one and two star three-days events). Energy requirements were calculated from the weights of horse and the work measured on the track (duration, velocity), the energy values of the rations having been measured (digestibility and indirect calorimetry according to Vermorel, Martin-Rosset, 1997b) and allowing for variations in live weight and body reserves (INRA, 1990-1997; Martin-Rosset and Vermorel, 1991; and Martin-Rosset et al., unpublished). The feed allowances were the same order of magnitude as the requirements calculated by the analytical method, allowing for ancillary effects. This encouraging result shows that the basis of the calculation used is satisfactory, and the energy allowances recommended in 1984 and updated in 1990 (Figure 4) are reasonable. These values are orders of magnitude which will need to be refined in the future, especially for intense effort, when more is known about the effects of mettle and age, the horse's ability to adapt to training and the environment, and the rider's proficiency.

The energy requirements for work are additional to those for maintenance. For practical reasons, and in particular to forestall rationing errors, recommended total daily energy allowances tables (maintenance plus work) have been drawn up for the situations most often encountered by sport and leisure horses of different sizes (Table 6a). These tables are used directly to calculate rations.

Table 6a. Recommended daily energy allowances (Mcal) for exercising sport and leisure horses (adapted from INRA, 1990).

Use	Live weight [1] kg			
	450	500	550	600
	NE (Mcal) per day			
Maintenance: Horse at rest[5]	8.99	9.73	10.47	1.20
Work				
Very light [2][4]	11.32	11.88	12.44	13.2
Light [2][4]	14.52	15.18	15.84	16.50
Moderate[2][4]	16.72	17.38	18.04	18.70
Intense [3]	15.35	15.95	16.55	17.16

NE : Net energy (Mcal) adapted from INRA energy requirement evaluation system (INRA, 1984 ; Martin-Rosset et al., 1994 ; Vermorel and Martin-Rosset, 1997)

[1] This allowance is for geldings and dry mares. For stallions, add 0.88 Mcal (1.10 Mcal for stallions of 550 to 600 kg)

[2] Horse assumed to work 2 h per day (average observed in horse training facilities)

[3] Horse assumed to work 1 h per day

[4] For short rides out, very light work is considered to be 1 h out and light work 2 h out. For long rides, light work is 2-4 h out and moderate work longer than 4 h

[5] Recommended allowance might be increased in resting horse from 5 to 15% depending on breed and intensity of work (INRA, 1984)

Table 6b. Recommended daily energy (Mcal) and nitrogen (MADC) allowances   for exercising sport and leisure horses of light breed (500 kg mature live weight[1]) (INRA).

Use	Daily allowances						Daily Feed* (kg DM**)
	NE (Mcal)	MDAC (g)	Ca (g)	P (g)	MG (g)	Na (g)	
Maintenance:							
Horse at rest	9.73	295	25	15	7	12	7.0-8.5
Work:							
Very light [2][4]	11.88	370	28	16	8	22	8.95-9.5
Light [2][4]	15.18	470	30	18	9	37	9.5-11.5
Moderate [2][4]	17.38	540	35	19	10	47	10.5-13.5
Intense [3]	15.95	490	35	19	10	40	10.0-12.0

*   The lower values are used with high proportion of concentrate in the diet and the higher with hay-based diet.
**   DM: dry matter.
[1]   These recommendations are suggested for geldings and mares. 0.88 Mcal NE and 30 g MADC are added daily for stallions.
[2]   We considered two hours of daily work (mean observed in riding school).
[3]   We considered one hour of daily work.
[4]   For short outside riding, very light and light work intensities are considered for one and two hours of exercise, respectively. For medium (2 to 4 hours) and long (> 4 hours) outside riding, light and moderate work intensities are considered.

For load pulling in draft horses, the total energy expense is the sum of energy expenses of locomotion and for load pulling. Pulling work is the result of the developed strength (F in kg) by the distance (d in m) expressed in kilogrameters. The strength of pulling the load and the resulting work during a given period decreases 60% when the velocity increases by 70 to 300 m/mn as measured by Gouin (1932) with half cold-blood horses (500 kg). The energy expenses of the horse rise linearly with the developed power (F x Velocity = kgm/s) and the length of work (Brody, 1945). For cold-blood (half-bred or draft horses), we have preferred to borrow energy requirements for work from former standards (Jespersen, 1949) by Olsson and Ruudvere (1955) because they were set up from feeding trials and practical observations, which were more reliable than assessments based on oxygen consumption as measured by Brody (1945) and Nadal'Jak (1961) on a very small numbers of horses. The additional energy requirements per hour are 0.77 Mcal for light, 1.23 Mcal for medium, 1.65 Mcal for hard and 2.38 Mcal for very hard work.

Pregnancy and Lactation Requirements

These are above maintenance and have been estimated by a factorial method (Martin-Rosset and Doreau, 1984) and revised by the same group (INRA, 1990). The increase in energy content of the fetus was calculated from the data of Dusek (1966), Den Engelsen (1966) and Meyer and Alshwede (1976) to estimate the amount of energy retained by the fetus (Table 7). The efficiency of energy utilization for pregnancy was estimated to be 25% (mean between those in cows and in sows because of the lack of measurements in mares). The amount of energy fixed in the conceptus (fetus, adnexa and placenta) and udder, calculated by multiplying energy content of fetus by 1.20 to take into account the energy fixed in the fetal adnexa, uterus and udder (Martin-Rosset and Doreau, 1984), increases daily from 152 kcal to 467 kcal/100 kg BW/day between the 8th and the 11th month of pregnancy.

Table 7. Weight gain and composition of the fetus*.

Months	Weight gain in % birth weight[1]	Gross energy Mcal/kg[2]	Protein content (%)[2]
8th ⎫		1.000	11.5
⎬ 19			
9th ⎭		1.100	13.0
10th	30	1.180	15.3
11th	31	1.280	17.1

*   From Doreau, unpublished and Martin-Rosset et al., 1994
[1]   Drawn from Dusek (1966) and Den Engelsen (1966)
[2]   Drawn from Meyer and Ahlswede (1976)

In lactation the amount of energy exported as milk was calculated thanks to better estimates of milk production and energy content (Table 8). Efficiency of energy utilization for milk production was estimated to be 65% from values obtained in the cow and the sow. The energy requirement for milk production is 682 kcal, 594 kcal and 572 kcal/kg for the first month, the second and third months and over the fourth month of lactation, respectively.

Table 8. Milk yield and composition.

Months	Milk yield (kg/100 kg LW)[1]	Milk composition	
		Energy Mcal/kg[2]	Protein content (g/kg)[2]
1st	3.0	0.575	24
2nd	2.5	0.500	24
3rd	2.5	0.500	24
4th	2.0	0.475	21

[1]   Doreau and Boulot (1989)
[2]   Doreau et al. (1988)
     In the first edition of the INRA recommendation (1984), the requirements were calculated with the available data (Martin-Rosset and Doreau, 1984), but the knowledge has increased a great deal since then.

As a result, the daily requirement for pregnancy (late pregnancy, 8th-11th months) and milk production (early lactation, 0-3 months) is 0.1-0.3 times and 0.9-0.6 times the maintenance requirements, respectively. However, the recommended energy allowances were estimated on the bases of many feeding trials carried out with mares of light or heavy breeds fed different diets, taking into account the possibility of body reserve utilization (Sutton et al., 1977; Henneke et al., 1981a and 1981b; Banach and Evans, 1981; Martin-Rosset and Doreau, 1984). Two levels of energy allowances, high level (HL) or low level (LL), were suggested according to the breed (heavy vs. light) and body condition score (Table 9).

Table 9. Feeding standards for mares and recommended nutrient allowances. Mare of light breeding (500 kg mature live weight) (INRA, 1990)*.

Physiological State	Daily Allowances							
	NE (Mcal/day)		MADC	Ca	P	Mg	Na	Daily feed[3]
	Low level[2]	High(g) level[1]	(g)	(g)	(g)	(g)	(g)	(kg DM)**
Dry mare or early pregnant mare	8.36	10.12	295	25	1	7	12	6.0-8.5
Pregnant mare								
8th & 9th months	9.02	11.00	340	29	18	7	12	6.5-9.0
10th month	10.34	12.54	460	38	26	7	12	7.0-10.5
11th month	10.56	12.76	485	39	28	7	12	7.5-11.0
Lactating mare								
1st month	19.58	23.54	950	61	55	10	15	12.0-15.0
2nd and 3rd month	16.72	20.24	800	47	40	9	14	10.0-15.0
4th month to weaning	13.42	16.50	660	39	32	8	13	8.0-12.5

* Live weight 24 hours after foaling;
** DM: dry matter
(1) For mares whose offspring are used in competition with the exception of the mares that are fat (body condition score (4) (INRA, 1990). Other mares (body condition score (2.5) and mares wintered outdoors or bred at three years old.
(2) Other cases.
(3) The lower values are used with proportion of concentrate in the diet and higher with hay-based diet values.

In **heavy mares**, energy allowances account for:

(1) 100 % (HL) or 80 % (LL) of total energy requirement in dry mares, pregnancy or late lactation (period 1)
(2) 110 % (HL) or 90 % (LL) of total energy requirement in early lactation (period 2)

These levels of energy allowances were suggested for two reasons :

(1) a moderate undernutrition in late pregnancy has no significant effect on the body weight and health of the foal at birth and its growth if the body condition score is good in early lactation.
(2) the effects of undernutrition in early lactation on the reproductive capacity of the mare are not well known. In light mares, energy allowances account for:
  (1) 110 % (HL) or 90 % (LL) in period 1
  (2) 120 % (HL) or 110 % (LL) in period 2

The suggested levels are higher for mares of light breeds to take into account the requirements of foals devoted to competition (bloodstock).

Growth Energy Requirements

These have been evaluated from feeding trials (Agabriel et al., 1984; Bigot et al., 1987; Micol and Martin-Rosset, 1995) in which NE intake, weight and weight gain of young horses were precisely measured, for three reasons.

(1) The energy requirement for maintenance (kg BW 0.75) varies with breed and growth rate. The variations in maintenance requirements due to breed are known only in the adult. Maintenance requirements account for 60-90% of total energy requirement of the growing horse.
(2) The amount of fixed energy per kg weight gain computed from the chemical composition of tissues was determined only in heavy breeds (Martin-Rosset et al., 1983).
(3) The efficiency of metabolizable energy utilization for growth is not known yet.

Energy allowances were computed according to body weight (BW) and weight gain (G) of young horses using a relationship established from the results of feeding trials carried out at INRA, and according to the following model drawn from the growing bull (Geay et al., 1978; Robelin, 1979) :

$$NE/kg\ BW\ 0.75/day = a + bG^{1.4}$$
a = coefficient of maintenance requirement
G = average daily gain; kg/day

The validity of the model and namely of the exponent '1.4' was checked in horses from the results of slaughter experiments in young heavy horses and extended to light breeds (Agabriel et al., 1984) (Table 10).

Table 10. Requirement for growth.

Energy : relationship between daily energy intake and live weight and growth in young horses of *light breeds* $UFC/day/kgW^{0.75} = a+bG^{1.4}$			Protein : relationship between daily protein intake and live weight and growth in young horses of *light breeds* $gMADC/day = aW^{0.75} + bG$		
Ages (months)	a	b	Ages (months)	a	b
6-12	0.0602	0.0183	6-12	3.5	450
18-24	0.0594	0.0252	18-24	2.8	270
30-36	0.0594	0.0252	30-36	2.8	270

a : coefficient of maintenance
G : average daily gain (kg/day) (Adapted from data of Agabriel et al., 1984 and Martin-Rosset et al., 1994)
UFC = Horse Feed Unit (2200 Mcal Net energy) in INRA energy requirement evaluation system
    (INRA, 1984 and 1990) *Relationships are different for heavy breeds (Agabriel et al., 1984)*

The relationships for the growing light-breed horse are given in Table 10. For 1 kg of daily gain, total requirements account for 10.78 Mcal at 250 kg BW and 14.08 Mcal at 350 kg BW. The increase deals with the maintenance requirements + 2.42 Mcal (8.36 Mcal at 250 kg BW compared to 10.8 Mcal at 350 kg BW) and with the cost of 1 kg of body gain + 0.88 Mcal (2.42 Mcal at 250 kg BW compared to 3.30 Mcal at 350 kg BW). At the same body weight, 300 kg, maintenance requirement reaches 9.46 Mcal but the cost of 1 kg of body gain increases with the level of growth, + 1.76 Mcal when growth requirement is calculated for 0.5 kg daily gain (1.10 Mcal) and compared with the requirement for 1 kg daily gain (2.86 Mcal). **Two levels of allowances** were suggested according to the goal of production (horses produced for school riding and hacking vs. bloodstock produced for competition) and the growth potential of the young horses (Table11) :

Table 11. Feeding standards for the growing horse. Recommended nutrient allowances for horses of light breeding (500 kg mature body weight) (INRA, 1990).

Age (months)	Mean body weight during the period (kg)	Growth		Daily allowances							Daily Feed * (kg/DM)**
		Level	Daily gain (g/d)	NE (Mcal)	MADC (g)	Ca (g)	P (g)	Mg (g)	Na (g)		
8 -12	320	Optimal	700-800	12.10	590	39	22	10	12		5.5-8.0
	280	Moderate	400-500	9.90	440	28	16	9	9		5.0-7.5
20 -24	470	Optimal	400-500	4.96	420	36	20	10	13		7.5-10.0
	440	Moderate	150-200	13.20	330	28	16	9	12		7.0-10.0
32 -36	490	Optimal	150-250	14.30	330	30	18	10	12		8.0-11.0
	470	Moderate	0-100	13.20	260	25	15	8	12		7.5-10.0

* The lower values are used with high proportion of concentrates in the diet and the higher values with hay-based diet.
** DM: dry matter

(1) optimal level for high growth (close to genetic potential) in young horses bred for competition:
(2) moderate level for limited growth in young horses bred for hacking but having compensatory growth on pasture (Figure 5).

Figure 5. Growth curve and management of young sport horses (Selle Français or Anglo-Arab) established from INRA long-term feeding experiments (INRA, 1990).

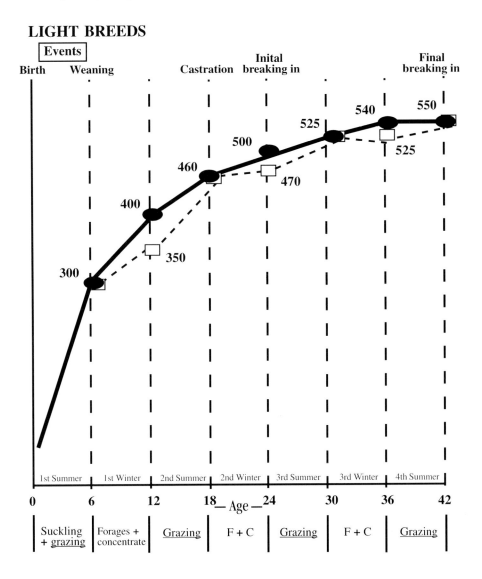

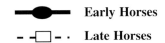

Protein Requirements and Recommended Allowances

On the basis of the nitrogen balance set up by Slade et al., (1970), and Prior et al., (1974), MAINTENANCE REQUIREMENTS have been estimated to be 2.4 g DCP (digestible crude protein)/kg $BW^{0.75}$; expressed in MADC, allowances are 10-15% higher, 2.8 g MADC/kg $BW^{0.75}$ because the DCP content of forages has been reduced by 10 to 20%. Daily allowances reach 295 g and 380 g MADC respectively for horses weighing 500 and 700 kg. Hence, recommendations are very close to those suggested by the NRC (1978) (2.8 DCP/kg BW $^{0.75}$). INRA recommendations cannot be compared to the new requirements suggested in crude protein (CP) by 1989 NRC. The former requirements were compiled by Olsson and Ruudvere (1955). However, they are lower than those calculated with the factorial method (Meyer, 1983) as this author took into account the possibilities of storing proteins.

All the feeding trials, digestion studies and nitrogen balances carried out show that the AA composition of the common diets for adult horses at maintenance is of no importance and that NPN (urea particularly) can be successfully used by the horse if the fermentable energy supply in the large intestine is sufficient.

Work

For work, protein expenses are unknown. They could increase more than energy expenses in relation to intensity and length of work if energy supply is sufficient to prevent body protein mobilization (Kellner, 1909). Thus, there is some reason to relate protein and energy supply for work to protein synthesis where the protein gain is positively related to the protein and energy contents of the diet. **A supply of 60 to 70 g MADC/UFC beyond maintenance requirements** seems satisfactory for the adult horse at work. A greater supply should be necessary for the two-year-old in training to enable increases in muscular mass, in the concentrations of myoglobin and in enzymes responsible for muscular metabolism as well as to prevent anemia. For the human athlete, protein supply should be increased by 20% at the beginning of training and by 100% during intensive training (Willmore and Freund Beau, 1984). There is no evidence to increase the protein or AA supplementation to equine athletes over what it is recommended for human athletes (Lawrence, 1998).

The protein requirements for pregnancy beyond maintenance were calculated by the factorial method (Martin-Rosset and Doreau, 1984). The total amount of daily protein was calculated by dividing the amount of protein fixed in the conceptus (Table 7) by the metabolic efficiency of DCP for that function. It has been stated at 55%, whereas Meyer (1983) and NRC (1978) assumed that it was 50 and 45% respectively. That efficiency is slightly lower than the value for pigs and ruminants (60%). The daily amount of protein fixed in the conceptus (fetus + adnexa + placenta + udder) reaches 5.0 g/100 kg BW and 21 g/100 kg BW between the 8th and 11th months of pregnancy. The recommended allowances given in the INRA table (1984-1990) account for 100% of the requirements (Table 9).

The protein requirements for milk production (above maintenance requirements) were also calculated by the factorial method on the bases of the amount of proteins expended daily in milk (Table 8) and of the metabolic efficiency of DCP estimated to be 55%. This efficiency is close to those suggested by Meyer (1983), 50 to 60%; Broster (1972) suggested true DCP to be 50-55%. The efficiency retained by NRC in 1978 and 1989 (69%) seems too high if we refer to the efficiency measured in ruminants (60% for the DCP). The protein allowances suggested account for 100% of the requirement (Table 9) as the influence of protein deficiency in the mare is not well known (Martin-Rosset and Doreau, 1984; Doreau et al., 1988). The protein allowances per kg of milk are 44 g, 38 g, and 36 g for the 1st month, 2nd and 3rd months, and 4th month of lactation, respectively.

Growth

Protein requirements of growing horses (above maintenance) were calculated from protein retention. The latter can be determined accurately from the composition of empty body weight and weight gain. This was defined on the basis of anatomical composition data on animals slaughtered at different ages (Martin-Rosset et al., 1983) after several feeding trials (Agabriel et al., 1984).

Protein requirements of heavy breeds were estimated by the factorial method because all the feeding trials conducted were designed to define the optimum CP content of the diet or the effect of feed protein quality on the growth of the young horse. The efficiency of MADC stated was 45% as suggested by the NRC (1978). It has been suggested to supply 3.5 g MADC/kg $BW^{0.75}$ for maintenance in the young horse, instead of 2.8 g MADC/kg $BW^{0.75}$ for maintenance in the young horse until one year of age, to take into account the faster turnover of body protein, 25% as measured in other species (Reeds and Harris, 1982).

In light breeds, protein requirements were calculated by using a mathematical model similar to that used for energy (Table 10) as there were enough INRA feeding trials (Agabriel et al., 1984; Bigot et al., 1987) designed to study effect of the CP content of the diet on average daily gain.

Two levels of allowances were suggested in growing horses as for energy considering the goal of production, genetic potential of the animal and the use of compensatory growth on pasture (Table 11).

We have kept allowances suggested by NRC (1978 and 1989) due to the work of Ott et al. (1979, 1981 and 1983) and confirmed by Saastamoïnen et al. (1993) for lysine, the single AA recognized as limiting at this moment even though threonine is questionable (Graham et al., 1993). Lysine is 0.6 % of DM in the diet for the 6-month-old foal and 0.5 % for the yearling and 0.4 % for the two- and three-year-old. Feed protein quality is of less importance as the growth potential decreases with age and the proportion of hay in the diet increases simultaneously.

Tables of recommended allowances (examples tables 6, 9 and 11)

The daily allowances are set up to allow an easy calculation of the rations. These tables are different for light and heavy breeds. There are no tables for ponies as there are not enough available data. The allowances are given for different body weights for light breeds (450, 500, 550, 600 kg) and heavy breeds (700 and 800 kg). In all the situations, the mare, stallion, gelding and the growing and fattening horse are considered in their own physiological situations where applicable and include maintenance, pregnancy, lactation, breeding, growth or fattening, rest and exercise. The allowances are given only for the winter period as it is the only period studied in breeding horses.

The recommended allowances suggested by INRA (1984) were revised by INRA (1990), mainly for mares and growing horses as there were more available scientific data to define requirements and new feeding strategies. They could be slightly revised in 2001 just to take into account the slight variation of NE content of barley, which has moved from 2.200 Mkcal/kg in 1990 to 2.250 Mcal/kg in 1997 (Vermorel and Martin-Rosset, 1997).

The range of amount of daily feed suggested in the tables does not represent the maximum amount of feeds that the horses are able to consume, specifically in growing horses (Bigot et al., 1987) or in mares (Doreau et al., 1988; 1990), but only the amount to cover the nutrient requirements. These amounts of daily feeds are consistent with the ingestibility of the forages measured by INRA, in trials especially designed to study ingestibility (Martin-Rosset and Doreau, 1984; Dulphy and Martin-Rosset, 1997a and 1997b) and because all the intakes were measured in feeding trials in animals fed different diets (Agabriel et al., 1982 and 1984; Bigot et al., 1987; Doreau et al., 1988 and 1990; Martin-Rosset and Doreau, 1984b and 1984c; Martin-Rosset et al., 1989). Average value by forage category is proposed by INRA (1990) in its former recommendations published by Martin-Rosset et al. (1994) (Table 12). The recent work of INRA (Dulphy and Martin-Rosset, 1997a and 1997b) supports the idea that there is no relationship between dry matter intake of forages and its chemical components such as CP, CF or NDF, which is consistent with the previous work of Martin-Rosset and Doreau (1984) and Cymbaluk (1990). Therefore, the use of average values proposed previously is still very relevant.

Table 12. Ingestibility of the main forages in growing horses and adult horses (INRA, 1990).

Feedstuffs	Intake [1] kg DM / 100 kg live weight
Fresh natural meadow	1.8-2.1
Hays : grassland-gramineous	1.7-2.1
hays : legumes	2.1-2.3
Straw	1.2-1.5
Maize silage well preserved	
25 % DM	0.9-1.2
30 % DM	1.2-1.6
Grass silage well preserved (natural meadow)	
25 % DM	1.2-1.5
35 % DM	1.4-1.7

DM: dry matter

[1] Maximum intake when the forage is offered alone ad lib. The range represents the average variation between animals according to the quality of the forage offered.

Proportion of concentrates in the diet depends on daily requirement to be covered, then on nutritive value of forages and intake capacity of the different types of equine with regard to their level of performance. Guidelines of concentrate proportions in the diet to meet the requirements are provided to the users in INRA systems (Table 13).

Table 13. Diet composition guidlines (INRA, 1990)

Type of animal	Physiological status Performance goal	Percentages Forages *	Concentrate	Comments **
Light breeds Mares	Pregnancy	50 to 95	5 to 50	Forages G ↗to VG //
	Lactation			Forages M ↗ to VG//
Stallions	Rest	85-95	5-15	Straw//+Forages M to VG//
	Service	60-70	30-40	Straw//+Forages M to VG//
Growing horses	1 year, optimal growth	40-65	35-60	Forages G to VG ↗or E//
	1 year, moderate growth	65-75	25-35	Forages G and VG↗
	2 years, optimal growth	75-80	20-25	Forages G to VG ↗or E//
	2 years, moderate growth	80-85	15-20	Forages M ↗ or G//
	3 years, optimal growth	80-85	15-20	Forages G//
	3 years, moderate growth	85-90	10-15	Straw ↗+ Hay G//
Heavy breeds - growing	1 or 2 years, optimal growth	40-60	40-60	Forages B ↗to VG//
	1 or 2 years, moderate growth	85-90	10-15	Straw ↗+ hay G//
- fattening	10 months, very high growth	30-50	50-70	Forages VG to E ↗
	12 months, high growth	40-70	30-60	Forages VG to E ↗
Exercising horses - light breeds - heavy breeds	Rest to intensive	35-95 55-95	5-65 5-45	Hay G to VG// + Straw// Hay M to G// + Straw//

M :Medium quality forages   * Straw is included in the indicated forages percentages
G :Good quality forages   ** Straw is indicated where relevant and its feeding level pointed out
VG :Very good quality forages   ↗ : Forages offered ad libitum
E :Excellent quality forages   // : Forages offered in restricted amount

These feeding standard tables also include the daily recommended ALLOWANCES FOR MACROMINERALS (Ca, P, Mg, Na) (Tables 6a-9 and 11). The recommendations for TRACE MINERALS and VITAMINS are given alongside in another table (Table 14) where they are expressed in amount per kg dry matter intake (DMI) for the range suggested in the feeding standard tables.

Table 14. Optimum nutrient concentration of horses per kg DM referring to DM intake in the INRA daily recommended allowances tables (INRA, 1990).

	Adults [1]		Mares [2]		Growing horses		
	Rest light work	Medium work mating	End of pregnancy	Early lactation	6-12 months	18-24 months	32-36 months
Energy							
UFC	0.45-0.70	0.50-0.75	0.50-0.70	0.60-0.80	0.65-0.95	0.60-0.90	0.60-0.80
Nitrogen							
MADC (g)	30-50	40-55	45-70	55-90	55-90	40-55	30-40
Macrominerals							
Calcium (g)	3.1	2.9	4.6	4.3	5.5	3.8	3.3
Phosphorus (g)	1.9	1.6	3.2	3.7	3.0	2.2	1.9
Magnesium (g)	0.9	0.9	0.8	0.7	1.6	1.1	1.1
Sodium (g)	2.7	3.2	1.6	1.2	1.8	1.6	1.4
Potassium (g)	4.0	5.0	4.0	3.0	3.0	1.8	1.4
Sulfur (g)	1.5	-	1.5	-	1.5	-	-
Trace elements [3]							
Iron (mg)	80-100	80-100	80-100	80-100	80-100	80-100	80-100
Copper (mg)	10	10	10	10	10	10	10
Zinc (mg)	50	50	50	50	50	50	50
Manganese (mg)	40	40	40	40	40	40	40
Cobalt (mg)	0.1-0.3	0.1-0.3	0.1-0.3	0.1-0.3	0.1-0.2	0.1-0.2	0.1-0.2
Selenium (mg)	0.1-0.2	0.1-0.2	0.1-0.2	0.1-0.2	0.1-0.3	0.1-0.3	0.1-0.3
Iodine (mg)	0.1-0.3	0.1-0.3	0.1-0.3	0.1-0.3	0.1-0.2	0.1-0.2	0.1-0.2
Vitamins [4]							
Vitamin A (UI)	3250	3750	4200	3500	3450	3500	3500
Vitamin D (UI)	400	600	600	850	500	600	600
Vitamin E (UI)	8	10	11	7	7	10	10
Thiamine B1 (mg)	2.5	2.5	2.5	2.5	1.7	2.5	2.5
Riboflavin B2 (mg)	4.0	4.0	4.2	4.2	2.8	4.0	4.0
Niacin (mg) pp	12	12	12.5	12.5	8.5	12	12
Pantothenic acid (mg)	5.0	5.0	5.1	5.1	3.3	4.8	4.8
Pyridoxine B6 (mg)	1.3	1.0	1.3	1.2	0.8	1.2	1.2
Choline (mg)	65	60	63	84	42	60	60
Folic acids (mg)	1.3	1.2	1.3	1.2	0.8	1.2	1.2
Cyanocobalamine B12 (µg)	13	12	13	8	8	12	12

[1] The diet nutritive concentration of an intensively exercising horse should be increased.
[2] Mares during the last 3 months of pregnancy and the first 3 months of lactation.
[3] The allowances for traces elements are those suggested by Meyer (1982) in Germany.
[4] The recommended allowances for vitamins are those suggested by Wolter (1975) in France.
Remarks : The single concentrations given for macrominerals, some trace elements and vitamins were calculated for average DM intake.

Consequently, formulating rations for horses is quite easy with the modern INRA systems. NE and MADC values of feeds are additive as there is no forage and concentrate interaction on organic matter digestibility of the ration (Martin-Rosset and Dulphy, 1987) and very limited effect of level of feeding (Martin-Rosset et al., 1990). These systems allow an easy formulation of rations by hand with a graphical method (INRA, 1990) which is necessary to understand, to teach and to advise on the approach to ration formulation. But a ration calculation software package, The Chevalration, developed by teachers (Tavernier and Arslanian) and INRA workers both involved in the preparation of the handbook entitled 'Alimentation des chevaux' (INRA, 1990), is most accurate for formulating well balanced rations.

Comparison of recommended allowances expressed in DE and NE

The aim of any feeding system is to meet the requirement of the animals for type and level of production using available feeds. The nature, chemical composition and nutritive value of these feeds may vary with climatic and economic conditions. Ingestibility and palatability of feed, intake capacity of the animal and substitution rate between forages and concentrates must also be considered.

The differences between the NE and DE systems have been previously pointed out and differences between MADC and DCP systems have been discussed as well. Prediction of feed energy or nitrogen value in NE with regard to MADC and INRA systems has been described previously from a scientific point of view and primarily for the sake of application.

Energy and nitrogen requirements of different types of equines (mares, growing or exercising horses) as well as the energy and nitrogen value of feeds and rations may vary between feeding systems.

As far as the requirements suggested by NRC (1989) and INRA (1984-1990) are concerned and expressed in each system relative to maintenance as 100% bases, the NRC requirements seem to be lower than INRA requirements for energy or nitrogen (Table 15) when comparing the relative difference between the two systems.

As pointed out in the review of Hintz and Cymbaluk (1994), effect of environmental factors might be involved also, and added to the observed discrepancies which are acquainted to coefficients set up in the factorial method to perform the requirements of NRC primarily and of INRA partially. In the case of INRA recommendations, the procedure to implement allowances incorporates the impact of environmental factors because INRA allowances have been set up on feeding experiments designed and performed namely for this purpose or they are based on the requirements established from the factorial approach. Cuddeford (1997) compared NRC and INRA requirements. For energy, comparison takes place on requirements expressed in DE in both NRC and INRA, assuming conversion ratios from NE to DE drawn from INRA feed evaluation systems. For nitrogen, Cuddeford (1997) matched INRA requirement expressed in MADC to NRC requirement expressed in DCP, assuming digestion coefficient of CP drawn from NRC (1978). Conclusions are that NRC requirement would be slightly higher than INRA requirement for energy and for protein namely for growth, and for pregnancy to a lesser extent. But NRC energy requirement for pregnancy and lactation would be similar.

Table 15. Energy or nitrogen requirements (500 kg horse) related to maintenance in DE and CP system (NRC, 1989) and NE and MADC systems (INRA, 1984-1990). Relative difference between the systems.

SYSTEMS / ANIMAL	Energy			Nitrogen		
	NRC	INRA	% NRC/INRA	NRC	INRA	% NRC/INRA
MATURE HORSE						
- Maintenance[1]	**100**	**100**	**100**	**100**	**100**	**100**
- Mare						
• Pregnant						
8th –9th month	111	108	103	122	126	97
10th month	113	123	92	124	170	73
11th month	120	125	95	132	180	73
• Lactating						
1st month	173	232	75	218	352	62
2nd –3rd month	173	200	87	218	296	74
4th month weaning	148	161	92	160	244	66
GROWING HORSE						
• Yearling (12 months)						
Rapid growth	130	130	100	146	227	64
Moderate growth	115	107	107	130	169	76
• Two-year-olds (24 months) (not in training)						
Rapid growth	148	161	92	157	156	100
Moderate training	113	142	80	120	122	98
WORKING HORSE						
Very Light	125	128	98	125	137	91
Light	-	163	-	-	174	-
Moderate	150	187	80	150	200	75
Intense	200	172	116	200	182	110

[1]Stricto sensu maintenance requirement in adult horse whatever the systems

Base 100 | INRA = 9.3 Mcal NE, 270 g MADC

NRC = 16.4 Mcal DE, 656 g CP

Concerning the net energy requirements for work and maintenance, Harris (1997) pointed out that INRA requirements cannot be compared with requirements calculated through a partitioning system for the INRA system would assume the same efficiency of feed energy for work as for maintenance. The NE INRA system does not assume that efficiency of feed energy is the same for work and maintenance. The requirements (e.g. allowance in INRA system) have been determined with long-term feeding experiments. Factorial method was only implemented at the beginning of the research to state the scientific basis of the energy requirement variations according to the duration, intensity and ancillary factors.

The INRA system is based on the concept that the NE value of nutrients for both maintenance and work (physical activity) depends on the free energy (ATP) produced by their oxidative catabolism (Vermorel and Martin-Rosset, 1997). This concept was already suggested by Armsby (1917) in his famous handbook entitled 'The Nutrition of Farm Animals.' INRA has taken into account the new figure, that muscular contraction requires ATP as well as the other maintenance energy expenditures.

More than a century ago, German and French workers (Wolff et al., 1888-1895; Grandeau Alekan, 1904) stated that the maintenance DE requirements of horses were much higher with a hay-based diet than with a hay and concentrate diet (Figure 6). On this basis, a NE system for work was established by these workers (Table 16). It compares very well with the NE system for maintenance (Table 16) proposed by INRA. It has been extensively confirmed by other workers to this day.

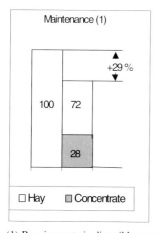
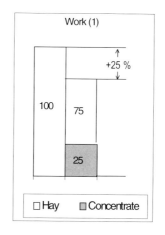

(1) Requirements in digestible energy

Figure 6 : Relative nutritive value of feeds at maintenance and work (adapted from Wolff et al., 1880-1895 and Grandeau et al., 1888-1904)

Table 16. Relative net energy value of feeds to barley for a horse at work and at maintenance.

	WORK Adapted from Wolff et al., 1888-1895	MAINTENANCE Adapted from INRA, 1984		
	Relative NE	Relative NE	NE (Mcal/kg DM)	UFC (/kg DM)
Maize	117	114	2.90	1.32
Barley	**100**	**100**	**2.55**	**1.16**
Oat	87	87	2.22	1.01
Horsebean	107	94	2.40	1.09
Pea	99	100	2.57	1.17
Lucerne hay	53	54	1.38	0.62
Grass hay	44	47	1.21	0.55

As a result, feeds can be substituted on their NE value for maintenance to cover the work energy expenditure. INRA allowances performed on such assumptions through long-term feeding experiments carried out for that purpose are relevant as pointed out by Harris (1997). This approach for determining energy work requirement through long-term experiments is still the most consistent. No country has available facilities to do otherwise even though INRA keeps in mind that energy requirements and allowances for intense work implemented in various types of high competition races, three-day events, and endurance rides require more refinement as pointed out by Southwood et al. (1993), Lawrence (1994), and Jones and Carlsson (1995).

As nutrient requirements as well as nutritive value of feeds and rations vary among feeding systems, it is essential to evaluate whether total energy and protein requirements of the different types of equines can be met by the same amount of feeds in balanced rations of the various feeding systems, namely NRC vs. INRA.

Frape (1997) compared applications of DE (NRC, 1989) and NE (INRA, 1990) systems to the formulation of simple balanced daily rations based on grass hay, barley grain and extracted soyabean meal for a horse 500 kg at maturity in regard to energy and protein and with feed intakes, kg DM/day provided by NRC (1978). Total dry matter intakes (TDMI) for each type of animal are stated to be the same in the two feeding systems for a type of animal. As a result, the hay to concentrate ratios vary to match energy and protein requirements in each system.

Dry matter intake for forage (DMIF) is lower with the NRC system for all types of equines, with the exception of the exercising horse where it is higher. The discrepancies range between - 9 to - 24% for breeding horses and + 9 to + 19% for exercising equines respectively. The inconsistency of differences between NRC and INRA may partially be due to differences in requirements between NRC and INRA. Descriptions of, and as a result requirements for, high and moderate work are different between the two systems (NRC, 1989 and INRA, 1990; Martin-Rosset et al., 1994). For example, light work in NRC (1989) recommendations corresponds to very light work in INRA 1990 recommendations.

As a result the appropriate procedure to compare the two systems that INRA proposes is as follows. Requirements for each type of equine are provided in each system according to performances level: DE, CP and DMI for NRC (1989) and NE, MADC; DMI for INRA (1990).

Good and poor (in some situations) quality hay-based diets were tested. Forages were supplemented with barley and extracted soybean meal in appropriate proportions to meet the energy and protein requirements of the different types of equines depending on their performance level. The characteristics of the four feeds used are given in Table 17.

Table 17. Chemical composition and nutritive value of feeds.

Feeds	DM g/kg	/kg DM							
		DE (Mcal)	NE (Mcal)	CF (g)	CP (g)	DCP (g)	MADC (g)	CP/D E	MADC/ NE
Good grass hay (n° 62)*	850	2.34	1.34	333	102	59	50	44	37
Poor grass hay (n° 70)*	850	1.69	0.92	382	76	37	31	45	34
Barley grain (n° 120)*	860	3.56	2.55	54	117	92	92	33	36
Soyabean meal extracted 48-50 (N° 130)*	883	4.20	2.40	39	545	496	496	130	207

* Tables INRA 1984

The energy value (DE or NE per kg DM) and nitrogen value (CP or MADC, per kg DM) of each diet and the amount of feed required to meet energy and nitrogen requirements were computed in each energy and nitrogen system. In a first approach, the hay to concentrate ratio was that proposed by NRC (1989). Then, in a second step, the hay to concentrate ratio tested was that proposed by INRA (1990).

For the lactating mare fed a hay and concentrate diet supplemented with the same concentrate percentage proposed by NRC, TDMI of rations are 10 to 13% higher with INRA rations than with NRC rations at 1st month of lactation, according to quality of hay. Thereafter differences come to be erratic (0.1 to 7.2%). For mares fed a hay and concentrate diet supplemented with the concentrate percentage proposed by INRA, TDMI of INRA rations are 14 to 18% higher than the NRC ration in early lactation, according to hay quality. The differences decrease as late lactation approaches. As a result DE provided by balanced rations calculated in the INRA system supplemented with concentrate percentage proposed by INRA are 5 to 19% higher than DE requirement (NRC) in the first and fourth months of lactation; NE provided by a NRC ration calculated in the NRC system supplemented by adequate concentrate percentage in the NRC approach are 3 to 10% lower than NE requirements (INRA) (Table 18).

Discrepancies are very difficult to interpret further, as there are strong interactions either with the relative NRC/INRA requirements (energy and protein), and the evolution of the requirement during lactation (Table 15), or with the hay and concentrate substitution rate on DM basis and to the respective protein/energy ratios of hay, barley and soybean (Table 17) in each system.

Table 18. Comparison of balanced rations calculated in either the NRC or INRA system balanced for energy and protein for the mare in the 1st month of lactation (BW = 500 kg).

		DM (kg)	C (%)	DE (Mcal)	CP (g)	NE (Mcal)
NRC	Hay (good)	5.00		11.70	510	6.70
	Barley	4.30		15.31	503	10.96
Requirements	Soyabean	0.70	50	3.94	382	1.68
28.3 Mcal DE		10.00		30.95	1395	19.34
1427 g CP				NE provided by NRC ration		= - 2.22 Mcal
2.0 < TDMI	< 3.0	(2.0 % BW)		NE requirements (INRA)		(- 10 %)
% BW						

		DM (kg)	C (%)	NE (Mcal)	MADC (g)	DE (Mcal)
INRA	Hay (good)	7.26		9.73	363	17.00
	Barley	4.27		10.88	392	15.20
Requirements	Soyabean	0.36	39	0.86	179	1.51
21.56 Mcal NE		11.89		21.47	934	33.71
950 g MADC				DE provided by INRA ration		= +5.41 Mcal
2.4 < TDMI	< 3.0	(2.4 % BW)		DE requirements (NRC)		(+ 19 %)
% BW						

For the pregnant mare fed a good hay and concentrate diet supplemented with the same concentrate percentage proposed by NRC, the TDMI differences between NRC and INRA rations are erratic, mainly in regard to early pregnancy. DE provided by INRA balanced rations are 5 to 7% lower (8th-9th month and 11th month, respectively) or 2% higher (10th month) than the NRC requirements. For mares fed only a good hay-based diet, TDMI of a INRA balanced ration are 4 to 11% higher than NRC ration from 8th-9th months to the 11th month. DE provided by this balanced good hay INRA ration calculated in the INRA system is 10% higher than DE NRC requirements in late pregnancy (10th-11th month) and the difference varies as the quality of hay decreases. But the differences become erratic as soon as rations are supplemented with concentrate percentage proposed by NRC.

For the yearling (12 months) fed good hay supplemented with the concentrate percentage proposed by NRC, TDMI differences between NRC and INRA balanced rations are erratic, ± 4 to 19% for rapid (0.750 kg/d) and moderate growth (0.450 kg/d). But DE provided by balanced good hay and concentrate ration calculated in the INRA system are 5 to 16% lower than the DE NRC requirements for moderate and rapid growth, respectively. Conversely, the NE provided by the NRC balanced diet calculated in the NRC system is 8 to 19% higher than INRA requirements for moderate and rapid growth (Table 19), respectively. And 60% concentrate for moderate growth suggested by NRC appears to be far too high in the INRA system; 40 % could be more appropriate.

For the two-year-old (24 months) not in training, fed good quality hay with 35% concentrate as proposed by NRC, TDMI differences are erratic, ± 6% between NRC according to the growth rate (+ 6% for INRA ration when 0.200 kg/d and - 6 % when 0.450 kg/d). But the DE provided by a balanced ration calculated in the INRA system are lower (- 4%) than DE according to NRC requirements, primarily for rapid growth. And NE provided by a balanced ration calculated in the NRC system are 5% higher for rapid growth and -5% for moderate growth.

Table 19. Comparison of rations calculated in each NRC and INRA system balanced for energy and protein for the growing horse.

Yearling 12 month - Rapid growth 0.750 kg/d - BW = 320 kg

		DM (kg)	C (%)	DE (Mcal)	CP (g)	NE (Mcal)
NRC		2.80		6.55	286	3.75
		3.54		12.6	414	9.03
Requirements		0.66	60	2.77	359	1.58
21.7 Mcal NE		7.00		21.92	1059	14.36
1083 g CP		(2.2 % BW)				
2.0 < TDMI % BW	< 3.0					

$$\frac{\text{NE provided by NRC ration}}{\text{NE requirements INRA}} = + 2.26 \text{ Mcal} \ (+ 19 \%)$$

		DM (kg)	C (%)	NE (Mcal)	MADC (g)	DE (Mcal)
INRA	Hay (good)	2.39		3.20	120	5.59
	Barley	3.12		7.96	287	11.10
Requirements	Soyabean	0.39	60	0.93	193	1.64
12.1 Mcal NE		5.90		12.82	600	18.33
590 g MADC		(1.8 % BW)				
1.7 <TDMI % BW	< 2.5					

$$\frac{\text{DE provided by INRA ration}}{\text{DE requirements NRC}} = - 3.37 \text{ Mcal} \ (- 16 \%)$$

As in the mare, interpretation of the discrepancies is confusing as to NRC/INRA requirements; NRC requirements are lower than INRA requirements (Table 15). In addition, there is a strong interaction with hay and concentrate substitution either on DM and/or energy. Hay to concentrate ratios based on TDMI % BW suggested by NRC 1989 are rather high and very close to the maximum feed intake capacity of the growing horse fed hay ad libitum supplemented with a very high percentage of concentrate.

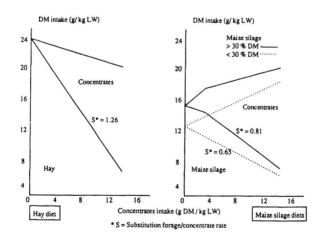

Figure 7. Effect of concentrate supplementation on forage and diet intake in young horses (12 months of age) (Agabriel et al., 1982 ; Martin-Rosset and Doreau, 1984b).

Forage-concentrate substitution on DM basis is well known in growing horses in the INRA system (Figure 7). Forage intake decreases in the yearling (12 months) when the amount of concentrate rises in the diet. The decreasing amount of forage DM intake, so-called substitution rate, is on average 1.26 with hay (NE = 1.14 to 1.34 Mcal NE/kg DM) but it reaches only 0.73 for maize silage (NE = 1.78 - 1.87 Mcal NE/kg DM). And the effect of substitution might rise with the age of the growing horse (Martin-Rosset and Doreau, 1984b). As a result, it becomes easier to understand NRC, assuming a high percentage TDMI/BW. For example, with a 2-3% BW for a yearling higher on average than INRA TDMI % (BW = 1.72-2.50%), the DE requirement can be easily reached with a high percentage of concentrate, even though amount of forage intake is restricted by substitution. Conversely, the amount of hay intake in the INRA ration is more restricted for the substitution ratio on an energy basis (and nitrogen basis as well) and should be much higher due to MADC/NE ratio of hay and soyabean meal. The figure could be discussed for the two-year-old as well. In the older, growing horse (2 or 3 years old), as requirements decrease, intake capacity increases and ingestibility of forage is not very limiting. As a result, the high proportion of concentrate suggested by NRC (35 %) is not appropriate in the INRA system.

For the working horse, direct comparisons are impossible as definitions of the work intensities are different between NRC and INRA. As a result, requirements are different (Table 15 and Figure 4). After attempting adjustments of INRA work intensities and requirements to NRC proposals, comparisons were attempted with good hay-based diets supplemented with the percentage of concentrate proposed by NRC. TDMI is higher with NRC rations than with INRA regardless of work intensity, but it was impossible to balance INRA rations for MADC requirements, as NRC percentages of concentrate are far too high to be consistent in the INRA system. However, DE provided by the INRA rations was 12 to 18% lower than DE NRC requirements.

The rations were recalculated in each system with a good hay diet supplemented with a lower percentage of concentrate and the addition of some wheat straw to fit the MADC requirements in the INRA ration. TDMI was 2 to 19% higher for INRA rations than for NRC rations.

Validity of the energy systems

Agreements or differences in feed allowances among systems, even with the same feeds and diets, do not allow conclusions to be drawn about the validity and accuracy of each system. Animals respond to inadequate rations through changes in production (milk, weight gain or performance), body weight and body composition.

In France, the validity of the NE (UFC) system was evaluated using two data files, feeding trials performed at INRA with different diets based on hay (Agabriel et al., 1982; Agabriel et al., 1984; Vermorel et al., 1984; Martin-Rosset and Doreau, 1984b; Bigot et al., 1987; Doreau et al., 1988a,b; Martin-Rosset et al., 1989; Doreau et al., 1991; Martin Rosset and Vermorel, 1991; Micol and Martin-Rosset, 1995), hay, maize silage, hay-straw and supplemented with concentrates containing cereals, soyabean meal, dehy-alfalfa and bran according to type of animal (mares, growing horses, exercising horses). Data concerning stallions were provided by bibliography (Axelsson, 1949; Jespersen, 1949) and survey in the stable of the French National Studs (n = 1600 stallions).

Calorimetry studies were carried out with horses at maintenance fed 12 different diets (Vermorel et al., 1996) and compared with results obtained in long-term feeding trials with horses fed the same diets at maintenance (Martin-Rosset and Vermorel, 1991) or in working horses (Vermorel et al., 1984; Martin-Rosset et al., 1989).

Tables of nutrient concentration in total diet using NE and MADC systems (Tables 20 -24)

Tables of daily nutrients (allowances) have been stated for horses of different weights in each breed: 400 to 600 kg for light breeds (namely sport and leisure horses) and 700 to 800 kg for heavy breeds. All types of horses are included (stallions, geldings, mares, growing horses and exercising horses) (Tables 20-24).

For horses at maintenance, effect of genetic type is included for the appropriate BW for sport horses (Tables 20 to 22) and for heavy horses (Tables 23-24). For bloodstock (400-500 kg), this calculation should be performed: maintenance energy requirement of gelding in stricto sensu maintenance + 10%; energy requirement for geldings at rest during the period of training is stricto sensu maintenance + 15%; for stallions out of breeding season stricto sensu maintenance + 20 %.

For mares, two levels of energy and protein allowances are provided in pregnancy and lactation according to the breed and the body condition score in each breed reached at the 8th month of pregnancy and at foaling.

For growing horses, allowances for young horses before weaning is not yet stated. But it could be determined in the very near future using a relationship between milk yield of the mare, determined with an original method (Doreau et al., 1986) and the average daily gain of the foal (Doreau et al., 1986), considering the effect of age at weaning (Warren et al., 1998) and the effect of concentrate supplementation of the foal on the growth (Martin-Rosset and Doreau, 1984; Breuer, 1998). After weaning, three categories of age are stated for sport-leisure horses and heavy breeds. Energy and protein allowances are provided for two growth rates in accordance with the growth curve and management stated for young light breed horses (Figure 5). Similar curves and management for young heavy horses have been performed (Micol and Martin-Rosset, 1995).

For working horses, descriptions of work intensities of light breeds are those described in Figure 4.

Lysine requirements are provided by NRC (1978 and 1989) and are 0.5% and 0.4% DM for yearlings and two-year-olds (and three-year-olds by extension). It is not stated for mature horses in Tables 20 and 24 as lysine requirements are not very well known. Ca, P, Mg, Na and K requirements are those drawn from NRC

(1978) and based on the work of Schryver and Hintz between the years of 1970 and 1978. Trace element and vitamin requirements are those drawn from the recommendations of Meyer (1982) and Wolter (1975), respectively. All these allowances are given in optimum concentration of TDMI (Table 14). TDMI proposed in these tables (20 to 24) are generally similar to or lower than those stated by NRC (1989), particularly for the growing horse. Diet proportions are given for each type of horse (Table 25). Ranges of concentrate and forage percentages are preferably provided as they take much more into account :

(1) concerning the feedstuffs - the types of forages and concentrates which are available to horse breeders and horsekeepers, the large variations of forage quality, possible use of some good straw in some situations;

(2) concerning the animal - the feed intake capacities of the different types of horses which have been measured in the interaction of body condition score according to INRA method (INRA - Institut du Cheval - InstitutElevage, 1997) based on slaughter experiments.

The concentrate percentages proposed by INRA are in most cases lower than those stated in NRC (1989) tables. The INRA proposals for energy, protein and feed intake capacity have been stated both on factorial and feeding experiments. Ingestibility of forages has been measured as well, and nutritive value of forages can be accurately predicted in the INRA system (Table 3 and appendix A, B, C).

Discussion - Conclusion

At the beginning of the 19th century, German and French workers carried out a lot of experiments which provided new promising scientific bases in equine nutrition. But after World War II scientific work declined tremendously in western Europe and to a lesser extent in eastern Europe and in the USA. As a result, horses were fed during this period using mostly the feed evaluation systems stated for ruminants which was inappropriate, as we know very well now.

In the 1960s, new research arose in the USA (H.F. Hintz was a pioneer in that field) and in the 1970s in Europe. The NRC of the USA published the first new feed evaluation systems for equines in the 1980s, the DE and DCP systems. These systems have been used to some extent in European countries. In 1989, NRC proposed a CP system instead of DCP system for evaluating nitrogen value of feeds.

In the 1980s and 1990s, INRA from France has proposed two new feed evaluation systems, the NE (UFC) and MADC systems. The difference in utilization efficiency of digestion end products involves large variations in the efficiency of DE or ME utilization and justifies expression of the energy value of feedstuffs in NE. The validity of the system was tested extensively.

In the MADC system, nitrogen value of feedstuffs in horses is linked to crude protein content (and NPN/N ratio), to digestibility of crude protein and respective proportion of crude protein digested either in the small intestine or in the large intestine. Total digestibility of CP has been extensively measured in equines for forages and to a lesser extent for concentrates. Proportion of CP digested in the small and large intestine is now much more well known thanks to experiments carried out with fistulated equines in the USA and Europe. Recent studies of Brazilian researchers support the coefficient of true N digestibilities stated in the MADC system in the small intestine. Amino acid requirements are mainly supplied by AA absorbed in the small intestine when horses are fed a high proportion of concentrates. At the same DCP content, concentrates provide 20 to 40% more AA than forages. The difference between forages and concentrates might be reduced as forages provide some essential AA and nitrogen sources, AA and/or ammonia to microbial population in the large intestine for synthesizing microbial protein. If the AA absorption in the large intestine should be considered very negligible, which might be the case, the MADC system would account for any amount of AA (and ammonia) which are used by microflora to satisfy its own nitrogen requirements in respect to amount of available energy (Glade, 1984; Reitnour, 1979-1980). Bacteria are known to be able to use AA (and NPN) for synthesizing their own protein which is then digested partially by microbial enzymes (Baruc et al., 1983). Net absorption of nitrogen is as high in the small colon as in the cecum (Glade, 1983). Amount of urea recycled is estimated (Prior et al., 1974) and efficiency of available urea or ammonia for microbial protein synthesis is estimated as well (Hintz and Schryver, 1972 ; Slade et al., 1973). Microbial protein accounts for 50 to 60% of total nitrogen in the feces (Meyer et al., 1963).

Energy and protein requirements and recommended allowances stated in the NE and MADC systems are consistent for the different types of equines as they have been extensively checked with long-term feeding experiments, but allowances for intense exercise in light horses need to be refined more accurately in relation to the training method.

These systems are now used in southern European countries where the INRA handbook has been translated into Italian and Spanish and in Northern European countries where the last meeting of the European Association of Animal Production (EAAP) was held in Iceland in 1993 and then in Norway in 1996. Eastern countries such as Poland and Romania are using them as well. In other countries the question is still open. For that purpose, the English version of the INRA handbook is in progress.

DAILY NUTRIENT ALLOWANCES

Key notes for using Tables 20 to 25

*	Body weight 24 h after foaling in good conditions
**	The lower values are used with high proportion of concentrates in the diet and the higher values are used with hay-based diet
(1)	Riding horses are mainly considered in the table (For bloodstock see paragraph 2.6)
(2)	For bloodstock, requirements have to be increased (see paragraph 2.6)
(3)	Includes 1 hour of daily light work for stallions managed in box or in paddock
(4)	This DM intake accounts for limited straw intake as well
(5)	Light breeds : - For mares whose offspring are used in competition with the exception of mares that are fat (BCS (4)) (INRA, 1990) - Other mares (BCS (2.5)) and mares wintered outdoors or mares 3 years old
(6)	Light breeds: other situations
(7)	Geldings and mares are covered by these allowances. For stallions add daily 0.88 Mcal and 30 g MADC
(8)	For 2 hours work (average daily work in riding school)
(9)	In case of working outside consider : - very light work = 1 hour riding out ; - light work = 2 hours riding out - for long hacking: light work = 2-4 hours riding out ; moderate work = riding out over 4 hours
(10)	For 1 hour of daily work
(11)	Not in training (for training see Figure 4)
(12)	Lysine: - 0.5 % DM 12 months - 0.4 % DM 24 months and 36 months
(13)	Heavy breeds: for lean mares (BCS < 2.5), mares managed outdoors
(14)	Heavy breeds : other situations
(15)	Hay cutting, wilting, etc.
(16)	Plowing soft land or harrowing
(17)	Plowing heavy land

Table 20: Daily nutrient allowances of horses 400 kg[1][2]

Animal	Weight (kg)	Daily gain (kg)	NE (Mcal)		MADC (g)	Lysine (g)	Ca (g)	P (g)	Mg (g)	Na (g)	K (g)	Vit. A (10^3IU)	DM** (kg)
Mature horses													
- MAINTENANCE[2]	400		7.89		230		20	12	5	16	27	22	6.0-7.5
- STALLIONS	400												
• Out of breeding season[3]			10.35		305		24	12	8	22	32	26	7.0-9.0[4]
• Breeding season = service													
Light			13.86		460		28	15	8	25	39	32	8.5-10.0[4]
Moderate			15.62		530		32	15	8	34	53	39	9.5-11.5[4]
Intense			16.94		600		34	15	8	36	56	42	10.5-12.0[4]
- MARES			LL[6]	HL[5]									
• Pregnant	400*												
Before 8th month (or dry)			7.04	8.80	255		21	11	5	10	25	26	5.0-7.5
8th-9th months			7.70	9.46	290		25	14	5	10	27	28	5.5-8.0
10th month			8.80	10.34	390		32	23	5	10	30	32	6.0-9.0
11th month			8.90	10.56	415		33	22	5	10	32	34	6.5-9.5
• Lactating													
1st month			16.50	19.58	780		57	49	8	14	35	40	10.0-13.0
2nd-3rd months			14.08	16.72	680		46	36	7	14	32	37	8.0-13.0
4th month to weaning			11.22	13.86	550		38	30	7	14	26	30	6.5-10.5
- WORKING HORSES[7]	400												
• Rest			8.26		240		20	12	5	16	27	22	6.0-7.5
• Very light[8][9]			10.56		330		22	12	6	20	32	26	7.5-8.5
• Light[8][9]			13.86		430		24	13	6	35	38	31	8.5-10.5
• Moderate[8][9]			16.06		490		29	15	8	45	55	41	9.5-12.5
• Intense[10]			14.52		450		29	15	8	38	50	38	9.0-11.0
Growing horses													
• Yearling : 12 months[12]													
- rapid growth	280	0.6-0.7	10.34		530	30	33	18	8	10	18	21	5.0-7.0
- moderate growth	260	0.3-0.4	8.58		360	28	31	12	8	8	17	19	4.5-6.5
• Two-year-old: 24 months[11][12]													
- rapid growth	390	0.3-0.4	12.76		340	29	32	18	8	11	13	25	6.0-8.5
- moderate growth	360	0.1-0.15	11.00		270	27	30	12	8	8	12	24	5.5-8.0
• Three-year-old: 36 months[11][12]													
- Rapid growth	400	0.1-0.15	11.66		270	31	26	12	8	10	11	27	6.5-9.0
- Moderate growth	380	0.0-0.1	11.44		200	29	24	13	8	11	10	25	6.0-8.5

Table 21: Daily nutrient allowances of horses 500 kg[(1)(2)]

Animal	Weight (kg)	Daily gain (kg)	NE (Mcal)		MADC (g)	Lysine (g)	Ca (g)	P (g)	Mg (g)	Na (g)	K (g)	Vit. A (10⁷IU)	DM** (kg)
Mature horses													
- MAINTENANCE[(2)]	500		9.29		270		25	15	7	18	31	25	7.0-8.5
- STALLIONS	500												
• Out of breeding season[(3)]			11.85		360		28	15	10	24	36	29	8.0-10.0[(4)]
• Breeding season = service													
Light			14.52		480		30	18	10	28	41	33	9.5-11.0[(4)]
Moderate			16.45		550		32	18	10	31	41	33	10.5-12.5[(4)]
Intense			17.60		620		34	18	10	39	61	46	11.5-13.0[(4)]
- MARES			LL[(5)]	HL[(6)]									
• Pregnant	500*												
Before 8th month (or dry)			8.36	10.12	295		25	15	7	12	29	30	6.0-8.5
8th-9th months			9.02	11.0	340		29	18	7	12	31	33	6.5-9.0
10th month			10.34	12.54	460		38	26	7	12	35	37	7.0-10.5
11th month			10.56	12.76	485		39	28	7	12	37	39	7.5-11.0
• Lactating	500*												
1st month			19.58	23.54	950		61	55	10	15	41	47	12.0-15.0
2nd-3rd months			16.72	20.24	800		47	40	9	14	38	44	10.0-15.0
4th month to weaning			13.42	16.50	660		39	32	8	13	31	36	8.0-12.5
- WORKING HORSES[(7)]	500												
• Rest			9.73		290		25	15	7	18	31	25	7.0-8.5
• Very light[(8)(9)]			11.88		370		28	16	8	22	36	29	8.5-9.5
• Light[(8)(9)]			15.18		470		30	18	9	37	53	39	9.5-11.5
• Moderate[(8)(9)]			17.38		540		35	19	10	47	60	45	10.5-13.5
• Intense[(10)]			15.95		490		35	29	10	40	55	41	10.0-12.0
Growing horses													
• Yearling : 12 months[(12)]													
- rapid growth	320	0.7-0.8	12.10		590	33	39	22	10	12	20	23	5.5-8.0
- moderate growth	280	0.4-0.5	9.90		440	31	28	16	9	9	19	22	5.0-7.5
• Two-year-old: 24 months[(11)(12)]													
- rapid growth	470	0.4-0.5	14.96		420	35	36	20	10	13	16	31	7.5-10.0
- moderate growth	440	0.15-0.25	13.20		330	33	28	16	9	12	15	29	7.0-9.5
• Three-year-old: 36 months[(11)(12)]													
- Rapid growth	490	015-0.25	14.3		330	38	30	18	10	12	13	33	8.0-11.0
- Moderate growth	470	0-0.10	13.2		260	35	25	15	8	12	12	31	7.5-10.0

Table 22: Daily nutrient allowances of horses 600 kg[(1)(2)]

Animal	Weight (kg)	Daily gain (kg)	NE (Mcal)	MADC (g)	Lysine (g)	Ca (g)	P (g)	Mg (g)	Na (g)	K (g)	Vit. A (10³IU)	DM[(*)] (kg)
Mature horses												
- MAINTENANCE	600		10.69	310		30	18	8	24	35	28	8.0-9.5
- STALLIONS	600											
• Out of breeding season[(3)]			13.42	420		30	18	9	27	40	33	9.0-11.0[(4)]
• Breeding season = service												
Light			15.18	500		36	21	10	30	45	37	10.5-12.0[(4)]
Moderate			17.25	570		36	21	11	34	63	47	11.5-13.5[(4)]
Intense			18.80	640		36	21	12	42	66	50	12.5-14.0[(4)]
- MARES	600*	LL[(6)] HL[(5)]										
• Pregnant												
Before 8th month (or dry)			9.46 11.66	340		30	18	7	15	32	34	6.5-9.5
8th-9th months			10.34 12.54	395		35	22	8	15	35	37	7.0-10.5
10th month			11.88 14.52	535		46	32	8	15	41	43	8.5-12.0
11th month			12.10 14.74	565		47	34	8	15	41	43	8.5-12.0
• Lactating												
1st month			23.10 27.72	1125		73	67	11	18	47	55	14.0-17.5
2nd-3rd months			19.58 23.54	960		57	48	10	17	44	52	12.0-17.5
4th month to weaning			15.62 19.14	780		47	38	9	16	36	42	9.5-14.5
- WORKING HORSES[(7)]	600											
• Rest			11.20	330		30	18	8	24	35	28	8.0-9.5
• Very light[(8)(9)]			13.20	415		33	19	9	25	40	33	9.5-10.5
• Light[(8)(9)]			16.50	510		33	21	10	40	46	37	10.5-12.5
• Moderate[(8)(9)]			18.70	580		38	22	11	50	65	49	11.5-14.5
• Intense[(10)]			17.16	530		42	22	11	43	60	45	11.0-13.0
Growing horses												
• Yearling : 12 months[(12)]												
- rapid growth	360	0.8-0.9	13.64	660	39	44	24	12	13	23	27	6.5-9.0
- moderate growth	320	0.5-0.6	11.44	510	35	33	18	11	11	21	24	6.0-8.0
• Two-year-old: 24 months[(11)(12)]												
- rapid growth	530	0.5-0.6	17.16	480	41	42	24	10	15	19	36	8.5-12.0
- moderate growth	500	0.25-0.35	14.96	390	38	33	19	10	13	17	33	8.0-11.0
• Three-year-old: 36 months[(11)(12)]												
- Rapid growth	580	025-0.35	16.72	390	44	37	22	10	15	16	39	9.0-13.0
- Moderate growth	550	0.-0.10	14.96	320	41	29	17	11	14	14	36	8.5-12.0

Table 23: Daily nutrient allowances of horses 700 kg[(1)(2)]

Animal	Weight (kg)	Daily gain (kg)	NE (Mcal)	MADC (g)	Lysine (g)	Ca (g)	P (g)	Mg (g)	Na (g)	K (g)	Vit. A (10³IU)	DM[**] (kg)
Mature horses												
MAINTENANCE	700		11.37	340		30	18	9	26	38	31	9.0-10.0
STALLIONS	700											
• Out of breeding season[(3)]			13.42	430		35	21	12	30	49	35	10.0-12.0[(4)]
• Breeding season = service												
Light			14.76	480		37	24	12	32	48	39	11.0-13.0[(4)]
Moderate			16.43	540		38	24	12	42	66	50	12.0-14.5[(4)]
Intense			17.50	590		42	24	12	45	70	53	13.0-15.0[(4)]
MARES			LL[(14)] HL[(13)]									
• Pregnant	700*											
Before 8th month (or dry)			9.24 11.44	380		35	21	7	17	32	34	6.5-9.5
8th-9th months			9.90 12.32	445		41	26	8	17	36	38	7.0-11.0
10th month			11.44 14.30	570		53	37	9	17	40	42	8.0-12.0
11th month			11.66 14.74	600		55	40	9	17	40	42	8.0-12.0
• Lactating												
1st month			23.54 28.6	1295		85	78	13	21	54	63	14.5-21.5
2nd-3rd months			19.80 24.2	1045		66	56	11	20	45	53	12.0-18.5
4th month to weaning			15.62 19.58	860		55	45	10	19	38	44	9.5-15.5
WORKING HORSES	700											
• Rest			11.95	350		30	21	9	7	38	31	9.0-10.0
• Light (6 h/d) [(15)]			16.06	510		38	25	10	35	49	40	11.5-13.0
• Moderate (5 h/d) [(16)]			17.60	550		42	26	13	46	69	52	12.5-15.0
• Intense (4 h/d) [(17)]			20.90	645		49	26	13	57	74	55	13.5-16.0
Growing horses												
• Yearling : 12 months[(12)]												
- rapid growth	450	0.9-1.0	15.18	750	43	57	22	13	16	26	29	7.5-9.5
- moderate growth	410	0.6-0.7	12.32	590	39	40	23	11	14	24	27	7.0-8.5
• Two-year-old: 24 months[(11)(12)]												
- rapid growth	600	0.5-0.6	15.40	570	44	45	26	12	17	20	39	10.0-12.0
- moderate growth	560	0.2-0.3	13.20	440	42	36	21	11	15	19	37	10.0-11.0
• Three-year-old: 36 months[(11)(12)]												
- moderate growth	640	0-0.1	13.42	380	46	36	21	12	16	16	40	11.0-12.0

Table 24: Daily nutrient allowances of horses 800 kg[1][2]

Animal	Weight (kg)	Daily gain (kg)	NE (Mcal)	MADC (g)	Lysine (g)	Ca (g)	P (g)	Mg (g)	Na (g)	K (g)	Vit. A (10³IU)	DM** (kg)
Mature horses												
- MAINTENANCE[2]	800		12.54	3.70		33	20	10	28	42	34	10.0-11.0
- STALLIONS	800											
• Out of breeding season[3]			14.74	440		37	23	14	32	48	39	11.0-13.0[4]
• Breeding season = service												
Light			15.99	490		40	25	14	35	52	42	12.0-14.0[4]
Moderate			17.67	555		41	23	14	46	71	53	13.0-15.5[4]
Intense			18.75	600		44	24	14	48	75	56	14.0-16.0[4]
- MARES			LL[14] HL[13]									
• Pregnant	800											
Before 8th month (or dry)			10.12 12.54	420		40	24	7	15	35	28	7.0-10.5
8th-9th months			11.00 13.64	495		46	30	8	17	39	41	8.0-11.5
10th month			12.76 15.84	630		61	42	10	20	44	46	9.0-13.0
11th month			12.98 16.28	660		62	46	10	20	44	46	9.0-13.0
• Lactating												
1st month			26.18 32.12	1470		98	89	14	24	61	71	16.0-24.5
2nd-3rd months			22.22 27.06	1180		76	64	13	23	51	60	13.5-20.5
4th month to weaning			17.38 21.78	930		62	51	11	22	41	47	10.5-16.5
- WORKING HORSES	800											
• Rest			13.17	390		33	20	10	28	42	34	10.0-11.0
• Light (6 h/d)[15]			17.16	510		38	25	11	38	53	43	12.5-14.0
• Moderate (5 h/d)[16]			18.70	550		43	28	14	45	74	55	13.5-16.0
• Intense (4 h/d)[17]			22.00	650		54	30	14	52	79	59	14.5-17.0
Growing horses												
• Yearling : 12 months[12]												
- rapid growth	500	1.0-1.1	17.16	820	48	57	32	14	18	29	33	8.5-10.5
- moderate growth	460	0.7-0.8	14.30	660	43	46	26	13	15	26	29	8.0-9.0
• Two-year-old: 24 months[11][12]												
- rapid growth	680	0.60-0.65	16.94	600	48	51	29	12	19	22	42	11.0-13.0
- moderate growth	640	0.30-0.40	14.74	490	46	42	24	12	17	21	40	11.0-12.0
• Three-year-old: 36 months[11][12]												
- moderate growth	730	0.00-0.10	14.96	410	50	37	24	13	16	18	44	12.0-13.0

Table 25: Nutrient concentration in total diets for horses (dry matter basis)

	NE (Mcal/kg)	MADC (%)	% MADC/NE	DM % BW	Diet proportions Concentrate %	Forage[1] %		Comments
Mature horse								
MAINTENANCE	1.20	3.5	29	1.4 – 1.9	0	100		Hays M to G//
STALLIONS	1.28 – 1.36	3.9 – 4.5	32	1.4 + 3.0	5 – 15 / 30 - 40	85 – 95 / 60 – 70	Out of br .season service	Hays M to VG // + Straw// / Hays M to VG // + Straw //
MARES								
• pregnant								
8 – 9th months	1.14 - 1.40	4.6	36	1.0 – 2.0				
10th month	1.16 – 1.41	5.4	42	1.1 – 2.3				Forages G↑ to VG//
11th month	1.16 – 1.42	5.6	43	1.1 – 2.4	5 - 50	50 - 95		Forages M↑ to VG//
• Lactating								
1st month	1.39 – 1.68	7.1	46	2.0 – 3.3				
2nd - 3rd months	1.32 – 1.60	6.6	45	1.7 – 3.3				
4th month to weaning	1.29 – 1.59	6.6	46	1.3 – 2.6				
WORKING HORSE								
Rest	1.25	3.7	31	1.3 – 1.9			Light breed	Hay G to VG// + straw//
Light work	1.37	4.3	31	1.6 – 2.6	5 – 65	35 – 95	Heavy breed	Hay M to G// + straw//
Moderate work	1.38	4.3	31	2.0 – 3.1	5 – 45	55 – 95		
Intense work	1.43	4.3	31	1.8 – 2.8				
Growing horse								
• Yearling : 12 months								
Rapid growth	1.75*-1.80**	8.7	49	1.7 – 2.5	35 – 60 / 40 - 60	40 – 65 / 40 - 60	Light breed / Heavy breed	Forages G↑ to VG// or E// / Forages G↑ to VG//
Moderate growth	1.59*-1.64**	7.2	45	1.7 – 2.5	10 – 35 / 10 - 15	65 – 90 / 85 - 90	Light breed / Heavy breed	Forages M↑ to G↑ / Straw↑ + Hay G//
• Two-year-old: 24 months								
Rapid growth	1.41**-1.76*	4.9	32	1.6 – 2.3	20 – 25 / 40 – 60	75 – 80 / 40 -60	Light breed / Heavy breed	Forages G to VG↑ or E// / Forages G↑ to VG//
Moderate growth	1.26**-1.62*	4.1	28	1.6 – 2.2	15 – 20 / 10 - 15	80 – 85 / 85 - 90	Light breed / Heavy breed	Forages M↑ or G// / Straw↑ + Hay G//
• Three-year-old: 36 months								
Rapid growth	1.51*	3.5	23	1.5 – 2.2	15 – 20	80 – 85	Light or	Forages G//
Moderate growth	1.20**-1.58*	3.1	22	1.6 – 2.2	10 – 15	85 - 90	Heavy breed	Straw↑ + Hay G//

(1) straw is included in the hay percentages for stallions   *   Light breed
(2) Straw is distinguished to indicate its specific mode of distribution   **   Heavy breed
M   Medium quality
G   Good quality
VG   Very good quality
E   Excellent
↑   Forages offered ad libitum
//   Forages restricted

APPENDIX

Routine laboratory methods for predicting the organic matter digestibility (and net energy) of FORAGES in the horse

A. from chemical composition
B. using enzymatic method
C. using near infrared spectrophotometry (NIRS)

A. from chemical composition

Prediction of organic matter digestibility (OMD) of forages in horses from chemical composition.

W Martin-Rosset, J Andrieu, M Jestin

INRA, Department of Animal Husbandry, Research Center of Clermont-Femand/Theix, 63122Saint-Genes-Champanelle (France).

INTRODUCTION

OMD coefficients of forages measured in vivo in horses are known to be the reference method, but they cannot be measured for all forages in view of predicting routinely their NE value in the UFC system. As a result, OMD must be predicted from chemical composition.

MATERIAL & METHODS

Botanical and crop characteristics of the 52 forages studied

types	species	number	1st cycle	regrowth
green forages	grassland	7	3	4
hays	grassland graminea	39	24	15
hays	legumes (lucerne)	6	0	6

Range of chemical composition and in vivo digestibility of the 52 forages

Ash (g / kg DM)	63-152
CP (g / kg DM)	55-214
CF (g / kg DM)	235-424
*CC (g / kg DM)	7-136
NDF (g / kg DM)	477-737
ADF (g / kg DM)	255-452
ADL (g / kg DM)	27-97
OMD (%)	41-66

* CC = water soluble carbohydrate

In vivo measurement of OMD

- 5-6 gelding horses 500 kg BW
- level of feeding : 1.2 times the maintenance requirement
- 2 weeks adaptation
- 6 days total feces collection

RESULTS

The OMD coefficient of all the forages can be predicted with only one relationship :

$$\text{OMD (\%)} = 67.78 + 0.07088\, CP - 0.000045\, NDF^2 - 0.12180\, ADL$$
$$RSD = \pm 2.5 \quad R^2 = 0.878 \quad N = 52$$

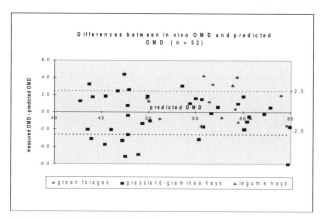

Differences between in vivo OMD and predicted OMD (n = 52)

♦ green forages   ■ grassland-graminea hays   ▲ legume hays

The average differences between OMD predicted with this model and OMD measured in vivo are :
- 1.9 for green forages
  0.0 for grassland and graminea hays
  +0.06 for legume

CONCLUSIONS

The prediction equation of forages OMD based on NDF, ADL and CP is much more accurate (RSD = ± 2.5) than the former INRA model 1984 using only CF as RSD ranged from 3.7 to 6.0 depending on the type of forage.

B. using enzymatic method

Prediction of organic matter digestibility (OMD) of forages in horses by pepsine cellulase method.

W Martin-Rosset, J Andrieu, M Jestin

INRA, Department of Animal Husbandry, Research Center of Clermont-Femand/Theix, 63122Saint-Genes-Champanelle (France).

INTRODUCTION

Chemical composition remains a medium quality estimator of OMD for horses especially when complex forages (natural grassland) are concerned. Enzymatic method using pepsine cellulase might be a suitable method.

MATERIAL & METHODS

Botanical and crop characteristics of the 52 forages studied

types	species	number	1st cycle	regrowth
green forages	grassland	7	3	4
hays	grassland graminea	39	24	15
hays	legumes (lucerne)	6	0	6

Range of chemical composition and in vivo digestibility of the 52 forages

Ash (g / kg DM)	63-152
CP (g / kg DM)	55-214
CF (g / kg DM)	235-424
*CC (g / kg DM)	7-136
NDF (g / kg DM)	477-737
ADF (g / kg DM)	255-452
ADL (g / kg DM)	27-97
OMD (%)	41-66

* CC = water soluble carbohydrate

In vivo measurement of dOM

- 5-6 gelding horses 500 kg BW
- level of feeding : 1.2 times the maintenance requirement
- 2 weeks adaptation
- 6 days total feces collection

Method of pepsine cellulase degradability of dry matter (CDMD) or of organic matter (COMD)

1) pre-treatment with pepsine in hydrochloric acid (0.2 % pepsin in 0.1 N HCL) in a water bath at 40°C for 24 h
2) starch hydrolysis in a water bath in the same mixture for exactly 30 min at 80°C
3) attack by cellulase (cellulase ONOZUKA R10) for 24 h in a water bath at 40° C, then filtration and rinsing.

RESULTS

OMD (%) can be predicted for all forages from CDMD (%) with only one curvilinear relationship set up for the grassland and graminea hays, provided a correcting effect (di) due to the type of forage :

$$OMD = -29.38 + di + 2.3032 \; CDMD - 0.01384 \; CDMD^2$$

$n = 52 \quad RSD = \pm 1.90 \quad R^2 = 0.927$

di = + 4.12 for green forages
di = 0 for grassland - graminea hays
di = - 2.61 for legumes hays

For green forages and legumes hays, OMD prediction must be restricted to the CDMD range between 51 to 69 %.

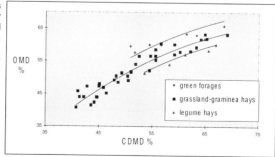

CONCLUSION

OMD of forages can be predicted routinely using pepsine cellulase method with a better accuracy (RSD = 1.90) than from chemical composition (RSD = 2.50) especially for complex forages.

C. using near infrared spectrophotometry (NIRS)

Prediction of organic matter digestibility (OMD) of forages in horses near infrared spectrophotometry (NIRS).

W Martin-Rosset, J Andrieu, M Jestin

INRA, Department of Animal Husbandry, Research Center of Clermont-Femand/Theix, 63122Saint-Genes-Champanelle (France).

INTRODUCTION

Routine laboratories need to use fast and accurate methods for predicting OMD and UFC value of forages. NIRS method might be the appropriate method.

MATERIAL & METHODS

Botanical and crop characteristics of the 52 forages studied

types	species	number	1st cycle	regrowth
green forages	grassland	7	3	4
hays	grassland graminea	39	24	15
hays	legumes (lucerne)	6	0	6

Range of chemical composition and in vivo digestibility of the 52 forages

Ash (g / kg DM)	63-152
CP (g / kg DM)	55-214
CF (g / kg DM)	235-424
*CC (g / kg DM)	7-136
NDF (g / kg DM)	477-737
ADF (g / kg DM)	255-452
ADL (g / kg DM)	27-97
OMD (%)	41-66

* CC = water soluble carbohydrate

In vivo measurement of OMD

- 5-6 gelding horses 500 kg BW
- level of feeding : 1.2 times the maintenance requirement
- 2 weeks adaptation
- 6 days total feces collection

NIRS method :
- forages were ground through a 0.8 mm screen
- dried again in an oven (40°C - one night)
- mononochromatic spectrophotometer (type 6500 Nirsystems)
- two reflectance spectra for each sample between 1100 to 2500 nm wavelengths
- calibration
- prediction

RESULTS

Accuracy of the calibration performed to predict OMD is very high

		52 forages		45 hays		39 hays (1)	
Results	Model (3)	R	SEC	R	SEC	R	SEC
Absorbance (derived functions)	stepwise	0.965	1.82	0.959	1.89	0.964	1.80
Absorbance (derived functions)	mpls	0.950	2.18 (2)	0.949	2.10 (2)	0.954	2.02 (2)

(1) grassland - graminea hays   (2) standard error of cross validation
(3) stepwise = method of multiple linear regression; MPLS = modified partial least squares regression

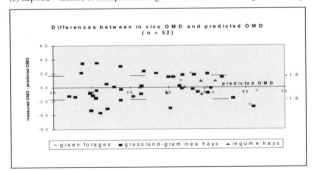

OMD can be predicted with a single calibration as far as the green forages and lucerne hays are included in the data bank to perform the regression for all the forages (n=52).

CONCLUSION Accuracy of OMD prediction with NIRS method is as high as OMD prediction with enzymatic method, RSD = 1.80 and 1.90 respectively and NIRS is a much faster method.

REFERENCES

Agabriel, J., Trillaud-Geyl, C., Martin-Rosset, W., Jussiaux, M., 1982. Utilisation de l'ensilage de maïs par le poulain de boucherie. Bull. Techn. CRZV Theix, INRA, 49, 5-13.

Agabriel, J., Martin-Rosset, W. and Robelin, J., 1984. Croissance et besoins du poulain. Chapitre 22. In : R. Jarrige, W. Martin-Rosset Editor « Le cheval » INRA Publications, route de St Cyr, 78000 Versailles. 370-384.

Almeida, F.Q., Valadares Filho, S.C., Donzele, J.L., Coelho Da Silva, J.F., Queiroz A.C., Leado M.I., Cecon P.R., 1999. Prececal digestibility of amino acids in diets for horses. In Proceedings 16th Equine Nutrition Physiology Symposium Raleigh, North Carolina USA. June 2-5, p. 274-277.

Anderson, C.E., Potter, G.D., Kreider, J.L. and Courtney, C.C., 1983. Digestible energy requirements for exercising horses. J. Anim. Sci.,   41:568-571.

Andrieu, J. and Martin-Rosset, W., 1995. Chemical, biological and physical (NIRS) methods for predicting organic matter digestibility of forages in horses. Proceedings 14th Equine Nutrition and Physiology Symp., Ontario, CA, USA, pp. 76-77.

Andrieu, J., Jestin, M. and Martin-Rosset, W., 1996. Prediction of the organic matter digestibility (OMD) of forages in horses by near infra-red spectrophotometry (NIRS). In: Proceedings of the 47th European Association of Animal Production Meeting. Lillehammer, Norway, August, 26-29, Abstract H 4.5, p. 299. Wageningen Pers Ed. Wageningen. The Netherlands.

Armsby, H.P., 1917. The nutrition of farm animals. Ed. McMillan Co, New York. 743pp.

Austbo, D. (1996). Energy and protein evaluation systems and nutrient recommendations for horses in the Nordic countries. In : Proceedings of the 47th European Associations of Animal Production Meeting. Lillehammer, Norway, August, 26-29, Abstract H4.4, p. 293. Wageningen Pers Ed. Wageningen, The Netherlands.

Axelsson, J., 1949. Standard for nutritional requirement of domestic animals in the Scandinavian Countries. In : Ve Congrès Int . de Zootechnie, Paris, Vol. 2, Rapports particuliers, pp. 123-144.

Banach, M.A., Evans, J.W., 1981. Effects of inadequate energy during gestation and lactation on the estrous cycle and conception rates of mares and on their foal weights. Proceedings 7th Equine Nutrition Physiology Symposium, 97-100.

Baruc, C.J., Dawson, K.A., Baker, J.P., 1983. The characterization and nitrogen metabolism of equine cecal bacteria. In : Proceeding 8th Equine Nutrition Physiology Symposium, Lexington, Kentucky, USA, April 28-30, p. 151-156.

Bigot, G., Trillaud-Geyl, C., Jussiaux, M. and Martin-Rosset, W., 1987. Elevage du cheval de selle du sevrage au débourrage : Alimentation hivernale, croissance et développement. Bul. Tech. Theix INRA. 69:45-53.

Bochröder, B., Schubert, R., Bodecker, D., 1994. Studies on the transport in vitro of lysine, histidine, arginine and ammonia across the mucosa of the equine colon. Equ. Vet. Journ., 26 (2), 131-133.

Borton, A., Anderson, D.L., Lyford, S., 1973. Studies of protein quality and quantity in the early weaned foal . In : Proceedings. 3rd Equine Nutrition Physiology Symposium Gainesville Florida USA, p. 19-22.

Breuer, L.H., 1968. Energy nutrition of the light horse. In : Proceedings 1st Equine Nutrition Physiology Symposium, Lexington, Kentucky, USA. pp. 8-9.

Brody, S., 1945. Bioenergetics and growth. Hafner Pub. Co. New-York, 1023 pp.

Broster, W.C., 1972. Protein requirements of mares for lactation. In: Handbuch der Tierernährung (Ed. W. Lenkeit, K. Breirem et, E. Crasemann), 2, 323-329. Paul Parey. Hamburg et Berlin.

Chenost, M. and Martin-Rosset, W., 1985. Comparaison entre espèces (mouton, cheval, bovin) de la digestibilité et des quantités ingérées des fourrages verts. Ann. Zootech., 34 :291-312.

Cordelet, C., 1990. Contribution à l'étude de l'alimentation azotée du cheval : utilisation de la fraction azotée dans le gros intestin. Mémoire Ingénieur CNAM, Dijon, 94 pp.

Cuddeford, D., 1997. Feeding systems for horses. Chap. 11. In : Feeding systems and feed evaluation models. Theodorou, M.K., France J., Ed. Cabi Publishing, Oxon, U.K., NY, USA, p. 239-274.

Cymbaluk, N., 1990. Comparison of forage digestion by cattle and horses. Can. J. Anim. Sci., 70, 601-610.

Den Engelsen, cited by Meyer and Alhswede 1976.

Doreau, M., Boulot, S., Martin-Rosset, W., Dubroeucq, H., 1986. Milking lactating mares using oxytocin : milk volume and composition. Reprod. Nutr. Develop. 26, (1A) 1-11.

Doreau, M. Boulot, S., Martin-Rosset, W., Robelin, J., 1986. Relation between nutrient intake, growth and body composition of nursing foal. Reprod. Nutr. Develop., 26, (2B) 683-690.

Doreau, M., Martin-Rosset, W. and Boulot S., 1988. Energy requirements and the feeding of mares during lactation : a review. Livest. Prod. Sci., 20 : 53-68.

Doreau, M., Bruhat, J.P. and Martin-Rosset, W., 1988. Effets du niveau des apports azotés chez la jument en début de lactation. Ann. Zootech., 37 :21-30.

Doreau, M. and Boulot, S., 1989. Recent knowledge on mare milk production. A review. Livest. Prod. Sci., 22 :213-235.

Doreau, M., Moretti, C. and Martin-Rosset, W., 1990. Effect of quality of hay given to mares around foaling on their voluntary intake and foal growth. Ann. Zootech., 39 :125-131.

Doreau, M., Boulot, S., Martin-Rosset, W., 1991. Effect of parity and physiological state on intake milk production and blood parameters in lactating mares differing in body size. Anim. Prod. 33, 111-118. P. 90-102.

Dulphy, J.P., Martin-Rosset, W., Dubroeucq, H., Jailler, M., 1997b. Evaluation of voluntary intake of forage trough fed to light horses. Comparison with sheep. Factors of variation and prediction. Livest. Prod. Sci., 52, 97-104.

Dusek, J., 1966. Notes sur le développement prénatal des chevaux (in Czech language). Ved. Pr. Vysk. San. Chov. Keni., Slatinany, 2 :1-25.

Frape, D., 1998. Equine Nutrition and Feeding. 2nd Ed. Blackwell. Science Ltd, London, pp.564.

Freeman, D.E. and Donawick, W.J., 1991. In vitro transport of cycloleucine by equine cecal mucosa. Am. J. Vet. Res., 52, 539-542.

Freeman, D.E., Kleinzeller, A., Donawick, W.J. and Topkis, V.A., 1989. In vitro transport of L-alanine by equine cecal mucosa. Am. J. Vet. Res,. 50, 2138-2144.

Futtermittel Tabellen, D.K.G. für Pferde, 1984. Ed. DLG Verlag. D6000 Frankfurt and Main.

Geay, Y., Robelin, J., Beranger, C., Micol, D., Gueguen, L. and Malterre, C., 1978. Bovins en croissance et à l'engrais. Chapter 11, 297-344. In : INRA, 1978 R. Jarrige Editor INRA Publications, Route de St Cyr, 78000 Versailles. pp 597.

Glade, M.J., 1983. Nitrogen partitioning along the equine digestive tract. J. Anim. Sci., 57 :943-953.

Glade, M.J., 1984. The influence of dietary fiber digestibility on the nitrogen requirements of mature horses. J. Anim. Sci., 58 :638-646.

Goodebee, R.G., Slade, L.M., 1979. Nitrogen absorption from the caecum of a mature horse. In : Proceeding 6th Equine Nutrition and Physiology Symposium, Texas, A.M. University, 75-76.

Gouin, R., 1932. Alimentation des animaux domestiques. Ed. Ballière, Paris, pp. 432.

Graham, P.M., Ott, E.A., Brendemuhl, J.H., Ten Broeck, S.H., 1993. The effect of supplemental lysine and threonine on growth and development of yearling horses. Proc. 13th Equine Nutrition Physiology Symposium. Gainesville : Florida USA, pp. 80-81.

Grandeau, L. and Alekan, A., 1904. Vingt années d'expériences sur l'alimentation du cheval de trait. Etudes sur les rations d'entretien, de marche et de travail. Ed. L. Courtier, Paris. p. 20-48.

Harris, P., 1997. Energy sources and requirements of the exercising horse. Ann. Rev. Nutr., 17, 185-210

Henneke, D.R., Potter, G.D., Kreider, J.L., 1981a. Rebreeding efficiency in mares fed different levels of energy during late gestation. In : Proceedings 7th Equine Nutrition Physiology Symposium, p.101-104.

Henneke, D.R., Potter, G.D., Kreider, J.L., 1981b. A condition score relationship to body fat content of mares during gestation and lactation. In : Proceedings 7th Equine Nutrition Physiology Symposium, p.105-110.

Hintz, H.F., 1968. Energy utilization in the horse. Proc. Cornell Nutr. Conf., pp. 47-49.

Hintz, H.F., Schryver, H.F., 1972. Nitrogen utilization in ponies. J. Anim. Sci., 34, 592-595.

Hintz, H.F., 1983. Nutritional requirements of the exercising horse. A review. In : Proceedings 1st Equine Exercise Physiology Symposium, Snow D.H., Persson S.G.B., Rose R., Editors, ICEEP Publications Cambridge, p. 275-290.

Hintz, H.F., Cymbaluk, N.F., 1994. Nutrition of the horse. Ann. Rev., 14, 243-267.

Hoffmann, L., Klippel, W. and Schiemann, R., 1967. Untersuchungen über den Energieumsatz beim Pferd unter besonderer Berücksichtigung der Horizontal bewegung. Archiv. Tierern., 17:441-449.

Hörnicke, H., Ehrlein, H.J., Tolkmitt, G., Nagel, M., Epple, E., Decker, E., Kimmich, H.P., Kreuzer, F., 1974. Method for continuous oxygen consumption measurement in exercising horses by telemetry and electronic data processing. In : Energy metabolism of farm animals, K.H. Menke et al., Ed. EAAP Publ. N° 14, 257-260, Stuttgart.

Hörnicke, H., Meixner, R. and Pollmann, R., 1983. Respiration in exercising horses. p.7-16. In : Equine Exercise Physiology. Editors. D.H. Snow, S.G.B. Persson et R.J. Rose, Granta Edition, Cambridge.

INRA, 1978. Alimentation des ruminants. R. Jarrige Editor INRA Publication Route de St-Cyr 78000 Versailles. pp. 597.

INRA, 1984. Le Cheval : Reproduction, Sélection, Alimentation, Exploitation. R. Jarrige et W. Martin-Rosset Editors. INRA Edition Route de St-Cyr 78000 Versailles. pp. 689.

INRA, 1984. Tables de la valeur nutritive des aliments pour le cheval. In : R. Jarrige, W. Martin-Rosset Editors. « Le Cheval » Reproduction - Sélection - Alimentation - Exploitation. INRA Publications, Route de St Cyr, 78000 Versailles, p. 661-689.

INRA, 1984. Tables des apports alimentaires recommandés pour le cheval. In : R. Jarrige, W. Martin-Rosset Editors. « Le Cheval ». INRA Publications,

Route de St Cyr, 78000 Versailles, p. 645-660.
INRA, 1988. Ruminant Nutrition. E. Jarrige Editor, INRA and John Libbey Ed.
    London or Paris. 471.
INRA, 1990. L'alimentation des chevaux. W. Martin-Rosset (Editor) INRA
    Publications, Route de St Cyr, 78000 Versailles. pp.232.
INRA, Institut du Cheval, Institut Elevage, 1997. Notation de l'état corporel des
    chevaux de selle et de sport. Guide pratique. Institut de l'Elevage Ed. 149
    rue de Bercy 75595 Paris Cedex 12. pp. 40.
Jackson, S., 1998. Myths and wives tales of feeding horses 'Some truth, some
    fiction'. In : Advances in Equine Nutrition. J. Pagan, Editor. Kentucky
    Equine Research. Nottingham University Press. U.K., p. 307-318.
Jarrige, R., Tisserand, J.L., 1984. Métabolisme, besoins et alimentation azotée du
    cheval.Chapter 18. In : R. Jarrige, W. Martin-Rosset Editors. « Le Cheval »
    INRA Publications, Route de St Cyr, 78000 Versailles, p. 277-302.
Jespersen, J., 1949. Normes pour les besoins des animaux : chevaux, porcs et poules,
    in : Vème Congrès International de Zootechnie, Paris, Vol. 2, Rapports
    particuliers, pp. 33-43.
Jones, J.H., Carlson, G., 1995. Estimation of metabolic energy cost and heat
    production during a 3-day-event. Equine Vet. J. 20 (Suppl.) 23-31.
Karlsen, G. and Nadal'Jak, E.A., 1964. Gas-energie Umstaz une Atmung Bei Trabern
    Während der Arbeit. Nonevodstvo i konnyi sport, 11 : 27-31.
Kellner, O., 1909. Principes fondamentaux de l'alimentation du bétail. 3ème Ed.
    Allemande, traduction, A. Grégoire, Berger Levrault, Paris, Nancy, pp. 288.
Kentucky Equine Research Inc. 1998. Advances in Equine Nutrition.
    Editor: J. Pagan. Nottingham Press Ed. Nottingham UK. pp. 566.
Knox, K.L., Crownover, D.C. and Wooden, G.R., 1970. Maintenance energy
    requirements of mature idle horses. In : Proceedings 5th Symposium Energy
    Metabolism of Farm Animals. Witznau (Suisse) Shurch, A., Wenk, C.
    Editors 181-184. Juris-Druck Verlag Zurich.
Kossila, V., Virtanen, R. and Maukonen, J., 1972. A diet of hay and oat as a source of
    energy digestible crude protein, minerals and trace elements for saddle
    horses. J. Sci. Agric. Soc. Finland, 44 : 217-227.
Lawrence, L., 1994. Nutrition and the athletic horse. In The Athletic Horse,
    Editors D.R. Hodgson, R.J. Rose, London Saunders, p. 205-30.
Lawrence, W., 1998. Protein requirements of equine athletes. In : J. Pagan Editor.
    Advances in Equine Nutrition. Nottingham Univ. Press. Ed., p.161-166
Lowman, R.S., Theodorou, M.K., Dhanoa, M.S., Hyslop, J.J. and Cuddeford, D.,
    1997. Evaluation of an in vitro gas production technique for estimating the
    in vivo digestibility of equine feeds. In : Proceedings of the 15th Equine
    Nutrition and Physiology Symposium. Forth Worth, Texas USA. p. 1-2.
Macheboeuf, D., Marangi, M., Poncet, C. and Martin-Rosset, W. 1995. Study of
    nitrogen digestion from different hays by the mobile nylon bag technique in
    horses. Annales Zootechnie, 44 219 (abstract).
Macheboeuf, D., Poncet, C., Jestin, M. and Martin-Rosset, W. 1996. Use of a mobile
    nylon bag technique with caecum fistulated horses as an alternative method
    for estimating pre-caecal and total tract nitrogen digestibilities of feedstuffs.
    In : Proceedings of the 47th European Association of Animal Production
    Meeting. Lillehammer, Norway, August, 26-29, Abstract H 4.9, p. 296.
    Wageningen Pers Ed. Wageningen The Netherlands.
Macheboeuf, D., Jestin, M., Andrieu, J. and Martin-Rosset, W. 1998. Prediction of

the organic matter digestibility of forages in horses by the gas test method. In : In vitro Techniques for Measuring Nutrient Supply to Ruminants. Occasional Publication N° 22 BSAS, Penicuik, Edinburgh, p.

Macheboeuf, D. and Jestin, M., 1998. Utilization of the gas test method using horse faeces as a source of inoculation. In : In vitro techniques for Measuring Nutrient Supply to Ruminants. Occasional Publication N° 22. BSAS, Penicuik, Edinburgh, p.

Mc Meniman, N.P., Elliot, R., Groenendyk, S., Und Dowsett, K.F., 1987. Synthesis and absorption of cysteine from the hindgut of the horse. Equine vet. J. 19, 192-194.

Martin-Rosset, W., Boccard, R., Jussiaux, M., Robelin, J. and Trillaud-Geyl, C., 1983. Croissance relative des différents tissus, organes et régions corporelles entre 12 et 30 mois chez le cheval de boucherie de différentes r races lourdes. Ann. Zootech., 32 :153-174.

Martin-Rosset, W., Andrieu, J., Vermorel, M. and Dulphy, J.P., 1984a. Valeur nutritive des aliments pour le cheval. Chapter 17. In : R. Jarrige, W. Martin-Rosset Editors. « Le Cheval » INRA Publications, Route de St Cyr, 78000 Versailles, p. 208-239.

Martin-Rosset, W., Doreau, M., 1984b. Consommation des aliments et d'eau par le cheval. Chapter 22. In R. Jarrige. W. Martin-Rosset Editor 'Le Cheval' INRA Edition Route de St-Cyr 78000 Versailles, p. 334-354

Martin-Rosset, W. and Doreau, M., 1984c. Besoins et alimentation de la jument. Chapter 23. In : R. Jarrige, W. Martin-Rosset (Editors). 'Le Cheval' INRA Publications, Route de St Cyr, 78000 Versailles, p. 355-370.

Martin-Rosset, W., Dulphy, J.P., 1987. Digestibility. Interactions between forages and concentrates in horses : influence of feeding level. Comparison with sheep. Livest. Prod. Sci., 17, 263-276.

Martin-Rosset, W., Tavernier, Vermorel, M., 1989. Alimentation du cheval de club avec un régime à base de paille et d'aliments composés. Proceedings 15e Journée Recherche Chevaline. Paris - 8 mars - CEREOPA Ed. 16 rue Claude Bernard 75231 Paris Cedex 05, p. 90-102.

Martin-Rosset, W., Doreau, M., Boulot, S., Miraglia, N., 1990. Influence of level of feeding and physiological state on diet digestibility in light and heavy breed horses. Livest. Prod. Sci., 25, 257-264.

Martin-Rosset, W. and Vermorel, M., 1991. Maintenance energy requirements determined by indirect calorimetry and feeding trials in light horses. Eq. Vet. Sci., 11 :42-45.

Martin-Rosset, W., 1993. Dépenses et apports énergétiques chez le cheval à l'effort. Sciences et Sports, 8, 101-108.

Martin-Rosset, W., Vermorel, M., Doreau, M., Tisserand, J.L. and Andrieu, J., 1994. The French horse feed evaluation systems and recommended allowances for energy and protein. Livest. Prod. Sci., 40 : 37-56.

Martin-Rosset, W., Andrieu, J. and Vermorel, M., 1996a. Routine methods for predicting the net energy value (UFC) of feeds in horses. In: Proceedings 47th European Association for Animal Production Meeting, Lillehammer, Norway, August, 26-29, Horse Commission. Session IV. Abstract H 4.1. p. 292. Wageningen Pers Ed. Wageningen The Netherlands.

Martin-Rosset, W., Andrieu, J. and Jestin, M., 1996b. Prediction of the digestibility of organic matter of forages in horses from the chemical composition. In: Proceedings of the 47th European Association of Animal Production

Meeting. Lillehammer, Norway, August, 26-29, Abstract H4.7, p. 295. Wageningen Pers Ed. Wageningen The Netherlands.

Martin-Rosset, W., Andrieu, J. and Jestin, M., 1996c. Prediction of the digestibility of organic matter of forages in horses by pepsin-cellulase method. In: Proceedings of the 47th European Association of Animal Production Meeting. Lillehammer, Norway, August, 26-29, Abstract H4.6, p.294. Wageningen Pers Ed. Wageningen The Netherlands.

Meixner, R., Hörnicke, H. and Ehrlein, H.J., 1981. Oxygen consumption, pulmonary ventilation and heart rate of riding horses during walk, trot and gallop. Biotelemetry, p. 6.

Meyer, H. and Ahlswede, L., 1976. Uber das intrauterine wachstum und die Körperzusammensetzung won Fohlen sowie den Hährstoffbedarf tragender stuten ubers. Tierernähr., 4, 263-292.

Meyer H., 1982. Ernährung des Pferdes. In : Lowe H. und Meyer H., Pferdezucht und Pferdefutterung. Eugen Ulmer Stuttgart. pp. 440.

Meyer, H., 1983. Protein metabolism and protein requirements in horses. In : IVème Symposium International Métabolisme et Nutrition Azotés. Clermont-Ferrand. M. Arnal, R. Pion, D. Bonin (Editors). Vol. 1. Les Colloques de l'INRA, n° 16, INRA Publications, Route de St Cyr, 78000 Versailles, p. 343-374.

Meyer, H., Nutrition of the equine athlete. In : Proceedings 2nd Equine exercise Physiology Symposium, Gillepsie J.P., Robinson N.E., Editors. ILEEP Publication, Davis, USA, p. 644.

Meyer, H., Radicke, S., Kienzle, H., Wilke, S. and Kleffen, D., 1993. Investigations on preileal digestion of oat, corn and barley starch in relation to grain processing. In : Proceedings 13th Equine Nutrition and Physiology Symposium, January 21-23. Gainesville. Florida, USA, p. 66-71.

Micol, D., Martin-Rosset, W., 1998. Feeding systems for horses on high forage diets in the temperate zones. Chapter 15. In : Proceedings IVth International Symposium Nutrition Herbivores. M. Journet et al. Editors Clermont-Ferrand. September 11-15, INRA Edition Route de St-Cyr 78000 Versailles, p.569-584.

Miller-Graber, P., Lawrence, L., Foreman, J., Bump, K., Fischer, M., Kurcz, 1991. Effect of dietary protein level on nitrogen metabolites in exercised quarter horses. In : Proceedings 3rd Equine exercise Physiology Symposium, ICEEP Publications, Uppsala, Sweden, p. 305-314.

Miraglia, N. and Olivieri, O., 1990. Statement and expression of the energy and nitrogen value of feedstuffs in Southern Europe In : Proceedings of the 41st European Association of Animal Production Meeting. Toulouse, France, Abstract p. 390. Wageningen Pers. Ed. The Netherlands.

Moore-Colyer, M.J.S., Longland, A.C., Hyslop, J.J. and Cuddeford, D., 1998. The degradation of protein and non-starch polysaccharides (NSP) from botanically diverse sources of dietary fiber by ponies as measured by the mobile bag technique. In : In vitro Techniques for Measuring Nutrients Supply to Ruminants. Occasional Publication N° 22, BSAS, Penicuik, Edinburgh.

Nadal'Jak, E.A., 1961. Gaseous exchange in horses in transport work at the walk and trot with different load and rates of movements. Gaseous exchange and energy expenditure at rest and during different tasks by breeding stallions of heavy draft breeds. Effect of state of training on gaseous exchange and energy expendure in horses of heavy draft breeds (in Russian). Nutr. Abstr. Reviews, 32, n° 2230-2231-2232 : 463-464.

National Research Council, 1978. Nutrients requirements of domestic animals , no. 6. Nutrients requirement of Horses. 4th Revised Edition. National Academy of Sciences, Washington, D.C. pp. 33.

National Research Council, 1989. Nutrients requirements of domestic animals, no. 6. Nutrients Requirements of Horses. 5th Revised Edition. National Academy of Sciences, Washington, D.C. pp. 100.

Nicoletti, J.N., Wohlt, J., Glade, M.J., 1980. Nutrition utilization by ponies and steers as affected by dietary forage rations. J. Anim. Sci., 51, suppl. 1, 25.

Olsson, N.A. and Ruudvere, A., 1955. The nutrition of the horse. Nutr. Abstr. Reviews, 25, 1-18.

Ott, E.A., Asquith, R.L., Feaster, J.P., Martin, F.G., 1979. Influence of protein level and quality on the growth and development of yearling foals. J. Anim. Sci., 49, 620.

Ott, E.A., Asquith, R.L., Feaster, J.P., 1981. Lysine supplementation of diets for yearling horses. J. Anim. Sci., 53, 1496.

Ott, E.A., Asquith, R.L., 1983. Influence of protein and mineral intake on growth and bone development of weanling horses. Proceedings 8th Equine Nutrition Physiology Symposium, Lexington, Kentucky, p. 39.

Pagan, J.D. and Hintz, H.F., 1986b. Equine energetics. II. Energy expenditure in horses during submaximal exercise. J. of Anim. Sci., 63, 822-830.

Pagan, J.D., Essen-Gustavsson, Lindholm, M., Thornton, J., 1987. The effect of dietary energy source on exercise performance in Standardbred horses. In : Proceedings, 2nd Exercise Physiology Symposium, Gillepsie J.R., Robinson N.E., Editions ICEEP Publications, Davis, USA, p. 686-701.

Pagan, J. ; 1998. Computing horse nutrition : how to properly conduct an equine nutrition evaluation. In Advances in Equine Nutrition J. Pagan Editor Nottingham Univers. Press Ed., p. 111-124.

Persson, S.G.B., 1983. Analyses of fitness and state of training. In : Proceeding 1st Equine Exercise Physiology Symposium. Snow D.H., Persson S.G.B., Rose R.J. Editors Granta, Editions Cambridge, p. 441-457.

Potter, G.D., Evans, J.W., Webb, G.W. and Webb, S.P., 1987. Digestible energy requirements of Belgian and Percheron horses. In : Proceedings of the 10th Equine Nutrition and Physiology Symposium. Fort Collins, Colorado USA, p.133-138.

Potter, G.D., Arnold, F.F., Householder, D.D., Hansen, D.H. and Brown, K.M., 1992a. Digestion of starch in the small and large intestine of the equine. In Proceedings 1st European Conference On Horse Nutrition, Hannover, Germany, p. 107-111.

Potter, G.D., Gibbs, P.G., Haley, R.G., Klendshoj, C., 1992b. Digestion of protein in the small and large intestine of equines fed mixed diets. In Proceedings 1st European Conference On Horse Nutrition, Hannover, Germany, pp. 140-143.

Prior, R.L., Hintz, H.F., Lowe, J.E. and Visek, W.D., 1974. Urea recycling and metabolism of ponies. J. Anim. Sci., 30 :565-571.

Reeds, P.J. and Harris, C.I., 1981. Protein turnover in animals: Man in his context. In: Nitrogen Metabolism in Man (ed. J.C. Waterlow and, J.M.L. Stephen). Applied Sciences Pub., London, p.392-402.

Reitnour, C.M., 1979. Effect of caecal administration of corn starch on nitrogen metabolism in ponies. J. Anim. Sci., 49, 988-992.

Reitnour, C.M., 1980. Protein utilization in response to caecal corn starch in ponies. J. Anim. Sci., 51, Suppl. 1, 218.

Robelin, J., 1979. Influence de la vitesse de croissance sur la composition du gain de poids des bovins : variations selon la race et le sexe. Ann. Zootech., 28 :209-218.

Robinson, D.W., Slade, L.M., 1974. The current status of knowledge on the nutrition of equines. J. Anim. Sci., 39, 1045-1066.

Saastamoinen, M.T., Kostinen, E., 1993. Influence of quality of dietary protein supplement and anabolic steroids on muscular and skeletal growth of foals. Anim. Prod. 56 (1) : 135-44.

Schmitz, M., Abrens, F., Und Hagemeister, H., 1990. Beitrag der Absorption von Aminosäuren im Dikdarm zur Proteinversorgung bei Pferd, Rind und Schwein. J. Animal Physiol. Animal Nutr. 64, 12-13.

Slade, L.M., Bishop, R., Moriss, J.G. and Robinson, D.G., 1971. Digestion and absorption of N-labelled microbial protein in the large intestine of the horse. Br. Vet.J., 127, XI, XII.

Slade, L.M., Robinson, D.W. and Casey, K.E., 1970. Nitrogen metabolism in non-ruminant herbivores. I - The influence of non-protein and protein quality on the nitrogen retention of a diet in mares. J. Anim. Sci., 30 :753-760.

Slade, L.M., Robinson, D.W., Al-Rabbat, F., 1973. Ammonia turnover in the large intestine. In Proceedings 3th Equine Nutrition Physiology Symposium, Gainesville, Florida USA, p. 1-12.

Smolders, E.A.A., 1990. Evolution of the energy and nitrogen systems used in The Netherlands. In : Proceedings of the 41st European Association of Animal Production Meeting. Toulouse, France, p. 386 (abstract). Wageningen Pers. Ed. Wageningen The Netherlands.

Southwood, L.L., Evans, D.L., Bryden, W.L., Rose, R.J., 1993. Nutrient intake of horses in Thoroughbred and Standardbred stables. Aust. Vet. J., 70 :164-168.

Staun, H., 1990. Energy and nitrogen systems used in northern countries for estimating and expressing value of feedstuffs in horses. In : Proceedings of the 41st European Association of Animal Production Meeting. Toulouse, France, Abstract p. 388.Wageningen Pers. Ed. Wageningen The Netherlands.

Stillions, M.C. and Nelson, W.E., 1972. Digestible energy during maintenance of the light horse. J. Anim. Sci., 34 :981-982.

Sutton, E.I., Bowland, D.P., Ratcliff, W.D., 1977. Influence of level of energy and nutrient intake by mares on reproductive performance and on blood serum composition of the mares and foals. Can. J. Anim. Sci., 57, 551-558.

Thomas, D.P. and Fregin, G.F., 1981. Cardiorespiratory and metabolic response to treadmill exercise in the horse. Pflügers Arch., 385 :65-70.

Tisserand, J.L., Martin-Rosset, W., 1996. Evaluation of the protein value of feedstuffs in horses in the MADC system. In : Proceeding of 47th Annual Meeting of European Association for Animal Production. Lillehammer August 25-29 Norway. Aleshat H.4.3. p. 293. Wageningen Pers. Ed. Wageningen The Netherlands.

Van Es, A.J.H., 1975. Feed evaluation for dairy cows. Livest. Prod. Sci., 2 : 95-107.

Vermorel, M., Jarrige, R. and Martin-Rosset, W., 1984. Métabolisme et besoins énergétiques du cheval. Le système des UFC. Chapter 18. In : R. Jarrige et, W. Martin-Rosset (Editors). « Le Cheval » INRA Publications, Route de St Cyr, 78000 Versailles, p. 237-276.

Vermorel, M., Martin-Rosset, W. and Vernet, J., 1991. Energy utilization of two diets for maintenance by horses : agreement with the new French net energy system. Eq. Vet. Sci., 11, 33-35.

Vermorel, M., Martin-Rosset, W., Vernet, J., 1997. Energy utilization of twelve forages or mixed diets for maintenance by sport horses. Livest. Prod. 57, 157-167.

Vermorel, M., Martin-Rosset, W., 1997. Concepts, scientific bases, structure and validation of the French horse net energy system (UFC). Livest. Prod. Sci., 47 : 261-275.

Vogelsang, M.M., Potter, G.D., Kreideer, J.L., Anderson, J.G., 1981. Determining oxygen consumption in the exercising horse. In : Proceeding 7th Equine Nutrition Physiology Symposium, Warrenton, Virginia, USA, p. 195-196.

Warren, L.K., Lawrence, L.M., Griffin, A.S., Parker, A.L., Barnes, T., Wright, D., 1998. The effect of weaning age on foal growth and bone density. In : Pagan Editor, KER Advances in Equine Nutrition Nottingham Press University, p. 457-460.

Willmore, J.H., Freund Beau J., 1984. Nutritional enhancement of athletic performance. Nur. Abs. Rev., Series A., 54, 1-16.

Wolff, E., 1888. Principes de l'alimentation rationnelle du cheval : nouvelles séries d'expériences réalisées à la Station d'Hohenheim en 1885-1886. Ann. Sci. Agron., 5ème année, 2, 336-339. Traduction par, M. Margottet de l'article original publié dans Landw. Jahrb., 1887, 3 :49-131.

Wolff, E., Siegling, Kreuzhage, C. and Riess, C., 1887. Versuche über den Einflusseiner verschiedenen Art der Arbeitsleistung auf die Verdauung des Futters sowie über den Verhaltens des Rauhfutters gegenüber den Kraftfutters zur Leistungsfähigkeit des Pferdes. Landw. Wjahrb., 3 :49-131.

Wolff E., Kreuzhage, C., 1895. Pferde Fütterungsversuche über Verdauuung und Arbeitsäquivalent des Futters. Landw. Jahrb., 24, 125-271.

Wolter, 1975. L'alimentation du cheval. 2e Edition. Vigot Frères Edition Rue de l'Ecole de Médecine Paris. pp. 177.

Wooden, G.R., Know, K.L. and Wild , C.L., 1970. Energy metabolism of light horses. J. Anim. Sci., 30 :544-548.

Wysocki, A. and Baker, J.P., 1975. Utilization of bacterial protein from the lower gut of the equine. In : Proceedings 4th Equine Nutrition and Physiology Symposium. California University. Pomona, Los Angelès, USA, p. 21.

Zuntz, N. and Hagemann, O., 1898. Untersuchungen über den Stoffwechsel des Pferdes bei Ruhe und Arbeit. Landw. Jahrb., 27, suppl. 3, 1-437.

VITAMIN REQUIREMENTS IN THE HORSE

KATHLEEN CRANDELL
Kentucky Equine Research, Inc., Versailles, KY

Vitamins may be required in minute amounts by the body, but they are undeniably necessary for normal metabolic and physiological processes. The discovery of vitamins and an improved understanding of their functions are among the most important achievements of this century. Most vitamin deficiency diseases are rarely seen today except in developing countries. What exactly are vitamins? The nature of vitamins is as varied as any group of nutrients. In general, vitamins are organic compounds which cannot be produced in adequate quantities in the body and must be obtained from food or the environment. More specifically, vitamins serve the similar purpose of maintaining normal body function.

Vitamins are necessary for optimal growth, health, feed conversion and reproduction. To further complicate this, vitamins are also necessary for proper physical performance in horses. The target tissues of the fat-soluble vitamins A, D, E, and K are skin, bones, muscles and blood. B vitamins serve as catalysts in the conversion of nutrients to tissues and the metabolism of fats, carbohydrates and protein to the final oxidation end products of CO_2 and H_2O (via the citrate cycle).

Table 1. Functions and deficiency signs of vitamins (BASF 2000).

Nutrient	Major Function	Signs of Major Deficiency
Vitamin A	Vision, mucous tissue integrity, immunity	Xerophthalmia, tissue keratinization, polyneuritis, hind leg paralysis, elevated cerebrospinal fluid pressure, ataxia, depressed immune system, reduced fertility
Vitamin D	Calcium and phosphorus balance	Rickets, osteomalacia, metabolism, bone calcification, immunity
Vitamin E	Intracellular respiration, antioxidant, membrane integrity	Encephalomalacia, depressed immune status, skin edema, steatitis, jaundice, liver necrosis, anemia, erythrocyte hemolysis, muscular dystrophy, fetal death, reduced fertility
Vitamin K	Blood coagulation	Prolonged blood clotting, low prothrombin, intramuscular bleeding, anemia, hemorrhage
Thiamine (B_1)	Metabolism of carbohydrates and proteins, nervous system	Loss of appetite, polyneuritis, skin edema, fatty liver, fatty heart, convulsions, cyanosis, gastrointestinal hemorrhage, diarrhea
Riboflavin (B_2)	Antioxidant, H-transfer, ligament integrity	Poor growth, shortened bones, curled toe paralysis, fused ribs, dermatitis, poor hair coat, seborrhea, photophobia, cataracts, anemia, stiff crooked legs, fetal death, reduced fertility, collapsed ovarian follicles, diarrhea, anal mucosa inflammation, ulcerative colitis
Pantothenic Acid	Conversion of amino acid groups as coenzyme A, skin integrity	Dermatitis in feet and mouth, blindness, goose stepping, demyelination of spinal cord, depressed immune system, decreased milk production, embryo detachment, diarrhea, GI ulceration, fatty liver

Niacin	Metabolism of carbohydrates, protein and fats, H carrying enzymes NAD and NADP, skin integrity	Bowed legs, diarrhea, general dermatitis of feet and mouth, anorexia, hind leg paralysis, GI ulceration
Vitamin B$_6$	Metabolism of proteins	Dermatitis around the eyes, fatty liver, ataxia, convulsions, diarrhea
Choline	Methyl donor group, phospholipids	Poor growth, fatty liver, enlarged spleen, abnormal gait, demyelinization of peripheral nerves, depressed immune system, reduced fertility, kidney damage
Vitamin B$_{12}$	Protein metabolism, transport of methyl groups	Anemia, poor growth, poor hair coat, fatty kidney, kidney damage, ataxia, uncoordinated hind legs, impaired thyroid, diarrhea
Folacin	Transfer of single carbon units in activated form for methylation reactions involving methyl donors such as methionine and choline, immunity	Poor growth, anemia, poor skin condition, reduced fertility
Biotin	Metabolism of fats, carbohydrates and proteins as coenzyme for CO$_2$ fixation and transcarboxylation	General dermatitis, dermatitis on feet, mouth and eyes, poor hair coat, weak-walled brittle hooves, spasms in hind legs, stiff gait, reduced fertility
Vitamin C	Antioxidant, hormone synthesis, conversion of vitamin D$_3$ to calcitriol, essential for bone calcification	Poor hair coat, depressed immune system, hemorrhage, delayed wound healing, degenerated enlarged adrenal, scoliosis, lordosis

Vitamin Requirements

Vitamin requirements, like those of other nutrients, are affected by age, reproductive status, amount of exercise and a variety of stresses such as gastrointestinal infections and intense muscular exercise. The need for vitamin supplementation also depends on the type and quality of the diet, the length of exposure to sunlight, the amount of microbial vitamin synthesis in the digestive tract, and the extent of vitamin absorption from the site of synthesis. Nonworking horses grazing high quality pastures are likely to need little or no vitamin supplementation because forages are a rich source of most fat- and water-soluble vitamins. Because many horses do not have the advantage of lush green pasture year-round, supplementation of vitamins becomes the responsibility of the care provider. Today, nearly all commercial horse feeds are fortified with vitamins (primarily fat-soluble vitamins) to supplement the low natural vitamin content of the grains.

Unlike research conducted on humans and food animals, actual vitamin requirements for horses have not undergone intensive scrutiny. Research into requirements is time consuming due to the lengthy period it takes to initiate obvious outward signs of vitamin depletion. Further, research may involve considerable suffering of the animal before death occurs, and many researchers are reluctant to put the animals through this suffering. Currently, function tests are used most frequently in vitamin studies in an attempt to detect the onset of vitamin depletion before it causes irreversible damage and suffering. With the use of function tests, it is becoming apparent that there is a clear distinction between the minimal requirement of a vitamin and an optimal requirement.

The minimum vitamin requirement is that quantity which has to be supplied daily to the animal to prevent or correct deficiency symptoms. These values were determined earlier in the twentieth century under experimental laboratory conditions; however, these values have only a theoretical value as far as practical animal feeding is concerned. Many times the research animals were maintained in a rigidly controlled environment, not in a natural situation. Animals were also not subjected to any form of physical exercise or reproductive effort, physiological states which may alter minimal vitamin requirements.

The optimum requirement is the quantity that promotes maximal growth rate, performance, health, feed utilization and body reserves. Experience indicates that the optimum supply is probably several times higher than the minimum vitamin requirement. Researchers are attempting to define the optimum requirement for many vitamins in the horse. For production animals, inadequate growth is the common measure of vitamin deficiency. Slowed growth can have adverse effects on the health and productivity of the animal. However, growth alone may be a misleading measure in the horse. Supplementation may be adequate for normal growth but not for optimal performance.

Symptoms of vitamin deficiency are not always visible. Suboptimal intake can occur frequently in practice and result in a nonspecific depression of performance, increased susceptibility to disease, reduced fertility, and shorter productive life. For these reasons, the objective in designing the ideal diet is to meet optimum requirements through vitamin supplementation.

Table 2. Levels of vitamin supplementation (BASF 2000).

Minimum supplementation	The vitamin quantity required for prevention and/or correction of deficiency symptoms (only of little practical value).
Optimum supplementation	The vitamin quantity required for achieving best performances (growth rate, feed conversion, health) can only be determined by using sensitive biological and physiological criteria.
Suboptimal supplementation	Supply of vitamins somewhere between minimum and optimum level (occurs frequently in practice without knowledge, resulting in nonspecific depression of production).

The National Research Council (NRC) last published Nutrient Requirements of Horses in 1989. The recommendations presented in this edition are restricted to the research that had been done for each nutrient before 1989. In compiling the requirements, many different recommendations had to be considered and a consensus was decided upon to establish requirements. The NRC vitamin requirements should be considered the minimum vitamin requirement under ideal circumstances to prevent clinical symptoms of vitamin deficiency. Since these requirements often differ drastically from the optimal vitamin requirements under true field conditions, the NRC recommendations should be considered only a part of the total vitamin requirement.

The most recent Nutrient Requirements of Horses (NRC, 1989) provides estimated requirements for vitamins A, D, E, thiamin and riboflavin, but not for the other vitamins. Deficiencies of other vitamins in the horse were not researched extensively prior to 1989. However, in designing diets for optimum health and well-being of horses, the gray zone between minimal and optimum requirement needs to be taken into consideration. For example, McMeniman et al. (1995) reported that exercising horses had lower peak plasma lactate concentrations and lower maximum heart rates when they received a diet containing a multi-vitamin supplement (A, D, E, K, thiamin, riboflavin nicotinamide, pyridoxine, pantothenic acid, biotin, choline, folic acid and cyanocobalamine), but the indices of performance were not improved. The fact that performance was not improved may be justification to not change the minimal requirement, but the changes in heart rate and lactate are sufficient to suggest that there can still be improvement with optimal levels.

Table 3. BASF recommendations for vitamin supplementation per 100 kg live weight/day.

Vitamin	Foals	Leisure Horses	Race & Breeding Horses
Vitamin A, IU	10,000-12,000	6,000-8,000	12,000-15,000
Vitamin D, IU	1000-1200	600-800	1200-1500
Vitamin E, mg	100-120	60-80	200-300
Vitamin K, mg (menadione)	3	2	3
Thiamin, mg	8-10	6-8	8-12
Riboflavin, mg	8-12	6-8	8-12
Vitamin B_6, mg	6	4	6
Vitamin B_{12}, ug	60-80	50-70	60-80
Biotin, ug	200-300	200*	200-300*
Folacin, mg	6	4	8
Niacin, mg	10-20	10-15	15-25
Pantothenic Acid, mg	8-10	6-8	10-12
Choline, mg	150-250	150-250	300-400
Vitamin C, mg	200	100	200-300
B-carotene, mg			400-500+

* for improvement of hoof health and integrity 15 - 20 mg/day for at least 6 months
+ for reproduction per horse per day, 4 weeks before birth to 10 weeks after

Aside from the basic and optimal vitamin requirements in the horse, there is increasing interest in using vitamins as healing foods. This is drastically changing the manner in which vitamin supplementation of horses is being viewed. Vitamin supplementation has evolved from meeting the basic physiological requirements of the animal to providing a food that will improve the quality of life. This is essentially the same development that occurred in human nutrition over the last ten years. Human supplementation with antioxidant vitamins (vitamins E and C, beta-carotene) and omega 3-fatty acids has become commonplace. The main justification for this supplementation is to improve the quality of life and to reduce risks associated with aging such as cancer, low immune competence, cardiovascular diseases, cataracts and renal failure. Such supplementation may improve the quality of life for horses as well. Supplement products found on the market frequently contain nutrients such as vitamins E and C, beta-carotene and omega 3-fatty acids.

Table 4. Factors that influence vitamin supplementation (BASF 2000).

Effect of feed	Level of other nutrients in the feed (protein, energy, minerals, drugs, etc.)
	Bioavailability of vitamins in feed ingredients
	Destruction of vitamins in feed due to high temperature/ pelleting, oxidation and the catalytic effects of trace minerals or the peroxiding effects of rancidifying polyunsaturated fats
	Binding of vitamins in feed
	Growth of fungi, bacteria and yeast
Effect of animal and metabolism	Race, breed and genetic variation
	Variation in carry through of vitamins from breeding stock to progeny
	Variation in the absorption of vitamins in the gut: Parasitic damage to intestine walls Low dietary fat levels to support optimum fat-soluble vitamin absorption Lack of bile salts to form micelles and other mechanisms involved in vitamin absorption Inadequate level of lipoproteins essential in vitamin transport
	Competition between vitamins that use similar absorption mechanisms (vitamins A and E)
	Influence of gut enzymes (lipase, thiaminase)
Effect of animal health and environment	Level of disease and other stress factors
	Malabsorption due to destruction of microvilli
	Endoparasites and ectoparasites
	Mycotoxins and peroxides are often responsible for malabsorption

Survey of Vitamin Research

Vitamin A
Because of the narrow range between deficiency and toxicity of vitamin A, recommendations for supplementation are very conservative in the 1989 NRC despite research indicating that the requirement should be two to five times the recommended level (Donoghue et al., 1981). Research done by Maenpaa and coworkers (1988b) also suggested that the recommended levels were insufficient when green forage was not available. In current research, vitamin A status was assessed by use of the function test. Relative dose response was found to be a more sensitive measure of vitamin A status than serum vitamin A (Greiwe-Crandell

et al., 1995). Depletion of vitamin A reserves was found within two months of a diet of hay and vitamin A free concentrate. Subsequent supplementation of vitamin A palmitate at two times the NRC recommended level was not adequate to completely replete stores of vitamin A in mares with no access to pasture (Greiwe-Crandell et al., 1997a). Mares with access to green pastures had adequate liver stores of vitamin A regardless of vitamin A supplementation. Additional vitamin A palmitate supplementation did not induce any excesses of vitamin A in liver or serum (Greiwe-Crandell et al., 1997a). Further investigation on ß-carotene found carotenes in grass readily available as a source of vitamin A, but synthetic ß-carotene was not readily absorbed. Use of synthetic ß-carotene as a sole source of vitamin A could not meet vitamin A requirements of horses and is not recommended (Greiwe-Crandell et al., 1997b).

Vitamin D
Focus in current vitamin D research has been in exercise effects on requirement. Horses with adequate access to sunshine each day may always be able to meet their vitamin D requirement, but many of the horses in intensive work are confined indoors for the majority of the day. Extensive bone remodeling was found in young horses undergoing race training; this resulted in changes in calcium, phosphorus and vitamin D levels in serum and bone (Nielsen et al., 1995). Logically, anything that stresses the bone structure of the animal (such as carrying the weight of the rider, jumping or intensive training) will increase the amount of bone remodeling, and therefore, may increase the need for vitamin D. An increase in vitamin D intake has been found to increase efficiency of calcium and phosphorus absorption in the intestinal tract (Cromwell, 1996), which would increase the supply of these nutrients as needed in times of stress due to exercise.

Vitamin E
The vitamin most closely focused upon in both equine and production animals during the last decade has been vitamin E. Continuing previous research investigating the best form to supplement vitamin E, Wooden and Papas (1991) studied absorption of four forms of tocopherol in long yearling Thoroughbreds (dl-alpha-tocopheryl acetate, d-alpha-tocopheryl acetate, d-alpha-tocopheryl polyethelene glycol (TPGS) and TPGS + d-alpha-tocopherol acetate). Researchers found d-forms resulted in higher plasma concentrations than equivalent amounts of dl-forms. Gansen et al. (1995) reported on the ability of the horse to absorb synthetic and natural sources of vitamin E. Natural vitamin E elicited serum vitamin E increases equal to that of three times the amount of synthetic E. A function test of vitamin E absorption, the OVETT (oral vitamin E absorption test), found that maximal absorption occurred when vitamin E was naturally consumed in one liter of grain as compared to an oral paste or stomach tube (Craig et al., 1991).

Horses receiving a diet containing about 3000 IU of vitamin E/day had higher serum alpha-tocopherol concentrations than horses receiving less than 800 IU vitamin E/day (Siciliano et al., 1997). However, there was no difference in the susceptibility to exercise-induced muscle damage as measured by increase in serum creatine kinase or aspartate aminotransferase activities after a strenuous exercise bout. Low vitamin E levels were reported in exercising horses kept in confinement and fed hay (Hall et al., 1991). Additional vitamin E supplementation is warranted for horses when pasture is not available. In the same study, no differences in serum alpha-tocopherol levels were found between yearlings and mature horses in training; both were low compared to horses not in training. The vitamin E requirement in exercising horses and its effect on performance and cellular peroxidation were investigated (McMenniman and Hintz, 1992). The suggested intake of vitamin E was adequate for the exercising horse at 42 IU/kg DM. This is well below the currently recommended level of 80 IU/kg DM. However, it can be safely assumed that as the workload or the duration increases, so may the requirement.

The increasing popularity of high fat diets for horses has brought forth concern for an increased requirement for vitamin E because of its cellular antioxidant properties, quenching free radicals produced by fat oxidation for energy. Horses consuming a diet consisting of 6% added soybean oil did not show any indication of interference with vitamin E status in non-exercising horses (Siciliano and Wood, 1993). Exercising ponies fed a diet providing 10% of dietary energy in the form of corn oil caused a challenge to the cellular antioxidant systems in the muscle, but no reduction in plasma or muscle vitamin E (McMeniman and Hintz, 1992).

Vitamin E status in reproductive horses has been investigated. Hoffman et al. (1999) reported an increase in serum IgG concentrations in foaling mares from 160 IU vitamin E/kg daily intake as compared to 80 IU. After foaling, suckling foals of mares supplemented with the 160 IU vitamin E had higher serum IgG concentrations, even though there was no difference at birth.

A deficiency of vitamin E may cause a variety of symptoms and pathological changes, which may include nutritional muscular dystrophy (weak and poorly oxygenated muscles) and poor immunity to diseases (recurrent colds and coughing) (Moore and Kohn, 1991). Low serum vitamin E and blood glutathione peroxidase values were found in horses with degenerative myeloencephalopathy (Dill et al., 1989; Blythe et al., 1991). Treatment of this disease and equine motor neuron disease with 6,000 to 10,000 IU of vitamin E per day has been successful in numerous cases (Marcella, 1997).

Vitamin K

There is currently no requirement for vitamin K in the horse because of microbial production in the intestinal tract and amounts present in forage. Recent research is focusing on the role of vitamin K in bone growth and the possible link to developmental orthopedic diseases. Siciliano et al. (1999b) reported an increase in vitamin K status with age because of increased forage consumption and consequent increase in microbial synthesis as foals are weaned from milk. Vitamin K status affects bone metabolism as well as blood coagulation. However, there was a decline in serum osteocalcin during the same period which reflects the slowing of bone growth. In further investigations on vitamin K in the young growing horse in training, Siciliano et al. (1999a) reported no effect of exercise training on vitamin K status as measured by hydroxyapatite binding capacity. During times of increased bone metabolism (onset of exercise training), the vitamin K requirement may increase due to an increase in osteocalcin production. Lower vitamin K status was related to an increased incidence of hip stress fractures in humans (Szulc et al., 1994).

Vitamin C
There is no NRC requirement for vitamin C in horses because of adequate hepatic production from glucose under normal circumstances. However, the requirement has been seen to increase in times of stress and disease when body production may not be able to meet the need. Foals exhibiting high levels of stress during weaning by confinement in stalls had lower than normal plasma ascorbate levels (Hoffman et al., 1995). Vitamin C supplementation (20 g per day) increased antibody response to vaccines in aged horses, especially those with pituitary dysfunction or Cushing's syndrome (Ralston, 1999). Lieb et al. (1995) investigated the effectiveness of a nutrient supplement which included vitamin C and niacin (as well as tyrosine) in alleviating the clinical signs of anhidrosis (inability to sweat). The supplement appeared to improve heat dissipation in non-exercised anhidrotic horses by increasing the amount of body sweat area.

B vitamins
The NRC suggests that B vitamins are usually supplied in adequate amounts for the mature horse with ingestion of quality forages and that there is utilization of B vitamins produced by the microflora of the cecum and colon. However, bacterial products can not supply all the requirements at all times because B vitamin deficiencies can be produced by feeding diets containing low concentrations of certain B vitamins. The two B vitamins that have minimal requirements according to Nutrient Requirements of Horses (NRC, 1989) are thiamin and riboflavin. Little work has been done to investigate requirements of these vitamins since the last publication.

The most significant research on B vitamins has been on biotin. Even when the minimal requirement to maintain normal body function is met in the horse, there have been anecdotal accounts of feeding additional biotin improving hoof quality. In 1995, Josseck et al. demonstrated significant improvement in hoof horn quality with daily supplementation of 20 mg of biotin. Supplementation only affected new hoof growth. Improvements were observed after nine months of supplementation but a measurable improvement in tensile strength took 19 months. Lindner et al. (1994) reported that biotin supplementation alone was not found to affect lactate concentrations during exercise.

Parker et al. (1997) investigated the use of supplemental niacin in exercising horses to increase lipolysis, thus allowing for improved utilization of FFA during submaximal and maximal exercise. Just as small amounts of niacin are vital to energy metabolism (component of NAD and NADP), excess amounts have been found to inhibit lipolysis and thereby affecting mobilization of FFA from adipose tissue. In this study 3 g of nicotinic acid were fed for six weeks which resulted in neither improvement of exercise parameters or in niacin status over control exercised horses.

Conclusion

Continued investigation into vitamin requirements in horses will broaden the understanding of the vital importance of these delicate organic compounds and to highlight the significance of optimal, not just minimal, nutrition in the equine.

References

BASF. 2000. Vitamins - One of the Most Important Discoveries of the Century. BASF Documentation DC 0002. Animal Nutrition 6th Edition.

Blythe, L.L., A.M. Craig, E.D. Lassen, K.E. Rowe, and L.H. Appell. 1991. Serially determined plasma alpha-tocopherol concentrations and results of the oral vitamin E absorption test in clinically normal horses and in horses with degenerative myeloencephalopathy. Am. J. Vet. Res. 52:908.

Craig, A.M., L.L. Blythe, K.E. Rowe, E.D. Lassen, and L.L. Walker. 1991. Evaluation of the oral vitamin E absorption test in horses. Am. J. Vet. Res. 52:912.

Cromwell, G.L. 1996. Metabolism and role of phosphorus, calcium and vitamin D3 in swine nutrition. In: M.B. Coelho and E.T. Kornegay (Ed.) Phytase in Animal Nutrition and Waste Management. BASF Reference Manual. 101.

Donoghue, S., D.S. Kronfeld, S.J. Berkowitz, and R.L. Copp 1981. Vitamin A nutrition of the equine: Growth, serum biochemistry, and hematology. J. Nutr. 111:365.

Gansen, S., A. Lindner and A. Wagener. 1995. Influence of a supplementation with natural and synthetic vitamin E on serum a-tocopherol content and V4 of Thoroughbred horses. Proc. 14th Equine Nutr. Physiol. Soc. Symp. 68.

Greiwe-Crandell, K.M., D.S. Kronfeld, L.A. Gay, and D. Sklan. 1995. Seasonal vitamin A depletion in grazing horses is assessed better by the relative dose response test than by serum retinol concentration. J. Nutr. 125:2711.

Greiwe-Crandell, K.M., D.S. Kronfeld, L.A. Gay, D. Sklan, W. Tiegs and P.A. Harris. 1997a. Vitamin A repletion in Thoroughbred mares with retinyl palmitate or ß-carotene. J. Anim. Sci. 7:2684.

Greiwe-Crandell, K.M., D.S. Kronfeld, L.A. Gay, D. Sklan and P.A. Harris. 1997b. Daily B-carotene supplementation of vitamin A depleted mares. Proc. 15th Equine Nutr. Physiol. Soc. Symp. 378.

Hall, R.R., R.W. Brennan, L.M. Peck, J.P. Lew and S.E. Duren. 1991. Comparisons of serum vitamin E concentration in yearlings and mature horses. Proc. 12th Equine Nutr. Physiol. Soc. Symp. 263.

Hoffman, R.M., D.S. Kronfeld, J.L. Holland and K.M. Greiwe-Crandell. 1995. Preweaning diet and stall weaning method influences on stress response in foals. J. Anim. Sci. 73:2922.

Hoffman, R.M., K.L. Morgan, M.P. Lynch, S.A. Zinn, C. Faustman, and P.A. Harris. Dietary vitamin E supplemented in the periparturient period influences immunoglobulins in equine colostrum and passive transfer in foals. Proc. 16th Equine Nutr. Physiol. Soc. Symp. 96.

Josseck, H., W. Zenker, and H. Geyer. 1995. Hoof horn abnormalities in Lipizzaner horses and the effect of dietary biotin on macroscopic aspects of hoof horn quality. Equine Vet J. 27:175.

Lieb, S., K. Bowker, G. Lester, S. Ralston and P.E. Ginn. 1995. Effect of a nutrient supplement on the respiration, pulse, temperature, sweating, and serum electrolyte and amino acid levels of anhidrotic horses. Proc. 14th Equine Nutr. Physiol. Soc. Symp. 87.

Lindner, A., P. von Wittke and M. Frigg. 1992. Effect of biotin supplementation on VLa4 of Thoroughbred Horses. J. Equine Vet Sci. 12:149.

Maenpaa, P.H., T. Koskinin and E. Koskinin. 1988a. Serum profiles of vitamins A, E, and D in mares and foals during different seasons. J. Anim. Science 66:1418.

Maenpaa, P.H., A. Pirhonen, and E. Koskinen. 1988b. Vitamin A, E, and D nutrition in mares and foals during winter season: Effect of feeding two different vitamin-mineral concentrates. J. Anim. Sci. 66:1424.

Marcella, K. 1997. E is for "ease": Vitamin E may ease the symptoms of motor neuron disease and other stressors. Equine Athlete. Nov/Dec p. 28.

McMeniman, N.P. J.R. Thorton, and K.F. Dowsett. 1995 Effect of grain type and vitamin supplementation on performance of exercising horses. Equine Vet J. Supplement 18:367.

McMeniman, N.P. and H.F. Hintz. 1992. Effect of vitamin E status on lipid peroxidation in exercised horses. Equine Vet. J. 24:482.

Moore, R.M. and C.W. Kohn. 1991. Nutritional muscular dystrophy in foals. Compendium for Continuing Education 13/3:476.

Neilsen, B.D., G.D. Potter, L.W. Greene, E.L. Morris, M. Murray-Gerzik, W.B. Smith, and M.T. Martin. 1995. Does the onset of training alter mineral requirements in the young racing Quarter Horses? Proc. 14th Equine Nutr. Physiol. Soc. Symp. 70.

NRC. 1989. Nutrient requirements of horses. (5th Ed.) National Academy Press, Washington D.C.

Parker, A.L., L.M. Lawrence, S. Rokuroda and L.K. Warren. 1997. The effects of niacin supplementation on niacin status and exercise metabolism in horses. Proc. 15th Equine Nutr. Physiol. Soc. Symp. 19.

Ralston, S.L. 1999. Management of geriatric horses. In: Advances in Equine Nutrition. Proc. 1999 Equine Nutr. Conf. for Feed Manufacturers. 57.

Siciliano, P.D. and C.H. Wood. 1993. The effect of added dietary soybean oil on vitamin E status of the horse. Proc. 13th Equine Nutr. Physiol. Soc. Symp. 3.

Siciliano, P.D., A.L. Parker, and L.M. Lawrence. 1997. Effect of dietary vitamin E supplementation on the exercising hrose. Proc. KER 1997 Equine Nutr. Conf. 33.

Siciliano, P.D., C.E. Kawcak, and C.W. McIlwraith. 1999a. The effect of initiation of exercise training in young horses on vitamin K status. Proc. 16th Equine Nutr. Physiol. Soc. Symp. 92.

Siciliano, P.D., L.K. Warren, and L.M. Lawrence. 1999b. Changes in vitamin K status over time in the growing horse. Proc. 16th Equine Nutr. Physiol. Soc. Symp. 94.

Snow, D.H. and M. Frigg. 1987. Plasma concentrations at monthly intervals of ascorbic acid, retinol, b-carotene and a-tocopherol in two Thoroughbred racing stables and the effects of supplementation. Proc. 10th Equine Nutr. Physiol. Soc. Symp. 55.

Szulc, P., M. Arlot, M.C. Chapuy, F. Duboef, P.J. Meunnier, and P.D. Delmas. 1994. Serum undercarboxylated osteocalcin correlates with hip bone mineral density in elderly women. J. Bone Min. Res. 9:1591.

Wooden, G.R. and A.M. Papas. 1991. Utilization of various forms of vitamin E by horses. Proc. 12th Equine Nutr. Physiol. Soc. Symp. 265.

MICROMINERAL REQUIREMENTS IN HORSES

JOE D. PAGAN

Kentucky Equine Research, Inc., Versailles, KY

Introduction

Minerals required in minute amounts by horses are specified as microminerals. These nutrients play an important role in a wide range of biochemical systems which affect virtually every metabolic function in the horse. With the exception of selenium, little attention has been given to microminerals in horse nutrition until shortly before the publication of the current NRC Nutrient Requirements of Horses in 1989. Since then, a great deal of research has been directed towards microminerals, particularly as they affect skeletal development in growing horses. Still, many questions remain unanswered about specific requirements for microminerals in many classes of horses. This paper will briefly review the principal functions of copper, chromium, iodine, iron, manganese, selenium and zinc and will highlight recent research that has been conducted with each micromineral in horses. It will also attempt to combine these data with practical experience to provide recommendations for inclusion of these minerals in horse rations.

Expressing Micromineral Requirements

The 1989 NRC expresses micromineral requirements as a concentration (ppm or mg/kg) of the dry matter intake of a total ration. While this is an easy way to remember requirements, it is fundamentally flawed because it presumes a single dry matter intake for a particular class of horse when in reality dry matter intake can range considerably depending on the overall energy density of the ration that is consumed. Since most of the micromineral requirements are not actually related to the energy density of the ration, a better way to describe micromineral requirements would be on a daily intake basis (mg/day) or on a body weight basis (mg/kg BW/day). This paper will attempt to describe requirements both in terms of concentration of diet dry matter (ppm) and on the basis of daily intake.

Copper

Copper (Cu) is essential for proper functioning of enzymes involved in the synthesis and maintenance of elastic tissue, mobilization of iron stores, preservation of the integrity of mitochondria, and detoxification of superoxide. Copper has received a great deal of attention since the last publication of the NRC because of its purported role in the pathogenesis of developmental orthopedic disease (DOD).

The 1989 NRC estimated that all classes of horses require 10 mg Cu/kg of dry diet. Pagan (1998a) and Hudson et al. (2000) both estimated that 500-600 kg

317

mature, idle horses require about 95-100 mg Cu/day (10.5-11 ppm at 9 kg/d dry matter intake), which is very close to the NRC. Other studies, however, suggest that the requirements of copper for young growing horses and broodmares are considerably higher, especially in certain breeds.

New Zealand researchers (Pearce et al., 1998a) studied the effect of copper supplementation on the incidence of DOD in Thoroughbred foals. Pregnant Thoroughbred mares (n=24) were divided into either copper supplemented or control groups. Live foals born to each group of mares were also divided into copper supplemented or control groups. The four treatment groups were: 1) mares supplemented with copper, but their foals were not supplemented; 2) both mares and foals were supplemented with copper; 3) mares were not supplemented, but their foals received supplementation; 4) neither mares nor foals received supplementation.

Supplemented mares received 0.5 mg Cu/kg body weight (BW) daily (~30 ppm at 9 kg/d dry matter intake) while copper supplemented foals received 0.2 mg Cu/kg BW from 21-49 days of age and 0.5 mg Cu/kg BW (~20 ppm at 5-6 kg/d dry matter intake) from 50 days to 150 days. Mares were supplemented for the final 13 to 25 weeks of gestation. At 150 days of age, the foals were sacrificed and an exhaustive postmortem examination was performed which included investigation of all limb and cervical spine articulations and examination of the physes from the proximal humerus, proximal and distal radius and tibia and distal femur, third metacarpus and third metatarsus. The number of articular and physeal cartilage lesions was noted for each treatment group along with a physitis score that was determined from radiographs of the distal metatarsus.

Copper supplementation of mares was associated with a significant reduction in the physitis scores (p<0.01), assessed radiographically, of the foals at 150 days of age. Foals from mares that received no supplementation had a mean physitis score of 6, while foals out of supplemented mares had a mean score of 3.7. A lower score indicates less physitis. When only foals were supplemented with copper, no significant effect on physitis scores was noted. There was a significantly lower (p<0.05) incidence of articular cartilage lesions in foals from mares supplemented with copper. However, no significant effects on articular and physeal cartilage lesions occurred in foals supplemented with copper.

Two North American dose-response studies examining the effect of increased dietary copper intakes on bone and cartilage abnormalities (Knight et al., 1990; Hurtig et al., 1993) found that the incidence of DOD decreased by increasing the copper content of the diet above NRC recommendations. In Knight's study, both mares and their foals received copper supplementation, so it is difficult to determine whether the effect resulted from supplementation of the mare or the foal. New Zealand research (Pearce et al., 1998a,b) would suggest that supplementation of the pregnant mare is more important than supplementation of the foal. Oral copper supplementation of mares in late gestation altered the copper balance in these horses and resulted in an increase in the foal's liver copper stores at birth. Increased liver copper stores of the neonate may be important for ensuring healthy development of the skeleton during the period of maximum postnatal growth.

The studies cited above certainly provide proof that copper supplementation of mares and their foals can play an important role in skeletal development. Copper is not, however, the only factor involved in the pathogenesis of DOD, and it is questionable whether the lesions produced by copper deficiency are the same as those most often seen in the field. Pagan and Jackson (1996) documented the incidence of DOD on a commercial Thoroughbred farm over a four-year period. Two hundred and seventy-one foals were monitored. DOD was diagnosed in 10% of the foals even though the horses were fed a ration that provided pregnant mares >250 mg Cu per day (~30 ppm at 9 kg/d dry matter intake) and foals began receiving copper supplementation at 90 days of age.

Copper deficiencies may either be primary in origin because of a lack of copper intake or induced (secondary) due to interactions with other substances in the ration. Zinc (Zn) and molybdenum (Mo) have often been implicated as minerals that can interfere with copper absorption in horses, but several studies have suggested that neither Zn (Coger et al., 1987; Young et al., 1987; Bridges and Moffitt, 1990; Pagan, 1998b) nor Mo (Cymbaluk et al., 1981; Strickland et al., 1987) affects copper utilization when fed at levels found in practical diets. Pagan (1998b) found significant negative correlations between true copper digestibility and the concentration of both crude protein and calcium in thirty different diets. These interactions may be particularly relevant where horses are fed predominantly legume forage.

Chromium

Chromium (Cr) is a component of glucose tolerance factor (GTF). GTF is thought to potentiate the action of insulin in Cr-deficient tissue (Mertz, 1992). Insulin has anabolic characteristics as it promotes glucose uptake by the cell, stimulates amino acid synthesis and inhibits tissue lipase. Chromium excretion is greater in athletic than in sedentary humans and the chromium requirement is increased by physical activity (Anderson et al., 1991). Chromium supplementation has increased lean body mass in humans and pigs and has resulted in a partitioning effect on nutrients which favors tissue anabolism and muscle protein accretion. In calves chromium excretion is greater during stress and chromium supplementation has resulted in a stimulation of the immune system and less mortality and morbidity in shipped feedlot cattle. The 1989 NRC does not include a recommendation for chromium supplementation in horses.

Pagan et al. (1995) reported that supplementing performance horses with 5 mg/day of chromium (0.5 ppm at 10 kg/d dry matter intake) in the form of chromium yeast had a beneficial effect on the response of horses to exercise stress. Horses were subjected to a standardized exercise test on a high speed treadmill and blood and heart rate were monitored. Horses receiving chromium cleared blood glucose following a meal more quickly than control horses and showed lower peak insulin values and lower cortisol levels. Chromium supplemented horses also had higher triglyceride values during exercise indicating perhaps more efficient fat mobilization. There was no difference in the heart rate in response to exercise between the two groups but peak lactic acid concentrations in the chromium supplemented group of horses were significantly lower than for the controls.

Iodine

Iodine (I) is an essential nutrient for reproduction and normal physiological function in the horse. Thyroxine (T_4) contains iodine and this hormone, along with triiodothyronine (T_3), has powerful effects on the overall health of the horse. These hormones influence nearly every process in the body, from heat regulation and feed utilization to proper bone growth and maturation. Based on data from other species, the 1989 NRC estimated that the iodine requirement of horses was 0.1-0.6 mg/kg of diet.

Nearly 75% of the iodine in an animal's body is in the thyroid gland. Iodine deficiency may result in goiter as the thyroid becomes enlarged in an attempt to produce adequate levels of thyroxine. In the horse, goiters often occur in the foal at birth. Foal goiter may result from a deficiency in iodine in the mare's ration during pregnancy or it may be caused by a goitrogenic substance. Symptoms of iodine deficiency may be a stillborn foal or a very weak foal that cannot stand and nurse. The foal may also have a rough haircoat, contracted tendons, angular limb deformities or other abnormal bone development. A Russian study (Kruzkova, 1968) indicated that mares which had shown anovulatory cycles responded to iodine supplementation.

While iodine deficiency is the primary cause of goiter in foals, excessive levels of iodine may also cause this condition. The maximal tolerable dietary concentration of iodine has been estimated to be 5 mg/kg (ppm) of dry matter (NRC, 1980), equivalent to 50 mg of iodine/day for a horse consuming 10 kg of dry matter daily. The horses most sensitive to high iodine levels are foals from mares who are supplemented with high levels of iodine. Iodine is concentrated across the placenta and in milk so that the fetus and nursing foal receive much higher concentrations than are present in the mare's ration. Therefore, goiters may be present in newborn foals while sparing the mother. A dietary intake of 83 mg I/day is the lowest level reported to have caused goiter in a horse more mature than a suckling foal (Drew et al., 1975).

Baker and Lindsey (1968) reported that goitrous foals were born on three farms which were feeding mares high levels of iodine. The incidence of goiter was proportional to the level of iodine fed and equaled 3% on one farm feeding 48-55 mg I/day, 10% on a farm feeding 36-69 mg I/day and 50% on another farm feeding 288-432 mg I/day. A neighboring farm which did not have any goitrous foals fed iodine at a rate of 6.3-7 mg I/day. In a separate paper, Sipple (1969) reviewed a case in which 11% of the foals born on a farm had goiters. Analysis of the diet revealed that the mares received between 160-400 mg I/day. Coincidentally, the author discovered that the manager of this farm was the brother of the manager of one of the farms in Baker's study in Florida. Apparently, the Florida horseman had prescribed the same iodine supplement for his brother's horses 1,000 miles away.

Drew et al. (1975) reported that on one stud farm in England four foals were born with greatly enlarged thyroids and leg weaknesses. One mare also had an enlarged thyroid. Feed analysis showed that the mares had received 83 mg I/day from a proprietary feed during pregnancy. The year before the introduction of this proprietary feed, the mares received a vitamin/mineral supplement which supplied about 12 mg I/day and there was no problem with goiter on the farm.

It appears from these reports that around 50 mg of dietary iodine is required in the daily rations of mares to produce any incidence of goiters in their foals. One other study (Driscoll et al., 1978) reported goitrous foals from mares receiving 35 mg I/day. There is some question, however, about what levels of iodine the mares in this study actually received. The authors reported that the mares were given 12 ounces per day of a supplement which was reported to contain 58 ppm iodine. The guaranteed analysis on the product's label stated that it contained 340 ppm iodine and independent analyses of the same product revealed that it contained at least 580 ppm iodine, a level 10-fold higher than reported in the paper. Using the manufacturer's guarantee, the mares would have received a total of 131 mg I/day and according to the independent analyses, a total of 212 mg I/day. These levels are within the ranges reported to produce goitrous foals in other studies.

Toxic dietary iodine concentrations may result from adding excessive supplemental iodine, such as from ethylenediamindihydroiodide (EDDI), to concentrates or from using feedstuffs high in iodine. A common feedstuff that may contain excess iodine is kelp (*Laminariales*), a specific family of seaweeds that may contain as much as 1,850 ppm iodine (NRC, 1989). Unfortunately, people have a tendency to classify all seaweeds as kelp just as the layman might consider every breed of horse a Thoroughbred. There are numerous other specific seaweeds, including *Fucaceae, Palmariaceae, Gigartinaceae, Bangiaceae,* and *Ulvaceae* that contain considerably less iodine than kelp.

Iron

Iron (Fe) is the trace mineral most often associated with exercise, even though its true relevance is questionable. A recent survey conducted at a California racetrack indicated that many trainers used some type of iron supplement (Carlson, 1994). This concern with iron stems from the well-known function of iron as part of the heme molecule. The first symptom associated with iron deficiency is anemia. The anemia associated with iron deficiency is hypochromic, microcytic anemia. There are few instances, however, when practical diets would result in iron deficiency anemia. In the previously mentioned study of Carlson, horses which received supplemental iron had iron levels which were in the normal range for adult horses. Very few of the unsupplemented horses examined had any evidence of anemia and those with resting hematocrits below 34% (defined as anemia) showed no evidence of impaired iron status. This scenario is frequently the case and it is rare that a horse with a lowered hematocrit responds to supplemental iron. Lawrence et al. (1987) failed to show an increase in hemoglobin, hematocrit or serum iron when ponies were supplemented with high levels of iron. More times than not, low hematocrits are an indicator of infection, low-grade systemic disease or even marginal B vitamin status.

Clinically significant anemia in the athletic horse is rare. Exceptions to this are severely parasitized horses, horses with gastric ulceration that leads to blood loss and perhaps horses which suffer from severe EIPH (exercise-induced pulmonary hemorrhage). Meyer (1987) suggested that the iron requirement of the 500 kg horse is 500, 600 and 1200 mg/day for light, moderate and heavy exercise, respectively.

High levels of iron supplementation may affect the availability of other minerals in the diet. Lawrence et al. (1987) reported that high levels of dietary iron supplementation (500 and 1000 mg/kg) depressed both serum and liver zinc. Pagan (1998b), however, failed to show any correlation between iron content and mineral digestion in diets with a wide range of iron content (127-753 ppm). Most of the iron in this study was not supplemented and was probably in the form of iron oxide.

Manganese

Manganese (Mn) is essential for carbohydrate and lipid metabolism and for synthesis of the chondroitin sulfate necessary for cartilage formation. The 1989 NRC based its recommendation of 40 mg Mn/kg of diet on research from other species. Pagan (1998a) estimated that the maintenance requirement for manganese in idle 500-600 kg horses equaled about 385 mg/d, or 43 ppm of the diet, when the horses consumed 9 kg of dry matter per day. Recently, Hudson et al. (2000) measured manganese retention when mature idle horses were fed four different levels of copper, zinc and manganese supplementation (0, 50, 100 or 200% of NRC) added to the same basal ration. The maintenance requirement for manganese calculated from linear regression of intake against retention equaled 540 mg per day or 60 ppm of diet (Figure 1). When data from these two studies were combined, true manganese digestibility was estimated to be 37%, endogenous losses equaled 151 mg/d (0.27 mg/kg BW), and the maintenance requirement equaled 409 mg/d or 45 ppm of diet.

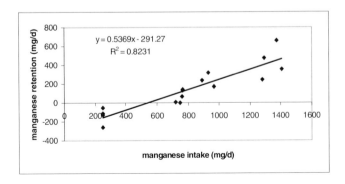

Figure 1. Relationship between manganese intake and retention.

Selenium

Selenium (Se) plays an important role in the maintenance of membrane integrity, growth, reproduction and immune response. A deficiency of Se in foals may produce white muscle disease, a myopathy which results in weakness, impaired locomotion, difficulty in suckling and swallowing, respiratory distress, and impaired cardiac function (Dill and Rebhun, 1985).

Strenuous exercise is known to induce oxidative stress, leading to the generation of free radicals. An increased generation of free radicals may induce lipid peroxidation and tissue damage in both the respiratory system and working muscle. This is particularly true if the animal has a deficient or impaired antioxidant status. Many antioxidants, including glutathione peroxidase (GSH-Px), are selenoproteins, making selenium an extremely important mineral for performance horses.

Although the Food and Drug Administration (FDA) has approved maximal selenium supplementation at 0.3 mg/kg of dry matter in complete feeds for cattle, sheep, and swine (FDA, 1987), selenium supplementation of equine feeds is restricted only by nutritional recommendations and industry practices (NRC, 1989). The selenium requirement for mature idle horses was estimated by the NRC to be 0.1 mg/kg of diet. This requirement is based on studies that evaluated the relationship between selenium intake and blood selenium in mature idle horses (NRC, 1989). Shelle et al. (1985) investigated the effect of supplemental selenium on plasma selenium and on glutathione peroxidase in Arabian and cross-bred horses subjected to a conditioning program. They reported that conditioning increased erythrocyte glutathione peroxidase activity and suggested that horses at high work intensities may have higher requirements for selenium than the 0.1 ppm requirement suggested by the NRC. Stowe (1998) has suggested that the appropriate concentration of selenium in the total diet of a horse is 0.3 ppm. This would mean that if a concentrate mix was 50% of the diet and the forage component of the diet supplied 0.1 ppm Se, the grain mix would need to supply roughly 0.5 ppm.

Recent research from the University of Kentucky (Janicki et al., 2000) demonstrated that foals from mares receiving 3 mg Se/d had higher concentrations of IgG at 2 wk (P<0.05) and at 4 and 8 wk (P<0.1) compared to foals from mares receiving 1 mg Se/d.

Selenium in forages and seed grains is normally present as organic selenium in the form of selenocystine, selenocysteine, and selenomethionine. Sodium selenite and sodium selenate are common inorganic sources of supplemental selenium for horses, and evidence in horses (Podoll et al., 1992) indicates there is no difference between them in potency as measured by blood selenium status. Measurement in laboratory animals, however, shows that organic plant sources of Se are more potent than inorganic (Frape, 1998). Pagan et al. (1999) measured selenium utilization by exercised Thoroughbreds in two diets containing 0.41 ppm selenium. In one diet, about 3/4 of the selenium was from sodium selenite and in the second diet it was from selenium-enriched yeast. The apparent digestibility of yeast selenium was greater than for selenite (57%

vs 51%). Selenium retention was 25% greater for yeast selenium than for selenite. Exercise increased urinary excretion of selenium more in the selenite group than in the yeast-fed group and plasma selenium remained higher in the selenium yeast group than in the selenite group. More research is needed to quantify the selenium requirements of horses at various intensities of exercise and to determine the effect of form of selenium on antioxidant status.

The maximal tolerable level of selenium in horses is estimated to be 2 mg/kg of diet (NRC, 1980). Acute selenium toxicity (blind staggers) is characterized by apparent blindness, head pressing, perspiration, abdominal pain, colic, diarrhea, increased heart and respiration rates and lethargy (Rosenfeld and Beath, 1964). Chronic selenium toxicity (alkali disease) is characterized by alopecia, especially of the mane and tail, as well as cracking of the hooves around the coronary band (Rosenfeld and Beath, 1964; Traub-Dargatz and Hamar, 1986).

Zinc

Zinc (Zn) is present in the body as a component of many metalloenzymes. The biochemical role of zinc relates largely to the functions of these enzymes. The 1989 NRC suggests that all classes of horses require 40 mg zinc/kg of dry matter. Pagan (1998a) estimated that idle, mature 500-600 kg horses require 257 mg/d, or 28.5 ppm zinc at 9 kg/d dry matter intake. Hudson et al. (2000) estimated that horses of this size require 230 mg Zn/d. Combining data from these two studies (Figure 2) yields a Zn requirement of 248 mg/d, or 27.5 mg Zn/kg diet. This requirement assumes a true zinc digestibility of 21% and endogenous losses of 52 mg/d (0.10 mg/kg BW). Pagan (1998b) evaluated interactions between zinc digestibility and a number of nutrients in 30 different diets. The only nutrient that was significantly correlated to zinc digestibility was magnesium. None of the trace minerals, including iron, affected zinc digestibility.

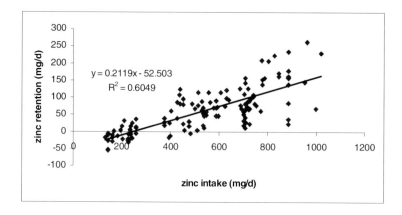

Figure 2. Relationship between zinc intake and retention.

The 1989 NRC noted two studies in which 40 mg Zn/kg diet resulted in acceptable growth rates and no deficiency symptoms in young horses. Ott and Asquith (1987), however, suggested that while yearlings grow satisfactorily with this level of supplementation, optimum bone mineralization may require intakes above NRC recommendations. The current NRC recommendations for zinc appear adequate for mature horses, but they may be low for growing horses.

Recommendations

Table 1 contains a comparison of the current NRC recommendations and those currently used by Kentucky Equine Research (KER). Both sets of recommendations are given as concentrations in the total ration for comparison purposes because this is how the NRC requirements are commonly expressed. The KER recommendations are listed as ranges to account for variability in dry matter intake within each class of horse. KER recommendations are generally more liberal than those of the NRC for several reasons. First, the NRC requirements are considered "minimum amounts needed to sustain normal health, production, and performance of horses." The KER recommendations are generally higher than minimal to account for potential metabolic differences among horses and to protect against possible interactions with other substances in the ration that might result in depressed digestibility or utilization. Second, there is current research which shows that some of the NRC recommendations are too low, particularly for pregnant mares and growing horses. Finally, the NRC provides no recommendation for chromium for any class of horse. Research in other species and with exercised horses suggests that chromium may be an essential mineral for all classes of horses. More research is certainly warranted in this area.

	Maintenance		Pregnant Mares		Lactating Mares		Growing Horses		Working Horses	
	NRC	KER	NRC	KER	NRC	KER	NRC	KER	NRC	KER
Iron	40	40	50	40-50	50	40-50	50	40-50	40	40-50
Manganese	40	40-50	40	40-60	40	40-50	40	60-80	40	40-60
Copper	10	10-15	10	15-25	10	10-15	10	20-30	10	10-15
Zinc	40	40-50	40	50-60	40	40-50	40	60-80	40	40-60
Chromium	None	0.1-0.3	None	0.2-0.4	None	0.2-0.4	None	0.3-0.4	None	0.4-0.5
Selenium	0.1	0.1-0.3	0.1	0.2-0.3	0.1	0.2-0.3	0.1	0.2-0.3	0.1	0.2-0.3
Iodine	0.1-0.6	0.1-0.2	0.1-0.6	0.15-0.3	0.1-0.6	0.15-0.25	0.1-0.6	0.15-0.25	0.1-0.6	0.15-0.25

Table 1. A comparison of 1989 NRC micromineral recommendations and those currently used by Kentucky Equine Research (KER). Recommendations are expressed as a concentration (mg/kg) of dry matter in the total ration.

References

Anderson, RA, Bryden, NA, Polansky, MM and JW Thorp. 1991. Effect of carbohydrate loading and underwater exercise on circulating cortisol, insulin and urinary losses of chromium and zinc. Eur. J. Appl. Physiol. 63:146.

Baker, HJ and JR Lindsey. 1968. Equine goiter due to excess dietary iodine. J. Am. Vet. Med. Assoc. 153:1618.

Bridges, CH and PG Moffitt. 1990. Influence of variable content of dietary zinc on copper metabolism of weanling foals. Am. J. Vet. Res. 51. 275-280.

Carlson, GP. 1994. Forty years later - can we diagnose and treat anemia yet? Proc. 16th Bain-Fallon Memorial Lectures. Aus. Eq. Vet. Assn. Queensland, Australia.

Coger, LS, HF Hintz, HF Schryver, and JE Lowe. 1987. The effect of high zinc intake on copper metabolism and bone development in growing horses. Proc. Equine Nutr. Physiol. Soc. Symp. 10: 173-177.

Cymbaluk, NF, HF Schryver, and HF Hintz. 1981. Influence of dietary molybdenum on copper metabolism in ponies. J. Nutr. 111:96-106.

Dill, SG and WC Rebhun. 1985. White muscle disease in foals. Comp. Cont. Ed.7:S627.

Drew, B, WP Barber, and DG Williams. 1975. The effect of excess iodine on pregnant mares and foals. Vet Rec. 97:93.

Driscoll, J, HF Hintz, and HF Schryver. 1978. Goiter in foals caused by excess iodine. J. Am. Vet. Med. Assoc. 173:858.

Frape, D. 1998. Equine Nutrition and Feeding. (2nd Ed.). Blackwell Science Ltd, UK.

Food and Drug Administration. 1987. Food additives permitted in feed and drinking water of animals. Fed. Reg. 52 (Part 573, No. 65):10887.

Hudson, CA, JD Pagan, KE Hoekstra, A Prince, S Gardner, and RJ Geor. 2000. The effects of intake level on the digestibility and retention of copper, zinc and manganese in sedentary horses. Proceedings of the KER Nutrition Conference for Feed Manufacturers, Lexington, KY.

Hurtig, MB, SL Green, H Dobson, Y Mikuni-Takagaki, and J Choi. 1993. Correlative study of defective cartilage and bone growth in foals fed a low-copper diet. Equine Vet. J., Suppl. 16:66-73.

Janicki, KM, LM Lawrence, T Barnes and CI O'Connor. 2000. The effect of dietary selenium source and level on broodmares and their foals. Proceedings of the KER Nutrition Conference for Feed Manufacturers, Lexington, KY.

Knight, DA, SE Weisbrode, LM Schmall, SM Reed, AA Gabel, LR Bramlage and WI Tyznik. 1990. The effects of copper supplementation on the prevalence of cartilage lesions in foals. Equine Vet. J. 22:426-432.

Kruzkova, E. 1968. Mikroelementy I vos proizvoditel'naja funkeija kobyl. Tr. Vses. Inst. Konevodstvo, 2:28 (as cited in Nutr. Abst. Rev. 39:807, 1968).

Lawrence, LM, EA Ott, RL Asquith and GJ Miller. 1987. Influence of dietary iron on growth, tissue mineral composition, apparent phosphorus absorption and chemical properties of bone. Proc. 10th Equine Nutr. Physiol. Soc. Symp.

Mertz, W. 1992. Chromium, history and nutritional importance. Biol Trace Elem. Res. 32:3.

Meyer, H. 1987. Nutrition of the equine athlete. In: Proc. 2nd International Conference on Equine Exercise Physiology. J.R. Gillespie and N.E. Robinson (eds).

National Research Council. 1980. Mineral Tolerance of Domestic Animals. Washington, D.C.: National Academy Press.

National Research Council 1989. Nutrient Requirements of Horses. Washington, D.C.: National Academy Press.

Ott, EA and RL Asquith. 1987. The influence of trace mineral supplementation on growth and bone development of yearling horses. Proc. Equine Nutr. Physiol. Soc. Symp. 10:185-192.

Pagan, J.D., S.G. Jackson and S.E. Duren. 1995. The effect of chromium supplementation on the metabolic response to exercise in Thoroughbred horses. Proc. Equine Nutr. Physiol. Soc. Symp.

Pagan, JD and SG Jackson. 1996. The incidence of developmental orthopedic disease on a Kentucky Thoroughbred farm. Pferdeheilkunde 12:351-354.

Pagan, JD. 1998a. Nutrient digestibility in horses. In: J.D. Pagan (Ed.) Advances in Equine Nutrition. Nottingham University Press, UK. 77-83.

Pagan, JD. 1998b. Factors affecting mineral digestibility in horses. Proceedings of the KER Nutrition Conference for Feed Manufacturers, Lexington, KY.

Pagan, JD, P Karnezos, MAP Kennedy, T Currier and KE Hoekstra. 1999. Effect of selenium source on selenium digestibility and retention in exercised Thoroughbreds. Proc. Equine Nutr. Physiol. Soc. Symp.

Pearce, SG, EC Firth, ND Grace and PF Fennessy. 1998a. Effect of copper supplementation on the evidence of developmental orthopaedic disease in pasture-fed New Zealand Thoroughbreds. Equine Vet. J. 30:211-218.

Pearce, SG, ND Grace, EC Firth, JJ Wichtel, SA Holle and PF Fennessy. 1998b. Effect of copper supplementation on the copper status of pasture-fed young Thoroughbreds. Equine Vet. J., 30:204-210.

Podoll, KL, JB Bernard, DE Ullrey, SR DeBar, PK Ku, and WT Magee. 1992. Dietary selenate versus selenite for cattle, sheep, and horses. J. Anim. Sci. 7 70:1965-1970.

Rosenfeld, I and OA Beath. 1964. Selenium: Geobotany, Biochemistry, Toxicity, and Nutrition. New York: Academic Press.

Shelle, JE, WD Vanhuss, JS Rook and DE Ullrey. 1985. Relationship between selenium and vitamin E nutrition and exercise in horses. Proc. Equine Nutr. Physiol. Soc. Symp.

Sipple, WL 1969. A veterinarian's approach to stud farm nutrition. Equine Vet. J.1:203.

Strickland, KF, M Woods and J Mason. 1987. Dietary molybdenum as a putative copper antagonist in the horse. Equine Vet. J. 19:50-54.

Stowe, HD 1998. Selenium supplementation for horse feed. In: J.D. Pagan (Ed.) Advances in Equine Nutrition. Nottingham University Press, UK. 97-103.

Traub-Dargatz JL and DW Hamar. 1986. Selenium toxicity in horses. Comp. Cont. Vet. Ed. (Equine) 8:771.

Young, JK, GD Potter, LW Green, SP Webb, JW Evans and GW Webb. 1987. Copper balance in miniature horses fed varying amounts of zinc. Proc. Equine Nutr. Physiol. Soc. Symp. 10:153-157.

MACROMINERALS - CALCIUM, PHOSPHORUS AND MAGNESIUM

HAROLD F. HINTZ
Cornell University, Ithaca, NY

The 1989 National Research Council (NRC) requirements for calcium, phosphorus and magnesium are the latest published guidelines for equine ration formulation. Since that time, research has led nutritionists to suggest changes in requirements.

Calcium

The calcium requirements were estimated for the 1989 NRC publication by the factorial method. The maintenance requirement was estimated and then estimates were added for the amount of calcium needed for fetal growth, milk production or growth of foal to provide requirements for pregnant mares, lactating mares and growing animals, respectively.

Maintenance

Estimates for maintenance requirements were, for the most part, based on measurements of endogenous losses in feces and urine. The losses were measured with the use of isotopes or by extrapolation from studies in which Ca balances were determined in animals fed a wide range of Ca intake. In the latter case, the endogenous loss was estimated by extrapolating the loss of Ca at zero Ca intake. The amount of dietary Ca needed to replace endogenous losses was calculated by assuming the efficiency of absorption was 50%. There are several potential sources of error in such an approach. The length of the balance studies was seldom greater than ten days. It is quite probable a horse can adapt to a lower level of calcium when fed over long periods.

The validity of the estimate of 50% efficiency of absorption is of concern. Many factors such as Ca:P ratio, form of Ca, level of Ca intake and presence of inhibitors such as phytate and oxalate can influence the efficiency of absorption.

For example as the dietary intake of Ca decreases, efficiency of absorption may increase. Absorptive rates of 70% or greater are often found when a readily available source of Ca was fed at near maintenance levels.

The 1989 NRC maintenance requirement for a 500 kg horse is 20 g of calcium. Pagan (1994) reported endogenous losses for Ca that were higher than estimated by NRC but he also found higher true digestibility. Therefore, the two factors combined resulted in maintenance requirements similar to those suggested by NRC.

It was thought that the estimates of the Ca maintenance requirement made by the factorial method were probably on the liberal side because of the potential adaptations that could be made by the horse over a long period. It was also thought that liberal estimates were better than conservative estimates, thereby decreasing the chance for the horse to suffer from Ca deficiency. The horse can easily adapt to moderate Ca excesses by limiting absorption and increasing urinary excretion.

A direct approach in which bone was actually evaluated would be preferred to the indirect approach used above. For example, it would be desirable to feed horses various levels of Ca for prolonged periods and measure bone mineralization over those periods to determine if bone integrity is maintained.

I found no studies that suggested horses at maintenance fed Ca intakes satisfying the 1989 recommendations were at risk of being Ca deficient, at least in cases where no inhibitors such as excess P, phytate and/or oxalate were present in the feed.

The increased use of commercial feeds over the last several decades has significantly decreased the incidence of Ca deficiencies. However, the potential for Ca deficiency cannot be ignored. Many case reports have demonstrated that severe Ca deficiency, hypocalcemia or nutritional secondary hyperparathyroidism still exist worldwide because of inadequate feeding practices (Richardson et al., 1991; David et al., 1997; Hintz, 1997a; Ramirez and Seahorn, 1997; Couetil et al., 1998; Hudson et al., 1999; Wisniewski et al., 1999).

It is frustrating that the importance of dietary Ca is still not understood by all those who care for horses, particularly when Ca supplements can usually be purchased inexpensively. The lack of attention to the Ca content of the diet has caused many horses to suffer, even in modern times.

Pregnant Mares
Calcium deposition in the fetus was estimated to be 11.1, 25.3 and 11.4 grams of calcium daily per kg of the mare's body weight during months 9, 10 and 11 of gestation, respectively (Meyer and Ahlswede, 1976; Drepper et al., 1982). The NRC decided to use a mean daily deposition rate of 15.9 mg of calcium/kg of mare body weight during the last three months. Again, as for maintenance, it was assumed the efficiency of absorption was 50%. The mean dietary calcium required for deposition was added to maintenance needs and divided by the mean digestible energy (DE) requirement for the same period. This value was then multiplied by the daily DE requirement for each month. The daily requirements for various weights of mares are shown in Table 1.

Table 1. Estimates of daily calcium requirements for pregnant mares (NRC).

Body Weight	Month of Gestation		
kg	9	10	11
400	28	29	31
500	35	35	37
600	41	42	44
700	45	46	49
800	48	49	52
900	51	52	55

The 1989 NRC estimates were slightly greater than those of 1978 and much greater than the estimates of 1949, 1961 and 1973 (Table 2). If the grain contained 3.3 Mcal/kg and hay contained 2 Mcal/kg of dry matter, it could be calculated that the diet should contain 0.45% calcium compared to 0.22% calcium recommended in 1949.

Table 2. Changes in estimates of daily calcium requirements by NRC[a].

Year	Edition	g/day
1949	1	16.5
1961	2	17
1973	3	24
1978	4	34
1989	5	35-37

[a]500 kg pregnant mare in late gestation.

Estimates by the French and German counterparts of NRC are similar to those of NRC (Table 3). Perhaps the similarity should be expected because the NRC committee used data from those countries when making its estimates, and the French and German committees used data from the United States. However, it is encouraging that three scientific committees arrived at similar results after independently studying the data that are available.

Table 3. Estimates of calcium requirements of mares in late pregnancy[a].

Source	g/day
NRC (1989)	35-37
Meyer (1994)	38
Martin-Rosset (1990)	38-39

[a]500 kg mare

Martin et al. (1996) studied calcium metabolism by a different method than balance studies. They measured changes in serum concentrations of calcium and parathyroid hormones in mares fed diets containing calcium concentrations below (0.35%) and above (0.55%) the NRC requirement of 0.45%. They concluded, "The practical importance of dietary calcium for mares during the last month of pregnancy has been reinforced by our findings." They found less extreme perturbations of serum total calcium, ionized calcium and parathyroid hormone in the mares that were fed 0.55% calcium than in mares fed 0.35% calcium. They suggested that the optimal concentration of dietary calcium for prepartum mares was closer to 0.55% than 0.35%.

Glade (1993) estimated metacarpal breaking strength (MBS) by transmission ultrasonics of mares during the last 12 weeks of gestation and for 40 weeks after parturition. MBS increased during the last 6-10 weeks of gestation in mares fed amounts of Ca similar to NRC recommendations but mares fed 20% less Ca than NRC recommendations did not have an increase in MBS. Furthermore, foals of the mares fed the lower level of Ca had thinner mid-cannon mediolateral diameters and mechanically weaker bones at birth than foals of control mares and the differences persisted for 40 weeks.

The studies by Martin et al. (1996) and Glade (1993) indicate that the pregnant mare probably needs a Ca intake at least as great as NRC recommendation but extrapolations as to the optimum level of Ca from their data are not possible because only two concentrations were used in each of the studies.

Lactating Mares

It was estimated that the milk of mares contains about 1.2 g of Ca per kg of fluid milk during early lactation (foaling to three months) and 0.8 g of Ca per kg of fluid milk during late lactation (three months to weaning) (Schryver et al., 1986). Similar data were recently reported by Grace et al. (1999a). NRC (1989) assumed the mare produced milk at a rate of 3% of body weight in early lactation and 2% in late lactation. Again, an absorption rate of 50% was assumed. The calcium requirements for a 500 kg mare would be 56 g of calcium/day in early lactation and 36 g of calcium/day in late lactation. NRC calculated that the typical diet of the mare should contain 0.47% and 0.33% Ca during early and late lactation respectively.

Glade (1993) reported that mares fed the NRC recommended levels of Ca gradually lost density during the first 12 weeks of lactation, but bone density started increasing at that time and was fully restored at 24 weeks post-parturition. However, mares fed 20% less Ca than recommended had not recovered bone density at 40 weeks after parturition even though the foals were weaned at 20 weeks.

Growing Horses

NRC assumed that the growing horse required about 16 g of Ca/kg of gain. A foal weighing 215 kg with an estimated mature weight of 500 kg and gaining 0.65 kg per day would require 29 g of Ca/day. A foal gaining 0.85 kg/day would need 34 g of Ca/day.

Grace et al. (1999b) measured the body mineral content of 21 foals at about 150 days of age. The foals had grazed on tall fescue pasture. The body levels of calcium, phosphorus and magnesium were similar to values previously reported by Schryver et al. (1974) and utilized by NRC in 1989.

Grace et al. (1999b) calculated that a 200 kg horse gaining 1.0 kg/day would require 28 g of calcium per day. The NRC estimate for the same weight of foal and rate of gain is 30 g of calcium per day. The estimate by Grace et al. (1999b) was lower than NRC values because they estimated the coefficient of absorption to be 70% compared to the 50% used by NRC.

It has been suggested that dietary cation-anion difference (DCAD) could influence calcium utilization and perhaps lead to a calcium deficiency in growing horses (Wall et al., 1997). However, Cooper et al. (2000) recently reported that horses consuming a lower DCAD can compensate for the increased urinary excretion of calcium by enhancing intestinal calcium absorption and thereby maintain calcium status. Young horses fed about 30 g of calcium per day were in positive balance and had no obvious calcium problems when fed high or low DCAD diets.

Young Horses In Training
NRC recommendation for young horses in training was based on extrapolations and the assumption that if the Ca:calorie ratio was maintained the increased feed intake needed to supply the extra energy for work would also supply the calcium needed for bone formation and for that lost in sweat (Hintz, 1997b).

Nielsen et al. (1997a) studied changes in the third metacarpal bone in Quarter Horses put into race training at 18 months of age. Bone density began to decrease at the onset of training and continued to do so until day 62 of training, remaining low through day 104 when it began to increase to day 244. Horses with greater cortical mass in the lateral and medial aspects of the third metacarpal relative to the palmar aspect at the commencement of training had fewer injuries. Could additional calcium help delay the loss of bone density?

Nielsen et al. (1997b) studied calcium balance in Quarter Horses about two years of age. Sixteen horses were started in race training. The trial was blocked into four periods of 28 days. The horses were galloped 5500 m on a track for the first week and 8250 m per week for the last three weeks of the period. During the third period, the horses galloped 4630 m per week and did a weekly sprint of 230 m. During the fourth period, the horses were galloped for 4585 m and sprinted for 275 m per week.

Balance studies were conducted for three days every 28 days. Eight horses were fed about 36 g of Ca per day during the last three periods and eight horses were fed about 30 g of Ca per day. The horses fed the higher Ca intake retained 4-6 more grams of Ca per day during the balance periods than the horses fed the lower intake. This study demonstrated that Quarter Horses, about 24 months of age and weighing around 485 kg, in typical race training required at least 36 g of calcium daily. This is slightly above the NRC estimate of 34 g per day. Nielsen et al. (1997b) recommended that diets for young horses in training should contain 0.4% calcium and that seems to be a prudent recommendation.

Phosphorus

Maintenance
The endogenous losses of phosphorus were estimated at 10 mg/kg of body weight and an absorptive efficiency of 35% was used because maintenance diets are likely to contain plant sources of phosphorus (NRC, 1989). Factors influencing phosphorus absorption include form of phosphorus. Phytate phosphorus, the type often found in grains, was considered not to be as effectively utilized as

inorganic phosphorus. The maintenance requirement of phosphorus of a 500 kg horse was estimated to be 14 g.

Pagan (1994) reported endogenous phosphorus losses slightly lower than NRC but he also reported lower true phosphorus availability. Therefore, the two factors combined for a daily phosphorus requirement similar to that of NRC.

The same potential problems suggested for the calcium requirements determined by the above method exist for phosphorus. Cymbaluk and Christison (1989) suggested that young horses could adapt to lower dietary phosphorus intake. Increased intake of phosphorus does not greatly decrease phosphorus utilization (Schryver et al., 1974), but environment might influence phosphorus utilization. Cymbaluk et al. (1990) reported that the true digestibility of phosphorus was greater in horses housed in a warm barn (mean temperature of 10.9 + 0.66°C) than in a cold barn (mean temperature of -5.2 + 1.72°C).

Age of the horse might also influence digestibility of phosphorus. Horses eight months of age were more efficient than horses 12 months of age (Cymbaluk et al., 1990).

Pregnant Mares
The phosphorus deposition in the fetus was estimated to be 9.4 g/day for a 500 kg mare during the last three months of gestation based on the body composition data of Drepper et al. (1992). An efficiency of absorption of 35% was assumed. Thus, a 500 kg mare would require about 27 g of phosphorus per day.

Lactating Mares
The phosphorus content of fluid milk in early lactation was 0.75 g/kg and decreased to 0.50 g/kg during late lactation (NRC, 1989). Similar values were reported by Grace et al. (1999a). An absorptive efficiency of 45% was assumed because a significant amount of the phosphorus in a ration for a lactating mare would probably come from inorganic sources, which perhaps would be used more effectively than plant sources of phosphorus.

Growing Horses
NRC (1989) suggested that growing horses with an expected mature weight of 500 kg would need 16 to 20 g of phosphorus per day. Grace et al. (1999b) suggested 21 g of phosphorus per day. Cooper et al. (2000) reported that horses about six months of age fed 17 g of phosphorus were in positive phosphorus balance and no bone problems were reported.

Magnesium

Maintenance
There are fewer measures of endogenous losses of magnesium and true digestibility of magnesium available than for calcium and phosphorus. NRC used a value of 6 mg/kg of body weight. Meyer and Ahlswede (1977) suggested a value of 7 mg/kg of body weight whereas Pagan (1994) reported 2.2 mg/kg of body weight. Hintz and Schryver (1972) reported true digestibility values of 40-60%.

Pagan (1994) reported an average of 52%. Thus the data of Pagan would indicate a requirement of 4 g/day for a 500 kg horse whereas NRC suggested 7.5 g/day and Meyer recommended 10 g/day. Further studies are needed to better define the magnesium requirements. However, there is no evidence that the 1989 recommendations are not adequate and the data of Pagan (1994) suggest they can be decreased.

Pregnant Mares
As with calcium and phosphorus, body composition data from the laboratory of Dr. Helmut Meyer (Drepper et al., 1982) were utilized to estimate requirements of mares. Deposition of magnesium was calculated to be 0.3 mg/kg of body weight of the mare and it was calculated a 500 kg mare would need about 9 mg of magnesium per day.

Lactating Mares
It was estimated that the magnesium concentration of milk averages 90 mg/kg during early lactation and 45 mg/kg during late lactation. Thus, a 500 kg mare would need about 11 g of magnesium during early lactation and 8.5 during late lactation.

Growing Horses
NRC estimated that growing horses with an expected mature weight of 500 kg would need about 4 g/day. Grace et al. (1999b) suggested 4.4 mg/day.

Discussion

Nutrient requirements of the macrominerals have apparently not received a great amount of attention in the last decade. Only the calcium requirements of the pregnant mare and of the young horse in training were questioned in the papers that were reviewed. Emphasis on the mineral needs of the performance horse is justified. Musculoskeletal injuries continue to be of major concern. Estberg et al. (1996) reported that 1.7/1000 California Thoroughbred race entrants were euthanized because of catastrophic musculoskeletal injuries. Bailey et al. (1998) reported a rate of 0.6 musculoskeletal injuries per 1000 entrants in flat races and 14/1000 entrants in steeplechases. More (1999) reported a high wastage of Thoroughbreds in Australia. Only 46% of the horses started at 2 or 3 years raced for two years after their first start. Removal from racing was linked to poor performance. The role of musculoskeletal problems in the poor performance remains to be determined.

Many factors such as accumulation of a large total of high-speed distances or rapid accumulation of high speed distances within a two-month period may increase the risk of musculoskeletal injuries (Estberg et al., 1996). Firth et al. (1999) suggested that imaging techniques could be developed to monitor bone changes and thus allow appropriate changes in training intensity to minimize damage. The effect of nutrition on bone changes such as bone density, bone stiffening and shock absorbing capability should receive greater attention using the new methods of bone evaluation.

However, answers about the role of nutrition will not come easily. Porr et al. (1998) reported that stall rest for 12 weeks decreased the bone mineral content of highly conditioned mature Arabian horses. Loss of BMC was not prevented by dietary calcium intake at twice the level recommended by NRC. In conclusion, I agree with Grace et al. (1999b) who stated that dietary requirements of minerals "need to be reviewed as new data come to hand because there is a dearth of information on the endogenous losses and coefficients of mineral elements." However, I think more studies using direct methods to measure the effect of mineral intake on bone integrity are required.

References

Bailey, C.J., Reid, S.W.J., Hodgson, D.R., Bourke, J.M. and Rose, R.J. 1998. Flat, hurdle and steeple racing: Risk factors for musculoskeletal injury. Equine Vet J. 30:498-503.

Cooper, S.R., Topliff, D.R., Freeman, D.W., Breazile, J.E. and Geisert, R.D. 2000. Effect of dietary cation-anion difference on mineral balance, serum osteocalcin concentration and growth in weanling horses. J Eq Vet Sci. 20:39.

Couetil, L.L., Sojka, J.E. and Nachreiner, R.F. 1998. Primary hypoparathyroidism in a horse. J Vet Internal Medicine. 12:45-49.

Cymbaluk, N.F. 1990. Cold housing effects of growth and nutrient demand of young horses. J Anim Sci. 68:3152-3162.

Cymbaluk, N.F. and Christison, G.I. 1989. Effects of dietary energy and phosphorus content on blood chemistry and development of growing horses. J Anim Sci. 67:951-958.

David, J.B., Cohen, N.D. and Nachreiner, R. 1997. Equine nutritional secondary hyperparathyroidism. Compendium Cont Educ Vet. 19:1380-1387.

Drepper, K., Gutte, J.O., Meyer, H. and Schwarz, F.J. 1982. Energie und Nahrstoffbedard landwirtschaftlicher Nutztiere. Nr. 2 Empfehlungen zur Energie- und Nahrstoffversorgung der pferde. Frankfurt am Main, Germany: DLG Verlag.

Estberg, L., Stover, S.M., Gardner, I.A., Drake, C.M., Johnson, B. and Ardans, A. 1996. High-speed exercise history and catastrophic racing fracture in Thoroughbreds. Amer J Vet Research. 57:1549-1555.

Firth, E.C., Delahunt, J., Wichtel, J.W., Birch, H.L. and Goodship, A.E. 1999. Galloping exercise induces regional changes in bone density within the third and radial carpal bones of Thoroughbred horses. Equine Veterinary Journal. 31:111-115.

Glade, M.J. 1993. Effects of gestation, lactation, and maternal calcium intake on mechanical strength of equine bone. J Amer College of Nutr. 12:372-377.

Grace, N.D., Pearce, S.G., Firth, E.C. and Fennessy, P.F. 1999a. Concentrations of macro- and micro-elements in the milk of pasture-fed Thoroughbred mares. Aus Vet J. 77:177-180.

Grace, N.D., Pearce, S.G., Firth, E.C. and Fennessy, P.F. 1999b. Content and distribution of macro- and micro-elements in the body of pasture-fed young horses. Australian Vet J. 77:172-176.

Hintz, H.F. 1997a. Straight from the horse's mouth: Nutritional secondary hyperparathyroidism still happens. Equine Practice. 19:5-6.

Hintz, H.F. 1997b. Calcium requirements of young horses in training.

Equine Practice. 19:6-7.

Hintz, H.F. and Schryver, H.F. 1972. Magnesium metabolism in the horse. J. Anim. Sci. 35:755.

Hudson, N.P.H., Church, D.B., Trevena, J., Nielsen, I.L., Major, D. and Hodgson, D.R. 1999. Primary hypoparathyroidism in two horses. Aust Vet J. 77:504-506.

Martin, K.L., Hoffman, R.M. and Kronfeld, D.S. 1996. Calcium decreases and parathyroid hormone increases in serum of periparturient mares. J Anim Sci. 74:834-839.

Martin-Rosset W. 1990 L'Alimentation des Chevaux. Paris, INRA.

Meyer, H. 1994. Energie-und Nährstoff bedarf landwirtschaftlichie Nutztiere. Frankfurt, DLG Verlag.

Meyer, H. and Ahlswede, L. 1976. The intrauterine growth and body composition of foals and the nutrient requirements of pregnant mares. Anim. Res. Dev. 8:86.

Meyer, H. and Ahlswede, L. 1977. Untersuchunger zum Mg-Stoffwechsel des Pferdes. Zentrabl. Veterinacrmed. 24:128

More, S.J. 1999. A longitudinal study of racing Thoroughbreds: Performance during the first years of racing. Aust Vet J. 77:105-112.

National Research Council: Nutrient Requirements of Horses, ed 5. Washington, DC, National Academy - National Research Council. 1989.

Nielsen, B.D., Potter, G.D., Morris, E.L., Odom, T.W., Senor, D.M., Reynolds, J.A., Smith, W.B. and Martin, M.T. 1997a. Changes in the third metacarpal bone and frequency of bone injuries in young Quarter Horses during race training: Observations and theoretical considerations. J Equine Vet Sci. 17:541-549.

Nielsen, B.D., Potter, G.D. and Greene, L.W. 1997b. An increased need for calcium in young racehorses beginning training. Proc. 15th Equine Nutr Phys Symp p.153-158.

Pagan, J.D. 1994. Nutrient digestibility in horses. Feeding the performance horse. Proc. KER Short Course for Feed Manufacturers. p. 127-136.

Porr, C.A., Kronfeld, D.S., Lawrence, L.A., Pleasant, R.S. and Harris, P.A. 1998. Deconditioning reduces mineral content of the third metacarpal bone in horses. J Anim Sci. 76:1875-1879.

Porr, C.A., Ott, E.A., Johnson, E.L. and Madison, J.B. 1997. Bone mineral in young Thoroughbred horses is affected by training. Equine Practice. 19:28-31.

Ramirez, S. and Seahorn, T.L. 1997. How to manage nutritional secondary hyper-parathyroidism in horses. Vet Medicine. 92:980-985.

Richardson, J.D. 1991. Two horses with hypocalcaemia. Vet Rec. 129:5.

Schryver, H.F., Hintz, H.F., Lower, E., Hintz, R.L., Harper, R.B. and Reid, J.T. 1974. Mineral composition of the whole body, liver, and bone of young horses. J. Nutr. 104:126.

Schryver, H.F., Oftedal, O.T., Williams, J., Soderholm, L.V. and IIintz, H.F. 1986. Lactation in the horse: The mineral composition of mare's milk. J. Nutr. 16:2142.

Wall, D.L., Topliff, D.R. and Freeman, D.W. 1997. The effect of dietary cation-anion balance on mineral balance in growing horses. Proc 15th Equine Nutr Phys Symp p.145-150.

Wisniewski, E., Krumrych, W., Gehrke, M. and Mazurek, P. 1999. Fibrotic osteodystrophy in the horse. Medycyna Weterynaryjna. 55:84-88.

MACROMINERALS - SODIUM, POTASSIUM AND CHLORIDE

L. JILL MCCUTCHEON

University of Guelph, Canada

The importance of minerals in the diet of horses is well recognized by horse owners and equine nutritionists alike. The type and quantity of minerals required are very diverse and essential minerals include the major or macrominerals and the trace or microminerals. This discussion will be restricted to three macrominerals, sodium, potassium, and chloride (Na+, K+, Cl-), that are essential to the physiological well-being of the horse. Whereas certain assumptions are often made about the requirements for these three minerals, the demands for these minerals can vary substantially depending on age and activity. The following paper will elaborate on some of the basic dietary guidelines for sodium, potassium, and chloride and, by describing the role and use of these macrominerals in the body, provide a broader perspective as to how these basic requirements for the horse may be altered in different circumstances.

Sodium, Potassium and Chloride in the Body

All essential minerals can be detected in the tissue and fluids of healthy animals and sometimes reflect whether the mineral status of the animal is adequate or deficient. Distribution of minerals in tissues is not uniform; some tissues contain much higher concentrations of specific elements that are essential to that tissue's function. Perhaps the best known example is the structural support provided by concentration of calcium and phosphorus in bone. Equally important, although less obvious, are the complex biochemical, physiological and neurological functions of sodium, potassium and chloride. These elements are often essential for movement of specific ions across the cell membrane, for stabilizing enzymes in their most active conformation, and for maintenance of ionic and osmotic balance of intra- and extracellular fluids.

Sodium and chloride are the major ions found in extracellular fluids. Plasma sodium concentration, normally maintained within narrow limits, is the primary determinant of alterations in plasma tonicity and is important in the maintenance of acid-base balance in body fluids. Generally, of the sodium present in the horse's body, approximately 45% is present in extracellular fluid, 10% in intracellular fluid, and the remainder is integral to bone structure and is not available, consequently, for exchange with fluid compartments.

Whereas sodium and chloride are found predominantly in extracellular fluids, 90% of the body's potassium is retained within cells and, in particular, in skeletal muscle (70-75%). The retention of potassium largely within cells and of sodium in extracellular fluids is accomplished primarily by the action of specific ATP-consuming pumps (Na+-K+-ATPase), located within the cell membrane, that pump potassium into cells in exchange for sodium ions. The high intracellular potassium concentration, established by the constant work of Na+-K+-ATPase, maintains the elevated resting membrane potential in excitable tissues such as nerves and muscle, and has a crucial role in the triggering of action potentials that initiate muscle contraction and nerve impulse transmission. During exercise, there is rapid outflow of potassium from skeletal muscle that results in a transient rise in the concentration of plasma potassium with the extent of this hyperkalemia reflecting the intensity of exercise. This rise in plasma potassium can augment blood flow to working muscle through dilation of local arterioles. Upon cessation of exercise, the high intracellular potassium concentration in muscle is reestablished by Na+-K+-ATPase pump activity.

Sodium, potassium, and chloride are absorbed from the gastrointestinal tract. Whereas sodium is actively moved across the intestinal wall by an energy-requiring process, chloride and potassium diffuse from ingesta within the gastrointestinal tract across the intestinal wall with the rate being determined by their relative concentration in ingesta vs. the mucosal cells of the intestinal wall. Thus, when absorption is passive as is the case for chloride and potassium, the concentration of these ions in the feed and the quantity of feed in the intestine can significantly affect the amount absorbed by the gut. Transepithelial sodium transport has been demonstrated to be greater in distal as compared to proximal colon whereas 52-74% of potassium has been reported to be absorbed prececally, with less absorption in the large intestine (Meyer, 1987; Clarke et al., 1992). A few other factors can affect absorption of minerals in the intestinal tract. In particular, the age of an animal (young animals are more efficient in absorbing essential elements than older animals) and the pH of the intestinal tract can affect the rate of absorption (Lewis, 1995).

Effects of Sodium or Potassium Deficiency

When an adequate supply is available, horses normally consume sodium, potassium, and chloride in excess of their dietary requirements. Deficiencies of these minerals can occur when the quantity or the quality of certain feedstuffs in the diet limits their availability and/or when there are greater dietary demands based on growth, lactation, or activity. Consequently, there is a greater risk of sodium deficiency occurring during lactation or when horses are in moderate to heavy work. Secretion of the hormone aldosterone will increase in response to a reduced sodium intake, resulting in greater absorption and retention of sodium, and thereby lowering its excretion in urine, milk and sweat. When sweat production is decreased in an attempt to retain sodium, there is a greater likelihood of hyperthermia during exercise, reducing the rate of work and, as a result, performance capacity. In addition, there is evidence that significant electrolyte losses and subsequent electrolyte imbalances contribute to the development of exertional rhabdomyolysis ("tying up") in some horses (Harris, 1988).

Table 1. Concentrations of macrominerals in feeds.

	Amount in various feeds (% commonly present in dry matter)		
	Alfalfa	Grasses	Grains
Sodium	0.03 - 0.25	0.01 - 0.25	0.01 - 0.1
Potassium	1.0 - 4.0	0.5 - 3.5	0.2 - 0.7

Data adapted from Lewis,1995.

Most practical diets contain a sufficient quantity of potassium as a result of its relatively high concentration in most forage (Table 1). As intake of potassium in excess of requirements is excreted in urine, excess potassium in the diet rarely poses any difficulties. In contrast, a potassium deficiency can cause significant problems. When compared to human sweat, equine sweat has a high potassium content. Therefore, limited forage intake, substantial sweat losses due to strenuous and repeated exercise and/or hot ambient conditions are all factors that could contribute to a potassium deficiency. Aldosterone secretion, stimulated by loss of sodium in sweat fluid, serves to enhance sodium retention, but this conservation of sodium is accomplished by increasing potassium excretion. Thus, endurance horses that sustain large sweat fluid losses and undergo a period of reduced or no feed intake are particularly susceptible to substantial potassium loss. Potassium losses can also be exacerbated by the use of diuretics, such as furosemide (Lasix), commonly administered to racehorses in an attempt to reduce exercise-induced pulmonary hemorrhage. Finally, increased fecal potassium loss occurs in horses with diarrheal diseases and is often aggravated by the decreased feed intake associated with illness.

Potassium deficiency is more likely to occur when horses are maintained on a high grain, low forage diet. For example, when horses were fed a diet consisting of 1/3 grass hay and 2/3 oats, a diet commonly fed to horses in training or racing, a net loss of potassium occurred (Young et al., 1989). As stated earlier, approximately 75% of the body's potassium is found in skeletal muscle and is essential to its function. Therefore, muscle weakness, fatigue, and exercise intolerance are likely when potassium deficiency occurs (Table 2). In addition, decreased feed and water intake occurs in horses with potassium deficiency, and will serve to further reduce potassium intake (NRC, 1989).

Table 2. Sodium and potassium deficiencies: Causes and effects.

	Common causes	Effects
Sodium deficiency	Insufficient salt available as supplement and/or in feed Profuse sweating	Reduced sweating and performance Decreased food and water intake Dehydration, weight loss, constipation Restlessness, licking, pica
Potassium deficiency	Profuse sweating can be accentuated by: - a high-grain (low K+) diet - diuretic - diarrhea	Impaired muscle function and early onset of fatigue Decreased food and water intake Weight loss

Data adapted from Lewis,1995.

Dietary Requirements for Sodium, Potassium and Chloride

Maintenance requirements (sedentary horses)
Maintenance requirements for sodium, potassium and chloride have been estimated to be 20-25, 30-48, and 75-80 mg/kg body weight (BW) per day, respectively. For a 450 to 500 kg sedentary horse, this would amount to approximately 10 to 15 g/day of sodium, 15 to 20 g/day of potassium and 35 to 40 g of chloride (Coenen, 1999; Meyer, 1987; Schryver et al., 1987; Groenendyk et al., 1988; Hintz and Schryver, 1976). Most forages are high in potassium (0.5-3.5% and 1-4% DM in grass and alfalfa hay, respectively) (Table 1). Therefore, when fed in adequate quantity (at least 1% BW or a minimum of 5 kg for a 500 kg horse), maintenance potassium requirements can be met. Less data is available regarding chloride intake and requirements for the horse. However, recent work by Coenen (1999) has provided some insight into chloride requirements (Table 3). This work also demonstrated that chloride intake can vary from approximately 5 mg/kg BW/day (when horses are fed washed hay) up to 286 mg/kg BW/day when fed a ration of grain and hay. Whereas sodium content of grass and alfalfa is often far below 0.5 g/kg dry matter (DM), chloride content of forage is usually higher (estimated to be 7 to 10 g/kg DM). As such, there is greater risk of a deficiency of sodium than of chloride or potassium based on dietary intake (Jansson, 1999). The apparent digestibility of sodium, potassium, and chloride is approximately 90, 80, and 90%, respectively, indicating the efficient utilization of these macrominerals by the horse. Quantities in excess of requirements are normally excreted in urine. Intake of sodium and chloride also increases calcium and phosphorus absorption and retention.
    Provision of sodium chloride requirements for nonexercised horses is often met by access to a salt block (when coupled with sodium intake in feed) (Schryver et al., 1987; Jansson et al., 1996). However, salt consumption by individual horses can vary significantly. In a study by Schryver and coworkers

(1987), voluntary salt intake from a block by mature unexercised horses averaged 53 g/horse/day but varied from 9 to 143 g/day between horses and from 5 to 200 g/day by the same horse on different days. Horses also preferred a grain mix without added sodium chloride when compared to a grain mix with 2 to 4% added salt. Other researchers reported that in 4 of 6 horses studied, voluntary daily intake of salt was less than 20 mg/kg/day (Jansson and Dahlborn, 1999).

Voluntary salt consumption normally increases in hot and humid weather and may increase if horses are on pasture during periods of plant growth (Lewis, 1995). A small number of horses, when restricted to box stalls and allowed free access to salt, consume excess quantities of salt as a result of boredom.

Table 3. Maintenance dietary requirements for macrominerals.

	Sodium (mg/kg/day)	Potassium (mg/kg/day)	Chloride (mg/kg/day)
Maintenance (minimum requirement)	10 - 15	20 - 25	20
Maintenance (recommended)	20 - 25	30 - 38	75 - 80
Late pregnancy, lactation and growth	25 - 30	40 - 55	85 - 95

Data adapted from Coenen,1999; Lewis,1995; Meyer, 1987; Schryver et al., 1987; Groenendyk et al., 1988; Hintz and Schryver, 1976.

Growth and lactation

During the first two months of life, a growing foal's requirements for sodium, potassium, and chloride can be met by adequate intake of the mare's milk. Subsequently, the growing horse should be provided with unlimited access to water and a salt supplement. Often this is a trace-mineralized salt that also contains significant quantities of copper, zinc, and selenium. It should also be recognized that although foals may start to eat solid feed within the first one to two weeks of life, many do not drink any water, relying completely on the mare's milk as its only source of fluids (Crowell-Davis et al., 1985). Thus, the lactating mare can sustain substantial losses of both fluid and electrolytes. As a foal can consume upwards of 25% of its body weight in fluid each day, a foal weighing between 80 and 100 kg could rely on its dam for as much as 20 to 25 liters of milk each day. Assuming a sodium and potassium concentration in milk of ~ 0.17 and 0.51 g/liter, respectively, the demands of the foal could result in the requirement for an additional 3 to 4 g of sodium and 10 to 12 g of potassium for the lactating mare. For both the mare and the growing horse, sufficient quantities of potassium and chloride can normally be provided by a diet that includes an adequate quantity of grass or hay. However, given the paucity of sodium present in many feeds, salt should be made available at all times and can be added to the grain at a rate of 0.5 to 1% DM to ensure that horses receive sufficient sodium and chloride.

Horses in Work - Meeting the Needs of the Equine Athlete
The muscular activity associated with exercise results in substantial energy expenditure. However, the process of conversion of metabolic energy to mechanical work is only 20 to 25% efficient with the result that 75 to 80% of the chemical energy is converted to heat within skeletal muscle cells (Hodgson et al., 1994; Jones and Carlson, 1995). Like human athletes, the principal mechanism of heat loss in the horse during exercise is evaporation, primarily in the form of sweating. Sweating rates of 35 to 50 ml/m^2/min have been measured for horses exercising on a treadmill in a laboratory (Hodgson et al., 1993; McCutcheon et al., 1995a; McCutcheon and Geor, 1996). For a 500 kg horse with a total surface area of 4.5 to 5.0 m^2, such sweating rates correspond to sweat fluid losses of 10 to 15 liters/hour; this is similar to estimated hourly losses during prolonged exercise in field conditions (Carlson, 1983). Therefore, prolonged or strenuous exercise results in significant fluid losses and, in addition, extensive losses of ions.

The predominant ions in equine sweat fluid are sodium, potassium, and chloride. Although the ratio of these ions present in equine and human sweat are similar, the absolute quantities are 4 to 5-fold higher in horses (Figure 1) (Costill, 1977; McCutcheon et al., 1995b; McCutcheon and Geor, 1998). Therefore, the need to ensure adequate replacement of sodium, potassium, and chloride losses is an even more urgent consideration for the equine athlete when compared to their human counterparts.

Several factors affect the extent of sweat fluid and ion losses associated with exercise. Sweating rate, which can be used to estimate the rate of fluid and ion losses, is largely determined by the duration and intensity of exercise undertaken. In racehorses, fluid losses over the course of a full day may be as high as 15 liters, and endurance horses participating in 80-160 km events often have a fluid deficit of 20 to 40 liters despite access to water during the ride (Carlson, 1987). Horses completing 45 km of trotting on a treadmill in cool conditions in a simulated endurance event were reported to have lost approximately 27.5 liters of sweat fluid and estimated sweat ion losses of 241 g in less than 4 hours (Figure 2) (Kingston et al., 1999). Despite the relatively short period of endurance exercise in this study, these losses are in excess of estimates of voluntary intake of these macrominerals in an unsupplemented diet.

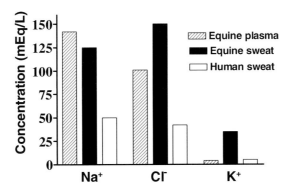

Figure 1. Comparison of the approximate concentration of sodium (Na+), chloride (Cl-), and potassium (K+) in equine plasma and in equine and human sweat, demonstrating the isotonic to hypertonic nature of sweat in horses compared to humans. (Data from McCutcheon et al., 1995b, McCutcheon et al., 1996; Costill,1977)

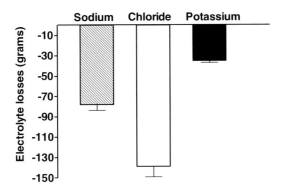

Figure 2. Electrolyte (sodium, chloride, potassium) losses in sweat in 6 Thoroughbred horses trotting for approximately 3 hours and 20 min (45 km) in moderate ambient conditions (20-24°C). (Data from Kingston et al., 1999)

In addition to exercise intensity and duration, other factors such as terrain, footing, the weight of the rider and tack, training status, and ambient temperature can contribute to the horse's work output and the need to increase dissipation of heat via sweating. Of these factors, the most important is the effect of the environment. Hot ambient conditions affect both the sweat ion composition and sweating rate (Table 4).

Table 4. Mean sweating rate of horses exercising on a treadmill at different exercise
intensities and ambient conditions.

Mean sweating rate (ml/m^2/min)		
Exercise intensity	Cool conditions	Hot conditions
Low (3.2 - 4.5 m/s)	21.1 ± 5.2	27.0 ± 6.2
High (9.0 m/s)	32.8 ± 5.1	52.3 ± 7.7

Data from McCutcheon and Geor,1996; McCutcheon and Geor, 1998.

Ion concentration of equine sweat is reported to be higher in hotter conditions; much of this increase is assumed to be a reflection of a higher rate of sweating under these more severe ambient conditions (McCutcheon et al., 1995b). Although sweat sodium concentration has been shown to decrease slightly in horses following training and in those acclimated to hot and humid conditions, this decline in the sweat sodium concentration was partially offset by higher sweating rates at a given body temperature (McCutcheon et al., 1999). The effects of higher ambient temperature on the extent of sweat fluid and ion losses was demonstrated in a group of six horses completing a treadmill speed and endurance test that simulated the speeds and distances required for each phase of an Olympic level three-day event in cool and hot ambient conditions. Total sweat fluid losses associated with the test in hot, dry conditions were almost double those measured in cool, dry conditions (Table 5) (McCutcheon and Geor, 1996).

Table 5. Calculated total sweat fluid and ion losses under different ambient conditions
during training and a treadmill simulated speed and endurance exercise test

	Daily training		Speed and endurance test	
	Cool	Hot	Cool	Hot
Total fluid losses (liters)	4.6	11.3	11.7	19.2
Total ion (Na+, K+ and Cl-) losses (g)	43.5	115.2	112.3	200.6

Cool = 20-22°C, 45-55% relative humidity; Hot = 33-35°C, 45-55% relative humidity;
Daily training = 1 hr of submaximal exercise. Data from McCutcheon and Geor,1996.

The various studies described in this section demonstrate the increased demand for sodium, potassium and chloride for horses in regular work. This additional demand will be particularly high in horses in heavy work during the summer months. For horses in training and competition, it is important to meet these needs as ongoing heavy sweat losses without adequate replacement can result in electrolyte imbalances and/or depletion that contributes to poor exercise performance.

Supplementation

Most forages are deficient in sodium and marginal in chloride. In contrast, forage is high in potassium and when fed in adequate quantities normally meets daily potassium requirements. However, as a result of the many factors that can affect intake and the rate of loss of sodium, potassium, and chloride, it is recommended that these electrolytes be provided to all horses. As electrolytes consumed in excess of requirements are excreted in the urine, access should be provided on a continuous (daily) basis. These minerals can be provided to horses in their ration or as a separate supplement. Salt supplements can be offered loose, top-dressed on the feed, or as a block. It should be noted that most salt blocks and other supplements typically do not contain potassium. Therefore, if the roughage portion of diet is restricted (less than 1% BW on a dry matter basis) and/or high sweat losses occur as a result of hot summer weather and moderate to heavy work, then potassium supplementation may be necessary. Sometimes, electrolytes are added to drinking water. However, horses should be introduced to electrolyte solutions gradually and water intake monitored as electrolyte solutions can result in reduced water consumption in some horses. Similarly, addition of electrolytes to feed should be approached conservatively to avoid decreased feed intake.

In addition to daily supplementation, administration of hypertonic electrolyte pastes before or during competition is widely practiced by endurance riders. Electrolyte pastes may contain just NaCl, or a 50:50 mixture of NaCl and KCl mixed with water. Experimentally, these pastes have been administered as 80 g each of NaCl and KCl or 90 g of NaCl. To date, there is limited data on the efficacy of these pastes. While such salt pastes provide a portion of the electrolytes lost in subsequent work, the primary purpose of the mixture is to stimulate the thirst response and encourage voluntary water consumption in order to maintain extracellular fluid volume during the ride and aid post exercise recovery of fluid losses. The work of Lindinger and Ecker (1995) reported that endurance horses with less pronounced fluid and electrolyte alterations during a competitive ride were more successful than those with greater alterations although recovery did not differ between the two groups. In work by Nyman and colleagues (1996), it was reported that horses offered 0.9% saline or given a salt paste during a 62 km ride had higher fluid intake during the ride and faster recovery of their body weight when compared to horses offered water. However, based on increases in plasma sodium concentration during the ride, the authors cautioned against the use of the salt paste. It should be emphasized that in the horse, the gastrointestinal tract is a very important source of both fluid and electrolytes utilized both during and after prolonged exercise (Schott and Hinchcliff, 1998). The importance of dietary fiber in increasing this gastrointestinal fluid reservoir underscores the need to maintain good quality roughage in horses training and competing in prolonged events. Depletion of this reservoir appears to be one of the main reasons for persistent body weight losses during recovery from prolonged endurance exercise (Hyppa et al., 1996; Schott et al., 1997).

Recommendations for Different Classes of Horses

Maintenance (sedentary horses)
Sedentary horses require at least 10 mg/kg BW of sodium and chloride daily (approximately 10 g for a 500 kg horse). In supplementing the diet, NRC (1989) recommendations suggest that chloride requirements are adequately met when sodium supplementation is provided as sodium chloride. Provision of at least 0.1% sodium in diet dry matter (0.25% common salt, which is 39% sodium and 61% chloride) is the minimal suggested requirement. However, at least 0.3% (0.75% common salt) would more adequately ensure provision of daily requirements of these macrominerals under any management conditions. In order to meet potassium requirements, provide forage at a rate of at least 1% BW on a dry matter basis. For a 500 kg horse, approximately 5 to 6 kg of hay (DM) will provide 75 to 150 g of potassium.

Growth and lactation
Similar recommendations for salt (0.3% sodium, 0.75% common salt in diet dry matter) will meet the needs of growing horses, lactating mares and those in late gestation. Again, this recommendation would include intake of adequate quantities of forage to supply potassium requirements. Top-dressing feeds made available in a creep provides a mechanism to ensure salt supplementation in foals that may not take advantage of a salt block.

Horses in work
As indicated earlier, horses in work have additional demands for sodium, potassium and chloride to replace loss of these ions in sweat fluid. There is some increase in intake of these ions as additional feed is provided to offset the additional energy consumption associated with exercise. However, salt and potassium intake in additional feed will not be sufficient to compensate for increased losses. A commercial electrolyte supplement can be provided to meet these additional needs; composition of such supplements should be carefully evaluated to ensure the specific needs of a particular horse or group of horses are met. Alternatively, top-dressing feed with 150 g of a 50:50 mixture of NaCl and KCl should meet the needs of most horses in moderate work (sweat losses of approximately 10 liters/day). During and following periods of intense work, particularly in hot ambient conditions, supplementation in excess of 150 g may be required (Table 6). Dividing this supplement between two or three grain meals each day will enhance the palatability of the ration.

Table 6. Estimated sweat fluid losses and sodium, chloride and potassium requirements (in grams) for a 500 kg horse at different levels of work.

Workload:	Non-exercised	Low workload	Moderate workload	Strenuous and/or prolonged exercise program
Sweat losses (liters per day):	0-5	5	10	20
Electrolyte requirements (grams) for a 500 kg horse				
Na+	10-15	25	45	90
Cl-	30-35	40	80	160
K+	15-20	30	45	80

Concluding Remarks

Sodium, potassium, and chloride are essential dietary requirements for all horses. Although present in grain and forage, the possible variation in quality, quantity and ratio of these two components of the diet leaves the potential for marginal or deficient intake of one or more of these minerals (particularly sodium). Furthermore, while access to salt should be provided at all times, dietary supplementation is a more accurate method of ensuring intake, as voluntary consumption of salt can vary significantly. Finally, loss of sodium, potassium, and chloride as a result of sweating will increase in hot weather and with the intensity and duration of exercise required by the horse. Therefore, supplementation of these minerals should be adjusted according to work required and ambient conditions.

References

Carlson, G. P. (1983) Thermoregulation and fluid balance in the exercising horse. Equine Exercise Physiology. D. H. Snow, S. G. B. Persson and N. E. Robinson. Cambridge, UK, Granta Editions: 291-309.

Carlson, G. P. (1987) Hematology and body fluids in the equine athlete: a review. Equine Exercise Physiology 2. J. R. Gillespie and N. E. Robinson (eds) Davis, California, ICEEP Publications: 393-425.

Clarke, L. L., M. C. Roberts, et al. (1992) Short-term effects of aldosterone on Na-Cl transport across equine colon. Amer. J. of Physiol. 262: R939-R946.

Coenen, M. (1999) Basics for chloride metabolism and requirement. Proc Equine Nutrition and Physiology Symposium. pp 353-354.

Costill, D. L. (1977) Sweating: its composition and effects on body fluids. Ann. N. Y. Acad. Sci.: 160-174.

Crowell-Davis, S. L., K. Houpt, et al. (1985) Feeding and drinking behavior of mares and foals with free access to pasture and water. J. Anim. Sci. 60: 883-889.

Groenendyk, S., P. B. English, et al. (1988) External balance of water and electrolytes in the horse. Equine Vet. J. 20: 189.

Harris, P. A. (1988) Aspects of equine rhabdomyolysis syndrome. Ph.D. Thesis. Churchill College, Univ. of Cambridge, Cambridge, England.

Hintz, H. and H. F. Schryver (1976) Potassium metabolism in ponies. J. Anim. Sci. 43: 637-643.

Hodgson, D. R., R. F. Davis, et al. (1994) Thermoregulation in the horse in response to exercise. Brit. Vet. J. 150: 219-235.

Hodgson, D. R., L. J. McCutcheon, et al. (1993) Dissipation of metabolic heat in the horse during exercise. J. Appl. Physiol. 74: 1161-1170.

Hyppa, S., M. Saastamoinen, et al. (1996) Restoration of water and electrolyte balance in horses after repeated exercise in hot and humid climates. Equine Vet. J. Suppl. 22: 108-112.

Jansson, A. (1999) Sodium and potassium regulation with special reference to the athletic horse. Ph.D. Thesis. Department of Animal Physiology, Swedish University of Agricultural Sciences. Uppsala, Sweden.

Jansson, A. and K. Dahlborn (1999) Effect of feeding frequency and voluntary salt intake on fluid and electrolyte regulation in athletic horses. J. Appl. Physiol. 86: 1610-1616.

Jansson, A., A. Rytthammar, et al. (1996) Voluntary salt (NaCl) intake in Standardbred horses. Pferdeheilkunde 12: 443.

Jones, J. H. and G. P. Carlson (1995) Estimation of energy costs and heat production during a three-day event. Equine Vet. J. Suppl. 20: 23-30.

Kingston, J. K., L. J. McCutcheon, et al. (1999). Comparison of three methods for estimation of exercise-related ion losses in sweat of horses. Amer. J. Vet. Res. 60: 1248-1254.

Lewis, L. D. (1995). Equine Clinical Nutrition: Feeding and Care. Media, PA, Williams & Wilkins.

Lindinger, M. I. and G. L. Ecker (1995) Ion and water losses from body fluids during a 163 km endurance ride. Equine Vet. J. Suppl. 18: 314-322.

McCutcheon, L. J. and R. J. Geor (1996) Sweat fluid and ion losses during training and competition in cool vs. hot ambient conditions: implications for ion supplementation. Equine Vet. J. Suppl. 22: 54-62.

McCutcheon, L. J. and R. J. Geor (1998) Sweating: Fluid and ion losses and replacement. The Veterinary Clinics of North America. Equine Practice: Fluids and electrolytes in athletic horses. K. W. Hinchcliff (ed) W. B. Saunders, Philadelphia, PA. 14: 75-96.

McCutcheon, L. J., R. J. Geor, et al. (1999) Sweating responses in horses during 21 days of heat acclimation. J. Appl. Physiol. 87: 1843-1851.

McCutcheon, L. J., R. J. Geor, et al. (1995a) Sweat composition and ion losses during exercise in heat and humidity. Equine Vet. J. Suppl. 20: 153-157.

McCutcheon, L. J., R. J. Geor, et al. (1995b) Sweat composition: comparison of collection methods and effect of exercise intensity. Equine Vet. J. Suppl. 18: 279-284.

Meyer, H. (1987) Nutrition of the equine athlete. In: Equine Exercise Physiology. J. R. Gillespie and N. E. Robinson (eds) ICEEP Publications, Davis, Calif. 644-673.

NRC (1989). Nutrient Requirements of Horses. 5th Edition. National Academy Press, Washington, D.C.

Nyman, S., A. Jansson, et al. (1996) Strategies for voluntary rehydration in horses during endurance exercise. Equine Vet. J. Suppl. 22: 99-107.

Schott, H. C. and K. W. Hinchcliff (1998) Treatments affecting fluid and electrolyte status during exercise. The Veterinary Clinics of North America. Equine Practice: Fluids and electrolytes in athletic horses. K. W. Hinchcliff (ed) W. B. Saunders, Philadelphia, PA. 14: 175-204.

Schott, H. C., K. S. McGlade, et al. (1997) Body weight, fluid, electrolyte, and hormonal changes in horses during and after recovery from 50- and 100-mile endurance rides. Amer. J. Vet.Res. 58: 303-309.

Schryver, H. F., M. T. Parker, et al. (1987) Salt consumption and the effect of salt on mineral metabolism in horses. Cornell Vet. 77: 122-131.

Young, J. K., G. D. Potter, et al. (1989) Mineral balance in resting and exercised miniature horses. Proc Equine Nutrition and Physiology Symposium. pp 79.

JAPANESE FEEDING STANDARDS

YO ASAI

Japan Racing Association, Tokyo, Japan

Introduction

The number of horses bred in Japan decreased sharply from over 1.5 million in the past to approximately 100,000 today. The situation in Japan, where over half of the horses being bred are light breed horses, is unlike that in any other country in the world. Many of the light breed horses (Thoroughbreds and Anglo-Arab horses) are being produced for use as racehorses. The 8,000 to 9,000 horses that are registered for breeding annually are the main source of racehorses.

Until now, there have been almost no textbooks dealing with the nutrition of horses written in Japanese, to say nothing of a feeding standard for horses designed specifically for Japan. Consequently, horse breeders who were eager to learn more about this subject had to utilize the NRC, but the majority of breeders were forced to rely on their experience and intuition. Furthermore, Japanese horse racing was closed to the rest of the world for many years, but with its recent internationalization the number of foreign-bred horses taking part in horse races in Japan is on the rise. This prompted those in the industry to think about the need to improve the ability and physical strength of racehorses born and raised in Japan. The consensus was that, in order to accomplish this goal, scientific breeding and training methods that reflect actual conditions in Japan must be actively introduced. The Japanese Feeding Standard for Horses was drafted as part of that initiative. In Japan, however, the number of universities and research institutions conducting research on horses is very small, so there was a dearth of research data on horses. Thus, foreign literature, including the NRC, was referred to when drafting the Japanese Feeding Standard. The first light breed horse feeding standard in Japan was published in 1998.

Basic Principles Behind Drafting of the Feeding Standard

Priority was given to the following principles when drafting the Japanese Feeding Standard for Horses:

1. Prepare a standard growth table in order to properly evaluate growth rates, which are said to be closely associated with the onset of bone diseases in growing horses.
2. Enhance instructions related to factors that affect nutritional requirements and feeding-related matters that require attention, using values obtained under average Japanese breeding conditions as the nutritional requirements.
3. Indicate nutritional requirements for nursing foals based on the results of experiments related to milk intake and changes in milk composition.

4. Enhance instructions regarding feeding management during grazing, because grazing is an important part of raising horses.
5. For performance horses, requirements are to be based on average nutritional intake by racehorses and information cited in the literature, because there are large individual differences in the amount of exercise performed by adult horses.
6. Indicate the procedure for determining feed amounts by adding a chapter entitled "How to Use the Japanese Feeding Standard."

Standard Growth

Based on the results of a growth study conducted in Japan (investigated in Hokkaido Prefecture where approximately 90% of Japanese light breed horses are produced), a standard growth table was included in the Feeding Standard. Shown in the standard growth table (Table 1) are body weight, withers height, heart girth and cannon bone circumference.

Table 1. Standard growth of Thoroughbreds in Japan.

	Male		Female		Daily gain (kg/day)
	Mean	Normal range	Mean	Normal range	
at birth					
BW (kg)	57	50-64	57	50-64	
1 month					1.33
BW (kg)	97	89-105	97	89-105	
WH (cm)	110	108-112	110	108-112	
HG (cm)	101	98-104	101	98-104	
CC (cm)	13.1	12.7-13.4	13.1	12.7-13.4	1.15
3 months					
BW (kg)	166	152-180	166	152-180	
WH (cm)	123	121-125	123	121-125	
HG (cm)	122	118-126	122	118-126	
CC (cm)	14.9	14.6-15.2	14.6	14.3-14.9	1.00
5 months					
BW (kg)	225	203-247	225	203-247	
WH (cm)	132	130-134	131	129-133	
HG (cm)	136	131-141	136	131-141	
CC (cm)	16.0	15.5-16.5	15.6	15.1-16.1	0.80-0.85
7 months					
BW (kg)	275	250-300	267	242-292	
WH (cm)	137	134-140	136	133-139	
HG (cm)	146	141-151	146	141-151	
CC (cm)	16.9	16.5-17.3	16.6	16.2-17.0	0.60-0.80
9 months					
BW (kg)	310	285-335	301	276-326	
WH (cm)	141	138-144	140	137-143	
HG (cm)	153	148-158	153	148-158	
CC (cm)	17.6	17.1-18.1	17.2	16.7-17.7	
11 months					
BW (kg)	335	309-361	328	302-354	0.40-0.50 [1]
WH (cm)	144	142-146	143	141-145	
HG (cm)	158	154-162	158	154-162	
CC (cm)	18.2	17.7-18.7	17.8	17.3-18.3	

13 months					
BW (kg)	364	344-384	364	344-384	
WH (cm)	148	146-150	146	144-148	
HG (cm)	163	159-167	163	159-167	
CC (cm)	18.5	18.0-19.0	18.1	17.6-18.6	0.40-0.70 [2]
15 months					
BW (kg)	408	379-437	408	379-437	
WH (cm)	151	148-154	149	146-152	
HG (cm)	168	162-174	168	162-174	
CC (cm)	18.8	18.4-19.2	18.3	17.9-18.7	
17 months					
BW (kg)	421	396-446	413	390-436	
WH (cm)	153	150-156	153	150-156	
HG (cm)	170	166-174	170	166-174	
CC (cm)	19.0	18.4-19.6	18.5	18.0-19.0	
19 months					
BW (kg)	439	415-463	429	405-453	
WH (cm)	154	151-157	154	151-157	0.10-0.30
HG (cm)	173	169-177	173	169-177	
CC (cm)	19.3	18.7-19.9	18.7	18.2-19.2	
27 months					
BW (kg)	451	425-477	433	410-456	
WH (cm)	159	156-162	158	155-161	
HG (cm)	175	171-179	174	170-178	
CC (cm)	19.5	19.0-20.0	18.8	18.3-19.3	

The following abbreviations were used : BW, body weight; WH, withers height; HG, heart girth; CC, cannon bone circumference.
1) Growth rate is slow in winter.
2) Growth rate is rapid in summer.

Nutrient Requirement of Horses

Daily nutrient requirements of horses in Japan are shown in Table 2 and digestible energy (DE) requirements compared with NRC are shown in Table 3.

Table 2. Daily nutrient requirements of horses.

Horse	Body weight -kg-	Daily Gain kg/day-	Digestible Energy Mcal/day- (Mcal/kg)	Crude Protein -g- (%)	Lysine -g- (%)	Ca -g- (%)	P -g- (%)	Mg -g- (%)	Vitamin A -1,000IU- (1,000IU/kg)
Growing horses									
Nursing foal	[Nutrient intake from mare's milk]								
2 months	130	1.15	8 [7.9]	330 [300]		23 [12]	13 [8]	3.4 [0.6]	
4 months	195	1.00	9 [5.9]	430 [220]		25 [7]	14 [5]	4.2 [0.3]	
Yearling – long yearling									
10 months	315	0.45	17.5 (2.50)	780 (11.2)	34 (0.49)	27 (0.39)	15 (0.22)	5.3 (0.08)	14 (2)
15 months	405	0.40	20.5 (2.30)	920 (10.3)	39 (0.44)	29 (0.33)	16 (0.18)	6.6 (0.07)	18 (2)
22 months (in training)	450	0.20	26.5 (2.65)	1,120 (11.2)	45 (0.45)	34 (0.34)	19 (0.19)	9.9 (0.10)	20 (2)
Mature horses									
Racehorses	455-475	-	27-35 (2.70)	1,300 (11.2)	46 (0.40)	40 (0.34)	29 (0.25)	15.1 (0.13)	22 (1.9)

Mare Late gestation	640	0.50	25 (1.80)	1,100 (7.7)	38 (0.27)	47 (0.33)	36 (0.26)	12.0 (0.08)	38 (2.7)
Early lactating	570		31 (2.40)	1,600 (12.5)	57 (0.45)	61 (0.48)	41 (0.32)	12.2 (0.10)	34 (2.7)
Late lactating	570		28 (2.20)	1,200 (9.4)	42 (0.33)	42 (0.33)	26 (0.20)	9.8 (0.08)	34 (2.7)

1) Values in parentheses show nutrient concentrations in total diets on dry matter basis.
2) Body weight and daily gain are based on the standard growth of Thoroughbreds in Japan (Table 1).
3) Requirements of crude protein, lysine, Ca, P, Mg and Vitamin A are referred to NRC (1989).
4) Values in racehorses are based on data obtained under Japanese conditions.

Table 3. DE requirement compared with Japanese Feeding Standard for Horses and NRC (1989).

	DE requirement (Mcal/day)		
		NRC	
Month of age	Japanese Feeding Standard	500 kg mature weight	600 kg mature weight
6		15.0	17.0
10	17.5		
12		18.9	22.7
15			
daytime grazing	20.5		
longtime grazing	30.5		
18		19.8	23.9
22			
in training	26.5		
24			
in training		26.3	32.3

A. Digestible Energy Requirements

1. Nursing Foals
    Digestible energy (DE) requirements for two- and four-month-old foals were indicated. The standard body weights were 130 kg and 195 kg, respectively. Maintenance DE requirements were calculated using the following formula (Pagan and Hintz, 1986), and 10% was added to correct for exercise.

    Maintenance DE (Mcal DE/day) = 0.975 + 0.021 x body weight (kg)

    Next, the amount of digestible energy required to increase body mass (daily gain, or DG, is 1.15 kg/day and 1.00 kg/day, respectively) was sought using the body composition of foals and how efficiently they utilize energy from protein and fat reserves. Digestible energy requirements, 8 Mcal/day for two-month-old and 9 Mcal/day for four-month-old foals, were calculated by combining the above results.

Meanwhile, the milk intake of foals was set at 16 kg/day at two months of age and 12 kg/day at four months of age from the results of trials to estimate milk yield. Gross energy (GE) was calculated from the milk composition at each stage (Table 5). As the digestibility of milk was set at 100%, the digestible energy amounts from milk are 7.9 Mcal/day and 5.9 Mcal/day, respectively. Therefore, two-month-old foals can obtain almost all their required energy from milk, but the energy from milk alone is insufficient for four-month-old foals.

2. Growing Horses
    Two to three month feeding and digestibility trials were conducted for 10-month-old horses (assuming winter when grazing is the mainstay of management, but horses cannot be expected to consume enough forage), 15-month-old horses (assuming summer when grazing is the mainstay of management and horses are expected to consume a substantial amount of forage) and 22-month-old horses (assuming winter when horses are in training). Digestible energy requirements for each stage were examined from the results of the trials. It was found that at all three stages DG increased at a regular rate when the amount of digestible energy consumed increased. From these regression equations, the DG appeared to be appropriate at each stage, namely 0.45 kg/day at 10 months old, 0.40 kg/day at 15 months old and 0.20 kg/day at 22 months old (Table 1). Digestible energy requirements were 17.7 (17.5 in requirement table) Mcal/day at 10 months old, 20.5 Mcal/day at 15 months old and 26.7 (26.5 in requirement table) Mcal/day at 22 months old. The digestible energy requirement of 15-month-old horses is for horses that graze only during the day (0830-1530 h, 7 hours). The digestible energy requirement of 15-month-old horses that graze from afternoon until the next morning (1530-0830 h, 17 hours) is 30.5 Mcal/day. Eighty percent or more of this requirement, however, can be consumed by grazing if the vegetation conditions are satisfactory.

3. Racehorses
    Large individual differences in the amount of exercise undertaken were observed and there were many areas that were unclear, so the digestible energy requirement was set at 27-35 Mcal/day using the results of feed intake surveys.

4. Mares
    Digestible energy requirements for late gestation, early and late lactation were indicated. Mean body weight at each stage was 640 kg, 570 kg and 570 kg, respectively. Using the results of feeding trials in Hokkaido, where late gestation is from winter to early spring, the digestible energy requirement for horses in late gestation was set at 25 Mcal/day. With the daily milk yields in early and late lactation set at 16 kg and 12 kg, respectively, the digestible

energy amounts required for lactation are 12.7 Mcal/day and 9.5 Mcal/day, respectively, when the digestible energy required to produce 1 kg of milk, 792 kcal, is totaled. The maintenance requirement of a horse weighing 570 kg as determined using an NRC formula, 18.5 Mcal/day, was added, leaving a digestible energy requirement of 31 Mcal/day for early lactation and 28 Mcal/day for late lactation.

B. Protein Requirements

1. Nursing Foals

The protein requirements of nursing foals are not necessarily clear. It is surmised from experience that the requirements of one- or two-month-old foals are met by the protein contained in the dam's milk, but in four-month-old foals the protein in the milk alone may not be sufficient. Assuming that the amount of digestible protein required to maintain two-month-old foals is 0.6 g/day per 1 kg of body weight and that protein makes up approximately 22% of their increase in body weight, the protein requirement of foals is 330 g/day (130 kg x 0.6 g/kg + 1,150 g x 0.22). Meanwhile, protein intake from milk is estimated to be approximately 300 g (16 kg/day x 1.9%) in two-month-old foals, and since the digestibility of the protein in the mother's milk is high, the protein from the milk alone is considered to be nearly sufficient in two-month-old foals. In four-month-old foals, however, the amount of protein provided from milk is approximately 220 g/day (12 kg/day x 1.8%), presumably leaving a deficiency of about 200 g.

2. Growing Horses

Protein requirements were calculated by multiplying the digestible energy requirements for 10-, 15- and 22-month-old horses by the coefficient in the formula for calculating protein requirements indicated in the NRC. The protein requirements were 780 g/day (45 x 17.5), 920 g/day (45 x 20.5) and 1,120 g/day (42 x 26.5) for 10-, 15- and 22-month-old horses, respectively.

3. Racehorses

The protein requirement was set at 1,300 g/day based on the requirements (1,312 g/day for intense working horses with a body weight of 500 kg) indicated in the NRC. The problem of overfeeding protein in racehorses is addressed later in this presentation.

4. Mares

According to the NRC, the protein requirement is found by multiplying the digestible energy requirement (Mcal/day) by 44. This formula was adopted and the protein requirement was set at 1,100 g/day (25 x 44). In addition, according to the NRC, the protein requirements of early (body weight of 500 kg) and late (body weight of 600 kg) lactating mares are 50 and 43 times the digestible energy requirement, respectively, so the protein requirements were set at 1,600 g/day (31 x 50) and 1,200 g/day (28 x 43), respectively.

C. Mineral Requirements

Calcium. Calcium requirements were calculated based on an endogenous loss (for maintenance) of 0.02 g/kg BW/day, 16 g required per 1 kg increase in body weight and an absorption efficiency of 50%. For instance, in the case of a growing 15-month-old horse (body weight 405 kg, DG 0.40), a calcium requirement of 29 g ((8.1 + 6.4)/0.5) was calculated based on an endogenous loss of 8.1 g and allowing 6.4 g for the increase in body weight. In the case of an early lactating mare (body weight 570 kg), a requirement of 61 g ((11.4 + 19.2)/0.5) was calculated based on an endogenous loss of 11.4 g and secretion into the milk of 19.2 g (milk yield of 16 kg x calcium concentration of 0.12%). Additionally, the calcium requirement for growing horses in training was sought in the same manner as that in the NRC: (calcium required for horses not in training) x (DE required for horses in training/DE required for horses not in training).

Phosphorus. Phosphorus requirements were calculated based on an endogenous loss (for maintenance) of 10 mg/kg BW/day, 8 g required per 1 kg increase in body weight and an absorption efficiency of 45%. For instance, in the case of a growing 10-month-old horse (body weight 315 kg, DG 0.45), a phosphorus requirement of 15 g ((3.15 + 3.6)/0.45) was calculated based on an endogenous excretion volume of 3.15 g and providing 3.6 g for the increase in body weight. In the case of a late lactating mare (body weight 570 kg), a requirement of 26 g ((5.7 + 6)/0.45) was calculated based on an endogenous loss of 5.7 g and secretion into the milk of 6 g (milk yield of 12 kg x phosphorus concentration of 0.05%). Additionally, the phosphorus requirement for growing horses in training was sought in the same manner as that in the NRC: (phosphorus required for horses not in training) x (DE required for horses in training/DE required for horses not in training).

Magnesium. Magnesium requirements were calculated based on an endogenous loss (for maintenance) of 6 mg/kgBW/day, 1.25 g required per 1 kg increase in body weight and an absorption efficiency of 40%.

Other minerals. For sodium, potassium, sulfur, iron, manganese, copper, zinc, selenium, iodine and cobalt, the requirements indicated in the NRC have been adopted. However, in the section entitled "How to Use the Feeding Standard," a note recommends 25-30 mg of copper and 100-120 mg of zinc be added per kg of feed in growing horses.

D. Vitamin Requirements

Vitamin A. The formula indicated in the NRC has been adopted: 60 IU/kg BW/ day for mares and 45 IU/kg BW/day for other horses.

Other vitamins. For vitamins D and E, thiamin and riboflavin, the requirements indicated in the NRC have been adopted.

Feeding Management and Feeding-Related Matters that Require Attention

This chapter includes matters related to feeding management that require attention at each stage of a horse's life. For instance, the section on mares discusses how to evaluate body condition scores, the importance of mineral supplementation in the last stages of gestation and the increase in requirements during lactation. The section on foals discusses the importance of ingesting colostrum, suckling behavior, the necessity and precautions of creep feeding and other matters. The section on growing horses discusses how to manage pastures, the advantages of extended grazing, the importance of preventing excessive weight gain during training, the importance of supplying minerals and other matters. The section on racehorses explains problems associated with giving horses too much concentrated feed, the utility of adding oils and fats to the diet, the importance of maintaining the proper balance of calcium and phosphorus, the importance of replenishing electrolytes and other matters.

Expected feed consumption by horses in Japan is shown in Table 4.

Table 4. Expected feed consumption by horses in Japan (% body weight)*.

	Forage	Concentrate	Total
Mature			
Maintenance	1.5 – 2.0	0 – 0.5	1.5 – 2.0
Mares, late gestation	1.2 – 2.0	0.5 – 0.8	1.7 – 2.2
Mares, early lactation	1.5 – 2.5	0.8 – 1.5	2.5 – 3.0
Mares, late lactation	1.5 – 2.0	0.5 – 1.0	2.0 – 2.5
Working horses			
Light – moderate work	1.0 – 2.0	0.5 – 1.5	1.5 – 2.5
Intense work	1.0 – 2.0	1.0 – 2.0	2.5 – 3.5
Young horses			
Nursing foal, 3-5 months	0.5 – 2.0	0.5 – 1.5	1.5 – 3.0
10 months	1.5 – 2.0	0.8 – 1.3	2.0 – 3.0
15 months	1.0 – 1.5	0.5 – 1.0	2.0 – 2.5
22 months (in training)	1.5 – 2.5	1.0 – 1.5	2.5 – 3.5

* Air-dry feed (about 90% DM)

How to Use the Feeding Standard for Horses

This section indicates the procedure for setting feed menus for growing horses and mares using nutritional requirement and feed composition tables.

Reference Materials

Tables of feed composition. These were prepared based on feed composition tables already published for domestic animals in Japan, and values in the feed composition tables listed in the NRC for feeds given to Japanese horses, or feeds that could possibly be given to them.

Characteristics of feeds used for horses. Lists nutritional features of each type of feed and how to use them.

Composition of milk in Thoroughbreds. Includes a milk composition table (Table 5) for mares bred under average Japanese breeding conditions. The state of changes accompanying the weekly course after gestation was nearly the same as that indicated in the NRC, but concentrations of fat, magnesium and calcium tended to be somewhat lower.

Growth of other breeds. Average body weight, withers height, heart girth and cannon bone circumference at different stages of growth are indicated for both light Hokkaido native horses and the heavy breed horses Breton and Percheron.

References. Literature and data referred to in the Japanese Feeding Standard for Horses are listed.

Conclusion

The environment in which horses are produced and raised in Japan is not necessarily ideal. Japan is a small country, the humidity is high and the weather is often unseasonable. The acidity of the soil is high and it lacks available minerals, because almost all organic soil in Japan is covered with volcanic soils. Consequently, it is difficult to cultivate quality grass. In such an environment, it is necessary to manage the feeding of horses based on scientific evidence in order to produce and raise "strong horses." However, there are a limited number of textbooks written in Japanese that could serve as a bible for those raising horses and there was no feeding standard that Japan could call its own, so it was imperative that one be prepared quickly.

   In order to complete a feeding standard in a short period of time, it was necessary to make it possible to use many formulas for calculating requirements adopted by the NRC. Since the basis of many of those formulas is body weight and digestible energy amounts, we sought the appropriate growth for horses being raised in Japan and the digestible energy requirements at each stage of a horse's life.

Horses in Japan also are affected by developmental orthopedic disease (DOD) such as osteochondrosis or epiphysitis, and potential problems related to nutrition and growth had been pointed out, so the demonstration of standard (appropriate) growth in Japan was very significant. Foals with a body weight higher than 100 kg at one month of age subsequently grow more and are more susceptible to DOD. Also, nursing foals with a lower-than-standard body weight that suddenly grow in their second year require attention.

The mainstay of feeding weanlings is grazing, which is greatly affected by seasonal change. In the latter half of this stage, training fitted to each horse is performed, so it is difficult to feed horses the exact required digestible energy. In addition, the amount of digestible energy given to horses must be raised or lowered as necessary. This is true for mares and racehorses. Therefore, those managers or trainers must have knowledge of nutrition in order to be able to give each horse the appropriate feed. One of the roles of the Japanese Feeding Standard for Horses is to supply them with that knowledge. The Japanese Feeding Standard for Horses was prepared based on the above strategy, but there are still many inadequacies, so a series of revisions will likely be required in the future.

Table 5. Composition of milk in Thoroughbreds (as liquid basis).

Time After Foaling	Total solids (%)	Energy (kcal/100g)	Protein (%)	Lactose (%)	Fat (%)	Ca (%)	P (%)	Mg (mg/kg)	Cu (mg/kg)	Zn (mg/kg)	Fe (mg/kg)
0 day (colostrum)	19.1	103.1	12.3	4.0	1.5	0.08	0.07	340	1.1	7.1	1.7
1 week	10.5	52.5	2.7	6.2	1.3	0.12	0.08	90	0.6	2.5	1.3
7 week	10.1	49.3	1.9	6.6	1.3	0.08	0.05	40	0.4	2.3	1.1
17 week	10.2	49.5	1.8	6.8	1.3	0.06	0.04	20	0.3	2.1	0.9

GERMAN FEEDING STANDARDS

MANFRED COENEN
School for Veterinary Medicine Hannover, Germany

Data about energy and nutrient requirements as well as feeding recommendations for the German horse industry are published by a particular committee set up by the German Society for Nutritional Physiology (GESELLSCHAFT für ERNÄHRUNGSPHYSIOLOGIE [GEH], 1994). Based on reviewed international literature, the tables cover energy, protein, macrominerals (Ca, P, Mg, Na, K, Cl) and trace elements (Cu, Zn, Fe, Mn, Se, I, Co) as well as vitamins (A, D, E, K, B_1, B_2, biotin, folic acid) for maintenance, exercise, reproduction and growth. The following description includes the current data about the requirement and recommendations in horse feeding as published in 1994 by the GEH. However, in some cases additional information published after 1994 may offer the opportunity for reassessing the requirements or to express more precise recommendations. Although this is an essential aspect in the discussion of a new edition of the feeding standards by NRC as well as anybody else, the presented calculations and tables are based on the latest official publication.

Energy and Protein

Maintenance

Digestible energy is still in use for horses comparable to most other countries (for France see Martin-Rosset, 2000). Digestible crude protein is established for protein requirement calculations.

The range of maintenance energy requirements is 0.48-0.62 MJ DE/kg $BW^{0.75}$ x d^{-1}, the standard value is 0.6 MJ DE/kg $BW^{0.75}$ x d^{-1} (143 kcal/kg $BW^{0.75}$ x d^{-1}). The differences to NRC levels are presented in Figure 1.

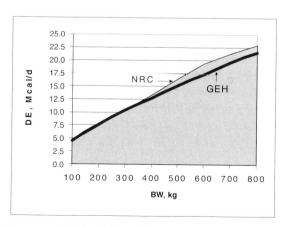

Figure 1. Digestible energy (DE, kcal/d) requirement for different body weights (BW, kg).

The endogenous N-losses and a utilization of digestible crude protein (DCP) of 80% (including a safety margin) yield 0.3 g DCP/kg BW$^{0.75}$ x d^{-1} (Figure 2). There are small differences between GEH and NRC values, if the latter are calculated by 0.6 g DCP/kg BW. The approximation of 40 g CP/kcal DE results in a bigger difference, with NRC data higher compared to those from GEH (see Table 1).

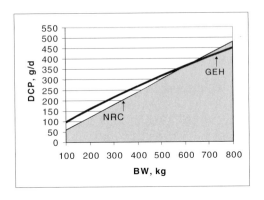

Figure 2. Digestible crude protein (DCP, g/d) requirement for different body weights (BW, kg).

Exercise

The DE required to produce kinetic energy is highly variable. Despite various influences like type of exercise, environment, capacity of the horse etc., in agreement with the NRC the following equation is used to derive the exercise- related DE requirement: DE (kJ/kg BW x h^{-1})=($e^{3.02+0.0065Y}$ - 13.92) x 0.441 (MJ instead of Mcal is used in Europe; 4.186 is the factor to convert Mcal to MJ).

For practical considerations it is crucial to have precise information about the exercise load on the one hand and not fail to recognize variable influences on the calculated requirement on the other hand. Particularly for horse keepers less experienced in animal production, a score system for evaluating the degree of exercise would be helpful. As a common practice low, medium or high intensity exercise creates an increase in relation to maintenance of 1.25, 1.25-1.5 and 1.5-2, respectively. These factors are also used by NRC.

In general, it is claimed that working muscle does not require additional protein. The elevated feed intake (increase in endogenous N-losses) and amino acid breakdown for energy metabolism and finally N-losses via sweat may change the DCP requirement. But as the DCP:DE ratio (g:MJ) in most feedstuffs is >5:1 (= >209 g DCP:1 Mcal), a sufficient DCP intake can be assumed even if the factorial approach may underestimate the DCP requirement. On the other hand, a protein supply approaching threefold the maintenance requirement is unfavorable. We recommend to keep the intake of DCP < 2 g/kg BW. Calculations for a 500 kg mature horse are presented in Table 1, comparing the data according to GEH and NRC, respectively.

Reproduction

The nonlinear curve for fetal growth reflects the accretion of energy and nutrients occurring mainly in the last months of pregnancy (Figure 3). From the growth curve and the body composition of the fetus, the energy and nutrient deposition as percentages of the total amount at birth can be calculated (Table 2). The estimation of a foal's birth weight is done by a regression equation (birth weight, kg = 0.45 * kg mare's $BW^{0.75}$), although this equation results in an over- and underestimation, respectively, of foals in small and large breeds, it is a helpful

Table 1. Daily energy and nutrient requirements for a 500 kg horse (GEH, 1994 vs. NRC, 1989).

BW, kg	DE, Mcal	DCP, g	Ca, g	P, g	Mg, g	Na, g	K, g	Cl, g
Maintenance, 500 kg BW								
GEH	15.2	317	25	15	10	10	25	40
NRC	16.2	361[1]	25	18	9.4	-	31.2	-
Exercise a) 60 min, 250 m/min								
GEH	19.8	595	26	15	11	28	35	68
NRC	20.5	451	25	18	9.4	-	31.2	-
Exercise b) 60 min, 250 m/min + 15 min 400 m/min + 5 min 600 m/min								
GEH	27.7	581	28	15	12	57	52	115
NRC	32.8	722	30	21	11.3	-	37.4	-
Reproduction, aa) pregnancy 8th month								
GEH	18.9	450	33	19	10	12	26	41
NRC								
Reproduction, ab) pregnancy 9th –10th month								
GEH	19.8	505	38	22	10	12	27	41
NRC	18.2-18.5	448	35	26	8.7	-	29.7	-
Reproduction, ac) pregnancy 11th month								
GEH	21.0	560	39	23	10	12	27	41
NRC	19.7	476	37	28	9.4	-	31.5	-
Reproduction, b) lactation, 3rd month								
GEH	29.6	720	55	42	13	14	45	36
NRC	28.3	785	56	36	10.9	-	46	-
Growth 3rd – 6th month								
GEH	15.1	580	37	27	5	6	13	20
NRC	14.4	396	34	19	3.7	-	11.3	-
Growth 7th-12th month								
GEH	15.8	540	30	21	7	7	16	26
NRC	15.0-17.2	413-473	29-36	16-20	4-4.3	-	13	-
Growth 13. -18. month								
GEH	16.2	485	30	20	8	8	19	31
NRC	18.9-21.3	468-526	29-34	16-19	5.6	-	18	-
Growth 19.-24. month								
GEH	16.7	445	29	19	9	9	22	34
NRC	19.8	491	27	15	6.4	-	21.1	-
Growth 25.-36. month								
GEH	17.7	415	28	18	10	10	24	38
NRC	18.8	400	24	13	7.0	-	23.1	-

[1] Assuming a crude protein digestibility of 55% accordingly (NRC, 1989).

Table 2. Intrauterine fetal growth, energy and nutrient accretion.

pregnancy, month	Accretion/month in % of total at birth				
	fetal BW	energy, protein	Ca, P	Mg, K	Na, Cl
up to 7th	17	10.5	10	15	16
8	18	14	14	15	21
9	19	21.5	22	21	19
10	23	23	36	22	28
11	23	31	18	27	16

Figure 3. Body weight (y) of the fetus in horses depending on day of gestation (x)

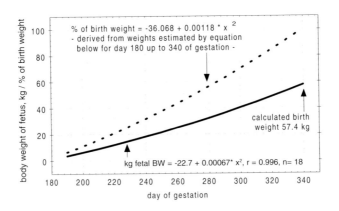

tool to predict energy and nutrient accretion during the last month of gestation. The pregnancy related requirement is calculated by the following equations: MJ DE/day = $0.004851 \times a \times BW^{0.75}$; DCP, g/day = $0.06055 \times a \times BW^{0.75}$; a is the relative accretion in each month of pregnancy as described in Table 2. A sufficient body condition is essential for breeding mares, and the plane of nutrition is important because accretion occurs not only in the pregnant uterus but also in maternal tissues.

To enable the mare to realize energy and protein deposition in maternal tissues, the additional requirement is calculated by: MJ DE/day= $BW \times 0.0164$; g DCP/day = $BW \times 0.085$. The requirement during lactation depends on milk yield (kg milk/kg $BW^{0.75}$: 1st month 0.14, 2nd month 0.17 and 5th month 0.12) and milk composition (NRC, 1989). The utilization of DCP and DE for milk production is set at 50 and 66%, respectively. Results of calculations for the requirements of pregnant mares are presented in Table 1.

Growth

A summary of data taken from the literature yields a model to predict a foal's actual BW in relation to mature BW depending on age; the regression equation (BW % = log (mature BW, kg) + 35469 x log (0.3345 x age, month + 1.1114); r = 0.989) describes the nonlinear growth curve which on average results in 72-57% of mature BW for a yearling (Figure 4).

The composition of the weight gain in the foal is calculated by use of the NRC (1978) equations for fat and protein (body fat % = 0.1388 x relative BW + 1.111; body protein % = 0.22 x (100 - body fat) and completed for macro-elements by literature data (Table 3).

The utilization of DE for growth is set at 60% as in other species, while for DCP, 50% is assumed for the 3rd - 6th month and 35% for older foals. Further information on amino acid turnover would present a more precise figure, taking into account what protein quality is needed. For a 6-month-old foal and a yearling, respectively, 2.3 and 1.88 g of lysine are recommended. Adapted from pigs the S-containing amino acids and threonine should be 0.6 in relation to lysine (=1).

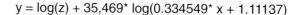

$$y = \log(z) + 35{,}469 * \log(0.334549 * x + 1.11137)$$

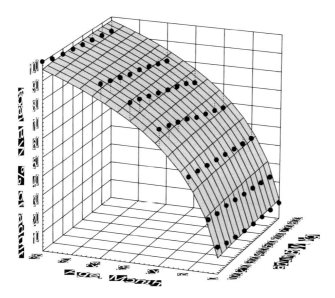

Figure 4. Body weight of foals (y, % of mature BW) depending on age (x, month) and mature body weight (z, kg).

Table 3. Estimated composition of foal's weight gain (per kg; mature BW 600 kg).

age month	protein	fat g	energy Mcal	Ca	P	Mg	Na g	K	Cl
3- 6	197	106	2.126	18	9	0.5	1.5	2.0	1.2
7-12	186	157	2.556	17	8	0.4	1.6	1.8	1.2
13-18	176	199	2.891						
19-24	170	229	3.153	15	8	0.4	1.7	1.8	1.2
25-36	165	257	3.392						

Table 4. Growth rate (GR, g/d) and growth related daily requirements for digestible energy (DE) and digestible protein (DCP) in foals.

age month	200[1] GR g	DE Mcal	DCP g	400[1] GR g	DE Mcal	DCP g	600[1] GR g	DE Mcal	DCP g
3 -6	344	1.28	134	656	2.37	256	984	3.49	388
7-12	242	0.99	126	440	1.93	230	560	2.38	298
13-18	132	0.68	64.6	264	1.33	132	429	2.07	216
19-24	87	0.48	41.1	175	0.94	84.0	295	1.55	143
25-36	44	0.26	20.0	110	0.69	50.9	181	1.02	85.4

[1] Mature body weight (kg)

For the derivation of complete requirements, the higher energy requirement for maintenance compared to adults is taken into account (month [M] 3rd-6th: 0.21; M 7th-12th: 0.167; M 13th-18th: 0.151; M 19th-36th: 0.143 Mcal/kg $BW^{0.75}$ x d^{-1}). For foals kept outside or in groups an extra 20% is added. The recommended relation DCP:DE Mcal of 20.9:1 is used to calculate maintenance requirements for DCP in foals.

Macroelements

Endogenous losses and assumed utilization are the basic data for calculation of macroelement requirements (Table 5). The utilization of calcium can be higher than 60% if the Ca intake is low. The utilization of phosphorus is rather low, recognizing the amount of P in concentrates which is mainly in the form of phytate-phosphorus. The principle to combine endogenous losses with utilization to calculate requirements fails in the case of chloride. Fecal losses are low and nearly completely independent of Cl intake. Intake covering endogenous losses is nearly sufficient to avoid metabolic alkalosis, the typical effect of Cl depletion in horses as in other animals, but an increase in pH and bicarbonate in blood is still present. At 80 mg/kg BW x d^{-1} normochloremia as well as no influence on acid-base balance can be expected; therefore requirement calculations are based on these data.

Table 5. Endogenous losses, utilization and derived daily requirement for macroelements.

		Ca	P	Mg	Na	K	Cl
endogenous losses	mg/kg BW x d[-1]	30	12	7	18	40	5
utilization	%	60[1]	40[2]	35	90	80	100
maintenance requirement	mg/kg BW x d[-1]	50	30	20	20	50	80[3]

[1] Lower if P intake is high
[2] Higher for P from inorganic sources, lower for P in phytate
[3] Less than 80 mg Cl/kg BW x d[-1] and even balancing endogenous losses is not sufficient to prevent changes in acid-base balance.

Table 6. Sweat composition.

element	g/l	element	mg/l
Na	3.1	P	<10
K	1.6	Zn	11
Cl	5.5	Fe	5
Ca	0.12	Cu	0.3
Mg	0.05	Se	traces

Table 7. Mean mineral accretion in fetus (F) and fetus plus adnexes (F+A) in 9th-11th month of gestation (mg/kg BW x d[-1]).

	mature body weight, kg							
	200		400		600		800	
element	F	F+A	F	F+A	F	F+A	F	F+A
Ca	18.1	19.9	15.2	16.7	13.7	15.1	12.7	14.0
P	9.6	10.6	8.1	8.9	7.3	8.0	6.8	7.5
Mg	0.36	0.40	0.31	0.34	0.28	0.31	0.26	0.29
Na	1.6	3.2	1.3	2.6	1.2	2.4	1.1	2.2
K	1.7	3.4	1.5	3.0	1.3	2.6	1.2	2.5
Cl	1.0	2.0	0.8	1.6	0.8	1.6	0.7	1.4

The sweat volume and sweat composition (Table 6) and mineral accretion in fetus and adnexes (Table 7) determine exercise- and pregnancy-related requirements, respectively. The same utilization is taken into account as for maintenance requirements. Although the development of the fetus is nonlinear, formulation of the pregnancy-related requirement is based on the average mineral accretion in the fetus plus adnexes for the 9th through the 11th month of gestation. By recognizing the adnexes, the net accretion in the fetus is elevated by 10% for Ca, P and Mg and by 50 % for the electrolytes. Finally the composition of foal's weight gain and growth rate (Tables 3 and 4) makes it possible to calculate the requirement for growing horses. The results for a 500 kg horse are included in Table 1.

Trace Elements

In general there is a lack of precise information on the trace element requirements in horses. The current recommendations are in part taken from other species. The usual dimension is mg/kg dry matter (DM), assuming a mean DM intake of 2 kg/100 kg BW x d^{-1}.

There are comparable recommendations for growing, breeding and exercising horses (Table 8). The rather high iron (Fe) accretion during pregnancy (75-185 µg/kg BW x $d^{-1)}$ as well as Fe output by milk contribute to levels of 80-100 mg/kg DM. The higher demand for copper (Cu) in foals compared to other horses is reflected by 12 mg Cu/kg DM.

Vitamins

Similar to trace elements, the database for vitamin requirement is incomplete; therefore the recommendations are assumptions and partly based on observations in practice (Table 8). For the fat-soluble vitamins, GEH recommends a broad range for horses. In part there is a higher intake proposed for exercising horses (Vitamins E, B_1). The results from experiments adding ß-carotene are controversial; therefore no clear level can be expressed as a special carotene requirement. Although in some cases supplemental vitamins or vitamin-like substances may be of benefit (folic acid - exercise, ascorbic acid - aged horses, carnitine - exercise), the experimental results do not provide rationale for clear recommendations.

Table 8. Recommendations for trace elements and vitamins in horse feeding.

		maintenance and exercise	breeding horses	foals
iron (Fe)	mg/kg DM	60 – 80	80	80 – 100
copper (Cu)	mg/kg DM	7 – 10	8 – 10	10 – 12
zinc (Zn)	mg/kg DM		50	
manganese (Mn)	mg/kg DM		40	
cobalt (Co)	mg/kg DM		0.05 – 0.1	
selenium (Se)	mg/kg DM		0.15	
iodine (I)	mg/kg DM	0.1 – 0.2		
vitamin A	IE/kg BW x d^{-1}	75	100 – 150	150 – 200
vitamin D	IE/kg BW x d^{-1}	5 – 10	15	15 – 20
vitamin E	mg/kg BW x d^{-1}	$1 – 2^1$	1	1
vitamin B_1	mg/kg DM		3^2	
vitamin B_2	mg/kg DM		2.5	
biotin	mg/kg DM		0.05	

[1]For strenous exercise 4 mg/kg BW x d^{-1}
[2]For strenous exercise 5 mg/kg DM

Table 9. Mean dry matter (DM) intake (kg/100 kg BW x d^{-1}).

BW, kg - mature -	maintenance	exercise	gestation	lactation	Growth, month of age 3 – 6	7 – 12	13 – 24
200	1.3 – 1.6	1.8 – 2.9	1.9 – 2.1	2.4 – 3.0	2.8 – 3.2	2.6 – 3.0	2.1 – 2.5
400	1.2 – 1.4	1.5 – 2.4	1.6 – 1.8	2.0 – 2.5	2.0 – 2.5	1.8 – 2.2	1.6 – 1.8
600	1.1 – 1.3	1.4 – 2.2	1.4 – 1.6	1.8 – 2.3	1.9 – 2.2	1.8 – 2.0	1.5 – 1.7
800	1.0 – 1.2	1.3 – 1.5	1.3 – 1.5	1.7 – 1.5	1.6 – 1.8	1.6 – 1.8	1.3 – 1.5

Feed and Water Intake

The average DM intake varies between 1.3 and 3.2 kg/100 kg BW x d^{-1}. The different values are summarized in Table 9. The mean water consumption is set at 3-3.5 l/kg DM (liters/100 kg BW x d^{-1}, foals 7-10, maintenance 3-5, lactating mares 8, exercise 3-10). However, environment, interindividual variation, and feed type can give reason for remarkable differences. In consequence, the best way to supply fresh water is freely, indoors as well as outdoors.

Basic Limits for Ration Design

To create a ration means to fulfill the following characteristics:
a) covering requirement
b) keep total DM within the margins for DM intake
c) have a ration which takes into account natural equine eating behavior
d) combine feeds in a compatible ration
e) keep the ration in a reasonable economical frame

The aspects related to a and b are described earlier in this paper. With regard to the natural behavior (~12-13 h/d spent on feed intake) we recommend 1 kg, but at least 0.5 kg hay or an alternate roughage per 100 kg BW x d^{-1}. Beside the horse's activity related to feed intake, the saliva production during chewing (mainly a function of chewing time/g feed) is a further reason to require a specific amount of roughage. Compared to grass or hay, the intake of concentrates results in lower saliva production, accelerated gastric fill without balanced gastric emptying and a rather high dry matter percentage in gastric contents (>30%). To prevent disturbances by this situation, the maximum amount of concentrate is 0.5 kg/100 kg BW x meal^{-1}; that means in cases of high energy requirement, 3 or more meals/day are necessary for a safe feeding system.

Selected Aspects for "Brain Storming"

Although nutrient requirements are closely examined before a new edition of feeding standards is published, some further points should be underlined. A system to predict the energy content of individual feedstuffs by chemical composition is an essential instrument in order to convert the knowledge about requirements into feeding practice. But there is a certain influence of the whole ration on the feeding value of a single component of the ration. A reasonable prediction (as well as an agreement as to what kind of calculation anybody should use) of feeding value for the whole ration and the isolated feed would offer remarkable benefit, particularly on the market of mixed feeds.

The described classes of exercise (low, medium and high) in terms of the x-fold of maintenance requirement are a rough but helpful approach to project energy and nutrient requirements for exercise. But they consider the intensity of exercise from the standpoint of requirement. In practice, however, horse keepers need help adapting feeding level to exercise intensity. This requires evaluation of requirement data in relation to duration and speed of exercise. This is very important because many people assume that their horse is in hard work.

More knowledge about digestion in the small and large intestine would be of interest, in particular the digestibility of protein and amino acids. The relation of ammonia in colon draining veins to peripheral venous blood is about 10:1; that means, the liver may be loaded by ammonia as a function of protein intake, protein quality and microbial fermentation in the large intestine.

Observations on the changes in bones in young horses during the first weeks of training stimulate discussion of the mineral supply, particularly Ca intake. However, at present it seems that elevated Ca intake alone is ineffective in minimizing the risk of injury in early training.

There are different results dealing with the anion-cation balance (focused on exercising horses). Possibly they can be introduced in the recommendation for electrolyte supply. The rather high potassium concentration in roughage from intensively used meadows as well as the possible link between glucose uptake by the muscle and the electrolyte homeostasis substantiate the interest in the effects of electrolyte balance in feed on the acid base balance in the horse.

Recent information on selenium in horses indicates that a low selenium intake is not simply related to muscle problems and that the selenium requirement may be overestimated by the current data. The big variation in blood selenium under identical feeding conditions poses the question for the optimal Se intake and the suitable tool for assessment of Se ingestion.

In general the vitamin intake is above the requirement if concentrates are in use. One question reflects a possible interaction between vitamin A and vitamin E in the case of vitamin A excess. A second one is related to the requirement of vitamin D. Some experimental data indicate a certain degree of independence of vitamin D intake for the horse. In consequence, it is doubtful if there is a remarkable vitamin D requirement in the horse; however, the typical recommendations are higher than necessary.

In general the supply of trace elements and vitamins is in excess of requirement in many cases. Limits for a reasonable intake would complete the formulation of recommendations.

References

The actual used data for requirements and feeding standards are published by: AUSSCHUSS FÜR BEDARFSNORMEN DER GESELLSCHAFT FÜR ERNÄHRUNGSPHYSIOLOGIE (1994): Empfehlungen zur Energie und Nährstoffversorgung der Pferde. DLG-Verlag Frankfut/Main, ISBN 3-7690-0517-1 (Committee for requirement formulations of the German Society of Nutritional Physiology: Recommendations for the energy and nutrient supply for horses).

Allen B V (1978): Serum folate levels in horses with particular reference to the English Thoroughbred. Vet. Rec. 103, 257 - 259.

Argenzio RA, Hintz HF (1971): Energy utilization in horses. Cornell Nutrition Conference, Buffalo, N. Y.

AUSSCHUSS FÜR BEDARFSNORMEN der Gesellschaft für Ernährungsphysiologie der Haustiere (1982), Nr. 2: Empfehlungen zur Energie- und Nährstoffversorgung der Pferde, DLG-Verlag, Frankfurt/M.

Barth KM, Williams JW, Brown DG (1977): Digestible energy requirements of working and non-working ponies. J. Anim. Sci. 44, 585 -589.

Bergner H, Ketz HA (1969): Verdauung, Resorption Intermediärstoffwechsel bei landwirtschaftlichen Nutztieren. VEB Deutscher Landwirtschaftsverlag, Berlin.

Bieberstein S von (1989): Prüfung von verschiedenen Mg-Verbin-dungen bei ruhenden und arbeitenden Pferden. Vet. Diss. Hannover.

Bostedt H (1977): Zur Klinik der ernährungsbedingten Muskeldegeneration bei Fohlen. Dtsch. tierärztl. Wschr. 84, 293 - 296.

Bouwman H, Van Der Schee W (1978): Composition and production of milk from Dutch warmblooded saddle horse mares. Z. Tierphysiol., Tierernährg. u. Futtermittelkde. 40, 39 - 53.

Breidbach S (1959): Über Zucht, Haltung, Leistungsfähigkeit und Einsatzmöglichkeiten von Kleinpferden in der Landwirtschaft der Bundesrepublik. Züchtungskde. 31, 241 - 250.

Brody S (1945): Bioenergetics and growth. New York, Reinhold Publ. Co.

Coenen M (1992): CI-Haushalt und CI-Bedarf des Pferdes. Habilitations-schrift, Tierärztliche Hochschule Hannover.

Crampton, E (1923): Rate of growth of draft horses. J. Agric. and Hort. 26, 172. Zit. nach Cunningham u. Fowler (1961), Bull. 546, Louisiana State University.

Crasemann: Hdb. Tierernährung 2, 674 - 689. Verlag Paul Parey, Berlin - Hamburg.

Cunningham K, Fowler S (1961): A study of growth and development in the Quarter Horse. Louisiana State University, Bull. 546.

Cymbaluk N, Christison J (1990): Environmental effects on thermoregulation and nutrition of horses. Vet. Clin. North America, Equi. Pract. 6, 356 - 371.

Donoghue S, Kronfeld DS, Berkowitz St J, Cop, RL (1981): Vitamin A nutrition of the equine: Growth, serum biochemistry and hematology. J. Nutr. 111, 365 - 374.

Doreau M, Boulot S (1989): Recent knowledge on mare milk production: a review. Livestock Prod. Sci. 22, 213 - 235.

Drepper K. (1980): Personal communication

Dusek J, Richter L (1972): Änderungen der Körpermasse von Stuten im Verlauf der Gravidität. Arch. Tierzucht 15, 361 - 366.

Dusek J (1976): Bewertung der Gewichtszunahme bei Fohlen. Bayer. Ldw. Jb. 53, 68-70.

Dyrendal S (1972): Arbeitsleistung, in: Lenkeit, W, Breirem K, Flade J (1957): Wachstum und Entwicklung beim Pferd. Tier-zucht 11, 162 - 170.

Geyer H, Schulze J (1994): The longterm influence of biotin supplementation on hoof horn quality in horse. Schweiz. Arch. Tier-heilkd. 136, 137-149.

Gibbs P, Potter D, Blake R, McMullan W (1982): Milk production of Quarter Horse mares during 150 days of lactation. J. Anim. Sci. 54, 496 - 499.

Gill RJ, Potter GD, Schelling GT, Kreider JL, Boyd CL (1983): Postpartum repro-ductive performance of mares fed various levels of protein. Proc. 8th Equine Nutr. Physiol. Symp., Lexington, Kentucky, 311 - 316.

Gürer C (1985): Untersuchungen zum K-Stoffwechsei des Pferdes. Vet. Diss. Hannover.

Gütte JO (1972): Energiebedarf laktierender Stuten, in: Lenkeit W, Breirem K, Crasemann E: Hdb. Tierernährg. 2, 393 - 398, Verlag Paul Parey, Berlin - Hamburg.

Harmeyer J, Twehues R, Schlumbohm C, Stader-Mann B, Meyer H (1992): The role of vitamin D on calcium metabolism in horses. Pferdeheilkunde. Sonderheft, 81 - 85

Heikens A (1992): Untersuchungen über den Se-Gehalt verschiede-ner Futtermittel sowie die Se-Versorgung von Wiederkäuern und Pfer-den in Ostfriesland. Vet. Diss. Hannover.

Hesse H. (1957.): Entwicklung und Wachstumsverlauf bei Kleinpfer-den. Z. Tierzüchtg. u. Züchtungsbiol. 70, 175 - 181.

Hintz HF, Roberts SJ, Sabin SW, Schryver HF (1971): Energy requirements of light horses for various activities. J. Anim. Sci. 32, 100 -102.

Hintz H, Hintz RL, van Vleck LD (1979): Growth rate of Thoroughbreds, effect of age of dam, year and month of birth and sex of foal. J. Anim. Sci. 48, 480 - 487.

Hoffmann L, Klippel W, Schiemann R (1967): Untersuchungen über den Energieumsatz beim Pferd unter besonderer Berücksichtigung der Horizontalbewegung. Arch. Tierern. 17, 411 - 449.

Jackson S, Baker J (1983): Digestible energy requirements of Thoroughbred geldings at the gallop. Proc. 8th Equine Nutr. Physiol. Symp., 113 - 118.

Jordan R, Myers MY (1972): Effect of protein levels on the growth of weanling and yearling ponies. J. Anim. Sci. 34, 578 - 581.

Knox KL, Crownover JC, Wooden GR (1970): Maintenance energy requirements for mature idle horses, in: Schürch A and C. Wenk C (ed.): Energy metabolism of farm animals, Juris-Verlag, Zürich.

Meadows DG, Potter GD, Thomas W, Hesby J, Anderson J (1979): Foal growth from mares fed supplemental soybean meal or urea. Proc. 6th Equine Nutr. Physiol. Symp. 14 - 16.

Menke K-H, Huss W (1980): Tierernährung und Futtermittelkde., Verlag Eugen Ulmer, Stuttgart.

Mertz W (1987): Trace elements in human and animal nutrition. 5. Aufl., Academic Press, San Diego.

Meyer H, Ahlsede L (1976): Über das intrauterine Wachstum und die Körperzusammensetzung von Fohlen sowie den Nährstoffbedarf tragender Stuten. Übers. Tierernährg. 4, 263 - 292.

Meyer H. (1983): Protein metabolism and protein requirement in mares. 4th International Symp. Protein metabolism in nutrition 1, 343 - 364.

Meyer H. (1990): Beiträge zum Wasser- und Mineralstoffhaushalt des Pferdes.
Fortschr. Tierphysiol. Tierernährg., Heft 21, Verlag Parey, Hamburg u. Berlin.
Meyer H, Stadermann B (1991): Energie- und Nährstoffbedarf
hochtragender Stuten. Pferdeheilkde. 7, 11- 20.
Meyer H. (1992): Pferdefütterung. Verlag Paul Parey, Hamburg u. Berlin, 2. Aufl.
Meyer H. (1994): Cu-Stoffwechsel und Cu-Bedarf beim Pferd. Übers.
Tierernährg. 22 (in Vorbereitung).
National Research Council (NRC) (1978): Nutrient requirements of horses.
4th edition, National Academic Press, Washington D. C.
National Research Council (NRC) (1989): Nutrient requirements of horses.
5th edition, National Academic Press, Washington D. C.
Nehring T. (1991): Einfluß von Futterart auf die Nettoabsorption von Calcium
sowie Magnesium und Phosphor beim Pferd. Vet. Diss. Hannover.
Neseni R, Flade E, Heidler G, Steger H (1958): Milchleistung und
Milchzusammensetzung von Stuten im Verlaufe der Laktation. Arch.
Tierzucht 1, 91 - 129.
Neuhaus U. (1960): Untersuchungen und Beobachtungen über Ge-winnung,
Eigenschaften und Zusammensetzung der Stutenmilch fürdie Ernährung
menschlicher Säuglinge. Züchtungskde. 32, 513 - 519.
Oftedal 0T, Hintz HF, Schryver HF (1983): Lactation in the horse: milk composition
and intake by foals. J. Nutr. 113, 2196 - 2206.
Olofsson NE, Thomke S (1963): Uppfödningsförsok med unghästar vid Ultuna
1952 - 1968 (Raising experiments with young horses at Ultuna,
1952 -1958). Lantbrukshögskonlans meddelen Ser. A, Nr. 1.
Olsson N, Ruudvere A (1955): The nutrition of the horse. Nutr. Abstr.
Rev. 25, 1 - 18.
Pagan J, Hintz H (1986): Equine energetics. 1. Relationship between body
weight and energy requirements in horses. J. Ani. Sci. 63, 815-821.
Pferdekamp M (1978): Einfluß steigender Proteinmengen auf den Stoffwechsel
des Pferdes. Vet. Diss. Hannover.
Ponen (1986): Vitamin E requirements of adult Standardbred horses
evaluated by tissue depletion and repletion. Equine. Vet. J. 18, 50 - 58.
Rambeck W (1994): Carry-over von Cadmium. Übers. Tierernährg. 22, 184 - 190.
Reed K, Dunn N (1977): Growth and development of the Arabian horse.
Proc. 5th Equi. Nutr. Phys. Symp. 76 - 98.
Roneus B0, Hakkarianen RVJ, Lindholm CA, Tyo JI
Saastamoinen M (1990): Heritabilities for body size and growth rate in young
horses. Acta Agric. Scand. 46, 377 - 386.
Saastamoinen M (1993): Feed energy and protein intakes of horses -review of
Finnish feeding trials. Agric. Sci. Finl. 2, 25 - 32.
Saastamoinen M, Koskinen E (1993): influence of quality of dietary protein
supplement and anabolic steroids on muscular and skeletal growth of
foals. Anim. Prod. 56, 135 - 144.
Sandersleben J von, Schlotke B (1977): Die Muskeldystrophie
(Weißmuskelkrankheit) bei Fohlen, eine offensichtlich im Zunehmen
begriffene Erkrankung. Dtsch. tierärztl. Wschr. 84, 105 - 107.
Schryver HF, Hintz HF (1972): Calcium and phosphorus requirements of
the horse. Feedstuffs 44, Nr. 28, 35 - 38.

Schryver HF, Hintz HF, Lowe JE, Hintz RL, Harper RB, Reid JT (1974): Mineral composition of the whole body, liver and bone of young horses. J. Nutr. 104, 126 - 132.

Schyver H. (1990): Mineral and vitamin intoxication in horses. Vet. Clin. North America - Equine Pract. 6, 295 - 317.

Schwarz FJ, Kirchgessner M (1979): Spurenelementbedarf und versorgung in der Pferdefütterung. Übers. Tierernährg. 7, 257 - 278.

Slade L., Lewis D, Quinn C, Chandler L (1975): Nutritional adaptation of horses for endurance type performance.Proc. 4th Equine Nutrition and Physiology Symposium, Davis, California, 114 - 128.

Teeter SM, Stillions MC, Nelson WE (1967): Recent observations on the nutrition of mature horses. Proc. 13th Ann. conv. Am. Ass. Equ. Pract., New Orleans, 39 - 49.

TIimm A. (1993): Einfluß der Gelatinefütterung auf das Skelett-, Haar-. Knorpel- und Hufwachstum bei Absatzfohlen. Diss. Vet, Hannover.

Topliff DR, Potter GD, Kreider JL, Creagor CR (1981): Thiamin supplementation for exercising horses. Proc. 7th Equine Nutr. Physiol. Soc. Symp. Warrenton, Va., 167 - 172.

Wooden GR, Knox KL, Wild CL (1970): Energy metabolism in light horses. J. Anim. Sci. 30, 544 - 548.

Zentek J (1991): Myopathien in einem Reitpferdebestand. Tierärztl. Praxis 19, 167 - 169.

Zuntz N, Hagemann O (1898): Untersuchungen über den Stoff-wechsel des Pferdes bei Ruhe und Arbeit. Landwirtsch. Jb. 27, Erg. Bd. 111.

THE EFFECTS OF INTAKE LEVEL ON THE DIGESTIBILITY AND RETENTION OF COPPER, ZINC AND MANGANESE IN SEDENTARY HORSES

C. A. HUDSON, J. D. PAGAN, K.E. HOEKSTRA, A. PRINCE, S. GARDNER, R. J. GEOR
Kentucky Equine Research, Inc., Versailles, KY

Few studies have directly evaluated the effects of intake level on the true digestibility and retention of copper (Cu), zinc (Zn) and manganese (Mn). Furthermore, the manganese requirements of horses have not been established (NRC, 1989). To address these questions, four mature Thoroughbred horses (mean age [± SD] 13 ± 2.2 yr; weight 559 ± 47 kg) were studied in a 16-week longitudinal experiment that consisted of 4 periods, each with a 23-day adaptation period followed by a 5-day complete collection digestion trial. In Period 1, horses were fed an unfortified diet (basal Cu, Zn and Mn intake) consisting of 6.8 kg of alfalfa cubes and 3 kg of sweet feed (45% cracked corn, 45% whole oats, 10% molasses). In Periods 2, 3 and 4, respectively, the diet was fortified with a supplement providing 50%, 100% and 200% of the National Research Council [NRC] (1989) requirements for Cu, Zn and Mn. During each adaptation period, the horses were housed in box stalls at night and given paddock turnout during the day. Muzzles were worn to prevent grazing. During the 5-day digestion trial, the horses were fitted with harnesses that permitted the complete and separate collection of urine and feces. Composite samples of feed, urine and feces were stored and subsequently analyzed for Cu, Zn and Mn contents by ICP radial spectrometry. For each mineral, and over the four levels of intake, external balance data were calculated. Linear regression analysis was performed to determine the relationship between intake level and mineral retention (retention = intake - fecal and urinary losses). The Lucas procedure was used to estimate the contribution of endogenous losses to the apparent digestibility of each nutrient. With this procedure, slope of the linear regression of nutrient retention against level of intake is the true digestibility and level of intake when y = 0 is taken as the maintenance requirement.

Data (mean ± SD, n = 4) for intake, fecal and urinary losses, and retention are presented in Table 1.

The linear regression equations describing the relationship between Zn and Mn intake and retention were: Zn: y = 0.235x - 53.1 (r^2 = 0.696, P < 0.001) and Mn: y = 0.537x - 291 (r^2 = 0.729, P < 0.001). On the other hand, the relationship between copper intake and retention was best described by a sigmoidal curve. In particular, Cu supplementation at 200% NRC resulted in only a small increase in Cu retention compared to supplementation at 100% NRC (see Table 1). From the Lucas procedure, estimates of true digestibility for Zn and Mn were 23.5% and 53.7%, respectively, while daily endogenous losses were 53 mg and 291 mg. Estimates of daily maintenance requirements for Zn and Mn were 230 and 540 mg, respectively.

Table 1. External balance for copper (Cu), zinc (Zn), and manganese (Mn) at four different levels of intake.

Mineral		Intake (mg/day)	Fecal losses (mg/day)	Urinary losses (mg/day)	Retention (mg/day)
Cu	Basal	53 ± 0	36 ± 3	0	17 ± 7
	50%	152 ± 4	101 ± 9	0	51 ± 9
	100%	187 ± 18	96 ± 23	0	90 ± 21
	200%	289 ± 29	186 ± 51	0	103 ± 45
Zn	Basal	145 ± 0	169 ± 74	0	-24 ± 22
	50%	450 ± 6	392 ± 95	0	58 ± 54
	100%	608 ± 33	513 ± 112	101 ± 7	92 ± 17
	200%	983 ± 32	832 ± 234	145 ± 4	176 ± 88
Mn	Basal	252 ± 0	389 ± 143	0	-137 ± 87
	50%	749 ± 21	668 ± 204	0	71 ± 77
	100%	885 ± 90	688 ± 175	0	198 ± 108
	200%	1,338 ± 62	904 ± 254	0	434 ± 176

The data for Zn true digestibility, endogenous losses, and maintenance requirements are consistent with previously published values (Pagan, 1998). However, the corresponding values for Mn were approximately 40% higher than previous estimates (Pagan, 1998). The sigmoidal relationship between copper intake and retention precluded use of the Lucas procedure for estimation of maintenance requirements. However, when Cu retention was regressed against the 50% and 100% NRC levels of supplementation (daily intake of approximately 150 mg and 190 mg, respectively), the estimated maintenance requirement was 100 mg/day which is in accord with published recommendations (NRC, 1989).

References

National Research Council. Nutrient Requirements of Horses, 5th edition. National Academy Press, Washington, D. C., 1989.

Pagan JD. Nutrient digestibility in horses. In: Advances in Equine Nutrition, edited by J. D. Pagan. Nottingham University Press, Nottingham, UK, 1998, pp 77-84.

APPARENT DIGESTION OF HAY/GRAIN RATIONS IN AGED HORSES-REVISITED

S. L. RALSTON[1], K. MALINOWSKI[1], R. CHRISTENSEN[1] AND L. BREUER[2]
[1]Rutgers, the State University of New Jersey, New Brunswick, NJ
[2]Purina Mills, Inc.

Research on digestion in aged (> 20 yrs old) Quarter Horse and Thoroughbred horses in the mid-1980s revealed that the population studied at that time had reduced apparent digestion of protein, phosphorus and fiber relative to younger horses fed pelleted alfalfa (Ralston et al., 1989). The digestive profile of the aged horses was very similar to that reported for horses which had had 90% resection of the large colon (Bertone et al., 1989a and 1989b).

In two subsequent digestion trials on aged horses conducted in New Jersey in the 1990s, the reductions of digestibility, especially in protein, were not as apparent. In trial 1, conducted in 1990, eight light horse mares > 19 years old were used in a simple crossover design to compare digestibility of a commercial textured sweet feed (OM) and one formulated specifically for aged horses (ES). The grains were fed in amounts calculated to provide 50% of the recommended caloric requirements (NRC, 1989) for maintenance, with the rest of the ration being provided as long stem hay. The mares were adapted to the rations for 2 weeks and then subjected to a standard 5-day digestion trial, during which intake was recorded and all feces were collected and weighed, with subsamples taken for nutrient analyses at a commercial laboratory (DHIA, Ithaca, NY). There were differences in apparent digestibility between the two rations. Protein, calcium and phosphorus intakes were higher in ES fed horses than in the OM. Mares on the OM were, as previously reported in aged horses, in low to negative phosphorus balance (-8.1%±4.3) but were in positive balance on the ES (13.3%±4.4). Calcium apparent digestion was also higher in ES fed horses (OM: 10±6; ES: 33±4, p<.05). Crude protein apparent digestion in OM fed mares was lower (OM: 63.9±1.0; ES: 71.5±0.7, p<.05); however it was still within the normal range of protein apparent digestion reported for horses. The differences between the two rations were probably due to the higher protein, calcium and phosphorus intake in the ES ration (Table 1).

In trial 2, conducted in 1995, 15 aged Standardbred mares were used in a trial comparing daily injections with equine somatotropin (eST). They were fed OM and mixed hay in amounts calculated to provide 120% recommended intakes (NRC, 1989). Results of this trial have been reported previously (Ralston et al., 1996). However, it is important to note that protein, phosphorus and calcium apparent digestion were within normal limits regardless of treatment (Table 2). The reduction in large intestinal absorption observed previously (Ralston et al., 1989) was not apparent, despite similar types of hay (timothy grass mix) and grain (OM).

Table 1. Total ration nutrient composition of rations fed in aged horse digestion trials (all values on dry matter basis).

Date	1989	1990 OM	1990 ES	1995
Crude protein	16%	10%	12%	13%
Crude fiber	22%	NA	NA	NA
Acid detergent fiber	NA	28%	32%	28%
Neutral detergent fiber	NA	49%	55%	39%
Calcium	1.2%	0.3%	0.5%	0.8%
Phosphorus	0.3%	0.3%	0.4%	0.3%

NA=not reported

Table 2. Covariately adjusted least squares means for percent apparent digestibilities of nutrients in aged mares given daily equine somatotropin (eST).[1]

Nutrient	eST (mg/day)			
Dry matter	56.6±1.8	61.5±2.0	63.8±0.04	0.23
Crude protein	64.8±2.5	70.3±2.9	69.5±0.31	0.23
Acid detergent fiber	33.9±2.8	30.3±3.2	41.8±2.4	0.04
Neutral detergent fiber	27.9±2.5	33.2±2.8	37.7±2.1	0.06
Calcium	53.3±5.7	46.9±6.6	54.1±5.6	0.69
Phosphorus	14.4±5.3	13.6±6.4	13.2±4.8	0.98

[1]Pretreatment (30 days prior to eST injection) was used as the covariate for day 22. N = 4, 3 and 5 for 0, 6.25 and 12.5 mg eST, respectively.

Comparing results of digestion trials which used different feeds, horses and conditions is somewhat risky. However, the same investigator (SLR) performed the trials and used exactly the same techniques in all 4 studies relative to the collection of feces, etc. The initial 1989 trial used straight alfalfa pellets as the basal ration which was higher in protein and calcium and equivalent in phosphorus relative to the rations used in subsequent studies (Table 1). If anything, the later studies should have shown lower digestion of these nutrients relative to the 1989 report. Both trials in the 1990s were conducted at the same facility and the phosphorus intakes in all trials were similar (Table 1). Therefore, the consistently positive phosphorus digestion in the 1995 study relative to the low to negative balance seen in OM horses in the 1989 and 1990 studies suggests a difference between the two populations of aged horses. One potentially significant difference between the studies was that the horses in both studies in the 1990s were in fair to good body condition with no major dental abnormalities, in contrast to the 1989 report, wherein 3 of the 7 aged horses were reported to have weight loss and/or poor dentition. In a field comparison of ES to a standard textured feed (Ralston and Breuer, 1996), it was noted that old horses in poor body condition received the most benefits from the "senior" formulation.

Another potential contributing factor to differences seen between the studies in aged horses may be improved gastrointestinal parasite control in the horses used in the later studies. Migrating *Strongylus vulgaris* larvae cause "verminous aneurysms" which result in thromboembolic lesions and scarring, especially in the large intestine (Drudge and Lyons, 1989). Effective, safe larvicidal anthelmintic administration was not a common practice until the late 1970s and early 1980s (Drudge and Lyons, 1989). The mares used in the original study had been born in the mid- to late-1960s and had not had larvicidal anthelmintics administered until already aged, if at all. The horses in the last two studies, born in the mid-1970s, had the benefit of larvicidal anthelmintic administration for most, if not all, of their lives. It is hypothesized that the reduction in apparent digestion of protein, phosphorus and fiber reported previously in aged horses may have been due in part to chronic parasitic damage to the large colons and/or abnormal dentition rather than caused by aging per se.

References

Bertone AL, Ralston SL, Stashak TS. Fiber digestion and voluntary intake in horses after adaption to extensive large colon resection. Am J Vet Res 50:1628-1632, 1989a.

Bertone AL, Van Soest PJ, Stashak TS. Digestion, fecal and blood variables associated with extensive large colon resection in the horse. Am J Vet Res 50:253-258, 1989b.

Drudge JH, Lyons ET. Internal parasites of equids with emphasis on treatment and control. Hoechst-Roussel Agri-Vet Co, Somerville, NJ, 1989.

Ralston SL, Breuer LH. Field evaluation of a feed formulated for geriatric horses. J Eq Vet Sci 16:334-338, 1996.

Ralston SL, Christensen RA, Malinowski K, Scanes CG, Hafs HD. Chronic effects of equine growth hormone (eGH) on intake, digestibility and retention of nutrients in aged mares. J Anim Sci 74 (Suppl. 1): 194, 1996.

Ralston SL, Squires EL, Nockels CF. Digestion in the aged horse. J Eq Vet Sci 9:203-205, 1989.

PATHOLOGICAL CONDITIONS

GASTRIC ULCERS IN HORSES: A WIDESPREAD BUT MANAGEABLE DISEASE

JOE D. PAGAN

Kentucky Equine Research, Versailles, KY

Every equine practitioner appreciates the delicate nature of the equine gut. Problems related to the small intestine and large intestine are well understood and routinely treated. What may be surprising to many is how often the stomach is affected.

Specifically, the incidence of gastric ulcers is extremely high, particularly in performance horses. This article will review why horses suffer from gastric ulcers and provide guidelines for their treatment and prevention.

Many studies since the mid-1980s have documented that gastric ulcers are commonplace in racehorses. An early postmortem study in Hong Kong (Hammond et al., 1986) of 195 Thoroughbred racehorses showed that 80% of the horses in active training had ulcers. The incidence of ulcers in horses retired from racing for one month or longer was 52%. Murray et al. (1989) examined the stomachs of 187 horses ranging in age from one to 24 years. Eighty-seven horses had clinical problems including chronic, recurrent colic, poor body condition or chronic diarrhea. One hundred horses had no clinical signs of gastrointestinal problems.

Figure 1.

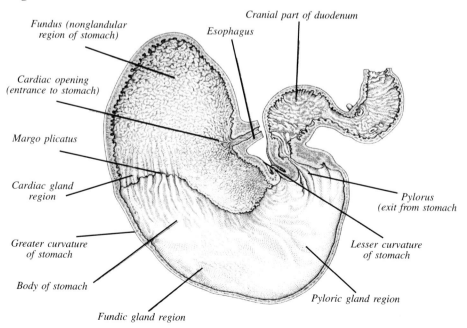

387

Ninety-two percent of the horses with clinical problems had gastric ulcers. Surprisingly, 52% of the horses displaying no clinical signs also had lesions. Racehorses in training had a higher incidence of ulcers (89%) than non-racers (59%). More recently, two studies evaluated the incidence of gastric ulcers in California racehorses. In one postmortem study (Johnson et al., 1994) of 169 horses in training, 88% of Thoroughbred horses in training had ulcers. A gastroendoscopic study of 202 Thoroughbred horses in training showed that 81% had ulcers (Vatistas et al., 1994).

Each of these studies produced remarkably similar results. 80-90% of racehorses in training have gastric ulcers. The vast majority of these lesions occur in the region of the stomach above the margo plicatus, with very few lesions in the glandular portion (Figure 1). The upper half of the horse's stomach consists of squamous epithelial cells that are very similar to the tissue found in the esophagus. Ulcers in this part of the stomach are more similar to esophagitis (heartburn) in humans than the ulcers that occur in the glandular region of the human stomach. It has also been determined that equine gastric ulcers are not caused by *Helicobactor pylori* bacteria, which are a common cause of ulcers in humans.

Gastric Acid is a Major Cause

Dr. M.J. Murray of the Marion duPont Scott Equine Medical Center in Leesburg, Virginia has proposed that the major cause of gastric ulcers in horses is prolonged exposure of the squamous mucosa to gastric acid.

Unlike the glandular portion of the stomach, this tissue does not have a mucous layer and does not secrete bicarbonate onto its luminal surface. The only protection that this portion of the stomach has from gastric acid and pepsin comes from saliva production. If adequate saliva is not produced to buffer the gastric acid and coat the squamous epithelium, gastric irritation occurs and lesions may develop.

The high incidence of ulcers seen in performance horses is a man-made problem resulting from the way that we feed and manage these horses, since ulcers are extremely rare in horses maintained solely on pasture. Horses evolved as wandering grazers with digestive tracts designed for continual consumption of forage. Meals of grain or extended periods of fasting lead to excess gastric acid output without adequate saliva production.

Horses secrete acid continually whether they are fed or not. The pH of gastric fluid in horses withheld from feed for several hours has consistently been measured to be 2.0 or less (Murray, 1992). Horses that received free choice timothy hay for 24 hours had mean gastric pH readings that were significantly higher than fasted horses (3.1 in fed vs 1.5 in fasted horses)(Murray and Schusser, 1989). Higher pH readings in hay-fed horses should be expected since forage consumption stimulates saliva production. Meyer et al. (1985) measured the amount of saliva produced when horses ate either hay, pasture or a grain feed. When fed hay and fresh grass, the horse produced 400-480 grams of saliva per 100 g of dry matter consumed. When a grain-based feed was offered, the horses produced only about half (206 g/100 g dry matter) as much saliva.

Grains and pelleted concentrates also increase the production of gastrin, a hormone that stimulates gastric acid production (Smyth et al., 1988). Hay fed alone affects gastrin production to a lesser degree. Therefore, horses that are fasted or that are fed high grain diets are more likely to produce more gastric acid with less saliva than horses offered free choice forage without grain.

It is easy to see why racehorses have such a high incidence of ulcers. Most horses in training are confined for most of the day and fed large grain meals. Often, racehorses are fasted for an extended period before exercise, allowing gastric acid to accumulate in the stomach. Intense exercise further increases the production of gastric acid so that the squamous mucosa of the stomach gets thoroughly bathed in acid during work.

Treatment and Prevention of Ulcers

Treating ulcers involves either inhibiting gastric acid secretion or neutralizing the acid produced. There are three classes of drugs that can be used to inhibit gastric acid secretion:

1)   Histamine type-2 antagonists (H_2 antagonists). H_2 antagonists act by competing with histamine for histamine type-2 receptor sites on the parietal cell, and therefore blocking histamine-stimulated gastric acid secretion. The two most popular H_2 antagonists used in horses are cimetidine (Tagamet) and ranitidine (Zantac).
2)   H+/K+ ATPase inhibitors. Direct inhibition of the proton pump can be achieved by substituted benzimidazoles. The only proton pump inhibitor licensed in the US and Europe is omeprazole.
3)   Prostaglandin analogues.

An alternative to suppression of acid production is to neutralize stomach acid and protect the squamous mucosa from exposure to acid. The natural buffering mechanism in the horse is from saliva production and indeed the most effective way to treat ulcers is simply to turn the animal out on pasture. In situations where this is not possible, administration of antacids may be a useful adjunct to acid suppression therapy in horses.

An Equine Antacid

For an antacid to work well in horses it must possess several properties. First, administering it should be easy, preferably by adding to the horse's feed. A small dose should neutralize a large amount of acid and it should coat and protect the mucosa from gastric acid irritation. In addition, antacids should be able to adsorb pepsin and other substances that can damage the mucosa. Finally, it should not affect fecal consistency and should create no harmful side effects.

Over the last three years, Kentucky Equine Research has worked to develop an equine antacid that would satisfy the above criteria. The first attempt at an equine antacid used ingredients that are common in human preparations. Unfortunately, when this type of preparation was fed at high doses to horses, they became ataxic and displayed signs that were similar to tying-up. Apparently, the horses were sensitive to magnesium hydroxide, one of the common antacids used for humans. Therefore, magnesium hydroxide was removed from the formula.

Another class of antacids contains aluminum. A concern with these is that the aluminum will interfere with phosphorus absorption. Although we found no detrimental effect of aluminum intake on phosphorus absorption in digestibility experiments (Pagan, unpublished, 1996), supplemental phosphorus (balanced with calcium) was added to the formula. Schryver et al. (1986) also showed that only small amounts of aluminum are absorbed by the horse and aluminum retention was not affected by level of aluminum intake. Therefore, it appears that aluminum-containing antacids can be safely fed to horses.

After a few more modifications, a commercial product was developed and patented. This equine antacid (Neigh-Lox) is in a palatable pelleted form that can be fed alone or mixed in feed. It contains a very fast acting antacid with 240 mequ/dose of acid neutralizing capacity. A four ounce (120 g) dose will neutralize about 6 hours of basal acid production. The recommended dosage for adult performance horses is 4 ounces per meal with a maximum daily intake of 16 ounces. Neigh-Lox also contains a compound that serves as a coating agent to protect the gastric mucosa. This ingredient has an astringent and anti-inflammatory action and pepsin binding capacity.

Since its development, Neigh-Lox has been field tested in hundreds of horses. Many horses that displayed signs of gastric irritation such as poor appetite, chronic colic, and sour disposition have shown immediate improvement after receiving only a few doses of Neigh-Lox. Since there are no studies to show that Neigh-Lox heals ulcers, it is recommended only as adjunctive therapy to acid suppressive drugs in horses that have been positively diagnosed with gastric ulcers. Neigh-Lox's most important role, however, may be in preventing ulcers from occurring in the first place. Antacid therapy along with a dietary program that emphasizes continual forage consumption may greatly reduce the incidence of gastric ulcers in performance horses and foals.

References

Hammond, C.J.. Mason, D.K. and Watkins, K.L. (1986) Gastric ulceration in mature Thoroughbred horses. Equine Vet J. 18, 284-287.

Johnson, W., Carlson, G.P., Vatistas, N., Snyder, J.R., Lloyd, K., and Koobs, J. (1994) Investigation of the number and location of gastric ulcerations in horses in race training submitted to the California racehorse postmortem program. 40th

Meyer, H., Coenen, M. and Gurer, C. (1985) Investigations of saliva production and chewing in horses fed various feeds. Proceedings of 9th ENPS, East Lansing, Mi., 38-41.

Murray M.J.and Schusser, G. (1989) Application of gastric pH-metry in horses: measurement of 24 hour gastric pH in horses fed, fasted, and treated with ranitidine. J. Vet. Intern. Med. 6, 133.

Murray, M.J., Grodinsky, C., Anderson, C.W., Radue, P.F. and Schmidt, G.R. (1989) Gastric ulcers in horses: A comparison of endoscopic findings in horses with and without clinical signs. Equine Vet J. Suppl 7, 68-72.

Murray, M.J. (1992) Aetiopathogenesis and treatment of peptic ulcer in the horse: a comparative review. Equine Vet J. Suppl. 13, 63-74.

Schryver, H.F., Millis, D.L., Soderholm, L.V., Williams, J. and Hintz, H.F.(1986) Metabolism of some essential minerals in ponies fed high levels of aluminum. Cornell Vet., 76, 354-360.

Smyth, G.B., Young, D.W. and Hammond, L.S. (1988) Effects of diet and feeding on post-prandial serum gastrin and insulin concentrations in adult horses. Equine Vet. J. Suppl 7, 56-59.

Vatitstas, N.J., Snyder, J.R., Carlson, G., Johnson, B., Arthur, R.M., Thurmond, M., and Lloyd, K.C.K. (1994) Epidemiological study of gastric ulceration in the Thoroughbred race horse: 202 horses 1992-1993. 40th AAEP Convention Proceedings, 125-126.

MANAGEMENT OF GERIATRIC HORSES

SARAH L. RALSTON
Cook College, Rutgers University, New Brunswick, NJ

Many geriatric horses (>20 years old) are able to maintain good to excellent body condition and health on normal maintenance rations. However, weight loss is not uncommon in old horses, especially in severe weather. Geriatric horses often suffer the adverse effects of irreparable dental abnormalities (tooth loss, wave mouth), chronic intestinal parasitism and pituitary dysfunction. If renal or hepatic function is reduced, tolerance of excess calcium and edible oils respectively may be adversely affected. Chronic pain associated with arthritic changes may exacerbate problems with appetite.

Evaluation

When confronted with a failing older horse, the first step should be to thoroughly evaluate what the horse is being fed, determine if there have been any recent changes in diet or environment and check the schedule of anthelmintic administration. Merely changing the horse to a better quality feed or hay will frequently solve the problem. Changes in environment can be particularly stressful to aged horses. Competition from new herd mates or loss of a herd companion can result in reduced intake and weight loss. Old horses are more sensitive to extremes of weather than are younger horses, regardless of body condition or pituitary/thyroid function (Ralston et al., 1988). Geriatric horses should have adequate shelter, though confinement to a stall can exacerbate orthopedic problems and stiffness. Intestinal parasitism may reduce digestive capabilities due to chronic mucosal damage and scarring.

The horse should be given a thorough physical examination, especially with regard to its teeth. Correctable dental abnormalities (sharp points, hooks, broken or infected molars) should be amended. Because most of these horses do not have much tooth growth left, overcorrection or aggressive floating should be avoided if possible (Scrutchfield et al., 1996). Before instituting dietary changes, blood should be drawn for blood chemistry and complete blood count to rule out medical causes of weight loss such as chronic infection, renal dysfunction, or hepatic failure. The standard indices for renal and hepatic function can be applied to the geriatric horse (Ralston et al., 1988). Chronic laminitis or infections, hyperglycemia and/or hyperinsulinemia following a glucose challenge, polyuria/polydypsia and hirsutism are suggestive of pituitary dysfunction (equine Cushing's disease), which is extremely common in geriatric horses (Ralston et al., 1988; Dybdal et al., 1994). If these signs are present the horse should be tested. A modified dexamethasone suppression test (MDST) is the single most sensitive and specific test for pituitary dysfunction (Dybdal et al., 1994). For the MDST a baseline blood sample is drawn followed by administration of 40 g dexamethasone/kg body weight IM between 12 PM and 4 PM. The second blood sample is drawn 20 to 24 hours later

394   *Management of Geriatric Horses*

(Scrutchfield et al., 1996). Cortisol concentrations exceeding 1 g/dl 20 to 24 hours after dexamethasone administration indicate pituitary dysfunction (Dybdal et al., 1994). A simple screening test for hyperglycemia/hyperinsulinemia can be employed to assess the need for the MDST. A blood sample can be obtained for glucose and insulin analysis before and then 1 to 2 hours after feeding the horse 3 lb of concentrate, preferably a sweet grain mix (Ralston, 1989-1998). If the values reported for the two samples differ by more than 100 mg/dl for glucose or by more than 200 IU/ml for insulin, these results are strongly suggestive of pituitary dysfunction (Ralston, 1989-1998).

Feeding Failing Older Horses

If there are no medical problems other than pituitary dysfunction and/or poor dentition, I recommend switching the horse to a ration formulated specifically for geriatric horses. All changes, however, should be done slowly. Long stem hay may still be offered as long as choke is not a problem. Most of the major feed companies now offer "geriatric" feeds (usually have the word "senior" or "vintage" in the product name) that are supplemented with water soluble vitamins and which contain 12-16% protein, <1.0% calcium, and 0.45-0.6% phosphorus. Most are designed to be "complete" feeds and contain at least 12% crude fiber. These feeds are usually either "predigested" or extruded to increase digestibility for the geriatric horse. Hay cubes can be used as a forage source if the horse has a problem chewing long stem hay. The hay cubes should be a mixture of grass hay or the whole corn plant and alfalfa, rather than straight alfalfa due to the high calcium content of alfalfa.

Cautions

Calcium intakes in excess of need result in high urinary calcium excretion in horses (Lewis, 1995). In my experience, there is an unusually high incidence of renal calculi in otherwise clinically normal geriatric horses fed straight alfalfa. Therefore alfalfa and beet pulp, both relatively high in calcium, should be used with caution in failing older horses. Sweet feeds (>3% molasses) may exacerbate glucose intolerance and also should be used with caution in horses with pituitary dysfunction. Hay cubes and pelleted/extruded geriatric feeds can be soaked in water to make a slurry if choke or impactions are a problem.

Supplements

Vegetable oil (1 to 2 cups per day) may be added to the ration for extra calories, but must be introduced slowly. Aged horses were documented to have lower plasma ascorbic acid than younger, healthy horses and the cause of this has yet to be determined (Ralston et al., 1988). However, vitamin C supplementation (10 gm twice a day) increased antibody response to vaccines in aged horses, especially those with pituitary dysfunction (Ralston and Quackenbush, unpublished data) and in my experience helps old horses with chronic infections.

Other Considerations/Treatments

If chronic pain due to arthritis appears to be a contributing factor to weight loss, the horse may be administered small amounts of anti-inflammatory or glucosamine/chondroitin sulfate compounds. Nontraditional therapies such as acupuncture have also been effective in some cases. Confinement appears to exacerbate stiffness and pain, so horses should be turned out as much as possible (Ralston, 1989-1998).

If pituitary dysfunction is present, consideration should be given to treating the horse with either cyproheptadine (0.6 – 1.2 mg/kg BW) or pergolide (0.005-0.01 mg/kg BW) in addition to dietary modifications (Beech, 1987). Although these drugs are not approved for use in horses, they can be effective in reducing clinical signs associated with pituitary dysfunction (Beech, 1987; Beech, 1995). Both drugs should be started at the lower dose and increased slowly over the course of two or three weeks. If signs of anorexia or depression develop, the dose should be reduced. Note that the cyproheptadine dose is based on metabolic weight, rather than the conventional body weight. Cyproheptadine causes fewer side effects and is less expensive than pergolide, but it is not as effective (Ralston, 1989-1998; Beech, 1995).

If renal or hepatic dysfunction are present, lower concentrations of protein (8-10%) and higher concentrations of carbohydrate should be fed (Lewis, 1995; Ralston, 1989; Ralston and Breuer, 1996). Grass hay (chopped or cubed), corn and/or barley are the feeds of choice. Beet pulp may be used as a roughage source for horses with hepatic disease, but should be avoided in horses with renal disease due to its high calcium content. Vegetable oil may be used as an additional calorie source for horses with renal disease but not for horses with evidence of hepatic dysfunction due to the danger of hyperlipidemia. Digestive aids such as yeast cultures may be of benefit.

Summary

Just because a horse is old does not mean it has to be thin and in poor health. With proper attention to dentition, ration and veterinary care, horses can maintain excellent body condition and health well past thirty years of age.

References

Beech, J. Pituitary tumors. In: Robinson, NE (ed). Current Therapy in Equine
     Medicine 2, Saunders, Philadelphia, pp182-185, 1987.
Beech, J. 1995. Personal communication.
Dybdal NO, Hargreaves KM, Madigan JE, Gribble DH, Kennedy PC, Stabenfeldt GH.
     1994. Diagnostic testing for pituitary pars intermedia dysfunction in
     horses. JAVMA 204:627-632.
Lewis, LD. 1995. Equine Clinical Nutrition. Williams and Wilkins, St. Louis., 1995.
Ralston SL, Nockels CF, Squires EL. Differences in diagnostic test results and

hematologic data between aged and young horses.
Am J Vet Res. 49:1387 -1392, 1988.

Ralston, SL. Digestive alterations in aged horses. J Eq Vet Sci. 9:203-205, 1989.

Ralston, SL. 1989-1998. Unpublished data.

Ralston, SL, Breuer LH. Field evaluation of a feed formulated for geriatric horses.
J. Eq. Vet. Sci. 16:334-338, 1996.

Scrutchfield, WL, Schumacher J, Martin MT. 1996. Correction of abnormalities of
the cheek teeth. Proc. 42nd AAEP. 42: 11-21.

GLUCOSE INTOLERANCE AND DEVELOPMENTAL ORTHOPEDIC DISEASE IN FOALS-A CONNECTION?

SARAH L. RALSTON

Cook College, Rutgers University, New Brunswick, NJ

Large foals that are growing rapidly are considered to be at increased risk of developmental orthopedic disease (DOD). A multifactorial problem, DOD includes problems such as osteochondritis dissecans (OCD), epiphysitis, flexural and angular limb deformities and perhaps wobbler syndrome (McIllwraith et al. 1991; Jeffcott, 1991; Jeffcott, 1996). Of these, OCD is probably the biggest problem in the equine industry in terms of lost revenue (Jeffcott, 1996). Genetics, nutrition and exercise all play a role in the incidence of DOD in horses. Recently a connection between high insulin concentrations, especially after feeding a meal of concentrates, and OCD has been reported (Ralston, 1996; Ralston et al., 1998). In this paper I will explore both the old and new theories regarding the causes of DOD and OCD in foals and provide recommendations, based on the most recent information, for feeding young horses.

General Factors

Hereditary predisposition to at least OCD is well documented in Standardbred and Swedish warmblood horses, with the incidence as high as 45% in some bloodlines (Grondahl, 1991; Schougaard et al., 1990). However the genetic defect that causes the growth associated problems in the horses has not been identified. Breeds selected for rapid growth maturation are at increased risk of developing problems, but it is not growth rate alone that causes the problem (Jeffcott, 1991). It is not always the most rapidly growing foal that develops DOD, but often the one with the most erratic growth rate (Jeffcott, 1991; Lewis, 1996).

Trauma due to excessive concussion, due either to obesity or forced exercise, may increase the incidence of DOD (Jeffcott, 1991; Lewis, 1996). Other reports, however, revealed that restriction of exercise adversely impacted bone growth and development in young horses. Turning the foals out in as large an area (either pasture or paddock) as possible for as long as possible is highly recommended. Ideally they should get 24 hour turnout. However, strenuous forced exercise, especially longeing in circles, should be avoided. Foals should not be allowed to become obese.

Nutritional Factors

MINERALS
Mineral imbalances have been well documented to cause DOD (Jeffcott, 1991; Lewis, 1996). Deficiencies of calcium, phosphorus and/or copper all result in defective bone maturation. Zinc toxicity and perhaps deficiency have also

resulted in lesions, though the effects of simple zinc deficiency are not well documented. The optimal intakes of copper and zinc for young horses have not been well defined. My current recommendations for acceptable ranges of mineral content of rations for foals less than 1 year of age are given in Table 1.

Table 1. Recommended concentrations of minerals* in rations fed to rapidly growing young horses.

Mineral	Range
Calcium	0.8 to 1.0%
Phosphorus	0.4 to 0.6%
Copper	10 to 15 mg/kg feed
Zinc	40 to 60 mg/kg feed

* Other minerals such as manganese, magnesium, selenium and iron
  are probably important, but there are no data available on requirements
  of young horses for these nutrients. There are no data on vitamin
  requirements either, though over 10 times the recommended level of vitamin A
  resulted in weakened bones in ponies (Donoghue et al., 1981).

PROTEIN
Excessive protein (greater than 16%) was incriminated as a cause of DOD in the 1970s but subsequent studies have not revealed a direct relationship between high protein rations and DOD (Lewis, 1996). Weanlings fed rations deficient in protein (less than 12%) had reduced growth rates and poor bone mineralization compared to weanlings fed rations that were higher in protein. Restricting protein in a rapidly growing foal's ration will not result in improved bone growth and may actually be detrimental to the animal.

CARBOHYDRATES
Rations providing over 100% of the National Research Council's recommended amounts of energy for rapid growth in foals may cause an increased incidence of DOD, especially if the ration contains more than 50% sweet feed (grain mix plus molasses) or other high sugar concentrate by weight (Lewis, 1996; Glade, 1986; Glade and Belling, 1986). High carbohydrate rations such as sweet feeds may contribute to the appearance of DOD, possibly related to the high blood glucose and insulin and low blood pH they cause for up to 4 hours after feeding (Glade, 1986; Glade and Belling, 1986). It is interesting that foals between 3 and 12 months of age have been documented to be relatively insulin resistant (Ralston, 1996; Krusic et al., 1997), since this is the period during which OCD lesions most commonly develop (McIllwraith et al., 1991; Jeffcott, 1996; Jeffcott, 1991). Pelleting and extrusion of concentrates appear to affect the availability of carbo-hydrates (Ralston, 1992; Ralston 1995-1999). Pelleted or extruded concentrates may cause lower glucose and insulin changes than textured feeds with the same basic formulation (Ralston, 1992; Ralston, 1995-1999). There may be a correlation between OCD and glucose intolerance (abnormally high blood

glucose and insulin after a meal of concentrates) in foals that are genetically predisposed to the problem (Ralston, 1996; Ralston et al., 1998). At Rutgers we are currently developing a patented glucose challenge that can be easily administered to foals < 3 mo old in order to identify those potentially at risk. These foals could then be fed specially formulated rations to reduce the risk of lesions appearing, since diet composition can dramatically affect glucose/insulin responses (Garcia and Beech 1986; Lawrence et al., 1993; Ralston et al., 1979).

Recommendations

1. Nursing foals should be introduced to concentrates when they are 1 to 2 months of age. The concentrate should contain 14 to 18% protein and have added calcium, phosphorus, copper and zinc (see Table 1) in a formulation designed specifically for growing horses. The higher percentages of protein and minerals should be used if only grass hay is available. The lower percentages can be used with legume or legume/grass mix hays. Concentrates should be fed at the rate of 0.50% to 1.0% of body weight, with the emphasis on maintaining lean body condition (ribs not visible but can be felt with mild pressure over the flank; loin, croup and neck have smooth outlines without creases or visible bony structures). If the lower amounts are fed (less than 0.5% body weight), addition of a balanced calcium/phosphorus/trace mineral mix may be necessary to maintain the proper mineral intake. Ideally the foals should be fed regulated amounts that are inaccessible to their dams or other foals twice a day. The mares should be fed the same concentrate if the foal has access to the mare's feed.

2. Don't let the foal get obese (obvious crease down the back, ribs cannot be easily felt) or excessively thin (ribs easily visible, hip bones prominent, hair coat dull and shaggy). If group feeding foals and mares, monitor their condition daily and feed any excessively thin or fat foals separately. Since pelleted and extruded feeds may cause lower glucose and insulin responses than do sweet feeds, the former two types of concentrate may be preferable to textured sweet feed mixes, especially in foals from bloodlines potentially predisposed to OCD.

3. Weanlings should be fed the same type of concentrate as when they were nursing and at the same rate as above and monitored carefully for signs of excessive weight gain or loss and DOD. From 0.25 to 1.0% body weight of a properly formulated concentrate divided into two or three meals a day, with free choice access to good quality mixed leume/grass hay or pasture, will maintain optimal growth rates of most light horse breeds while reducing the risk of DOD. The goal is to maintain steady growth, avoiding sudden increases or decreases, and to maintain good, but not fat or thin, body condition. Plain white or trace mineral salt and a good, clean source of water should be available free choice at all times.

4. If signs of epiphysitis (enlargement at the growth plates above the fetlocks and/or knees associated with lameness and reluctance to exercise) or other deformities (contracted tendons, angular deformities) appear, the ration is probably not properly balanced. The amount of concentrate fed should be TEMPORARILY reduced while the total ration's nutrient content is assessed. Any deficits or excesses should be corrected and a properly balanced ration reintroduced as soon as possible. Starving foals (feeding only grass hay and oats for a prolonged period of time, resulting in weight loss, poor growth and rough-looking hair coats) will not correct the problem on a long-term basis.
5. Yearling rations can be reduced to 12%-14% protein with lesser concentrations of minerals, but still above that usually found in mixes formulated for adult horses. Maintaining yearlings on the weanling rations will not hurt and may help, especially if horses are still growing fairly rapidly.

References

Donoghue S, Kronfeld DS, Berkowitz SJ, Copp RL. 1981. Vitamin A nutrition of the equine: Growth, serum biochemistry, and hematology. J. Nutr. 111:365-372.

Garcia MC, Beech J. 1986. Equine intravenous glucose tolerance test: Glucose and insulin responses of healthy horses fed grain or hay and of horses with pituitary adenomas. Am J Vet Res 47:570-572.

Glade MJ. 1986. The control of cartilage growth in osteochondrosis: a review. J Eq Vet Sci 6:175-187.

Glade MJ and Belling TH. 1986. A dietary etiology for osteochondrotic cartilage. J Eq Vet Sci 6:151-155.

Grondahl AM. 1991. The incidence of osteochondrosis in the tibiotarsal joint of Norwegian Standardbred trotters. J Eq Vet Sci. 11:273-274.

Jeffcott L.B. 1991. Osteochondrosis in the horse - searching for the key to pathogenesis. Equine Veterinary Journal 23: 331-338.

Jeffcott LB. 1996. Osteochondrosis-An international problem for the horse industry. J Eq Vet Sci 16:32-37.

Krusic L, Krusic-Kaplja A, Cestnik V, Snoj T, Pogacnik A, Pangos S, Gatta D, Moni P. 1997. Insulin response after oral glucose application in growing Lipizzaner foals. Proc 15th Eq Nutr Physiol Symp. (Fort Worth, TX):397-403.

Lawrence LM, Soderholm LV, Roberts A, Williams J, Hintz H. 1993. Feeding status affects glucose metabolism in exercising horses. J Nutr 123:2152-2157.

Lewis LD. 1996. Feeding and Care of the Horse, 2nd edition. Williams and Wilkins, Philadelphia.

McIlwraith CW, Foerner JJ, Davis DM. 1991. Osteochondritis dissecans of the tarsocrural joint: results of treatment with arthroscopic surgery. Eq Vet J 23:155-162.

Ralston SL. Effect of soluble carbohydrate content of pelleted diets on postprandial glucose and insulin profiles in horses. Pferdeheilkunde(September, 1992):112-115, 1992.

Ralston SL. Hyperglycemia/hyperinsulinemia after feeding a meal of grain to young horses with osteochondritis dissecans (OCD) lesions. Pferdeheilkunde (May, 1996):320-322, 1996.

Ralston SL. 1995-1999. Unpublished data.

Ralston SL, Van den Broek G, Baile CA. 1979. Feed intake patterns and associated blood glucose, free fatty acid and insulin changes in ponies.
J. Anim. Sci. 57: 815-821.

Ralston SL, Black A, Suslak-Brown L, Schoknecht PA. Postprandial insulin resistance associated with osteochondrosis in weanling fillies.
J Animal Science 76 (supplement 1):176, 1998.

Schougaard H, Falk-Ronne J, Philipsson J. 1990. A radiographic survey of tibiotarsal osteochondrosis in a selected population of trotting horses in Denmark and its possible genetic significance. Equine Vet J. 22:288-289.

NUTRITIONAL SUPPORT OF THE SICK ADULT HORSE

RAY J. GEOR

Kentucky Equine Research, Inc., Versailles, KY

Among hospitalized human patients, malnutrition has been associated with increased infectious morbidity, prolonged hospital stay, and increased mortality. Conversely, several studies of critically ill human patients have shown that nutritional support can attenuate body weight loss, reduce complication rates in surgical patients, and shorten the period of hospitalization. Recognition of the beneficial effects of nutritional intervention in human patients has spurred interest in veterinary clinical nutrition. Over the past 10 to 15 years, the nutritional management of hospitalized animals has received increasing attention (Donoghue, 1992).

Clinical experience suggests that nutritional intervention can improve clinical outcomes in diseased horses, particularly neonates. However, it should be noted that there have been no controlled studies of the relationship between nutritional support and clinically important endpoints (e.g. surgical complication rate, duration of hospitalization, mortality). Furthermore, there are few data on the effects of specific disease conditions on the nutritional requirements of horses. As a result, recommendations for the nutritional management of sick horses have largely been extrapolated from data in other species.

This paper considers: (1) the metabolic effects of starvation and disease; (2) clinical assessment of nutritional status and the need for nutritional intervention; (3) dietary management of certain disease conditions; and (4) methods for delivery of nutritional support to anorectic or hypophagic horses. Several other excellent reviews of equine clinical nutrition are available (Ralston, 1990; Burkholder and Thatcher, 1992; Naylor, 1992; Rooney, 1998).

Metabolic Effects of Starvation and Disease

To appreciate the importance of nutritional support in disease states, it is useful to distinguish the metabolic effects of fasting (or starvation) and severe illness. Fasting and starvation imply a lack of nutrient intake. To compensate, the body attempts to reduce energy expenditure and directs substrate stores to essential functions. The initial energy sources are stored liver glycogen and, to a lesser extent, glucose derived from amino acids. However, liver glycogen stores are quickly depleted, usually within 24 to 36 hours of the onset of a fast, necessitating an increase in gluconeogenesis (glucose synthesis in the liver) to meet the obligate needs of the central nervous system and red blood cells. The carbon skeletons from glycerol, lactate, and some amino acids (particularly alanine) are used for gluconeogenesis. Use of amino acids for glucose synthesis mandates the breakdown of body proteins (proteolysis), particularly those in muscle. Simultaneously, there is increased mobilization of fatty acids from fat stores in adipose tissue and increased utilization of fatty acids for energy (in all tissues but the central nervous system and red blood cells). Studies in man have demonstrated

403

that fat supplies approximately 80 to 85% of the body's energy during prolonged fasting, with the remainder derived from metabolism of protein, either direct oxidation of carbon skeletons via the Krebs cycle or utilization of glucose synthesized from amino acids.

The principal mechanism of adjustment to prolonged fasting or starvation is a change in hormone balance. In particular, there is a sharp decrease in insulin production. As well, the muscle and adipocytes become somewhat resistant to the action of insulin (i.e. whatever insulin available becomes less effective in promoting cellular uptake of nutrients for protein synthesis and lipogenesis). Decreased insulin activity, coupled with increased synthesis of counterregulatory hormones such as glucagon and cortisol, promotes fatty acid mobilization and the catabolism of muscle protein. An additional hormonal change facilitating adjustment to starvation is a decrease in synthesis of the thyroid hormone triiodothyronine (T_3), thus resulting in a lowered metabolic rate and daily energy requirement.

In several species, a further important adaptation to starvation is increased synthesis and utilization of ketones. Ketogenesis in the liver is favored under conditions of increased fat mobilization and a low insulin-to-glucagon ratio. Ketones are an important energy source during starvation and other negative energy states. In particular, the brain and other nervous tissues adapt to use of ketones and, as a result, become less dependent on glucose for energy. This reduction in glucose demand allows for a decrease in the rate of proteolysis, thus preserving lean body mass. Although plasma ketone concentrations increase during feed deprivation in horses (Rose and Sampson, 1982), the increase is small in comparison to man, suggesting that ketones are less important as an energy source in this species during prolonged fasting.

The body's response to severe stress, such as sepsis, major surgery and trauma, is much different than its response to starvation. Whereas adaptation to starvation is associated with a decrease in metabolic rate, severe stress induces a hypermetabolic state that results in rapid breakdown of the body's reserves of carbohydrate, protein and fat. The term "septic autocannabalism" has been coined to describe the metabolic response and wasting that accompanies severe sepsis in humans (Michie, 1996). Hyperglycemia with insulin resistance, hyperlipidemia, profound negative nitrogen balance, and diversion of protein (particularly glutamine) from skeletal muscle to the splanchnic tissues are prominent features. Unlike starvation, protein catabolism remains unchecked, resulting in severe wasting of lean body mass. These responses are mediated in part by marked increases in the counterregulatory or "stress" hormones (glucocorticoids, epinephrine and glucagon) and several of the inflammatory cytokines (low molecular weight peptides that evoke a number of varied reactions in the body), including interleukin-1 (IL-1) and tumor necrosis factor (TNF). This severely catabolic state quickly results in malnutrition that, in turn, impairs host defense, delays tissue repair, and increases risk of mortality.

The metabolic effects of fasting and, to a much more limited extent, illness have been studied in horses (Morris et al., 1972; Naylor et al., 1980; Sticker et al., 1995a and 1995b). As in other species, fasting results in linear increases in the plasma concentrations of glycerol and nonesterified fatty acids (NEFA), reaching a plateau about three days after the start of food deprivation. These changes are consistent with an increase in lipolysis and suggest greater use of fat during fasting. Some of the mobilized fat is re-esterified in the liver and transported to the peripheral tissues as very low-density lipoproteins (VLDL). As a result, serum triglyceride concentrations are also increased in fasting horses. Pregnancy, lactation and obesity accentuate the effect of fasting on plasma triglyceride concentrations. Furthermore, the increase in plasma triglycerides is typically more pronounced in sick, hypophagic horses (e.g. pneumonia, diarrhea, renal failure), perhaps reflecting the increased metabolic demands associated with the primary illness.

The term hyperlipemia has been used to describe a disease syndrome in ponies, miniature horses and, more rarely, horses that is characterized by a marked increase in plasma triglycerides (often >1000 mg/dl), cloudy serum or plasma, and fatty infiltration of the liver and kidneys. Affected animals usually have a primary illness that results in reduced feed intake (hypophagia), but the clinical manifestations of hyperlipemia and fatty liver often overshadow the primary problem. Increased metabolic demands associated with the primary illness tend to exacerbate the increase in fat mobilization associated with hypophagia. Furthermore, the increased metabolic demands of pregnancy or lactation intensify lipolysis. In obese ponies and miniature horses, insulin resistance is another underlying factor. Under conditions of rapid and sustained lipolysis, the liver's capacity for lipoprotein (VLDL) synthesis is overwhelmed and fat is deposited in the liver. Affected horses and ponies require aggressive nutritional support and, even then, the prognosis for recovery is guarded to poor.

Effects of Malnutrition on Host Defenses

Ultimately, all body systems are adversely affected by undernutrition. However, organs and cells with high rates of metabolism are more rapidly impaired by nutrient imbalances. The intestinal tract is particularly susceptible to the effects of anorexia. Enterocytes (the cells that line the intestinal mucosa) have an extremely high metabolic rate and a very short lifespan, with an average turnover rate of three days. Enterocytes participate in the digestion and absorption of nutrients and also provide a physical barrier that prevents entry of bacteria into circulation. Because many of the nutrients required for synthesis of new enterocytes are taken directly from the intestinal lumen, even short periods of food deprivation result in mucosal atrophy and impaired digestion and absorption. Perhaps more importantly, the mucosal barrier is compromised and there is increased risk of sepsis as a result of bacterial translocation.

Cells of the immune system also have high metabolic demands. In several species, including the horse (Naylor and Kenyon, 1981), 3 to 5 days of complete feed deprivation severely compromises cellular and nonspecific immune function, thus rendering the animal more susceptible to infection. This decrease in immune function, together with breakdown of the intestinal mucosa barrier, helps to explain the decreased survival rate in malnourished critically ill humans.

What are the Nutrient Requirements of Sick Horses?

The short answer to this question is "we don't know." As previously indicated, data from human and animal studies have demonstrated that metabolic rate is increased with certain conditions. For example, severe trauma increases energy expenditure (metabolic rate) by a factor of 1.3 to 1.4. Similarly, the metabolic rate of patients with sepsis or a major burn can be up to 1.4- to 1.7-fold higher when compared to resting healthy humans. Indirect calorimetry has been used to measure energy balance in a group of healthy neonatal foals and a small number of foals that were either premature or diagnosed with neonatal maladjustment syndrome (NMS) (Ousey, 1992; Ousey et al., 1996). Interestingly, the metabolic rate of these compromised neonates was reduced by approximately 50% compared to healthy, age-matched counterparts. As the premature and NMS foals were recumbent during the measurement periods, the most likely cause of the low energy expenditure was inactivity. Despite their low rates of metabolism, these foals were in negative energy balance because of very low milk intake (either by bottle or nasogastric tube). Energy deficits are probably greater in foals with more severe illnesses such as septicemia, diarrhea, and pneumonia as a result of increased metabolic demands (hypermetabolic states).

Pagan and Hintz (1986) reported that the resting energy expenditure (REE) of horses in metabolism stalls could be estimated from the formula: REE = 21 kcal (BWkg) + 975 kcal. Thus, for a 500-kg horse, REE would be approximately 11.5 Mcal/day, 30% lower than the requirement for maintenance under field conditions (16.4 Mcal/day for a 500-kg horse). Although data on the energy expenditure of injured or sick adult horses are not available, it has been proposed that the multipliers used in human medicine be applied to the horse. Therefore, a stall rested 500-kg horse that has generalized infection or is recovering from major surgery would have energy requirements of 16 to 20 Mcal/day (1.5 to 1.8 x 11.5 Mcal/day). Given the large increase in protein catabolism during severe illness or injury, it follows that protein requirements are also increased in sick patients. Rooney (1998) has suggested that 5 g of protein be provided per 100 kcal, i.e. 800 g of crude protein for a diet containing 16 Mcal digestible energy. This represents about a 25% increase over the NRC (1989) maintenance protein requirement.

It is probable that the requirements for other nutrients (e.g. micro- and macrominerals) are also altered during illness. However, in the absence of data, feeding levels should be based on the horse's current body weight and the recommended maintenance requirements. It should be emphasized that for horses that are completely anorexic and in need of "involuntary" tube feeding, it is often difficult to meet maintenance requirements. Nonetheless, delivery of even 60 to 70% of maintenance requirements is likely to be beneficial in sick horses. The primary goal is to limit catabolism and further loss of body weight.

In human medicine, there is currently intense clinical interest in the therapeutic role of specific nutrients, including arginine, glutamine, omega-3 fatty acids (from fish oil), and ribonucleic acid (a nucleotide). As each of these nutrients is important in the immune response, it is proposed that administration of these nutrients to critically ill patients will result in up-regulation of immune function and reductions in the morbidity and mortality associated with infection. A recent meta-analysis that addressed whether enteral nutrition with these "immuno-enhancing" nutrients benefits the critically ill demonstrated significant reductions in infection rate and length of hospital stay in patients receiving enteral supplements containing arginine with or without glutamine, nucleotides, and omega-3 fatty acids (Beale et al., 1999).

Of these nutrients, glutamine has been the subject of most clinical investigation. Glutamine is the most abundant amino acid in both plasma and the free intracellular amino acid pool in skeletal muscle, and is important in several key metabolic processes of immune cells and enterocytes. Because most tissues have the ability to synthesize glutamine, it is defined as a nonessential amino acid. However, marked decreases in plasma and skeletal muscle glutamine concentrations have been reported in a variety of catabolic states (Vinnars et al., 1975; Jackson et al., 1999). In horses, as in man, plasma glutamine concentrations decrease following viral infection (Routledge et al., 1999), a finding consistent with glutamine's proposed role in immune cell metabolism. The decrease in plasma glutamine concentrations during illness is temporally related to an increase in splanchnic amino acid uptake, perhaps reflecting increased intestinal demand for glutamine. Indeed, the small intestine is the principal organ of glutamine consumption (Nappert et al., 1997). These findings have led to the suggestion that glutamine may behave as a "conditionally essential" amino acid during severe illness, i.e. endogenous supply fails to meet increased demands (Jackson et al., 2000).

Several investigations in animals have indicated that oral glutamine supplementation improves growth and repair of the small intestinal mucosa and helps maintain intestinal immune function (Nappert et al., 1997). Glutamine supplementation also increases renal arginine production, another amino acid important for the body's response to injury (Welbourne, 1995). Human studies have demonstrated that glutamine supplementation (oral or intravenous) enhances in vitro measures of immune function, for example, bacterial killing function of neutrophils (Saito et al., 1999). Furthermore, in randomized clinical trials, patients receiving glutamine had reduced incidence of severe infection, decreased length of hospitalization (Jones et al., 1999), and a significant improvement in six-month survival (Griffiths, 1997). Taken together, these findings indicate that glutamine supplementation is beneficial in sick patients. Whether or not glutamine is similarly effective in the treatment of equine disease conditions is not known and requires study.

Selection of Patients Requiring Nutritional Support

Assessment of body condition and consideration of clinical history are important in identifying horses in need of nutritional support (Donoghue, 1992). Well-conditioned horses that are not pregnant (last trimester) or lactating can likely withstand a short period (2 to 4 days) of partial or complete anorexia without adverse affects. On the other hand, regardless of the duration of illness, thin horses (condition scores of 1 to 3) or those that have sustained a substantial loss of condition (10% or more of body weight) are candidates for immediate nutritional support. Conditions of adult horses associated with rapid weight loss include sepsis/endotoxemia, deep-seated bacterial infections (e.g. pulmonary abscess, pleuropneumonia, abdominal abscesses), diarrhea, severe trauma, surgery (e.g. colic surgery involving bowel resection), and intestinal disorders characterized by protein loss and nutrient malabsorption. Over-conditioned (scores of 7 to 9) horses, ponies, and miniature horses, when sick, can be poorly tolerant of even short periods of anorexia; in these animals, plasma or serum should be visually appraised and triglyceride concentrations measured. Hypertriglyceridemia (>500 mg/dl) and/or recognition of cloudy, opaque plasma or serum suggest hyperlipemia and risk of development of liver dysfunction from hepatic lipidosis. Neonates are also poorly tolerant of short periods of undernutrition. Particularly during the first week of life, neonates have negligible energy stores, and negative energy balance can quickly result in hypoglycemia, generalized weakness and death.

For many disease conditions, most horses maintain a reasonably good appetite. However, there are some circumstances in which some change in diet is indicated. These include inflammatory airway disease, chronic obstructive pulmonary disease (COPD), laminitis, renal failure, hepatic disease, hyperadrenocorticism (Cushing's syndrome), chronic diarrhea, and small intestinal malabsorption syndromes. Horses that have had portions of the small or large intestine removed also may require dietary adjustments. For athletic horses that have sustained minor injuries or are recovering from elective orthopedic surgery or bacterial or viral infections, a major consideration is a reduction in energy intake. During convalescence, and until resumption of training, most of these horses can be maintained on a diet composed primarily of hay (fed at 1.5 to 2.0% of body weight per day), a vitamin/mineral supplement, and loose salt (or access to a salt block). To facilitate the return to full grain feeding upon return to training, it may be desirable to continue feeding small amounts of grain (1 to 2 kg per day).

Nutritional Management of Selected Conditions

Respiratory diseases:
In horses with inflammatory airway disease and COPD, disease exacerbations are related to exposure to airborne allergens (dusts, molds, fungi) and toxins (endotoxin). Changes in feeding and housing management are essential for long-term control of these conditions. Ideally, affected horses are housed outside or in well-ventilated barns with ample daily turnout. If housed indoors, bedding must be dust-free, e.g. shredded paper, wood chips. Hay is a primary source of allergens, including molds and fungi. The hay should either be removed from the diet or only fed after thorough soaking (completely immersed in a tub of water for a minimum of 5 minutes). Hay cubes are an alternate fiber source, but these must also be soaked in water prior to feeding.

Gastrointestinal disorders:

Gastrointestinal disorders can be categorized into those affecting the small intestine (protein-losing enteropathies/malabsorption syndromes; "short-bowel syndrome" following surgical resection of intestine) and those involving the large intestine (colitis/diarrhea; colonic or cecal impaction; colon resection). With small intestinal diseases, the primary goal is to optimize large bowel digestive function. This can be achieved by feeding highly digestible fiber sources such as leafy alfalfa, beet pulp or soybean hulls, with a reduction in grain feeding. Offering small grain meals (i.e. <1 kg) may minimize the risk of passage of undigested starch into the hindgut. Ponies fed a complete pelleted feed were able to maintain body condition following resection of less than 50% of the small intestine (Tate et al., 1983). However, ponies fed the same diet after a more extensive resection (60 to 80% of the small intestine) suffered rapid and substantial weight loss. Clinical experience has also indicated difficulty in maintaining body condition of horses following resection of 50% or more of the small intestine (and horses with malabsorption syndromes). However, some horses can be maintained on a diet of highly fermentable fiber (beet pulp and alfalfa) and fat (rice bran and corn oil).

Diarrhea in the adult horse is usually due to large intestinal dysfunction (e.g. colitis secondary to *Salmonella* spp. or clostridial infection). In the acute phase, many affected horses are hypophagic and in need of enteral nutritional support (see below). Small intestinal digestive and absorptive function is probably well-maintained. Therefore, low residue diets with highly digestible sources of carbohydrate, protein and fat should be fed. As appetite improves, the diet should initially consist of hay and other sources of fermentable fiber. Yogurt cultures or commercially available bacterial preparations are often administered to aid reestablishment of normal microbial flora. There should be a gradual increase in grain feeding over a 1 to 2 week period.

Laminitis:

A common cause of laminitis is overconsumption of starch (grains and lush pasture). In this circumstance, a large amount of undigested substrate passes into the hindgut and undergoes bacterial fermentation. Therefore, a primary goal in the prevention of recurrences is elimination of grain from the diet. These animals should be fed hay, a vitamin-mineral supplement and, if needed, a fat supplement such as rice bran or corn oil. However, in many ponies and some horses, chronic active laminitis is associated with obesity and gradual weight loss is necessary. Older horses with hyperadrenocorticism (Cushing's syndrome) are also prone to laminitis. In addition, persistent hyperglycemia and insulin resistance are common features. Accordingly, nutritional management of affected horses should also emphasize restriction of grain intake.

Hepatic disease:

With severe liver dysfunction, hypoglycemia may develop because of depressed gluconeogenesis. Therefore, the diet should contain highly digestible starches to decrease reliance on hepatic gluconeogenesis for maintenance of glucose homeostasis. Abnormalities of hepatic liver protein metabolism contribute to the neurologic signs (hepatoencephalopathy) that frequently occur with acute hepatic failure. In particular, there is a decrease in the conversion of ammonia to urea. The resultant hyperammonemia contributes to dysfunction of the central nervous system. The body's ammonia load comes from two major sources: 1) bacterial synthesis in the large intestine; and 2) catabolism of amino acids. Therefore, to minimize colonic ammonia production, a low-to-moderate protein diet should be fed. In addition, maintaining positive energy balance will help to minimize protein catabolism. Ideally, the protein source should have a high branched chain to aromatic amino acid ratio. A 50:50 mix of ground corn cobs and sorghum meets this need and provides adequate protein to meet protein requirements. Processed corn and molasses can be added to provide glucose. Small amounts of this concentrate should be fed several times daily. Frequent feedings may facilitate maintenance of glucose homeostasis and prevent surges in colonic ammonia production. Legume hays, oats, and soybeans should be avoided because they contain a low branched chain to aromatic amino acid ratio. Grass hay should be fed as the fiber source.

Renal failure:
Chronic renal failure in horses is associated with hypercalcemia and azotemia (accumulation of urea), both the result of inadequate renal excretion. Phosphorus excretion can also be impaired and sodium deficits can develop because of poor renal conservation. In general, feedstuffs high in protein (legumes, soybeans), phosphorus (wheat bran), and calcium (legumes, calcium-containing supplements) should be avoided. On the other hand, hypoproteinemia can develop with chronic renal failure, necessitating an increase in the level of protein feeding (e.g. soybean meal). These animals also benefit from an increase in the energy density of the diet (fat supplementation).

Delivery of Nutritional Support to Anorectic or Hypophagic Horses

The method of nutritional support provided is primarily dependent on (1) whether or not the horse will voluntarily consume feed; (2) the nature and duration of the illness; and (3) economics. Adult horses can tolerate 2 to 3 days of feed deprivation providing hydration is maintained (either by water consumption or by intravenous administration of fluids). However, for illness of longer duration, some form of nutritional support is required. Nutritional support can be delivered by the enteral or parenteral (intravenous feeding) routes, and enteral feeding is either voluntary or involuntary. Voluntary feeding is by far the least invasive and should be encouraged. Donoghue (1992) has suggested that if an animal consumes at least 85% of its optimal intake, no other form of nutritional support is required.

Voluntary feeding can be encouraged by aggressive management of the primary problem, provision of a selection of feedstuffs (the "cafeteria" approach), and perhaps use of drugs purported to stimulate appetite (Naylor, 1999). In many sick horses, a reduction in appetite, at least in part, can be attributed to fever, pain and/or endotoxemia. In this regard, treatment of the primary problem is essential for restoration of normal appetite. Relief of pain and fever by administration of nonsteroidal anti-inflammatory drugs (e.g. phenylbutazone, flunixin meglumine) may improve feed intake in some horses. Although hospitalized horses should be offered feeds similar to those fed at home, it is often necessary to provide a variety of feedstuffs for encouragement of intake. Highly palatable feeds such as fresh grass, leafy hays, and small amounts of grain or bran mash containing some grain can be offered. Only small amounts of these feeds should be offered initially to avoid problems such as diarrhea and laminitis (founder); it is generally recommended that increases in grains or concentrates should be limited to < 0.5 kg per day for a 500-kg horse. In healthy horses, addition of molasses to grains improves voluntary intake and this approach has been successful in some sick horses. The feeding of oils will increase energy intake in the face of suboptimal feed consumption. Naylor (1999) has reported that the administration of diazepam (10 mg IV for a 500-kg horse) can stimulate appetite in some sick horses. This author's clinical impression is that diazepam is ineffective as an appetite stimulant.

Horses that remain anorexic (or intake is persistently inadequate) require enteral or parenteral feeding. Although total parenteral nutrition is frequently used in sick neonatal foals, it can be cost prohibitive in mature horses (see below). Therefore, "involuntary" enteral feeding is the most cost-effective means for delivery of nutritional support in anorexic horses. It should also be noted that, as with feed deprivation and starvation, a prolonged period of total parenteral nutrition is associated with atrophy of the intestinal mucosa. Thus, another advantage of enteral feeding is maintenance of intestinal health and function. Besides anorexic horses, enteral feeding is indicated in horses that have a good appetite, but where the presence of oral, pharyngeal or esophageal problems precludes voluntary feed consumption e.g. a severely fractured jaw; soft tissue injuries involving the pharynx or esophagus (although some of these horses will do well if fed a sloppy gruel of ground complete feed or alfalfa meal). Horses with pharyngeal paralysis or similar neurologic dysfunction should not be allowed to eat because of the risk of feed aspiration.

Enteral feeding can be delivered via a nasogastric tube or a surgically-positioned esophagostomy tube. Use of esophagostomy tubes is reserved for cases in which the nasal passages, pharynx, or proximal esophagus must be bypassed, e.g. injuries to the head and neck. This approach is associated with a high complication rate and should only be used when tube feeding is anticipated for 10 days or more (Stick et al., 1981). Use of nasogastric tubes for force feeding can also be problematic. Repeated passage of a tube is traumatic and, in most cases, it is preferable to leave the tube indwelling. The tube is secured by placement of tape butterfly sutures between the tube and the false nostril, and by taping the tube to the cheek and gullet straps of the horse's halter. Between feedings the tube should be closed with a plastic syringe case or rubber stopper to prevent gastric distention from air. Indwelling tubes are also traumatic and can cause rhinitis, pharyngitis, and ulceration of the esophageal mucosa. To minimize these complications, a smooth pliable tube should be used. Even then, tubes tend to stiffen when exposed to digestive fluids and I recommend use of a new tube every 3 to 4 days during prolonged tube feeding. Large-bore tubes (14F or ~12.5 mm internal diameter) are required for passage of most enteral diets, although slightly smaller tubes are suitable for administration of commercially available liquid diets (e.g. Osmolite-HN, Ross Laboratories).

Enteral Diets

Several enteral diets have been fed (via nasogastric tube) to sick horses, ranging from slurries of pelletized feeds or alfalfa meal to commercially available liquid diets designed for use in human patients. These liquid diets have been classified into three categories: 1) Blender diets - finely ground whole food suspended in water; 2) Composition diets - composed of highly digestible protein, carbohydrates, and fats; and 3) Elemental diets - diets containing small peptides and/or free amino acids rather than whole protein. Elemental diets are very expensive and rarely used in horses. Table 1 shows the ingredients for preparation of two enteral diets that can be used for nutritional support of horses. Note the number of batches required to meet the daily energy requirements of a 500-kg adult horse. Table 2 shows the calculated nutrient profiles of these diets as well as a commercial human diet that has been used in horses (Sweeney and Hansen, 1990). Currently, there is no commercially available enteral diet specifically designed for use in horses.

Table 1. Ingredients for two enteral diets, one made of alfalfa meal-casein-dextrose, the other a slurry of ground complete feed pellets, oil and water.

Alfalfa/casein/dextrose slurry**	Pellet-vegetable oil slurry**
454 g alfalfa meal	454 g pelleted complete feed†
204 g casein*	46 g (50 ml) corn oil
204 g dextrose	2-3 L water
52 g electrolyte mixture¶	DE per recipe = 1.76 Mcal
5 L water	
DE per recipe = 2.77 Mcal	
DE = digestible energy	

* Casein (Sigma Chemical Co.) or dehydrated cottage cheese (American Nutritional)
¶ Electrolyte mixture - 10 g NaCl, 15 g $NaHCO_3$, 75 g KCl, 60 g K_2HPO_4, 45 g $CaCl_2$, 25 g MgO
† e.g. Phase IV, Kentucky Equine Research, Inc.
** Approximately 6 batches of the alfalfa/casein/dextrose mixture and 9 batches of the pellet and vegetable oil slurry are required to meet the maintenance energy requirements of a 500-kg adult horse (DE = 16.4 Mcal/day).

Table 2. Nutrient profiles for three enteral diets supplying maintenance energy requirements for a 500-kg adult horse.

Nutrient	Requirements*	Naylor Diet**	Pellet Slurry Diet¶	Osmolite-HN†
Energy (DE), Mcal	16.4	16.4	16.4	16.4
Protein, g	656	1710	505	688
Calcium, g	20	81	25	11.7
Phosphorus, g	14	41	23	11.7
Sodium, g	8.2	16.2	10	14.4
Potassium, g	25	159	37	24.2
Magnesium, g	7.5	97	8	4.7
Copper, mg	82	35	193	23.6
Zinc, mg	328	100	535	264
Iron, mg	328	984	518	211
Selenium, mg	0.82	0.9	2.1	-

* From Nutrient Requirements of Horses, ed 5, National Research Council, 1989
** Diet developed by Dr. Jonathan Naylor and composed of alfalfa meal, casein, and dextrose
¶ A slurry of complete feed pellets (Phase IV, Kentucky Equine Research, Inc.) and corn oil
† Ross Products Division, Abbott Laboratories, Columbus, Ohio

Use of pelleted complete horse feeds allows the clinician to select the diet based on the age and physiologic state of the animal. In addition, these feeds are nutritionally balanced (providing protein, energy, fiber, vitamins and minerals), easily procured, and less expensive than other hand-prepared enteral diets. Pellets must be pulverized in a kitchen blender. Just prior to feeding, the ground feed is mixed with water. Addition of corn oil is also recommended to increase the energy density of the diet. Soybean meal or whey protein can be added if a higher protein diet is needed. A disadvantage of this approach is that large volumes of water are required to reduce the viscosity of the slurry mixture, thus reducing the energy density of the diet. In most cases, 2 to 3 liters of water is mixed with each 1 lb of ground pellet. Even then, this mixture is difficult to administer by gravity flow and tube blockage is a frequent occurrence.

An alternative diet that has been used for enteral nutritional support of horses is composed of alfalfa meal, casein or dehydrated cottage cheese, dextrose, and an electrolyte mixture [the "Naylor diet"](Naylor, 1977; Naylor, 1999). Although the individual components of this diet are relatively inexpensive, these costs do not account for the labor input required for procurement and preparation of the mixture. One advantage of this diet is the provision of fiber (alfalfa meal) which helps to support large colon function. Osmolite-HN (Ross Products Division, Abbott Laboratories, Columbus, Ohio), a liquid diet designed for use in hospitalized human patients, has been used as the sole source of nutritional support in sick horses (Sweeney and Hansen, 1990). Primary components of this diet include soy protein, starch hydrolysates (maltodextrins, oligosaccharides), and vegetable oils. Each liter contains approximately 1 Mcal of energy, of which

64% is derived from carbohydrate sources, 9% from fat, and the remainder from protein. The fat is predominantly medium-chain triglycerides (MCT) prepared from the fractionation of coconut oil. MCTs do not require micelle formation for absorption in the small intestine. Furthermore, the medium length fatty acids (6 to 10 fatty acids) are water soluble and gain direct access to the portal circulation, thus providing a readily available energy source. Use of starch hydrolysates reduces the osmolality of the diet. In fact, Osmolite-HN is iso-osmolar (~300 mOsmol/L), whereas preparations containing dextrose or glucose tend to be hyperosmolar, a factor that can contribute to digestive upsets. The main disadvantage of this diet is expense; to meet the maintenance energy requirements of a 500-kg horse, the cost per day of treatment is $70 to $80. As an alternative, 60 to 70% of energy requirements could be met by this human enteral diet, with the remainder provided by slurries of alfalfa meal (e.g. 3 to 4 feedings of 1 lb alfalfa meal mixed with 2 to 3 L of water).

Feeding Regimens and Clinical Monitoring

Regardless of the diet chosen, the most important consideration is a gradual increase in the rate of feeding. For horses that have been anorectic for five days or more, a reasonable goal for the first day of nutritional support is provision of 25% of the target feeding volume. The amount fed can then be gradually increased over the next 2 to 4 days. As mentioned, it may not be necessary to feed 100% of requirements and, in many horses, delivery of 75-80% of needs will maintain body weight. Repeated clinical assessment will provide the best indication of tolerance to tube feeding. Vital parameters (temperature, pulse and respiration), gastrointestinal motility, and fecal volume and character should be evaluated two to three times daily. Horses on liquid diets commonly develop low-grade diarrhea, although the volume of feces produced depends on the diet. On the other hand, signs of colic, abdominal distention, poor gastric emptying (residual feed can be easily evacuated through the nasogastric tube at the time of the next feeding) and gastrointestinal motility indicate intolerance to the diet and the need for a reduction in the level of feeding. Given the high carbohydrate content of these enteral diets, laminitis is another potential complication. The character and strength of digital pulses should be assessed regularly, particularly during the early phase of feeding.

In general, tolerance to liquid diets is best when small feedings are delivered frequently. When feeding slurry diets to an adult horse, a reasonable target is to feed 2 to 4 liters every 2 to 4 hours. If the horse tolerates this feeding regimen, the volume can be gradually increased and the feeding frequency decreased. However, a 500-kg horse has an average stomach capacity of 7 to 9 liters and single feedings should not exceed ~6 liters. Even then, this volume should be administered over a 10- to 15-minute period. Slurry diets can be administered using a large funnel and gravity or by use of a pump. Individual feedings of commercial liquid diets (e.g. Osmolite-HN) can be administered through gravity feeding sets (and a small bore tube), with or without use of an infusion pump. With these diets, it is feasible to administer feedings over a 30- to 60-minute period. In all circumstances, the proper position of the tube should be verified before each feeding. The tube must be flushed with warm water before and after each feeding. Recumbent horses should only be fed if kept in sternal position. Tube obstruction is common with the use of slurry diets. One method for clearing obstructions involves passage of a polyethylene tube inside the feeding tube to the site of obstruction. A 60-ml syringe is attached to the free end of the polyethylene tube and the obstruction is vigorously flushed with water.

The horse's daily water requirements should be calculated. If needed, additional water can be administered via the feeding tube. Even with a nasogastric tube in place, horses are able to safely consume water. Therefore, good-quality water should be available at all times. Horses with pharyngeal or esophageal dysfunction are the exception to this rule; their water requirements should be administered via the feeding tube.

The duration of enteral nutritional support will depend on the nature of the primary illness. Horses with pharyngeal or esophageal injury or reversible neurologic dysfunction may require a prolonged period of tube feeding. If possible, body weight should be measured every two to three days during the period of nutritional support. Maintenance of body weight will give the best guide as to the effectiveness of enteral nutrition. An increase in appetite will accompany improvement in clinical condition. When voluntary consumption is not precluded by the primary condition, the horse's appetite should be assessed regularly by offering palatable feeds such as fresh cut grass or leafy alfalfa hay. Even when the horse is willing to eat, it is best not to abruptly stop enteral nutritional support. Rather, the nasogastric tube can be removed to allow voluntary feed consumption, but replaced two to three times daily to allow administration of the enteral diet. The volume and frequency of tube feeding can be gradually reduced as voluntary intake increases, with discontinuation of enteral support when feed intake reaches 70 to 80% of maintenance requirements.

Parenteral Nutrition

Enteral feeding is contraindicated in horses with severe small intestinal diseases, such as anterior enteritis or following surgical correction of a strangulating obstruction. In these cases, intravenous feeding (parenteral nutrition) is the only option available for nutritional support. A complete discussion of parenteral nutrition is beyond the scope of this article. Furthermore, the expense of total parenteral nutrition often precludes its use in horses. Nonetheless, it should be emphasized that the administration of glucose-containing fluids can provide considerable energy. One to two liters of a 5% dextrose solution will provide 50 to 100 g of glucose (170 to 340 kcal). Over a 24-hour period, a 5% dextrose solution administered at a rate of 2 liters per hour will provide approximately 8 Mcal, or ~50% of maintenance energy requirements. Depending on requirements for other types of IV fluids, it may be possible to administer a larger quantity of dextrose. However, even at 1 to 2 liters per hour of 5% dextrose, some sick horses develop hyperglycemia and glucosuria, necessitating a decrease in the rate of glucose administration. This hyperglycemic state possibly reflects decreased peripheral glucose uptake as a result of mild insulin resistance. As a general rule, IV glucose should be used as the sole source of nutrition for no more than 2 to 3 days.

References

Beale RJ, Bryg DJ, Bihari DJ. Immunonutrition in the critically ill: a systematic review of clinical outcome. Crit Care Med 27: 2799-2805, 1999.

Burkholder WJ, Thatcher CD. Enteral nutritional support of sick horses. In: Current Therapy in Equine Medicine 3, NE Robinson (ed), W. B. Saunders, Philadelphia, 1992, pp 722-731.

Biolo G, Toigo G, Ciocchi B, Situlin R, Iscra F, Gullo A, Guarnieri G. Metabolic response to injury and sepsis: Changes in protein metabolism. Nutrition 13 (Suppl): 52S-57S, 1997.

Cooney RN, Kimball SR, Vary TC. Regulation of skeletal muscle protein turnover during sepsis: mechanisms and mediators. Shock 7: 1-16, 1997.

Donoghue S. Nutritional support of hospitalized animals. J AM Vet Med Assoc 200: 612-615, 1992.

Heyland DK. Nutritional support in the critically ill patient: A critical review of the evidence. Critical Care Clinics 3: 423-440, 1998.

Griffiths RD. Outcome of critically ill patients after supplementation with glutamine. Nutrition 13: 752-754, 1997.

Jackson NC, Carroll PV, Russell-Jones DL, Sonksen PH, Treacher DF, Umpleby AM. The metabolic consequences of critical illness: acute effects on glutamine and protein metabolism. Am J Physiol 276: E163-E170, 1999.

Jackson NC, Carroll PV, Russell-Jones DL, Sonksen PH, Treacher DF, Umpleby AM. Effects of glutamine supplementation, GH, and IGF-1 on glutamine metabolism in critically ill patients. Am J Physiol 278: E226-E233, 2000.

Jones C, Palmer TE, Griffiths RD. Randomized clinical outcome study of critically ill patients given glutamine-supplemented enteral nutrition. Nutrition 15: 108-115, 1999.

Madison JE. Hepatic encephalopathy - current concepts of the pathogenesis. J Vet Int Med 6: 341-353, 1992.

Michie HR. Metabolism of sepsis and multiple organ failure. World J Surg 20: 460-464, 1996.

Morris MD, Zilversmit DB, Hintz HF. Hyperlipoproteinemia in fasting ponies. J Lipid Res 13: 383-389, 1972.

Nappert G, Zello GA, Naylor JM. Intestinal metabolism of glutamine and potential use of glutamine as a therapeutic agent in diarrheic calves. J Am Med Assoc 211: 547-553, 1997.

Naylor JM. Nutrition of the sick horse. J Equine Med Surg 1: 64-70, 1977.

Naylor JM. Nutritional management in disease. In: Current Therapy in Equine Medicine 3, NE Robinson (ed), W. B. Saunders, Philadelphia, 1992, pp 736-7341.

Naylor JM. Feeding the sick horse. Proceedings of the British Equine Veterinary Association Specialist Days on Behaviour and Nutrition, pp. 87-90, 1999.

Naylor JM, Kenyon SJ. Effect of total calorific deprivation on host defence in the horse. Res Vet Sci 31: 369-372, 1981.

Naylor JM, Kronfeld DS, Acland H. Hyperlipemia in horses: Effects of undernutrition and disease. Am J Vet Res 41: 899-904, 1980.

NRC (1989) Nutrient Requirements of Horses, 5th ed, National Research Council, National Academic Press, Washington, DC, USA.

Ousey JC. Thermoregulation and energy metabolism in the newborn foal. PhD Thesis, University of Nottingham, UK, 1992.

Ousey JC, Holdstock N, Rossdale PD, McArthur AJ. How much energy do sick neonatal foals require compared with healthy foals? Pferdeheilkunde 12: 231-237, 1996.

Pagan JD, Hintz HF. Equine energetics. 1. Relationship between body weight and energy requirements in horses. J Anim Sci 63: 815-821, 1986.

Ralston SL. Clinical nutrition of adult horses. Vet Clin North Am: Equine Pract 6: 339-354, 1990.

Rooney DK. Clinical nutrition. In: Equine Internal Medicine, WM Bayly and SM Reed (eds). WB Saunders Co, Philadelphia, PA, 1998, pp 216-250.

Rose RJ, Sampson D. Changes in certain metabolic parameters in horses associated with food deprivation and endurance exercise. Res Vet Sci 32: 198-202, 1982.

Routledge NBH, Harris RC, Harris PA, Naylor JRJ, Roberts CA. Plasma glutamine status in the equine at rest, during exercise and following viral challenge. Eq Vet J Suppl 30: 612-616, 1999.

Saito H, Furukawa S, Matsuda T. Glutamine as an immunoenhancing nutrient. J Parenter Enteral Nutr 23 (5 Suppl): S59-S61, 1999.

Stick JA, Derksen FJ, Scott EA. Equine cervical esophagostomy: complications associated with duration and location of feeding tubes. Am J Vet Res 42: 727-732, 1981.

Sticker LS, Thompson Jr DL, Bunting LD, Fernandez JM, DePew CL. Dietary protein and(or) energy restriction in mares: Plasma glucose, insulin, nonesterified fatty acid, and urea responses to feeding, glucose and epinephrine. J Anim Sci 73: 136-144, 1995a.

Sticker LS, Thompson Jr DL, Bunting JM, Fernandez JM, DePew CL, Nadal MR. Feed deprivation in mares: Plasma metabolite and hormonal concentration responses to exercise. J Anim Sci 73: 3696-3702, 1995b.

Sweeney RW, Hansen TO. Use of a liquid diet as the sole source of nutrition in six dysphagic horses and as a dietary supplement in seven hypophagic horses. J Am Vet Med Assoc 197: 1030-1032, 1990.

Tate LP, Ralston SL, Koch CM, et al. Effects of extensive resection of the small intestine in the pony. Am J Vet Res 44: 1187-1191, 1983.

Vinnars E, Bergström J, Fürst P. Influence of the post operative state on the intracellular free amino acids in human muscle tissue. Ann Surg 182: 665-671, 1975.

Welbourne TC. Increased plasma bicarbonate and growth hormone after an oral glutamine load. Am J Clin Nutr 61: 1058-1061, 1995.

FEEDING PRACTICES

STUDIES OF PASTURE SUPPLEMENTATION

D. S. KRONFELD[1], W. L. COOPER[1], K. M. CRANDELL[1], L. A. GAY[1],
R. M. HOFFMAN[1], J. L. HOLLAND[1], J. A. WILSON[1], D. J. SKLAN[2],
P. A. HARRIS[3], W. TIEGS[3]

[1] Virginia Polytechnic Institute and State University, Blacksburg, VA
[2] Hebrew University, Rohovot, Israel
[3] WALTHAM Centre for Pet Nutrition, Leicestershire, UK.

Pastures provide the habitat, including much of the food, for about 80% of commercially active horses in Virginia. Pasture composition varies with type of soil, botanical composition, and season of the year. It is unlikely to provide energy and nutrients in optimal amounts and proportions for growth and reproduction. Therefore, most pastures require supplementation to provide diets with energy sources and essential nutrients in optimal ranges.

Although pasture is regarded as the main source of roughage or fiber for grazing horses, the actual fiber content fluctuates through the seasons and is marginal or too low in rapidly growing plants and in frost killed plants. At these times, overloads of soluble carbohydrates favor production of lactic acid rather than acetic acid, hence decrease metabolic efficiency (seen as a growth slump) and increase the risk of digestive upsets (osmotic diarrhea, colic, enteritis), metabolic disorders (laminitis, rhabdomyolysis, osteochondrosis). Thus, a flexible pasture supplement should provide a full spectrum of carbohydrates to promote a microbial population capable of adapting readily to changes in pasture composition.

In mature pastures, in contrast, too much fiber lowers the intake of sufficient food energy. This potential weakness is exacerbated by a supplement high in fiber (~2 Mcal/kg), so fats or oils (~9 Mcal/kg) are needed to raise the energy density to about 3.0 Mcal/kg, which is usual for a supplement.

For flexibility, a pasture supplement's contents of protein, vitamins and minerals should be designed using *sensitivity analysis* to cover wide ranges of forage composition and supplement:forage ratios.

Studies

- **Pasture composition.** Ranges of nutrients in pastures and hays from central and north-central Virginia were determined throughout the year to assess seasonal changes.

- **Carbohydrates.** The typical DHIA Forage Lab proximate analysis of carbohydrates was considered unsuitable for our hindgut fermenter. *Hydrolyzable carbohydrates* (CHO-H) were analyzed enzymatically on 130 pasture samples and 30 grains. Now CHO-H can be predicted by a simple regression on nonstructural carbohydrates (NSC).

- **Corn oil.** Preference tests demonstrated higher *palatability* (voluntary acceptability) of corn oil than other vegetable oils and animal fats. Tests of spontaneous activity and reactivity demonstrated *calmness* to be improved by dietary corn oil and mixtures of soy lecithin and corn oil.

- **Spring slump.** A slump in growth rate in March-April can be seen in data from Ontario, KY and VA for pastured yearlings fed typical supplements — corn, soybean meal and molasses (starch-and-sugar, SS). This slump can be prevented by feeding our fiber-and-fat supplement (FF). The slump may be attributed to overloads of hydrolyzable and rapidly fermented carbohydrates.

- **Glycemic index.** Exaggerated responses of blood glucose and insulin to a grain meal have been found in horses with developmental orthopedic disease (DOD) by Dr. Sarah Ralston (Rutgers). The glycemic index is much lower for FF than SS, so the risk of DOD should be lower with FF than with SS.

- **Vitamins.** Vitamin A status was determined by serum retinol concentration and, more sensitively, by a relative dose response (RDR) test. Vitamin A depletion developed in pregnant mares during winter, and was counteracted by supplementation with vitamin A at twice the NRC requirement (but not by a water-dispersible form of ß-carotene). Vitamin A deficiency resulted in low birth weight foals and more frequent retained placentas and contracted tendons. Other vitamins are included at twice NRC minimum requirements.

- **Minerals.** Analysis of forage samples, initially 20, then 33, finally 130, determined the range of mineral contents; P, Zn, Cu and Se were frequently marginal or deficient. The mineral premix was designed, using *sensitivity analysis*, to provide 1.5- to 3-times the NRC minimum requirements, taking into account the variation found in 130 forage samples and a range of supplement:forage intakes of 25:75 to 50:50. These forages were grasses or grass and legume mixtures with no more than 33% legumes. (A different mineral premix would be needed for alfalfa hay.)

- **Protein.** Initial protein content was simply the optimal value, 50 g/Mcal DE, to provide maximal growth rate determined by fitting a parabolic curve to data in the literature. Subsequent studies have tested fortification with lysine and threonine of 20 g/Mcal protein, which would minimize contamination of the environment.

- **Markers.** A limiting factor in pasture research and supplement design is the imprecision of estimating pasture intake (PI) by difference. Better estimates may be obtained by marker methods that measure fecal output

(FO) and digestibility (D): $PI = FO/(1 - D)$. We have developed a chromic oxide method for measuring FO. Currently being tested to measure D are alkanes, yttrium, and NIRS predictions from feeds and from feces.

FEED TYPE AND INTAKE AFFECT GLYCEMIC RESPONSE IN THOROUGHBRED HORSES

J.D. PAGAN[1], P.A. HARRIS[2], M.A.P. KENNEDY[3], N. DAVIDSON[2] AND K.E. HOEKSTRA[1]
[1]Kentucky Equine Research, Inc., Versailles, KY
[2]WALTHAM Centre for Pet Nutrition, Leichestershire, UK
[3]KER/WALTHAM Fellow

A 6 x 6 Latin square design experiment was conducted to determine glycemic response in horses fed six different feeds at 3 different levels of intake. Six Thoroughbred geldings were fed mixed grass hay and one of six diets: cracked corn, whole oats, sweet feed (45% cracked corn, 45% whole oats, and 10% molasses), sweet feed + 10% corn oil, alfalfa forage, or a low starch, high fermentable fiber mix (25% rice bran, 25% soy hulls, 25% wheat bran, and 25% soaked beet pulp). Horses were randomly assigned low (.75 kg), medium (1.5 kg), or high (2.5 kg) intake levels at each feeding during six 3-day test periods. Diets were fed at medium intake (1.5 kg/feeding) twice each day and each horse received 5.45 kg hay per day during transitions between test periods, during which time treatments were altered for the subsequent period. Horses were given access to free exercise on pasture during the day, although they were not allowed to graze. On test days, morning feeding levels equaled 750, 1,500, or 2,500 g of the treatment diets in every case except sweet feed + corn oil. Horses on this diet received 750, 1,500, or 2,500 g of sweet feed with an additional 75, 150, or 250 g corn oil, respectively. Blood samples were taken prior to the morning feeding on test days to determine baseline glucose values and at 30-min intervals following feeding until glucose levels returned to or dropped below baseline for 120 min. The morning allotment of hay was fed following completion of sample collection. Area under the curve, mean glucose (mg/dl), peak glucose (mg/dl), and time to peak glucose (min) were determined. Plasma glucose concentrations were statistically analyzed by the general linear model procedure for analysis of variance. Period, horse, day, diet, and intake were included in the model. Using area under the curve for whole oats at medium intake (1.5 kg) as a standard of reference, a glycemic index was generated from area under the curve for all diets and intake levels.

Area under the curve indicated differences in glycemic response between low (.75 kg) and high (2.5 kg) intake levels of all diets combined (Table 1). Sweet feed and whole oats demonstrated the greatest glycemic response, while alfalfa and sweet feed + corn oil provided the lowest response. Plotting the glycemic index by feed and level of intake revealed an appreciable drop in the index for whole oats fed at 2.5 kg, compared to that at 1.5 kg and relative to glycemic indexes generated for other feeds and intake levels (Figure 1).

Table 1. Area under the curve, mean glucose, peak glucose, and time to peak for all diets and intakes.

	Area under curve	Mean glucose (mg/dl)	Peak glucose (mg/dl)	Time to peak (min)
Dietary Treatment				
Sweet Feed	2,073[d]	99.6[c]	108.5[b]	148[a]
Whole Oats	1,602[cd]	99.2[c]	102.7[ab]	142[a]
Cracked Corn	1,438[c]	97.3[b]	105.6[b]	153[a]
High Fiber Mix	1,378[bc]	99.3[c]	106.8[b]	108[a]
Sweet Feed + Oil	898[ab]	96.9[b]	105.8[b]	238[b]
Alfalfa Forage	733[a]	94.5[a]	99.1[a]	135[a]
SEM	177	.63	2.1	20
Statistical Significance	.01	.01	.05	.01
Intake Level (all dietary treatments combined)				
.75 kg	1,087[a]	97.1	101.5[a]	120[a]
1.5 kg	1,428[ab]	98.4	106.5[b]	165[b]
2.5 kg	1,546[b]	98.0	106.2[ab]	178[b]
SEM	125	.45	1.5	14
Statistical Significance	.05	NS[e]	.05	.05
Diet x Intake Interaction				
Statistical Significance	NS[e]	NS[e]	NS[e]	NS[e]

[abcd] Treatments lacking a common superscript differ (P < .05)
[e]   Not significant

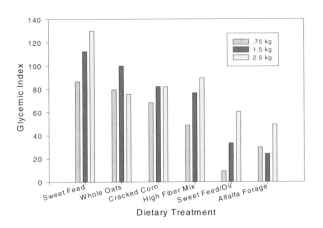

Mean glucose (mg/dl) was highest for sweet feed, whole oats, and the low starch, high fiber mix and lowest for the alfalfa diet. Peak glucose (mg/dl) was similar for all diets except alfalfa forage. Time to peak glucose (min) was greatly increased in the sweet feed + corn oil diet, while the remaining diets demonstrated similar responses. Increasing the level of intake from .75 kg to 1.5 kg per feeding increased time to peak glucose by 45 ± 14.1 min. Results of this study indicate that different grain diets demonstrate different glycemic responses and adding fat reduces both area under the curve and peak glucose values, as measured within this experimental design. More research is required to determine the relevance of glycemic response in predicting the effects of different feed ingredients on a horse's performance or behavior.

FEEDING AND MANAGEMENT PRACTICES IN THE UK AND GERMANY

PAT HARRIS

WALTHAM Centre for Pet Nutrition, Leicestershire, UK

A major part of this paper has been kindly reproduced with permission from the article "Review of equine feeding and stable management practices in the UK concentrating on the last decade of the 20th century" as published in Equine vet. J. Suppl 28 p 46-54.

UK

Overview

Equine feeding and stable management practices for horses kept in the UK vary greatly, and it is probable that almost any permutation of regimen could be found somewhere. Unfortunately, there is uncertainty about the number of horses in the UK and very limited data are available on the ways horses are being fed and managed. This paper reviews some of the information that is available and provides an outline of some of the factors influencing the practices used. To a certain extent, the way UK horses are fed and managed primarily reflects the purpose for which they are kept (e.g. racing Thoroughbred vs. native breeding pony), their location (urban vs. rural), the time of year, their breed and age and the owner's financial situation. In very general terms, the various ways that horses can be kept and managed fall between two extremes: the professional riders, owners or trainers who tend to keep horses and ponies in large barns or stable yards managed by themselves or a head stable person and the amateur competitors or leisure riders who tend to keep their horses at a livery yard or at their own home or the home of a friend. Many of these animals are kept under part- or full-time do-it-yourself (DIY) conditions. Common feeding practices range from feeding traditional home cereal-based mixes to feeding coarse mixes or pelleted manufactured feeds. Very few people, however, feed just a simple single grain or compound feed and roughage diet. Many add other feedstuffs and supplements including soaked sugar beet or straight molasses, primarily as palatability enhancers. Other common additions include cod liver oil, various types of vegetable oil, carrots, one or more vitamin and mineral mixes, herbal mixes and certain agents with ergogenic or performance-enhancing claims. Especially for horses used for competition purposes, provision of supplementary salt is common, either by means of a salt block or lick or as an addition to the feed. The soaking of hay, for a variable time period, is another common practice. Although the type and amount of feed fed fundamentally varies according to the workload of the horse, there tend to be marked seasonal variations in feeding practices due to the weather and the availability of pasture.

Introduction

This century has seen the horse change from having important roles in the military, in agriculture and as transportation to becoming part of the expanding leisure industry. This alteration in the fundamental role of the horse has in turn resulted in changes both to the ways horses are kept and managed as well as to who keeps and manages them.

An extensive horse census has not been carried out in the UK since the 1930s when it was estimated that there were 1,278,341 horses (Mellor et al., 1999) . In the 1997 Ministry of Agriculture, Fisheries and Food (MAFF) survey (Anon, 1997), which recorded the number of holdings on which horses were kept and their numbers, there were around 245,000 horses kept on about 39,000 agricultural holdings throughout the UK.   This would be an underestimation of the total number of horses in the UK, considering the change in the role of the horse in the last few decades. Records do not take into consideration the number of horses kept on land that is used for purposes other than agriculture. In a fairly large survey carried out also in the mid-1990s for the British Equestrian Trade Association (BETA) by the Produce Studies Group, the estimated number was about twice that of MAFF or 500-600,000 horses (Anon-BETA, 1996; see Table 1). This estimate was revised in the more recent survey to reflect a horse population of one million (Anon-BETA, 1999).

An extensive recent survey (Mellor et al., 1997; Mellor et al., 1999) suggested that in Scotland and northern England more than three times the number of horses were present than recorded by MAFF's annual census conducted in the same area at the same time. The authors implied that if the same ratio were to apply to the whole of the UK the number of horses would be around 800,000.

Even less published information is available on the ways in which horses are being fed and managed throughout the UK. This review therefore brings together some of the information that is available and looks at some of the factors that may influence feeding and management. This review highlights the lack of scientifically reliable information on the feeding and management of the UK horse population, which continues to play an important, if changed, role in our society. Currently, for example, one of the main sources of illustrative information is surveys that have been carried out by various organizations in different segments of the horse industry. The findings from these surveys obviously will have been influenced by the sample groups used. For example, those who ride mainly for pleasure are less likely to feed the recommended amounts of cereal-based feed in the summer when grass is available than those who ride mainly for competitive reasons. For this reason, a guide to the various surveys used to illustrate parts of this paper, and their target groups, are given in Table 1. Even the descriptors of the various equestrian activities used in the surveys are not equivalent.

Table 1. The percentage of the respondents, in a number of surveys, who participated in various equestrian activities.

Activity	Anon-			
	BETA 1996[a]	Equestrian 1995[b]	Eventing 1996[c]	Horse 1997[d]
Riding for pleasure/hacking	74		77	24
Having riding lessons	53			
Exercising the horse	36			
Schooling (the horse, not the rider)	25			
Show jumping	18	37	74	17
Trekking	16			
Dressage	12	35	74	17
Hunting	8	28		
Hunter trials	8			
Eventing (general)	5	29	77	2
Horse trials- unaffiliated				11
Eventing (three-day event)			38	
Eventing (one-day event)			85	
Point-to-point			16	4
Showing			44	20
Shows		42		
Driving			8	6
Sponsored riding	5			
Polo			8	2
Endurance riding			6	6
Others (UK chasers, driving, racing, etc.)	5			
Breeding				6
Buying/selling			35	11

[a] Survey carried out by the Produce Studies Group. This interviewed over 3,000 households with over 7,000 individuals and interviewed 500 horse owners/riders (maximum of one person per household).

[b] Survey carried out on behalf of the IPC Equestrian Group. The results were taken from an Omnibus Study (2,000 face to face interviews of adults aged 16+ years who were selected in a minimum of 130 sampling points at the end of 1994) together with desk research and a self-completion questionnaire sent to 3,000 members of a database of people who had bought or sold equestrian goods or services recently (553 questionnaires returned). 68% of respondents rode for pleasure and 28% worked with horses. For questions such as "where do you keep your horses/ponies" and "approximately how much have you spent in the last 12 months on your horses/ponies," the source of the information was 515 horse/pony owners.

[c] Reader survey was carried out for IPC Magazines Ltd. The data for this survey was gathered by means of a self-completion questionnaire included in an issue of Eventing magazine (August 1996). 245 replies were processed. Just over 3/5 of this sample rode for pleasure; 1/3 were actively involved with equestrian organizations/events, 1/5 worked with horses.

[d] Reader survey was carried out for IPC Magazine Ltd. The data for this survey was gathered by means of a self-completion questionnaire inserted into the launch issue of Horse Magazine (July 1997). A random sample of 1023 completed returned questionnaires was processed.

Factors Influencing Feeding and Management Practices in the UK

Today there is no general feeding or management practice for horses in the UK. It is probable that almost any permutation of feeding regimen can be found somewhere in the UK. In principle, the way UK horses are managed reflects a combination of the purpose for which they are kept, where and how they are kept, time of year, breed and age, and financial circumstances. An indication of how these factors may influence feeding and management practices is provided below.

The purpose for which horses are kept

As in other countries, a Thoroughbred in race training is likely to be kept in an individual stall within a larger complex and is often clipped and rugged with little or no access to grazing. Such an animal would typically be fed several meals a day of an energy-dense, usually cereal-based feed with relatively little of its energy or other nutrient supply coming from forage. This contrasts markedly with the native breeding pony, such as the New Forest pony, which receives little attention from man, eats primarily forage and eats for survival.

Horses kept for hacking or general purpose riding commonly spend part or most of their time out at grass and endurance horses are often fed high forage-based diets supplemented with vegetable oil. Many top dressage and show horses are kept stabled for most if not all of the day. Such general descriptions could continue and yet not apply to many individuals within any particular group. It is also not possible to categorize, even in the broadest sense, according to the purpose for which horses are kept, as not all horses are registered with either a breed society or a discipline-oriented association. It has been estimated that about 600,000 riders take part in some form of competitive riding, although 60% were participants in local competitions only (Anon-BETA, 1999). A proportion of owners and riders compete at a local (unregistered) and/or national (registered) level in a number of disciplines using the same horse. For many uses, such as hunting and polo, there is no formal registration scheme. Many horses are used for hunting during the winter and then for a different purpose in the summer. The Anon-BETA survey of 1999 estimated that 10% of riders hunt. Horses used for competitions or more intensive pastimes such as hunting tend to be kept stabled for prolonged periods of time and to be fed, when in work, some form of concentrate feed in addition to forage. An indication as to the size of the equine competitive industry in the UK is given in Table 2.

In the survey carried out in Scotland (Mellor et al., 1997) the most popular equestrian activities were hacking, involving 24% of the horses, breeding (20%) and use in riding or pony club events (18%). Only 2% of horses kept by respondents were used for endurance riding, and a similar number was kept for point-to-point racing. In the recent BETA survey, it was estimated that about 30% of animals are ponies (Anon-BETA, 1999). In this survey the main uses to which horses were put varied according to whether the respondents were private householders, riding schools and livery yards, breeders or trainers. Riding or hacking, however, was the main category of use overall.

Table 2. Indication of the purpose for which some horses were kept
in the UK in 1996.

Discipline/Use		~ Numbers
Horse Trials	Registered	7,500
	Advanced	772
Dressage	Registered	5,700
Show Jumping	Registered	16,200
British Endurance	Registered	950
Horse Driving Trials	Registered	630
Polo	Estimated	7-8,000
	High Goal	1,500
Race Training	Estimated	12-13,000
Thoroughbred mares covered each year including Ireland primarily for racing purposes		22,000
Number of foals born in UK and Ireland to Thoroughbred mares - primarily for racing purposes	Estimated	12,000

Where and how horses are kept

1. Urban vs. rural setting. Horses kept in urban settings are less likely to spend prolonged periods in large paddocks grazing than in rural areas (although it will also depend on other factors discussed in this section). At one extreme of this spectrum would be the urban pony kept in Dublin and at the other extreme would be ponies freely roaming areas such as the New Forest or Exmoor. Unfortunately, no survey has been carried out throughout the UK on this aspect. In one survey (Anon-Equestrian, 1995), it was estimated that about 7% of UK households had a member who rode, although only 19% of these riding households actually owned the horse or pony they rode (~340,000 households) and these were generally located in rural areas. In more recent surveys (Anon-BETA, 1996 and 1999), only an estimation of the geographical distribution of riders (not necessarily owners) was made as shown in Table 3.

Table 3. Regional distribution of riders compared with the national distribution (Anon - BETA survey, 1999).

Region	1999	
	Horse riders %	Population %
Scotland	8	9
North	21	25
W. Mids/ Wales	14	14
E. Mids/ E. Anglia	11	11
South West	9	8
London and South East	37	33

2. DIY vs. livery. The manner in which horses used for riding purposes are kept and managed in the UK falls broadly between two extremes. Firstly, professional riders, owners and trainers tend to keep horses and ponies either in large barns or stable yards managed by themselves or by a head stable person with help from a number of stable lads or grooms. Conversely, there are amateur competitors or leisure riders who tend to keep their horses at a livery yard or at their own home or the home of a friend. Many of these animals are kept under part- or full-time do-it-yourself (DIY) conditions. Full-time DIY would mean that the owner or keeper is responsible for feeding, turnout, grooming and riding. Full livery, on the other hand, means that the livery yard owner or operator does all of this with the owner or rider being able to turn up, usually at his or her convenience, to ride or to compete. The most commonly provided service among professionals who keep other people's horses in the Anon-BETA 1999 survey was full livery. In this survey it was estimated that 83% of all horses and ponies were owned and kept by private households and 7% were owned by private households, but kept by livery yards.

In the Scottish survey (Mellor et al., 1997), 50% of owners kept their horses on private premises and 50% on shared premises. Whether this would hold true for the rest of the UK, especially the more urban areas, is not known. The results from two surveys shown in Table 4 provide differing data, with 64% keeping their horses on their own land in the 1995 survey and 18% in the 1997 survey. The apparent decreasing trend in the number of people who keep horses on their own land probably reflects the demographic characteristics of people included in each survey and/or the question asked rather than a trend with time.

In the Scottish survey (Mellor et al., 1997), 29% of horses were stabled most of the time and a further 2% were permanently stabled and never given access to grazing.

Table 4. The percentage of respondents to two separate surveys who indicated where their horse(s) were kept and the question that was asked in each survey with respect to this.

Where kept	Anon -Equestrian 1994	Anon -Horse 1997
Livery	12	24
Rent land/stabling	12	20
Own land/stabling	64	18
Livery part-time	5	5
Livery full-time	5	4
Other	7	-
Question asked	Where do you keep your horse/pony?	Where horse(s)/pony(s) usually kept?

3. Single vs. multiple. This section also reflects whether the horse or pony is kept at the person's own home or not. When kept at home, there is an increased likelihood that just one or a few horses will be kept, often with a similar small number of stables. In one survey carried out in the mid-1990s (Anon-Equestrian, 1995) of those households that did own their own horse or pony, 70% owned only one. In a survey carried out into racehorse establishments in Ireland, 57% of the horses were kept in individual single story loose boxes (usually attached to each other in some linear configuration and often sharing air space) and 36% were housed in American-style horse barns (Townson et al., 1995).

Time of year

Each discipline tends to have a competitive season, although for disciplines such as show jumping and dressage the increasing availability of suitable indoor venues has extended the season to almost the whole year. Eventing under the BHTA, for example, runs from March to mid-October although unaffiliated horse trials start in the autumn. Hunting, however, tends to be an autumn and winter sport only. Some horses used only for this purpose may be given the summer off; others may be given a short break and then used for a different purpose.

Seasonal variations in feeding practices occur, mainly due to the weather, and tend to reflect the usage of the animal and the availability of pasture. In the leisure and breeding industries, many animals will be wintered out, often with protective clothing, although many will be brought in at night and fed. Native animals more commonly winter out, often with the provision of supplemental forage. In the Scottish survey (Mellor et al., 1997) 10% of the animals were permanently kept at pasture (~10,000 horses). In the later BETA survey (Anon-BETA, 1999) it was estimated that around 145,000 horses were kept in the open all year. In the riding schools and livery yards, 23% of horses that were owned by these establishments were stabled all year and 16% kept in the open all year as compared with 29 and 11% of those horses that they kept for others. The most common practice was to keep horses stabled in the winter only.

In the leisure industry, horses used for hacking purposes generally receive more supplemental feed in the winter months than in the summer months when pasture is plentiful. This is reflected in the figures on production of compound feeds for horses supplied by the Ministry of Agriculture, Fisheries and Food (MAFF) statistics (Commodity and Food) Branch A (Harris, 1997a). In 1994, for example, 9.1, 8.9, and 8.8 thousand tonnes (1000 kg) of feed were reported to have been produced in May, June and July respectively, compared with 10, 13.8, 14.9, 15.8 and 15.2 for August, September, October, November and December, respectively. Animals in competitive work and lactating mares tend to continue to be fed intensively throughout the summer months.

Breed, age and individuality

The breed distribution of all horses in the UK is not known but in both the Scottish (Mellor et al., 1997) survey and the Anon-BETA 1996 survey, as shown in Table 5, the most common breeds were Thoroughbred and Thoroughbred crosses.

Table 5. A guide to the breed characteristics of horses in Great Britain based on the Anon-BETA 1996 survey.

Breed	Number x 10^3
Thoroughbred/Thoroughbred crosses	216
Pony - Welsh	87
Pony - Shetland	18
Pony - Other	75
Arab/Anglo Arab	38
All others (includes "don't know" responses)	132
Total	566 *

* The figure actually given in the survey report is 565,000

In the Scottish survey these represented a slightly lower percentage (30 vs. 38%). Reference has been made previously to the different practices commonly used for the native pony considered to be more hardy than, for example, the less robust Thoroughbred. This is, once again, a generalization. Native ponies kept for showing purposes may be managed intensively and some individual Thoroughbreds may be kept fairly extensively. Breed and individual differences in ability to maintain weight or become overweight will obviously affect feeding and management practices. An individual's susceptibility to certain clinical disorders, especially laminitis, COPD, spasmodic colic and equine rhabdomyolysis syndrome, may also influence the way individuals are fed and managed.

In the Scottish survey (Mellor et al., 1997) the mean (± s.d.) age of the population was 11 ± 7.5 yrs with an equal sex ratio. Anon-BETA 1999 estimated that about 46% of horses and ponies were mares or fillies and amongst the private householders the average horse was about 12 years old. Breed and age together have some influence on the way the animal is kept. In a survey carried out on a number of Irish studfarms (O'Donohue et al., 1995) it was found that mares and young stock usually had access to pasture on almost all farms (n=46) surveyed. In general, animals were housed only during winter and then normally only overnight. The housing regimen varied with the studfarm involved, the age of the animal and time of year. For example, during the latter part of the winter on three farms the animals were not housed at night, yet on 41 studfarms the young stock were housed at night. One studfarm sold all foals during early winter and on another farm the animals tended to be housed almost completely during the latter part of their first winter. Few surveys are available which evaluate, in practice, the amounts and types of feeds fed to animals at various ages. This Irish survey is one of the few to have been published on breeding and young stock providing details on nutritional and management practices.

It is currently accepted by many that horses are living and being kept longer than horses would have been towards the beginning of this century. However, no scientifically published survey has been carried out into any associated changes in nutritional and management practices, although there are increasing numbers of commercially available compound manufactured diets for the senior horse.

Financial circumstances

Few substantial data are available on the cost of owning and competing a horse. The costs vary according to the type of horse kept (e.g. hardy native vs. sensitive Thoroughbred) and the purpose for which it is kept (e.g. general hacking vs. three-day eventing), a variability reflected in Table 6. The costs of keeping horses according to one survey (Anon-Horse, 1997) in which 24% of owners used their horses at least in part for hacking, 11% for unaffiliated horse trials and only 2% for BHTA horse trials, are much lower than those reported in

Table 6. Categorized spending on horse care according to the respondents
of four surveys.

Average amount spent £/year[a]	Anon-Equestrian (1995)[b]	Horse Survey (1997)	BETA (1996)[c]	Eventing Survey (1996)[d]
Livery	1,717	1,199		2,254 (27)
Stabling	916	958	505	1,202 (24)
Transport	936	657		995 (46)
Feed	1,172	463	345	965 (56)
Tack	+ Equipment 715	414	115	566 (62)
Shoeing	-	404	195	556 (67)
Vets	620	352	100	479 (66)
Insurance	533	310	90	521 (53)
Equipment	see Tack	211	see Tack	257 (57)
Competing	560	187	70	515 (60)
Security	150	173		305 (13)
Feed Supplements	see Feed	87		305 (13)
Other			Paid help £105	538 (14)
			Dewormers £50 Stud fees £20 Others £25	

Figures in parentheses = proportion spending on this category
[a] Of those stating that they spent on this category
[b] Given as the average spent per person in the last 12 months on their horses/ponies
[c] Based on total expenditure across 565,000 horses and each category includes a varying proportion
   of horses on which nothing was spent, i.e. average spent per horse, not average amount where spent
[d] Given as the average spent per person in the last 12 months on their horses/ponies includes some
   expenditure on feed, bedding and other services in bedding costs

another survey (Anon-Eventing, 1996). In this later survey, 85% of
the respondents were competing at the one-day event level and 38% at the
three-day event level. However, it is interesting that, although 85% said that
they were involved with one-day eventing, only 67% reported as having spent
money on shoeing. This may reflect that those completing the questionnaire
failed to fill out this section or were not aware of how much they spent. Again,
this highlights the lack of verified information available. In one survey
(Anon-BETA, 1996) it was noted that the cost estimates varied considerably both
within and between the different categories of owners. Another complicating
factor could be the respondent's understanding of the question and exactly what
the various categories referred to (e.g. what each respondent considered to be
stabling and/or livery costs).

An alternative way of looking at the situation was reported in 1999. A survey
suggested that the total spent in the UK on direct expenditure on horses was
£1200m (with a grand total of £1940m), £330m of which was associated with
feeding costs (Anon-BETA, 1999).

In the report on lifestyle in one survey (Anon-Equestrian, 1995), it was stated that the sample used was significantly more wealthy than the national average and 62% lived in a detached house compared to the national average of 15%. However, this survey did not evaluate availability of disposable income and what proportion of this disposable income was spent on the horses. Another survey reported that riding had equal appeal across social classes, split evenly between the middle and working classes (Anon-Gallup, 1994). Many owners keep their horses in the best possible conditions that their disposable income allows and for some this means that they take no holidays and buy few new clothes. Any spare money is spent on the horse.

Feeding and Stable Management Practices

There are few published reports of example feeding programs available in the scientific literature for the UK and Ireland and very few have been published in the last decade of this century (Harris, 1997a; O'Donoghue et al., 1995), although examples are provided in a number of the nutrition texts.

The preceding section described some of the general factors that may influence the way horses are kept and managed; some key aspects of management and feeding practices will now be explored.

Types of bedding used

Table 7. Percentage of respondents who used various types of bedding according to three surveys.

Types of bedding used	Anon-Equestrian* (1995)	Anon-Horse** (1997)	Anon-Waltham*** (1997)
Straw	58	41	68
Shavings	48	32	24
Shredded Paper	7	3	4
Hemp	2		
Peat	1		
Rubber Matting			4
a Other	7	(24)	

a   Other included wood chippings, sawdust and rubber matting.
*   Numbers did not add up to 100% in this survey as published. Those included in the other group may have included people using hemp, peat or rubber matting.
**   In this survey, some respondents noted that they used more than one type of bedding
***   In this survey, details of the predominant or sole bedding used by the respondents were requested.

Straw is not highly regarded as a forage in the UK, especially for young animals with an immature hindgut. It is believed to increase the risk of impactions and chronic obstructive pulmonary disease (COPD). However, as shown in Table 7, straw remains the most frequently used bedding, which may reflect tradition and its availability, relative low cost, and relative ease of disposal, as well as the anthropomorphic perception of straw providing a deep, warm and comfortable bed. In the survey and assessment of racehorses in Ireland carried out in 26 establishments, 65% used straw (twelve yards used straw only); 26% used wood shavings (four yards used wood shavings only) and 9% used shredded paper.

Choice of feed/feed supplier

In one survey (Anon-BETA, 1996) it was estimated that around 65,000 horses were owned by professionals and about 500,000 were owned by private households. What proportion of the decisions over the choice of feed and feed supplier was made solely by the owner is shown in Table 8.

Table 8. A guide as to who makes the sole decision on the choice of feed given to their horse and the feed supplier of this feed according to the Anon-BETA 1996 survey.

	Private owners	Professional owners - horses owned by them	Professional owners - horses kept by them on behalf of others
Choice of feed	59	97	63
Feed supplier	58	98	68

Among private owners, shared decisions (with the keeper) accounted for between 15 and 20% of all decisions.

Types of feed fed

Oats may be the traditional cereal fed to working horses (Harris, 1997), but in the UK oats are also considered to be a psychologically heating feed. Therefore, horses predominantly in light work are more likely to be fed cool mixes or non-heating products which are oat-free. This is illustrated by the low number of horses in a small survey (Table 9) being fed oats (Anon-Waltham, 1997 unpublished data).

Table 9. The percentage of respondents who stated they fed each type of feed listed below according to two surveys.

Types of feed	Anon-Equestrian (1995)	Anon-Waltham (1997)
Hay/hay substitute	85	86
Grass	76	85.5
Premixed feed	72	66
Traditional feed	48	17% fed oats
Other	4	&16% fed barley

Traditional feed in the Anon-Equestrian 1995 survey is taken to refer to an oat or other cereal-based diet.

Ninety percent of the respondents described their winter feeding program, 78% of whose horses were in light work, 19.5% in medium work and 2.5% in hard work. All horses were 4-20 years old inclusively and ranged in size from 13.2 hands and above, with 50% greater than 14.2 hands and less than 16 hands.

Barley is generally fed rolled or cooked to increase small intestinal digestibility. **Maize** is generally micronized or flaked before feeding. It is not commonly fed cracked. Traditionally, **wheat** has not been fed as a straight cereal to horses in the UK. However, the more modern cooking processes allow it to be used as a high energy feed in compound coarse mixes and home-mixed cereal rations.

Appropriately treated **soya bean** may be added in some form or another as a source of essential amino acids. The expelled meal and full-fat meal are often fairly unpalatable and therefore the full-fat flake, which has usually been micronized, tends to be the form that is sometimes used as a top dressing to balance "cereal straights."

The use of premixed or compound manufactured feeds has increased in the UK over the last few years (Harris, 1997a; 1998). The O'Donohue survey (1995) suggested that, even on studfarms, there is a greatly reduced dependence on completely home-compounded diets in the 1990s compared with the 1960s (McCarthy, 1975). The compound feed market is becoming increasingly fragmented with the development of many smaller niche markets such as the veteran animal and the convalescent animal. The major companies tend to have a wide range of feeds for different life stages and exercise types and offer the feeds in a number of different forms. This may reflect the feeding trends within the growing leisure section of the horse market in which the individual owner wishes to feed a diet specifically targeted to its animal's needs. However, the bulk of the feed sold still tends to be the general maintenance or light work cube and mixes.

Most feed manufacturers guarantee their products are free from caffeine and theobromine and are therefore suitable for animals competing under FEI or Jockey Club rules.

For maintenance, light work or breeding, **nuts, pellets/cubes** varying in size from 2 to 15 mm diameter are still very popular. Forty-six percent of the respondents in one survey (Anon-Waltham, 1997; unpublished data) fed cubes, but often in combination with coarse mixes (sweet feeds, a mixture of processed cereals often combined with molasses or glucose syrup and other ingredients such as grass nuts, sugar beet pellets and protein, mineral and vitamin balancing pellets).

Alfalfa hay is not as commonly fed in the UK as in countries such as the United States because it is usually imported. Its regular use is limited to some of the racing yards and breeding studs. However, alfalfa chaffs are being fed increasingly as part of the diet and are often mixed with the concentrate portion in order to provide bulk and to slow the rate of eating. Seventy-four percent of the respondents in one survey (Anon-Waltham, 1997; unpublished data) fed chaff (short chopped hay, straw, alfalfa or mixes).

Chaff is commonly added to concentrate in the UK. The amounts vary but commonly about 0.5 kg per feeding is included. There are four main types of chaff available in the UK: molassed straw chaffs which contain around 40-

60% molasses and straw, (+/- limestone); high temperature dried alfalfa with 10-20% molasses; a straw and alfalfa 50:50 mix; non-molassed pure hay and straw chaffs (see Table 5). Homemade chaff is still fed, but mainly in larger yards.

Another common component of horse feeds in the UK is sugar beet pulp (a by-product of sugar beet processing). It is usually molassed and is largely purchased as dehydrated shreds or compressed pellets which are then soaked before use. In one survey (Anon-Waltham, 1997; unpublished data) 47% of respondents regularly fed soaked sugar beet pulp. Wheat bran usage appears to be on the decrease, although it is still fed under more traditional feeding regimens (this is a personal observation and due perhaps, in part, to the increased awareness of the reversed calcium to phosphorus ratio). Only 11% of the respondents regularly fed bran in the Anon-Waltham 1997 survey.

Grass is a major component of the diet of most horses according to the surveys shown in Table 9.

In the Scottish survey (Mellor et al., 1997), 69% of horses grazed at least half of their time, with 10% permanently grazing and only 2% never grazing. Pastures vary in their composition not only from region to region but also from season to season. Calcium levels, for example, can decrease significantly during the winter when the calcium to phosphorus ratio may become reversed. Certain geographical areas may have pastures deficient in certain trace elements, particularly copper, zinc, manganese and/or selenium. Therefore, it is important to appreciate which areas are likely to be deficient, especially if the pasture is the main source of nutrients (Harris, 1997). Standard grazing levels vary; the commonly recommended rule of thumb is two acres for the first animal and then one acre for each subsequent animal, depending obviously on whether the pasture is the sole source of nutrients.

Feeding of supplements or feed additives

Very few horsemen provide their animal with a simple, single cereal or compound feed and roughage diet. Separate feedstuffs and various supplements and additives are commonly provided. In one survey (Anon-Equestrian, 1995), the respondents were asked whether they used feed additives or supplements. Fifty percent of respondents regularly fed supplements and were more likely to be males in the younger age groups, 36% said supplements were fed occasionally and 10% did not feed supplements. Various feed additives and supplements can be given to horses. One of the difficulties with surveys is defining what an individual considers to be a supplement or a feed additive. Carrots and apples are commonly given to horses as succulents, but are they considered to be supplements or normal feed components? Sixty-two percent of the respondents in the Anon-Waltham 1997 survey said they fed succulents. Salt is often available as a lick or rock in the horse's box or manger and is usually required to provide the sodium and chloride necessary for the horse, especially those which sweat extensively and regularly. Some respondents may have included salt as a supplement and others may have included salt as part of the normal diet. This highlights once again the need for a properly constructed survey.

The more commonly accepted supplements range from the broad spectrum vitamin and mineral supplements to those providing one or a selected few specific nutrients. Supplemental vegetable oil is often fed as an alternative or additional energy source (Harris, 1997b). In addition, herbal mixes, digestive aids, ergogenic aids, calming agents and coat conditioners are also available. Many supplements are, unfortunately, given with little knowledge of their suitability to be fed with the basal diet constituents or with other supplements in use. Often there is little understanding of their efficacy. This can sometimes lead to oversupplementation of certain nutrients with possible interactions and interferences (Harris, 1997a). Treats are often given to horses in the UK but the extent may depend on what is classified as a treat. Carrots, apples and mints are often used as rewards. Specifically manufactured treats for horses often include herbs or peppermint and are readily available.

Hay feeding practices

A variety of **hay** types is available in the UK, including meadow, seed and legume hays. Hays will differ in the number and type of grass species present, which in turn will affect the protein, mineral and energy levels, as will the stage of growth when harvested and the area where it was grown.

Hay quality can be variable in the UK due to unpredictable weather conditions; hay often tends to be dusty with a high fungal spore count. Several different methods have been used to deal with this, and there has been an increase in the use of alternative forage sources for horses, in particular big bale **haylage** (>40% DM) and **silages** (<40%DM). **Barn dried hay** is made from grass that is allowed to wilt in the field for two to three days when weather conditions are conducive and then loosely packed into special buildings where air of a particular temperature and humidity is blown through it for eight to ten days before it is baled.

Both commercial and home-produced haylages are made. The homemade varieties tend to vary considerably in composition. In the commercial production of haylages, plants are wilted to approximately 50-60% DM and placed into semi-permeable plastic packages where mild anaerobic lactic fermentation occurs, which stabilizes at a pH of approximately 4.5 -5.5. These haylages tend to have a protein level of 9-12 % and a DE value of 9-11 MJ/kg. Because of the high moisture content of the haylages, the fiber level is comparatively low (MAD ~32 -36%), so if fed at the same inclusion rate as hay may potentially result in problems due to a lowered fiber intake. There are concerns with feeding haylage, especially with respect to potential clostridial activity and lowered overall fiber intake, but the demand for haylage appears to be increasing. Once again, the extent of the increase depends on the participants in any particular survey. For example, only 3% of the 86% of respondents that fed forage in one survey (Anon-Waltham, 1997; unpublished data) fed haylage.

The practice of soaking hay to increase the likelihood of fungal spores adhering to the forage and being ingested rather than inhaled is popular. In the Anon-Waltham 1997 survey, 28% of respondents soaked their hay. The water used for soaking must be fresh, and although there is still some discussion over exactly how long soaking should be carried out, the author recommends soaking no longer than one hour because of the potential negative effect of prolonged soaking on the soluble carbohydrate and nitrogenous content of hay (Warr and Petch, 1992; Harris, 1997a). In the Irish studfarm survey, only 2 of the 26 establishments routinely soaked their hay, but all of the trainers surveyed replied that they would soak hay for individual animals when required and for the whole yard if the quality of the hay necessitated such a procedure.

In 1994, Brown and Powell Smith reported that many people in the UK preferred to feed hay on the stable floor. This was endorsed by Townson's survey of a number of Irish racing stables (Townson et al., 1995) which noted that 57% of the stables fed hay from the floor. However, in one survey (Anon-Waltham, 1997; unpublished data) of those who fed hay, 90.5% fed from a hay net. Suggested disadvantages of hayracks or nets high off of the ground included an increased risk of particles getting into the eyes and nose and feet getting caught in the rack or net. Eating from a high rack or net is thought to be an unnatural feeding position and suggested by some to affect muscles and nerve function adversely (Hintz, 1997). Hayracks at chest height are thought to increase the risk of injuries, decrease the space available within the stable and are costly. Although hay feeding on the floor induces a more natural feeding posture, it increases wastage, partially by increasing the risk of contamination with feces and urine. In addition, there is an increased risk of parasite egg ingestion.

Forage to concentrate ratios

Information for this composite table (Table 10) has been provided from a number of horse feed manufacturers in the UK in order to obtain a representative picture (Baileys, Dodson & Horrell, Spillers and Winergy).

Table 10. A guide to the amounts of forage and concentrates fed in the UK according to a number of nutritionists.

This is only a very approximate guide, as the actual ratios fed to a horse will vary according to a number of factors including the individual horse, the relative work intensity for that individual and therefore energy requirement, the rider/owner's requirement with respect to body condition and type of riding, the forage that is available, the actual forage and the concentrates being fed, the breed, the environmental conditions and management practices. With respect to young stock, growth rates, sales preparation, race training, and work intensity must also be considered.

Type of Animal	% of Body weight on an as fed basis Figures in bold italic refer to ratios of feed		
	Forage	Concentrates	Total
Mature horses at maintenance	2	0	2
	1.5 -1.8	0.2 - 0.5	2
	1.5 -2	0 - 0.5	1.75 - 2
	95	*5*	
Mares - late gestation	1.2	0.8	2
	1.4 -1.8	0.26 - 0.6	
	1.5-2.0	0.25-0.75	2 - 2.25
	non-TB 80	*non-TB 20*	
	TB 40	*TB 60*	
Mares - early lactation	1.5	1	2.5
	1.5	1.5	3
	1.5 - 1.75	1 - 1.5	2.5 - 3
	non-TB 80	*non-TB 20*	
	TB 40	*TB 60*	
Mares - late lactation	1.4	0.6	2
	1.5 - 1.75	0.75 - 1.0	2 - 2.5
	1.25 - 1.75	0.75 - 1.25	2 - 2.5
	non-TB 90	*non-TB 10*	
	TB 50	*TB 50*	
Light work	1.4	0.6	2
	1.5 - 1.8	0.2 - 0.6	2
	1.5 - 2.0	0.1 - 0.75	1.75 - 2.25
	90	*10*	
Moderate work	1.5	1	2.5
	1	1	2
	1 - 1.5	0.5 - 1.0	2 - 2.5
	80	*20*	
Intense work	1.25	1.25	2.5
	1 - 1.25	1.25 - 1.5	2.5
	0.75 - 1.5	1.25 - 1.5	2 - 2.75
	50	*50*	
Dressage	-	-	2.5
	1 - 1.5	0.5 - 1	2
Racing - flat	0.6	1.4 - 1.9	2 - 2.5
	1-1.25	1.25 - 1.5	2 - 2.5
	40	*60*	
Racing National Hunt	0.6	1.4 - 1.9	2.5
	1.25	1.25	2 - 2.5
	50	*50*	
Show jumping	1.5 - 1.75	0.75 - 1	2.5
	1.2	0.8	2
	80	*20*	
Eventing	1.5	1	2.5
	1.25 - 1.5	1 - 1.25	2 - 2.5
	70	*30*	
Hunting	1.5 - 1.75	0.75 - 1.0	2.5
	1 - 1.2	0.8 - 1	2
	50	*50*	
Weanling - 6 months (TB)	0.75	1.75	2.5
	40	*60*	
Weanling - 6 months (non-TB)	1	1.5	2-2.5
	50	*50*	
Yearling - 12 months (TB)	1.2	0.8	2
	1 - 1.2	0.8 - 1.0	2
	50	*50*	
Yearling - 12 months (non-TB)	1.2	0.8	2
	1 - 1.2	0.8 - 1.0	2
	1.25 - 1.5	0.5 - 1.0	2 - 2.25
	70	*30*	
Long Yearling (TB)	1.2	0.8	2
	1.25 - 1.5	1 - 1.25	2 - 2.5
	50	*50*	
Long Yearling (non-TB)	1.2	0.8	2
	1.4	0.6	2
	70	*30*	
Two-year-old (TB)	1.4	0.6	2
	1.4	0.6	2
	1.0 - 2.0	0.5 - 1.0	2 - 2.5
	40 Racing	*60*	
Two-year-old (non-TB)	1.4	0.6	2
	1.5	0.5	2
	1.25 - 1.75	0.5 - 0.75	2 - 2.25
	80	*20*	

Frequency of non-forage/roughage feeding

Most owners or managers feed at least once or twice a day when concentrate or hard feed is given. For animals with high energy requirements (such as horses in some racing, hunting or eventing yards), feed may be divided into three or more meals a day in order to provide sufficient energy while minimizing the risk of digestive disturbance. In one survey (Anon-Waltham; 1997; unpublished data) in which 78% of the horses were in light work, 14% were fed once a day, 80% were fed twice a day and 6% were fed three times a day.

Examples of Feeding Practices

A survey of a few top horses in a number of disciplines suggested that these animals, apart from the advanced dressage horses, were being fed less energy than the NRC requirements (Hollands, 1995). Only the advanced dressage and novice event horses were being fed slightly more than the NRC requirements for protein, and all were being fed far more than the NRC requirements of calcium, magnesium, phosphorus and potassium. Sodium intakes tended to be at or slightly lower than NRC recommendations, although the survey did not include the effects of any additional vitamin or mineral supplementation. Copper and iron levels tended to be higher then recommended. Iodine and cobalt levels were far higher than NRC stated levels as were vitamin A and D levels. Vitamin E levels prior to supplementation were approximately the same as NRC recommendations as was manganese, zinc and selenium content in most cases.

The results of a survey carried out on the feeding practices of a number of studs in Ireland are given in Tables 11-13. The mean nutrient intakes (using estimated pasture intakes) were compared against calculated NRC requirements for a pregnant mare, a ten-month-old foal and a 17-month-old yearling. For the mare and foal, the nutrient intake exceeded requirements in all cases. In the weanling, mean intakes of protein and lysine were deficient at the lower dry matter intake and marginal at the higher intake.

Table 11. Concentrate feedstuffs used on a number of Irish stud farms and percentage of stud farms on which feed was routinely used (O'Donohue et al., 1995).

Type of concentrate feedstuff	Pregnant mare	Short yearling	Long yearling
	% of farms n = 46	% of farms n = 45	% of farms n = 44
Straights			
Barley	6.5	6.7	2.3
Dried sugar beet pulp	21.7	20	11.4
Maize flaked	4.3	4.4	6.8
Molasses	17.4	17.8	20.5
Oats	84.8	82.2	72.7
Wheat bran	47.8	42.2	47.7
Grass meal	2.2	2.2	2.3
Linseed/flaxseed	2.2	4.4	2.3
Milk	0	0	2.3
Soya bean meal	28.3	20	20.5
Commercially compounded feeds			
Complete feeds	0	0	2.3
Oat replacer feeds	60.1	60	78.6
Balancers/conditioner	21.7	28.9	36.4

Pregnant mare expected foaling date April 1, as fed February 1.
Short yearling colt born April 1, as fed 10 months of age on February 1.
Long yearling colt born April 1, as fed 17 months of age on September 1.

Table 12. Mean (+/- SEM) weight of concentrate feed fed to a pregnant mare (expected foaling date April 1) or a lactating mare (foaled January 15) based on questionnaire results from a number of Irish studs (O'Donohue et al., 1995).

Month	Pregnant Mean +/- SEM (kg/day) based on air-dried feed	Lactating Mean +/- SEM (kg/day) based on air-dried feed
August	0 +/- 0	-
September	0.3 +/- 0.19	-
October	0.9 +/- 0.28	-
November	3.5 +/- 0.42	-
December	5.0 +/- 0.37	-
January	5.6 +/- 0.32	7.7 +/- 0.41
February	6.0 +/- 0.29	7.7 +/- 0.4
March	6.1 +/- 0.29	7.5 +/- 0.4

Table 13. Mean (+/- SEM) weight of concentrate feed fed to a weaned foal
or to a yearling (colt born on April 1) based on questionnaire results from
a number of Irish studs (O'Donohue et al., 1995).

	Month	Mean +/- SEM (kg/day) based on air-dried feed	Number of stud farms from which feed weights were collected. Includes farms which did not feed concentrates.
Weaned foal	August	2.5 +/- 0.3	6
	September	2.1 +/- 0.22	32
	October	2.6 +/- 0.18	42
	November	3.2 +/- 0.17	44
	December	3.7 +/- 0.17	44
Yearling	January	4.0 +/- 0.21	43
	February	4.3 +/- 0.22	45
	March	4.2 +/- 0.25	41
	April	3.4 +/- 0.28	41
	May	1.4 +/- 0.27	44
	June	1.2 +/- 0.22	44
	July	2.4 +/- 0.36	42
	August	5.7 +/- 0.42	41
	September	7.9 +/- 0.46	45
	October	8.5 +/- 0.41	42

Comparison with Germany

There are approximately 1.5 million riders and 600,000 horses in Germany. In 1994 the German Sports Club registered 665,000 members (35% male, 65% female) in German riding clubs, and it has been estimated that the number of non-organized riders is similar to this. In addition, there are around 800,000 hobby riders taking part in vaulting, riding school activities and pleasure riding. The popularity of western and Iberian riding continues to increase.

The number of Thoroughbred broodmares in Germany was about 2,500 in 1990 with approximately 2,400 races held on a total of 49 racetracks. In 1994, the estimated number of Thoroughbred broodmares was 7,000 and there were close to 3,000 races held. Unlike the UK and Ireland, trotting races are very popular in Germany. The most important breed in Germany is the German riding horse. The numbers of these have also increased. There were around 53,000 registered broodmares and 2,000 stallions in 1990 with around 80,000 and 3,000 respectively in 1994. All German riding horse breeding stallions must be licensed and have undertaken and passed a performance test at the age of 3 or 4 years.

A fairly recent survey of over 200 people, predominantly private horse owners, representing 735 horses was carried out in Germany. This survey involved participants from all over West Germany, although the majority came from the northern areas. Some of the findings are shown in Tables 14 and 15. The survey covered a wide number of breeds, ranging from the Shetland to the Hanovarian. As in the UK, most of the riders were female. In the UK survey, riding was most popular for women between 16-24 years of age, while in the German survey around 78% of the female participants were between 21 and 40. A slight increase in the number of male riders in the UK had been seen between 1992 and 1994, mainly due to an increase in riders between the ages of 25 and 34. In Germany, around 70% of the male participants were between 31 and 50 years of age.

Table 14. How horses were kept according to a recent survey in West Germany.

Livery	22%
At a riding club	13%
Individual stabling	54%
Misc. (including agricultural farms)	11%

Table 15. Age and sex distribution of horse owners surveyed recently in West Germany.

	Total Number	
Age	Male	Female
Up to 20	2	18
21-30	6	82
31-40	17	52
41-50	20	18
50 and up	8	3

In a survey completed in Germany, it was estimated that around 50% of horses were kept in privately owned small stables. Around one-third were kept at riding clubs and pension stables, 10% at farmers and around 15% at neighbors or friends. Although riding clubs (where a number of animals are kept at one establishment and managed by one individual) do appear to be much more common in Germany than in the UK, it was still estimated that only 11% of horses resided in stables with more than 30 horses. There are around 6,000 riding clubs in Germany. Horses are commonly kept in stalls and not turned out to pasture unless they are mares and foals or young stock under the age of three. The young stock often live in group stalls during the winter and are let out to pasture if the ground is suitable or turned loose into the indoor arena for exercise. Horses in competition training tend to be kept inside and take most, if not all, of their exercise in the indoor arena, although many barns or yards, especially if run by the top riders and trainers, do also have a small outdoor sand arena. Loose schooling in the indoor arena is quite common especially with the young stock. Deep littering is perhaps more commonly practiced in Germany than in the UK and straw is therefore the most common bedding material used, although shavings are becoming increasingly popular.

According to one German survey, about 60% of owners were responsible for feeding horses, but only around 15% of those owners with horses in riding clubs or pension stables have this responsibility. This is similar to UK statistics.

In Germany most of the feedstuffs seem to be bought via agricultural trading companies, although veterinarians are a fairly common source especially for the speciality feeds.

Oats tend to be the major cereal fed to horses in Germany. It has been estimated that oats have around 19% of the feed market share, roughages around 43% and compound manufactured feeds around 35%. Of the processed feeds, pellets have been suggested to have the largest share of the market value (see Table 16 for a comparison with the estimated expenditure in the UK). The rule of thumb for feeding, which is often quoted, is 1 kg oats and 1 kg of roughage per 100 kg body weight. In the larger liveries, riding clubs and stables, oats may be kept and distributed via silos rather than individual bags. Straw is more commonly fed to horses (especially warmbloods) often as a means of lowering the nutrient intake whilst still providing bulk and fiber. Horses in hard competition are commonly fed a competition coarse mix or pellet in addition to the oats. Alfalfa and chaff are fairly uncommon feedstuffs in Germany. The use of salt blocks as well as the addition of salt to the feed appears to be far less commonly practiced in Germany than in the UK.

Table 16. Distribution according to product type according to a UK survey (Anon- BETA, 1999).

Feeds	Estimated direct expenditure per year (£m)
Manufactured feeds	165
Oats and other "straights"	25
Vitamins, minerals and other	30
Hay and/or haylage	110

A guide to the recommended roughage to concentrate feeding ratios is given in Table 17. Concentrate feeds in Germany can be taken to mean vitamin and mineral supplements or similar feedstuffs rather than energy providing, non-forage feedstuffs as referred to in this table.

Acknowledgement

The author thanks the various discipline governing bodies for information supplied by them for Table 2 and the various providers of the Anon survey results.

Thanks are particlarly given to the various nutritionists working for a number of major UK feed manufacturers who assisted with the compilation of Tables 10 and 17, including Liz Bulbrook of Baileys, Teresa Hollands of Dodson & Horrell, Ruth Bishop of Spillers plus Annette Sommerhoff and Linda Mars of Winergy.

Table 17. A guide to the amounts of forage and concentrates fed in Germany.

This is only a very approximate guide, as the actual ratios fed to a horse will vary according to a number of factors including the individual horse, the relative work intensity for that individual and therefore energy requirement, the rider/owner's requirement with respect to body condition and type of riding, the forage that is available, the actual forage and the concentrates being fed, the breed, the environmental conditions and management practices. With respect to young stock, growth rates, sales preparation, race training, and work intensity must also be considered.

Type of Animal	% of Body weight on an as fed basis		
	Forage	Concentrates	Total
Mature horses at maintenance	1.5	0.25	1.75
Mares - late gestation	1-1.5	0.75-1.25	1.75 - 2.75
Mares - early lactation	1-1.5	0.8 - 1.5	1.8 - 3.0
Mares - late lactation	1-1.5	0.75 - 1.25	1.75 - 2.75
Light work	1-1.5	0.5- 0.75	Up to 2.25
Moderate work	1- 1.5	0.5 - 1.0	Up to 2.5
Intense work	1-1.5	0.7 - 1.75	Up to 3.25
Racing - flat	0.8 - 1.25	1.25 -1.5	2- 2.75
Weanling - 6 months (non-TB)	1	1.0 - 2.0	2.0 - 3.0
Yearling - 12 months (non-TB)	1-1.5	1.0	2-2.5
Long Yearling - (non-TB)	1-1.5	0.5 - 0.8	1.5-2.3
Two-year-old - (non-TB)	1.5 -2.0	0.5-1.0	2 -3.0* depends if in work

References

Anon (1997) The Agricultural and Horticultural Census conducted annually by the Ministry of Agriculture, Fisheries and Food.

Anon-BETA (1996) Survey undertaken by the Produce Studies Group.

Anon-BETA (1999) Survey undertaken by the Produce Studies Group.

Anon-Equestrian (1995) Survey undertaken for the IPC Magazines Ltd.

Anon-Eventing (1996) Survey undertaken for the IPC Magazines Ltd.

Anon-Gallup (1994) Gallup Omnibus Survey.

Anon-Horse (1997) Survey undertaken for IPC Magazines Ltd.

Brown, J.H. and Powell Smith, V. (1994) Horse and Stable Management, Blackwell, London. p 16.

Harris, P.A. (1997a) Feeds and feeding in the United Kingdom. In: Current Therapy in Equine Medicine, Ed: N.E. Robinson, W.B. Saunders, Washington. pp 121-124.

Harris, P.A. (1997b) Energy sources and requirements of the exercising horse. Annual Reviews of Nutrition 17. p 185-210.

Harris, P.A. (1998) Developments in equine nutrition: Comparing the beginning and end of this century. J. Nutr. 128, 2698S-2703S.

Hintz, H.F. (1997) Hay racks vs feeding hay on the stall floor. Equine Pract. 19, 5-6.

Hollands, T. (1995) Feeding the three-day event and dressage horse. In: Recent Advances in Equine Nutrition. Proceedings from the KER Equine Nutrition Conference. pp 163–177.

McCarthy, D.D. (1975) Survey of Thoroughbred Stud Farms in Ireland. An Foras Taluntais, Dublin. pp 1-145.

Mellor, D.J., Love, S., Gettinby, G. and Reid, S.W.J (1999) Demographic characteristics of the equine population of northern Britain. Vet. Rec. 299-304.

Mellor, D.J., Love, S., Reeves, M.J., Gettinby, G. and Reid, S.W.J. (1997) A demo graphic approach to equine disease in the northern UK through a sentinel practice network. Epidemiol. Sante Anim. 31-32.

O'Donohue, D.D, Smith F.H. and Strickland, K.L. (1995) Feed intakes on Irish Thoroughbred stud farms. In: Proceedings of the Equine Nutrition and Physiology Society, California. pp 229-234.

Townson, J., Dodd, V.A. and Brophy, P.O. (1995) A survey and assessment of racehorse stables in Ireland. Irish vet. J. 48, 364-372.

Warr, E.M. and Petch, J.L. (1992) Effects of soaking hay on its nutritional quality. Equine vet. Educ. 5, 169-171.

FEEDING PRACTICES IN THE UNITED STATES

STEPHEN DUREN

Kentucky Equine Research, Inc., Versailles, KY

Introduction

The population of horses in the United States as of January 1, 1999 was estimated at 5.32 million. Represented in these 5.32 million are horses of many different sizes and breeds and intended uses. From a practical standpoint, the physiological functions of horses can be roughly divided into the following categories: maintenance, work (performance), growth, breeding stallions, pregnancy and lactation.

The Nutrient Requirements of Horses (NRC, 1989) lists nutrient requirements based on physiological function and mature body weight. It is clear that horses with different physiological functions require nutrients at different levels in the diet. For example, the daily nutrient requirements of growing horses are certainly different from horses performing light exercise. On the other hand, horses with identical physiologic function and body weight would share a common daily nutrient requirement. Therefore, an 1100 lb (500 kg) mare in her tenth month of pregnancy would have the same daily nutrient requirements whether the mare lived in New Jersey or California. Although the nutrient requirements are the same for horses living in different areas of the United States, the feedstuffs used to satisfy these requirements and the feeding strategies are quite different.

Different Regions - Different Ingredients

Climatic conditions dictate which feed ingredients are commonly grown in different areas of the United States. Specifically, temperature, humidity, annual rainfall, and soil conditions influence crop production. In many animal production systems, the ability to grow an ingredient is the primary factor in determining its inclusion in a diet. Feeding horses in the United States is somewhat different from other animal production units in that most horse operations do not grow their own feeds. The one notable exception to this statement is pasture. Grain concentrate and stored forage are routinely purchased from suppliers that may have obtained the ingredients out of the immediate geographical area. For example, it is now common to find hay being fed in the East that was grown in the West or grain concentrates blended in the Midwest that are fed on the West Coast. Despite the ability to transport specific ingredients anywhere in the United States, many regional differences do exist in ingredient availability. These regional differences in feed ingredients influence price of the horse feed and ultimately influence which ingredients are commonly used in horse diets. The following is a brief discussion of some notable differences in ingredients used in different areas of the United States.

Northeast United States

Climatic conditions in this region favor the production of grass, both as pasture and as stored hay. Unfortunately, many horse operations are starved for space and pasture availability is limited or nonexistent. Common grasses utilized for hay production include timothy and orchardgrass. Legume plants, including alfalfa and clover, can also be grown in this region. Most of the pure alfalfa hay finds its way into diets of ruminant animals with some mixed legume/grass hay being utilized for horse feed. The vast majority of hay being fed to horses in this region is some type of grass hay. Table 1 presents an average nutrient profile for hay being fed in the region. Stage of maturity at harvest significantly influences nutrient profile of the hay. Therefore, the values presented in Table 1 can vary from year to year based on weather conditions that may delay harvesting. In years when hay quality suffers due to a poor growing season, it is common for horsemen to supplement their hay program with alfalfa pellets or cubes from the western United States or from Canada.

Table 1. Nutrient profile of northeast hay (dry matter basis)

Type of Hay	Crude Protein (%)	Nutrient NDF (%)	Ca (%)	P (%)	Cu (ppm)	Zn (ppm)
Legume Hay	17.5	46	1.25	0.25	9	22
Mixed - Mostly Legume	15.5	52	1.10	0.25	9	24
Mixed - Mostly Grass	11.5	61	0.74	0.23	8	25
Grass	10.5	63	0.58	0.22	9	27

The type and amount of grain fed in the Northeast varies greatly. Of the straight grains fed to horses, oats is the traditional favorite. Small amounts of other straight grains or grain byproducts are also fed and include corn, wheat bran, flaxseed, soybean meal and sugar beet pulp. Most horse owners have abandoned the practice of mixing straight grain on the farm to form their horses' grain diet. Many choose to feed commercial grain concentrates that are fortified with protein, vitamins and minerals. The notable exception to this trend is the Thoroughbred racing industry where mixing of many different ingredients is still popular. The physical form of the grain concentrates being fed is a mixture of textured sweet feeds, pellets and extruded products. As a rule, the amount of molasses being used on textured feeds in the Northeast (7-10%) is greater than commonly used in the West (4-7%). Complete feeds containing the forage, grain and vitamin/mineral portions of the diet are popular with Northeast horse owners. Unfortunately, most horse owners ignore the feeding instructions and feed these products at low levels that do not provide an adequate diet.

Southeast United States

High heat, high humidity and a long growing season are common in the Southeast. This type of climate favors growth of tropical and subtropical grasses. Grasses such as bahiagrass, coastal bermudagrass and pangolagrass make up a large percentage of the forage consumed by many horses in this area of the country. Southeastern states that are situated farther north have many varieties of cool-season grasses; the most common are fescue and bluegrass. High humidity is not conducive to hay production since it is difficult to get the forage plants properly dried prior to baling. Therefore, grazing is the primary method of utilizing natural forages. Grazing can normally be sustained from March through November during most years. During the time of year when stored forage must be fed, coastal bermudagrass and forage imported from other areas of the country are fed. Alfalfa production is generally considered to be low in the Southeast. However, peanut hay, another legume plant, is harvested and can be fed to horses. Table 2 presents an average nutrient profile of forages commonly fed in the Southeast. Again, growing conditions and stage of maturity at harvest will influence the nutrient profile of the forage.

Table 2. Nutrient profile of southeast forage (dry matter basis)

Type of Plant	Crude Protein (%)	Nutrient NDF (%)	Ca (%)	P (%)	Cu (ppm)	Zn (ppm)
Bahiagrass	11	75	0.29	0.25	10	34
Coastal Bermudagrass	12	73	0.36	0.27	6	25
Peanut Hay	10.5	61	1.23	0.16		
Bluegrass Hay	9	63	0.26	0.25	9	27
Fescue Hay	12	65	0.43	0.32	28	35

The Southeast has grains and grain byproducts that are unique. Sorghum, cottonseed meal, cottonseed hulls, cane molasses, rice bran and peanut hulls are some products often used in the Southeast. Sorghum can be utilized in horse feed once it is processed to increase digestibility. Unfortunately, palatability problems have been reported with the use of sorghum in horse feed. Cottonseed meal can be utilized as a protein source. It is lower in lysine than soybean meal and can contain gossypol, a substance that can interfere with the absorption of other minerals. If cottonseed meal is utilized in the diet, lysine should be added and care should be taken to insure it is a variety with a low gossypol content. Cottonseed hulls and peanut hulls are two byproducts that contain high levels of fiber. These fiber sources have a high lignin content and are low in digestibility. Rice bran is a byproduct of the rice milling industry. When heat stabilized immediately after separation from the rice kernel, rice bran is a high fat energy source for horses. The popularity of feeding stabilized rice bran to horses has caught on throughout the United States.

Midwest United States

The Midwest typically has a longer growing season than the extreme Northeast and less rain and humidity compared to the Southeast. Different areas in the Midwest rely on irrigation to grow crops, while other areas have enough rainfall to sustain crop production. These differences in growing conditions translate into many different types of forages and grains that can be grown in the region. In the upper Midwest, the predominant forage is grass pasture or grass hay. Farther south, heat and lack of rainfall make it more difficult to grow cool season grasses for utilization by horses. In these areas alfalfa can be grown with the help of irrigation systems. Throughout the Midwest, pastures consisting of native grasses are common. Some of the common grass types include brome, reed canarygrass, orchardgrass, ryegrass, and crested wheatgrass. It is not uncommon for these native pastures to have a low protein and energy content due to increased plant maturity at the time of grazing. Legume plants including alfalfa, clover, and birdsfoot trefoil are common and can be grown in the Midwest.

Corn is the king of grain in the Midwest. Corn is a high energy, low fiber, low protein grain that is common in horse diets. The corn kernel is typically processed prior to inclusion into horse feed to maximize small intestine digestion of starch. From time to time concern is expressed over the potential for mycotoxins in corn. However, the incidence of moldy corn poisoning in horses is limited compared to the number of horses who eat at least a portion of their grain diet as corn on a daily basis. The level of mycotoxins is elevated in corn that has been stressed by drought or excessive moisture. Further, the incidence of mycotoxins is more common in corn screenings from damaged corn kernels than from high quality, whole kernel corn. The other type of grain produced in abundance in the Midwest is the soybean. Soybean meal is the most common protein source used in horse diets in the United States. Soybean meal contains an excellent amino acid profile and is quite palatable. Soybean hulls are an excellent source of highly digestible fiber and also are used in horse diets. Finally, soybean oil is a common source of dietary fat in horse diets. Leaving all the components of the soybean together is also common by utilizing full-fat, roasted soybeans in horse diets.

Western United States

Many inland areas of the western United States depend on irrigation to grow crops. Irrigation allows the West to grow extremely high quality hay. Drying of hay is made simple with high temperatures and low humidity. Hay grown in the Western United States is exported to many horse production units throughout the country and the world. Of the hay grown in the West, alfalfa is the most common. Three cuttings of alfalfa per year are the norm in Washington, four cuttings are common in Idaho and up to seven or eight cuttings are possible in the desert Southwest. Most of the highest quality alfalfa finds its way into the dairy market, but significant amounts of alfalfa hay are fed to horses. Other types of hay that can be grown in the West include timothy, oat, orchardgrass and a variety of native grasses. One particular area in Washington is famous for high quality timothy hay, while several areas in California produce large amounts of oat hay. Oats are used as a cover crop and the entire plant is harvested in the late milk or dough stage for production of hay. Another type of hay that is very popular in the western United States is mixed (alfalfa/grass) hay. The most popular grass to mix with alfalfa is orchardgrass. Table 3 presents an average nutrient profile for forage fed in the western United States.

Table 3. Nutrient profile of western forage (dry matter basis)

Type of Hay	Crude Protein (%)	Nutrient NDF (%)	Ca (%)	P (%)	Cu (ppm)	Zn (ppm)
Alfalfa	21.0	34	1.40	0.21	9	20
Alfalfa/Orchardgrass	15.0	52	0.90	0.35	9	20
Timothy	11.0	61	0.51	0.30	7	25
Oat	9.5	63	0.32	0.25	6	22

The most predominant grains grown in the western United States are barley, wheat and canola. Barley is a high energy, moderate protein, low fiber grain. It has an energy value intermediate between corn and oats. Like corn, barley is processed prior to being fed to horses to maximize the digestion of starch in the small intestine. In the West, barley is utilized in horse feed to the same degree corn is utilized in horse feeds in the Midwest. The appearance of corn in horse feeds in the West is less than in the Midwest, Southeast or Northeast. Wheat is also grown in large amounts in the West. Processed wheat in the form of wheat bran and wheat middlings are common components of pelleted grain concentrates for horses. Due to availability wheat is often a low cost energy source for equine diets. Canola meal is a protein source that is unique to the West. Canola meal typically is lower in protein quantity and quality compared with soybean meal. However, a combination of canola and soybean meals can be used as an economical protein source for horses.

Feeding Strategy

Just as the ingredients commonly fed to horses differ from region to region in the United States, so do the feeding strategies. Guidelines for both the total amount of feed (hay and grain) consumed by horses and the ratio of forage to grain in the diet are provided by the Nutrient Requirements of Horses (NRC, 1989). The amount of feed a horse can reasonably consume in a day is a finite amount. The estimates set forth in the NRC (1989) are reasonable (Table 4). However, the ratio of forage to concentrate can vary significantly depending on the availability, palatability and, most importantly, energy content of the forage. For example, if one compared an 1100 lb (500 kg) performance horse in light work housed in Virginia with the same horse housed in California very different strategies can be used to feed the same horse. The NRC (1989) suggests in its Table 5-4 a total feed intake (90% DM) of 1.5 to 2.5% of body weight. Further, Table 5-2a in the NRC (1989) publication suggests a hay to concentrate ratio of 65% hay to 35% grain. This ratio works extremely well (Figure 1) when the horse is eating approximately 2% of its body weight as timothy hay and a grain concentrate formulated for use in Virginia.

Table 4. Expected feed intake and diet proportions for different classes of horses as established by NRC, 1989.

Class of Horse	Diet Proportion		Feed Intake
	Hay	Grain	% of B.W.
Maintenance	100	0	1.5 - 2.0
Stallions	70	30	1.5 - 2.0
Pregnant Mares			
9th Month	80	20	1.5 - 2.0
10th Month	80	20	1.5 - 2.0
11th Month	70	30	1.5 - 2.0
Lactating Mares			
Early	50	50	2.0 - 3.0
Late	65	35	2.0 - 2.5
Working			
Light	65	35	1.5 - 2.5
Moderate	50	50	1.75 - 2.5
Intense	35	65	2.0 - 3.0
Growing			
Weanling	30	70	2.0 - 3.5
Yearling	40	60	2.0 - 3.0
2-yr-old	65	35	1.75 - 2.5

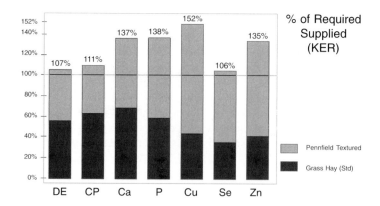

Figure 1. Diet for 1100 lb (500 kg) horse performing light exercise. Diet consists of 14.5 lb of timothy hay and 7 lb of Pennfield Textured Sweet Feed.

A diet for the same horse in California being fed alfalfa hay likely would depend more on the contribution of the high-energy alfalfa and less on grain concentrates. Figure 2 is an example of a diet featuring the horse eating 1.8% of his body weight in feed (90% DM) with the diet consisting of 83% forage and 17% concentrate.

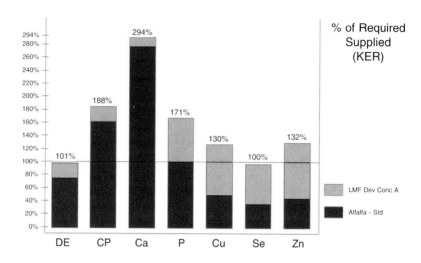

Figure 2. Diet for 1100 lb (500 kg) horse performing light exercise. Diet consists of 16.5 lb of alfalfa hay and 3.5 lb of LMF Textured Sweet Feed.

An even more likely diet scenario (Figure 3) would be the California horse eating a total of 2% of its body weight, consisting of 95% alfalfa and 5% supplement pellet. This diet consists of 20 lb of alfalfa hay and 1 lb of supplement pellet. The supplement pellet is added only to adjust vitamin and mineral levels in the diet.

These same western United States feeding scenarios in which the use of forage is maximized while grain intake is minimized also are common for pregnant and lactating broodmares and young growing horses. Table 5 reflects realistic diet proportions and daily feed intakes for horses consuming legume (alfalfa) forages. As the performance horse example illustrated, the availability of different feed ingredients in different regions of the country and underlying differences in feeding strategies make it difficult to utilize a fixed forage to grain ratio. The use of high energy forage in the western United States dictates a necessary change when considering an appropriate hay to grain ratio in a horse's diet.

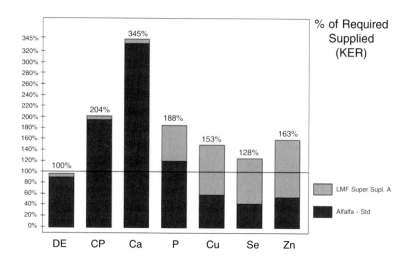

Figure 3. Diet for 1100 lb (500 kg) horse performing light exercise. Diet consists of 20 lb of alfalfa hay and 1 lb of LMF Supplement Pellet.

Table 5. Expected feed intake and diet proportions for different classes of horses consuming legume (alfalfa) forages.

Class of Horse	Diet Proportion		Feed Intake
	Hay	Grain	% of B.W.
Maintenance	95	5	1.5 - 2.0
Stallions	80	20	2.0 - 2.5
Pregnant Mares			
9th Month	95	5	2.0 - 2.5
10th Month	85	15	2.0 - 2.5
11th Month	80	20	2.0 - 2.5
Lactating Mares			
Early	80	20	2.5 - 3.0
Late	85	15	2.5 - 3.0
Working			
Light	85	15	2.0 - 2.5
Moderate	80	20	2.25 - 3.0
Intense	70	30	3.0 - 3.5
Growing			
Weanling	60	40	3.0 - 3.5
Yearling	70	30	3.0 - 3.5
2-yr-old	70	30	2.25 - 3.0

FEEDING PRACTICES IN AUSTRALASIA

PETER J. HUNTINGTON

Kentucky Equine Research, Inc., Versailles, KY

Introduction

Australia and New Zealand are great horse producing countries. Australia has the second largest horse population in the western world and New Zealand horses have long plundered Australia's richest races. There are about 36,000 active Thoroughbred racehorses in Australia competing at 400 racetracks. Over 18,000 Thoroughbred foals are born each year while there are about half that number of Standardbreds. Other popular breeds include Arabians, Quarter Horses, warmbloods and the various pony breeds. Equestrian sports are very popular and Australia has won the team three-day event at the last two Olympics Games. Horses are bred and trained in a variety of conditions and climatic regions, and can be kept outdoors all year round. In the North Island of New Zealand, pastures remain green and of high quality nearly year-round, and breeding of Thoroughbred and Standardbred horses is a very popular hobby and business. New Zealanders have excelled at three-day eventing in recent years and the New Zealand bred horse is renowned for versatility and toughness. This paper describes some of the common feeding practices in Australia and New Zealand, particularly where they differ from those used in North America or Europe. If further information is required, a practical 240 page review of feeding horses in Australia has just been published (Kohnke et al., 1999).

Forages

Pastures

Snow falls very rarely in areas where horses are kept, so horses can graze year-round. Most grazing horses are not stabled at night and have access to pasture all day. Overstocking, understocking and poor pasture management practices are common, so there is a great variation in pasture quality and quantity. In recent years, research and extension activities have been directed to getting more out of pastures and reducing the costs of supplementary feeding. In southern temperate areas, heat and lack of rain cause pastures to dry off over summer and autumn, which creates a need for supplementary feeding of horses at this time of year. Irrigation is used in some areas to maintain growth and green pasture during dry periods. In northern parts of the country, subtropical or tropical conditions prevail so the growing season occurs throughout summer. In northern Australia, introduced tropical grasses such as kikuyu, buffel and seteria are common. These grasses have high oxalate levels that interfere with calcium digestibility and can lead to the development of nutritional secondary hyperparathyroidism (big head) and various bone disorders. In areas of North Queensland, pasture species such as *Morinda* spp. or *Neptunia amplexicaulis* can accumulate selenium and cause

461

selenium toxicity. Australia has an abundance of poisonous plants with regional distributions, although most of the areas that are intensively stocked with horses are relatively free of poisonous plants.

The North Island of New Zealand is a warmer, higher rainfall area and is host to the main Thoroughbred breeding studs. The South Island of New Zealand has traditionally been the main breeding area for Standardbred horses. Pastures are mainly ryegrass and clover on both islands. South Island pastures may experience isolated snowfalls from May through September. Poisonous plants are not a concern in New Zealand horse pastures. Pastures rarely dry off over summer in the North Island; however, they do dry off in the South Island. Mares are often managed without supplementary grain as the high quality pastures supply enough energy and protein year-round, but mineral requirements are not met.

Hay and Chaff

Australian horses are fed forage in the form of hay, chaff, pellets and cubes. Lucerne (alfalfa) is the most common hay fed to horses, followed by grass (meadow) hay and oaten hay. Lucerne hay is grown in all states and is widely available, whereas grass hay is made in southern temperate areas. Lucerne and grass hay is usually packed in small bales and the unit of feeding is a biscuit (flake). Unfortunately, wide variation in the density of hay and the width of a biscuit means that the weight of a particular biscuit can vary from one kg to three kg. Under those circumstances, nutritional evaluation can be inaccurate unless the weight is actually measured. The value of grass hay can vary widely according to the composition of the grasses, time of haymaking and other factors. A high clover content will increase the nutrient content of the hay and good quality clover hay can approach lucerne hay in nutritious value and will be more palatable. However, it is hard to get clover hay year-round. Some studs use hayracks in the paddock for ad lib hay feeding and others use large round bales for ad lib feeding of hay to mares. In northern Australia the only hay available is lucerne and this can create problems for some performance horses with excess protein in the diet. The climatic conditions at the time of haymaking are usually favorable and Australian hay is generally free of dust, molds and other contaminants. Prices for hay vary substantially according to its type, quality, balance between supply and demand, and transport costs. Oaten hay is fed less commonly because production is lower and storage is more difficult. It is fed in sheaves or bales. Because hay quality is generally high, there is almost no use of haylage for feeding horses, although it is a common dairy feed.

Australian horses are fed large quantities of chopped hay or chaff. Chaff is made in commercial mills that produce 40 kg bags or it can be produced on the farm using a small-scale chaff cutter. Many horses are fed a mix of oaten or wheaten chaff and lucerne chaff with their grain. Immature oats or wheat crops are cut for chaff several weeks before harvest. The timing varies each season but is described as the "early heading stage" or no more than 7 days after flowering. At this stage you can squeeze "milk" out of the head of the grain and the nutrient content of the stalk is higher than at harvest. It is made into sheaves, which are stocked for drying in the paddock over several weeks, then cut into 5 mm segments. Most commercial chaff is "steam cleaned" to remove dust and allow easier cutting. Some chaff is rough cut and contains portions of longer fiber. High quality chaff should not contain any formed grain, but chaff often does contain grain.

The principal advantage of chaff is that it can be mixed in with the concentrate portion of the feed and the horse consumes forage with the concentrate. This can slow down the intake of concentrate and prevent starch overload in the large intestine. Horse owners commonly overestimate the weight of chaff fed because it is very light and may weigh only 300g/2 liters for oaten or wheaten chaff and 250g/2 liters for lucerne chaff. There is considerable debate as to the merits of oaten chaff over wheaten chaff and vice versa.

Because chaff is so popular, use of forage cubes or pellets is not as common as in other countries. The use of lucerne pellets instead of chaff has increased recently on studs and spelling operations, where it can be stored in silos. This can reduce the cost and handling required compared to the use of chaff in a bag. Forage cubes have been developed only in recent times and have not made any significant impact on feeding practices. Unlike countries in the Northern Hemisphere, beet pulp and soy hulls are not available as alternative forms of digestible forage. However, alkali-treated sugarcane bagasse is used as a cheap fiber source in some commercial feeds made in North Queensland. In many Southeast Asian countries, all horse feed is imported. This makes forage very expensive, and there is a tendency to feed less than is desirable for digestive health. Compressed hay is imported, but bagged chaff is a more popular feed because it is easier to handle for shipping. In India fresh lucerne and very poor quality grass hay are the popular forages.

In New Zealand, lucerne and oaten chaff are the main forms of forage added to grain. Meadow and lucerne hay are fed in similar quantities to stabled performance horses and horses in the paddock. The haymaking season is shorter and more difficult in New Zealand, and therefore a great deal of care is needed to ensure hay is cured correctly. Because it is harder to make high quality grass hay in New Zealand, haylage is becoming popular and is made commercially or "on farm." Average intakes for performance horses would be 2.5 kg of meadow or lucerne hays with approximately 300 to 400 grams of oaten or lucerne chaff. Access to pasture by racehorses varies from nil to about 8 hours, if available. Most racing stables have access to pasture and would often cut about 6 kg of wet pasture for stabled horses.

Concentrates

Grains

Australians have traditionally fed more "straights" than premixed feeds, although this pattern is changing with the development of better quality feeds and recognition of the performance, convenience and value these products can offer. Oats are by far the most common grain fed to horses based on safety, price and the fact that there is no need for further processing. Oats are grown in many of the cereal growing areas of the country and are often stored in a silo for bulk handling. Australian oats are usually lower protein than many Northern Hemisphere varieties averaging under 9% CP (as fed). Corn or maize is higher priced than oats and cannot compete on a cost/MJ basis, but it is often necessary to increase the energy density of the ration. It is usually fed in amounts less than 2 kg and is most popular with racehorse trainers. Corn is fed cracked, steam flaked or extruded. Unfortunately, extruded corn is not always readily eaten by racehorses when they are under pressure and for that reason most trainers still feed cracked corn even though extruded corn is more digestible. Barley is perceived by many to be a "non-heating" feed and is fed either steam rolled, micronized, extruded or boiled, but again it is usually fed in small amounts. It is a popular feed in showing circles where it is often the first choice grain. Sorghum is an economical grain grown in Queensland but is not widely used. Triticale, a wheat/rye hybrid, is a recent addition to the grain menu for some Australian horses. It is reported to give good results by users, but there is no published information on its digestibility.

Bran has been a popular ingredient, particularly for combination in a wet feed (bran mash) with various supplements, but its use is now declining. Pollard is popular among showing circles with people wanting to put condition on horses without feeding extra grain. Rice pollard is also used by some owners for conditioning, due to its high fat content.

Commercial Feeds

The range and quality of commercial feeds available in Australia have increased dramatically in recent years. Feed mills produce pelleted, textured and extruded feeds, but there is still a substantial prejudice against pellets by horsemen. Until recently, nutritionists who did not have equine nutrition experience have inappropriately formulated many processed feeds. In addition, the mineral and vitamin premixes added to cheap feeds have usually been inadequate for use in high performance horses or fast growing horses without extra supplementation. Many of these feeds are not used according to instructions, but are diluted with other grains, which diminishes the value of the vitamin and mineral premixes in the feeds. Some popular feeds are designed to be used as concentrates at for example 2 kg/d, and the user adds his own grain according to needs. These feeds are a simple and economic method of feeding but still give the user control over grain and energy intakes. Some lower energy textured feeds contain lucerne or oaten chaff as a source of fiber. These are popular for horses in light work. Recently feeds for performance and growing horses have been produced containing high levels of fat and specialist feeds have been developed for horses predisposed to tying up or behavioral problems.

Protein Supplements

In racing horses, protein supplementation is usually provided by feeding lupins, sunflower seeds, tick beans and peas. These also contain a higher energy content than traditional grains, but they are fed in cups rather than dippers. Linseed meal and cottonseed meal were traditional protein supplements, but these have declined in use while canola meal use has increased. Lupins are becoming increasingly popular as they represent good value, and they are a palatable energy and protein source with high digestible fiber content. However, lupins have low levels of methionine and tryptophan. The contention that linseed meal or sunflower seeds makes horses "look better" is likely due to the high oil content, as both provide relatively poor quality protein. On the breeding farms the value of soybean meal is increasingly being recognized and it is fed as a soybean meal or full fat soybean meal. However, some breeders still use protein supplements with lower quality protein (e.g. sunflower seeds, cottonseed meal). Copra (coconut) meal has become a popular supplement in recent years.

Where horses have access to the good pasture in New Zealand, protein supplementation is not the major limiting factor to growth as good pastures vary between 16% and 28% protein. Commercial feeds tend to be formulated with lower protein content in New Zealand than similar target feeds in Australia, where pasture quality is not as good year-round. Protein supplements given to racehorses are mainly full fat soya, peas and sunflower seeds, even though sunflower seeds are very expensive.

Supplements

Australian horse trainers are major users of supplements even when using fortified prepared feeds. There are many brands and types of feed supplements marketed for horses and these promise a variety of benefits. There is a strict registration process but despite this, many products provide spurious claims or contain inadequate supplementary minerals or vitamins. Some formulations are quite dated, and they do not provide sufficient supplementary minerals and vitamins to meet requirements. It is common for a product to contain a particular nutrient, but not contain enough to make a meaningful contribution to balancing the horse's diet. Overuse of supplements is common and many horses are fed five or six supplement products including several sources of the same nutrient. Iron supplements are still common despite the fact that all diets contain adequate iron from natural sources. Most owners have little concept of the mineral and vitamin needs of their horses and labels are often difficult to understand; hence the choice of supplements becomes difficult. Most supplements are powders although liquid electrolytes, vitamins, buffers and iron supplements are available. Some products are presented in a pelleted form to enhance intake of the supplement. Many of these products contain protein supplements in addition to minerals and vitamins and these products are widely used in feeding horses on studs, but are less popular in racing circles.

Feeding Practices with Racehorses

A survey on the feeding of racehorses in Sydney was published some years ago (Southwood et al., 1993). The results emphasized the dependence on straight grains in the early 1990s as only one of every 50 stables fed a commercial sweet feed as the only concentrate feed. However, there has been a significant change in the last 10 years with a dramatic increase in the use of premixed feeds, especially sweet feeds, but no published figures are available. Some trainers feed a premixed feed as part of the grain mix, but still add other grains and supplements. The average amount of concentrate fed was 7.8 kg by Thoroughbred and 7.7 kg by Standardbred trainers in this survey, and my experience is that this ranges from 4-9 kg, depending on workload. Trainers in tropical areas tend not to work their horses as hard and feed them less total grain accordingly; however, I believe hot and humid conditions can increase energy requirements. While research has been conducted into the impact of cold on energy needs, nothing has been done on the impact of heat. Oats is still the major grain fed with corn being the next most popular grain in both home-mixed and premixed feed diets. Most trainers feed both grains, but some horses with behavioral or tying up problems are fed oat-free diets. There are several commercial feeds that are oat-free for horses prone to these problems. Extruded corn is available as a separate ingredient and as a component of several brands of feed; however, some horses leave the extruded corn, as it is not as palatable as cracked corn.

The average daily forage intake of a 3.3 kg (Thoroughbred) and 4.1 kg (Standardbred) in the Southwood survey is less than the recommended minimum intakes, particularly for Thoroughbreds. When part of the forage intake is chaff, its intake is ensured when horses eat the grain, whereas hay intake is optional so it is possible that a combination of chaff and hay allows lower total forage intakes. The average chaff intake was 1 kg (Thoroughbred) and 2.2 kg (Standardbred) with 88% of trainers feeding lucerne chaff at an average of 33-45% of chaff intake. Standardbred trainers used chaff as the principal forage source while Thoroughbred trainers used hay, but very few feed ad lib hay. The concern of trainers is that if they feed ad lib hay, horses will eat too much hay and too little grain. However, the biggest hay feeders also entice their horses to eat the most grain. Some trainers feed less than 2 kg forage per day and it should not be surprising that many of these trainers complain about the difficulties in getting horses to eat well enough when they are in full work. One benefit of lower forage intake will be reduced weight handicap related to the weight of the large intestinal contents.

In this survey Thoroughbred trainers fed an average of 129 MJ/day and Standardbred trainers 132 MJ/day based on average body weights of 490 kg and 440 kg, respectively. There was significant variation in daily intake that presumably relates to workload, as would the increased energy intake in Standardbred horses. It appears that Thoroughbred trainers work horses for a shorter duration than trainers in the USA or Europe. Trainers fed more crude protein than NRC recommendations with an average of 1450 g or 12% dietary protein.

The mean calcium and phosphorus intakes were slightly higher than NRC recommendations and only 20% of trainers used a calcium supplement, yet approximately 30% of diets required additional calcium. However, the increased use of premixed feeds containing calcium appears to have reduced the incidence of this problem. Iron intakes were over 200% of requirements, yet iron supplements and injections were popular. Some trainers fed potassium deficient diets, presumably due to low forage intakes. Salt was fed by 25% of trainers with higher intake (70 g) by Standardbred horses than Thoroughbred horses (30 g). Electrolyte supplements were fed by only 44% of trainers, so many horses in this study would have been electrolyte deficient.

About 20% of trainers fed glucose while only 14% added fat. The average amount of oil fed was less than 125 ml. It is common to feed relatively small amounts of added fat as oil, although sunflower seeds are a significant source of fat in most stables. In addition, the new premixed feeds for racehorses also contain considerable quantities of added fat. Injectable vitamins were used more commonly than oral vitamins, although 25% of trainers fed extra vitamins. Many top-dress vitamin supplements to a fully fortified feed.

Most trainers thought the major problem with feeding racehorses is getting them to eat enough as they reach racing condition or after gallops and races. This is one reason for the low forage intake; however the low forage intake may also lead to inappetence as could the feeding of large meals twice daily, amino acid imbalances or B vitamin deficiencies. Ninety percent of trainers changed feeding prior to racing with a decreased forage intake, increased grain intake and increased supplement intake being common practice. Seasonal variations were also evident with increased salt and electrolyte intake in summer and greater maize and barley intake in winter.

Tying up and excitability are other feed related problems that are addressed by feed substitution, reduced feed intake and supplementation. Most trainers feed insufficient forage and this may be related to the high incidence of tying up in Australia. It is rare to find trainers who feed ad lib hay. Inadequate electrolyte supplementation may also contribute to the incidence of tying up. Some trainers do not feed lucerne as they consider it may lead horses to tie up and make them "thick in the wind."

Most New Zealand trainers feed three times a day, with some feeding four times a day. Owners and trainers tend to feed through conscience and have difficulty understanding that more feed does not necessarily mean horses will run faster. Some dramatic improvements have been experienced by identifying energy requirements and lowering grain intakes to be more in line with the horse's weight and workload. Most commercial trainers do not have this problem, as their experience has identified maximum intakes. Again, premixed sweet feeds have become very popular in recent years. More horses have access to grass during the day in New Zealand than is the case in Australia.

Feeding on Stud Farms

Feeding practices on studs vary widely throughout Australia. The size of the stud and the number of horses, climate, pasture development, irrigation, breed, location and commercial status all influence the feeding practices. Low agistment (boarding) rates mean that Standardbred studs only feed hay, oats or cheap pellets, apart from unusual circumstances. Relatively few studs weigh young horses, so assessment of growth rate is subjective. Breeders are now realizing the problems created by pushing large amounts of grain into young horses. However, demands by yearling purchasers for large, fat weanlings and yearlings at sales force many breeders to overfeed young horses.

Most studs have silos for bulk storage of oats, pellets or extruded feeds and many use feed mixers to mix grain, supplements and chaff with or without molasses. This means the use of premixed sweet feeds is reduced, although 1 kg per day balancer pellets or 2 kg per day sweet feed concentrates are popular and economic methods of feeding. The availability of oats reduces, and the price increases as you move further north in Australia. Therefore, a proportionally greater use of prepared feeds on breeding farms in Queensland occurs. The use of powdered supplements for feeding horses on studs has dropped dramatically in recent years as breeders have recognized the value of balancer pellets and fully fortified feeds. Extruded feeds are effective and popular on studs, as the same growth or conditioning outcome can usually be achieved with reduced intakes of feed and greater digestive safety.

As horses are not boxed routinely, they are fed in groups rather than individually and it is impossible to control individual intakes. On some farms feed is poured onto the ground, but it is more common to put feed into feed bins on the ground with one feeder per horse. Of course, pecking order effects will influence choice of feeder and feed intake. Weanlings tend to be fed 2-3 kg concentrates (1% body weight) with some chaff, lucerne hay if required and 24 hour access to pasture. This means the concentrate portion of the diet is 50% or less which is lower than recommended by the 1989 NRC. If pastures are lacking in quality or quantity, the shortfall is made up by lucerne or more rarely clover hay. Satisfactory growth rates are achieved under this regime although most farms do not weigh horses, and there has been no published compilation of growth data. Some farms increase grain/concentrate intake as the weanling/yearling grows and appetite increases while others tend to maintain yearlings on the same intake as weanlings. Obviously, pasture quality and weather conditions play a big part in determining whether growth will be maintained on lower intakes of grain. During spring it is often possible to cut grain out entirely and just maintain yearlings on a mineral supplement and pasture, particularly with fillies. Total concentrate intake for yearlings ranges from 0 to 5 kg. During sale preparation, breeders often add corn, barley, sunflower seeds and oil to the basic mix. If rapid weight gain is needed, yearlings can be fed up to 10 kg concentrates per day (over 2% body weight) although some fat fillies are maintained on almost no grain. Feed intake during this period depends upon the need for increased body condition, amount of exercise, size, energy density of the feed and feeding practices of the individual breeder/manager. Some breeders present fat yearlings and try to sell by the pound while others have more athletic types. During yearling preparation, some yearlings are consuming over 3% body weight in concentrate and forage.

Generally, more attention is paid to detail in feeding the growing horse than the mare. Creep feeding is relatively uncommon, but foals have the opportunity to eat from the mare's feed bin. While it is usual to feed supplementary minerals to young horses, many breeders are not aware of how important it is for the late pregnant mare. Some farms just feed oats or poorly fortified pellets to mares, without paying any attention to mineral intakes. One important consideration with mares is that owners expect mares to be in fat condition when they return from stud. To meet the owner expectations, mares need to be fatter than is necessary for reproductive performance and this often leads to a degree of overfeeding. Mares tend to remain on the stud where the stallion resides for several months until they are 45 days in foal. This means some mares are overfed so that the lightest mare in the group can be maintained in the desired condition. After weaning, many mares are run without any 'hard' feeding until winter arrives or they reach the last third of pregnancy. There is increasing recognition of the value of feeding extra minerals in the last third of pregnancy and many breeders also increase the grain intake at this time. Grain intakes range from 1 kg to 5 kg in the last trimester. Maiden mares are often fed so they foal in fat condition. Lactating mares can consume vast amounts of pasture, and under ideal conditions can maintain body weight on pasture alone. However, it is more typical to have Thoroughbred mares fed 3-6 kg grain to meet the significant energy and protein needs of lactation. If pasture is limited this is supplemented with up to one-half bale of lucerne or clover hay (10 kg) per day. As the foal gets older, it eats more of the mare's feed so intakes are maintained until weaning, even though the mare's nutritional needs are declining.

Feeding on studs in New Zealand has changed dramatically over the last 10 years mainly through the unfortunate experiences of overfeeding, when it was common to sell to the Australian market by the pound. The downturn in returns for the New Zealand yearling led studs to investigate the importance of nutrition. Seasonal pasture analysis is now common and structured feeding programs are justified by nutritional benefits and cost. Most of the Thoroughbred studs in New Zealand are feeding commercially prepared steam flaked or extruded textured feeds to their sale yearlings. The convenience and labor savings to the commercial studs together with the definable nutritional contributions are the main reasons for this switch in feeding practices. Many mares are not fed apart from haylage in winter and supplementary minerals in late pregnancy.

Feeding of Performance and Leisure Horses

There is a wide variety of feeding practices used for performance, stock and pleasure horses in Australia. Some horses can be maintained on grass year-round or are supplemented with hay in winter. Others are stabled and fully fed using a mix of commercial feeds, chaff and hay.

A recent survey was conducted into the feeding practices of horses competing at high levels in the three Olympic disciplines (Owens, 1998). This covered six dressage horses, five show jumping horses and 11 three-day event horses and involved measurements of feed intake and estimations of energy expenditure over a 5-day period. Average daily feed intakes and digestible energy intakes varied considerably with surprisingly low intakes for some show jumping and dressage horses (Table 1). Based on calculations of average dietary intakes, one-half of the horses were sodium deficient and six were fed diets deficient in trace minerals and vitamins.

Table 1. Body weight, digestible energy (DE) intake and dry matter intake of elite performance horses.

Type of Horse	n	Body Weight (kg)	DE Intake (MJ)	Dry Matter Intake (% B wt)
Dressage	6	570 - 760	75 - 130	1.0 - 1.8
Showjumping	5	487 - 555	60 - 140	1.1 - 2.5
Eventing	11	483 - 615	100 - 150	.5 - 2.5

There was a wide variation between predicted energy needs using the NRC equation or that of Pagan and Hintz (1986) and actual intakes. Equations used underestimated the energy needs of three-day event horses and produced underestimates for 60% of the show jumping and dressage horses. This may be due to poor feeding management or mismatching work and feed during the 5-day observation period. This period was not long enough to detect changes in body weight that reflected a real difference between energy intake and requirements. This study showed a suprising level of ignorance about basic principles of feeding and nutrition among elite riders. Horses managed by these riders showed greatest variation in feed and energy intake and energy requirements.

The concentrate and forage intakes of horses in various classes is shown in Table 2. The advent of well-formulated premixed feeds has made it easier to feed horses well; however many riders still make mistakes. That these riders can achieve good performance from their horses is a testimony to the marvelous athletic talents of the horses.

Table 2. Expected feed consumption by horses.

Horse	% of body weight forage	concentrate	% of diet forage	concentrate
Performance Horses				
Spelling	1- 2	0 - 1	0 - 50	
Light Work	1 - 2	0 - 1	50 -100	0 - 50
Moderate Work	1 - 2	0.5 - 1	50 - 80	20 - 50
Hard Work	1.5 - 2	0.5 - 1.5	0 - 80	20 - 50
Racehorse	0.5 - 1	1 - 2	30	70
Mares				
Maintenance/ Early Pregnancy	1 - 2	0 - 1	50 -100	0- 50
Late Pregnancy	1 - 2	0 - 1	70	30
Early Lactation	1 - 3.5	0 - 1.5	60	40
Late Lactation	1 - 2	0 - 1	65	35
Young Horses				
Weanling	1 - 2.5	0 - 1.25	50	50
Yearling	1.5 - 2	0 - 1	65	35
Sales Yearling	0.5 - 1	0.25 - 2	40	60

References

Kohnke J, Kelleher F and Trevor-Jones P. (1999) Feeding Horses in Australia - A Guide for Horse Owners and Managers, Rural Industries Research and Development Corporation Publication 99/49 240.

Owens E. (1998) Evaluation and Monitoring of the Nutritional Status of Australia's Equine Athletes, Australian Sports Commission Report.

Pagan JD and Hintz HF. (1986) Energy energetics. II. Energy expenditure in horses during submaximal exercise. J. Anim. Sci. 63:822.

Southwood LL, Evans DL, Bryden WL and Rose RJ. (1993) Nutrient Intake of horses in Thoroughbred and Standardbred stables. Aust Vet J 70:164-168.

Southwood LL, Evans DL, Bryden WL and Rose RJ. (1993) Feeding practices in Thoroughbred and Standardbred stables. Aust Vet J 70:184-185.

THE PERFORMANCE HORSE

THE EFFECT OF LEVEL AND TYPE OF DIETARY FIBER ON HYDRATION STATUS FOLLOWING DEHYDRATION WITH FUROSEMIDE

L.K. WARREN, L.M. LAWRENCE, T. BARNES, D. POWELL AND S. PRATT
University of Kentucky, Lexington, KY

The loss of fluid and electrolytes in sweat during endurance exercise can be extensive, resulting in dehydration and poor performance if these losses are not replaced. Feeds high in fiber, especially those high in soluble fiber, possess a high water-holding capacity. Therefore, feeding strategies that optimize the level of dietary fiber and fiber type have the potential to benefit endurance horses by creating a reservoir of fluid and electrolytes in the hindgut that could be used to replace sweat losses.

To determine whether the level of dietary fiber and fiber type could affect plasma volume in response to dehydration, three diets were fed to six mature Thoroughbred horses in a 3 X 6 Latin rectangle experiment. Previous research has shown that fluid and electrolyte losses in response to a diuretic are similar to those lost in sweat during 2-3 h of moderate intensity exercise. Therefore, furosemide was used in the current study to simulate exercise-induced sweat loss. All diets contained similar dry matter (DM), energy, protein and electrolyte content, but differed in total dietary fiber (TDF), as well as soluble (SDF) and insoluble dietary fiber (IDF) components. The three diets were: 1) HIGH-HIGH (high TDF (65%), high SDF (12%)); 2) HIGH-LOW (high TDF (61%), low SDF (7%)); and 3) LOW-LOW (low TDF (48%), low SDF (7%)). In each 10 d period, diets were fed for 9 d. Water consumption and fecal moisture content were determined on d 6. On d 10, plasma volume (PV) was determined using a dye-dilution technique before and 4 h after the administration of furosemide (1 mg/kg BW IM). In addition, blood samples were obtained before and at hourly intervals for 6 h after the administration of furosemide for the determination of packed cell volume (PCV), plasma total protein (TP) and electrolyte (Na^+, K^+ and Cl^-) concentration. Body weight (BW) was recorded before and at 4 h and 6 h after furosemide administration (and corrected for fecal loss) to assess fluid loss in response to the diuretic.

Water consumption was greater (P<.01) when horses received diets high in TDF (HIGH-HIGH: 3.4±.2 L/kg DM; HIGH-LOW 3.3±.2 L/kg DM) compared to the diet low in TDF (LOW-LOW: 2.6±.2 L/kg DM). Diet affected (P<.01) fecal moisture content, with the HIGH-HIGH diet having the greatest fecal moisture, followed by HIGH-LOW and LOW-LOW. The decline in PV following dehydration with furosemide was similar for all diets, averaging 5.2±1.3 ml/kg BW (10.3%). While no differences in plasma TP were observed between diets following furosemide administration, TP concentration was higher (P<.05) prior to dehydration when horses received the LOW-LOW diet compared to the HIGH-HIGH diet. Horses receiving the LOW-LOW diet also had a higher (P<.05) PCV before dehydration and 1 h after furosemide administration compared to horses fed the HIGH-HIGH and HIGH-LOW diets. Before dehydration, BW was greater (P<.05) when horses received diets high in TDF (HIGH-HIGH and HIGH-LOW) compared to the diet low in TDF (LOW-LOW). However, loss of BW in response to furosemide was greater (P<.05) when horses received the HIGH-HIGH diet (4.8±.2%) compared to the HIGH-LOW (3.8±.3%) and LOW- LOW (3.8±.3%) diets.

The greater loss of BW in response to dehydration without a proportional loss of PV when horses received the HIGH-HIGH diet suggests that a diet high in SDF may provide the horse with a source of dispensable water in the hindgut during dehydration.

METABOLIC EFFECTS OF WARM-UP ON EXERCISING HORSES

S. ROKURODA, L. LAWRENCE, S. PRATT, L. WARREN, D. POWELL AND A. CRUM
University of Kentucky, Lexington, KY

Athletes commonly warm-up prior to strenuous exercise. Warm-up may provide several benefits including enhanced blood flow to working muscles which could increase oxygen and substrate availability during the subsequent exercise. Little research has been conducted in the area of equine warm-up protocols, particularly in regard to identification of ideal warm-up procedures for various activities. In a previous study, warm-up intensity (walking vs. a combination of walking and trotting) did not affect heart rate or plasma lactate accumulation during a step-wise exercise test. However, it was concluded that the step-wise exercise test may not have been a suitable model as the initial steps were conducted at a very low intensity and may have acted as additional warm-up, thereby equalizing the warm-up treatments. Two additional studies were subsequently conducted to evaluate the effects of two warm-up intensities on response to shorter term, more intense exercise tests.

Five Thoroughbred geldings were used. All exercise tests were conducted in the morning, after a 12-16 h fast. In each experiment, the effects of two warm-up procedures were compared. The low intensity warm-up consisted of 15 min of walking and the high intensity warm-up consisted of a sequence of walk, trot, gallop and walk, but also lasted 15 min. Treatments were assigned to horses using a two period cross-over design in both experiments, so that all horses completed both warm-up treatments.

In experiment A, horses performed a 70 s exercise test following the warm-up period. The speed of the treadmill was increased during the last 30 s of the warm-up. The 70 s test consisted of 50 s of gallop at 8 m/s on a 10% grade, followed by 20 s at 8.5 m/s on a 10% grade. Heart rate and hematocrit were higher during warm-up ($P < .05$) when horses completed the high intensity warm-up but there were no treatment differences for these variables during the exercise test. Heart rate at the end of the exercise test averaged 196.8 b/min when horses completed the low intensity warm-up and 196.6 b/min when they completed the high intensity warm-up prior to the exercise test. Warm-up intensity affected plasma glucose and free fatty acid responses during the warm-up period, and glucose concentrations were higher ($P <.05$) during the exercise test when horses performed the high intensity warm-up. Higher glucose concentrations may suggest enhanced glucose availability as a result of catecholamine activation of hepatic glycogenolysis in response to high intensity warm-up. Plasma lactate concentrations at the end of the exercise test were lower ($P < .05$) when horses performed the high intensity warm-up. Lower lactate accumulation during the exercise test may suggest that high intensity warm-up promoted oxygen availability or aerobic energy production during short term high intensity exercise (70 s).

In experiment B, horses performed a 3.5 min exercise test consisting of 30 s at 6.5 m/s, 60 s at 7.0 m/s, 90 s at 7.5 m/s and 30 s at 8.0 m/s, all on a 10% grade. Warm-up effects were similar to experiment A except that plasma lactate responses were not significantly different. In addition, rectal temperature at the end of a 5 min recovery period was higher (P < .05) when horses completed the high intensity warm-up before the exercise test than when they completed the low intensity exercise test.

THE EFFECT OF LONG-TERM DIET RESTRICTION ON THYROXINE AND TRIIODOTHYRONINE CONCENTRATIONS AND METABOLIC RESPONSES IN HORSES FED AND FASTED PRIOR TO EXERCISE

D. POWELL, L. LAWRENCE, T. BREWSTER-BARNES, B. FITZGERALD,
L. WARREN, S. ROKURODA, A. PARKER AND A. CRUM
University of Kentucky, Lexington, KY

The results of a previous experiment indicated that the source of the calories can have an effect on the metabolic responses of feed restricted exercising horses. Because thyroid hormone can have a positive response to feeding a meal and exercise has been suggested to help maintain thyroid hormone production during negative caloric balance, this study was conducted to determine if horses consuming a restricted diet with 70% of the calories coming from the concentrate source, were better able to maintain thyroid hormone concentrations during exercise when fed prior to exercise.

All horses were previously determined to have normal thyroid functional capacity via thyrotropin releasing hormone response testing. Initial body weights and condition scores were obtained on each horse prior to treatment assignment. Eight Thoroughbred geldings were randomly assigned to one of two treatments: treatment 1, a calorie restricted diet containing 70% of the caloric need from the roughage source [RHR]; treatment 2, a calorie restricted diet containing 70% of the caloric need from the concentrate source [RHC]. The diets were designed to be adequate in grams of crude protein, calcium and phosphorus for horses undergoing moderate work. The diets were adjusted weekly in order for each horse to maintain an approximate 1.0 kg body weight/d weight loss. The horses were exercise conditioned for 55 d prior to exercise testing. The conditioning program consisted of progressive increases in work bouts until by day 22, all horses were working 3 d/wk on a high speed treadmill set at a 6% grade for 28 min and 2 d/wk for 30 min at a strong trot in a 60' round pen. At the end of the conditioning period, each horse performed two step tests: one, following a 12 h fast and one, 2 h following a meal of 2.0 kg of concentrate (a time previously determined to give an adequate thyroid hormone response to feeding). The exercise test consisted of a step test at a 10% grade, where the speed was increased every 2 min beginning at 2 m/s and ending at fatigue, determined as the point at which the horse could not keep up with the speed of the treadmill. Blood samples were taken pre-exercise, at the end of each 2 min step change, and during a 10 min recovery. The samples were analyzed for T_4, T_3, glucose, insulin, free fatty acid (FFA) and lactate concentrations. Heart rates and time to fatigue were also recorded. The exercise tests were separated by a 6 or 7 d interval during which time all horses continued their conditioning program.

Initial and final condition scores were 5.5 and 4.75 for the RHR group and 5.2 and 4.5 for the RHC group, respectively. Both treatment groups lost 8% of their initial body weight by the end of the study. Fed horses receiving the RHR dietary treatment had the greater T_4 response (time*diet*feeding state, P<.05) and a decline in glucose concentrations (time*diet*feeding state, P<.05). In response to fasting, horses receiving the RHR diet had a greater FFA response (time*diet*feeding state, P<.05) and numerically longer times to fatigue. Horses receiving the RHC dietary treatment tended to show an increase in T_3 concentrations during the step test (time*diet, P<.08) and tended to have lower heart rates when fasted (time*diet*feeding state, P<.1). Insulin concentrations declined during the step test when horses were fed (P<.05), but were not affected by diet. Plasma lactate concentrations increased (P<.01) in response to exercise, but were not affected by diet or feeding state.

These data suggest that exercise and calorie source may be important to hormonal regulation and energy metabolism in horses subjected to long-term restriction of calorie deficient diets.

STUDIES OF FAT ADAPTATION AND EXERCISE

D. S. KRONFELD[1], K. M. CRANDELL[1], S. E. CUSTALOW[1], P. L. FERRANTE[1],
L. A. GAY[1], P. GRAHAM-THIERS[1], K. A. KLINE[1], L. E. TAYLOR[1], J. L. HOLLAND[1],
J. A. WILSON[1], D. J. SKLAN[2], P. A. HARRIS[3], W. TIEGS[3]
[1]Virginia Polytechnic Institute and State University, Blacksburg, VA
[2]Faculty of Agriculture, Hebrew University, Rohovot, Israel
[3]WALTHAM Centre for Pet Nutritioon, Leicestershire, UK

The concept of fat adaptation - the combination of interval training and feeding a high fat diet to improve mitochondrial oxidation of fatty acids and to spare the use of glycogen and glucose - was introduced in 1973 to prevent tying-up in racing sled dogs. It was soon found also to improve stamina.

Subsequent studies of fat adaptation in horses have confirmed the protective effect against certain types of exertional rhabdomyolysis and suggested improvement of sprinting ability as well as stamina.

The balance of a ration is measured in terms of metabolic efficiency - the maximum output of the desired product (e.g., milk, work), and the minimum generation of undesired products (e.g., feces, acid, heat). For digestive and metabolic efficiency, fat > starch > fiber = protein.

In horses, optimal fat content is 40 g/Mcal DE of diet (not just the concentrate) to maximize muscle glycogen content, according to our parabolic analysis of data in literature.

Studies

- **Palatability.** Preference tests demonstrated higher palatability (voluntary acceptability) of corn oil compared to other vegetable oils and animal fats.

- **Calmness.** Spontaneous activity (pedometers) and reactivity (spook tests) were lower in horses when fed diets fortified with corn oil or mixtures of soy lecithin and corn oil.

- **Acid-base responses.** Fewer hydrogen ions and lower levels of carbonic acid were produced in fat-adapted horses during incremental exercise tests and repeated sprints. Increases in blood lactate contributed to acidosis during rest-work transition (sprints 1-3) but not during work adapted phase (sprints 3-9). The blood pH was dependent more on carbon dioxide tension, that is, respiratory regulation, than on blood lactate accumulation, that is, metabolism.

- **Lactate threshold.** The speed at which blood lactate concentration began to increase sharply during an incremental exercise test was higher in fat adapted horses (interval trained for 11 weeks while fed a 12%

481

fat diet) than in controls. Unlike maximal tests, such as VO_2max determination, this submaximal test could be used to predict the metabolic potential of an unfit horse and to monitor fitness of horses in training.

- **Metabolic regulation.** In fat adapted horses, blood lactate concentration was higher during anaerobic work but lower during aerobic work. These results suggest that glycolysis is enhanced during anaerobic work, which would enable sprinting, but suppressed during aerobic work, which would confer stamina.

- **Respiratory load.** Lower CO_2 production in fat adapted horses during strenuous exercise would reduce the mechanical effort of respiration and the stimulus for panting. Thus fat adaptation may be beneficial in horses with mildly compromised pulmonary functions (EIPH, COPD), especially in hot and humid conditions.

- **Calcium.** Bone mineral density (BMC) increased about 6% in 12 weeks interval training in horses fed twice the NRC requirement but not in those fed the NRC recommendation, which appears to be inadequate. On the other hand, dietary calcium did not affect a similar loss of BMC during de-training. Since BMC correlates with bone strength and elasticity, these results suggest that BMC loss during lay-ups may increase the risk of skeletal injury upon resumption of training. In these experiments, changing chloride, hence dietary cation-anion difference (DCAD), had little effect on BMC.

- **Protein.** Lowering dietary protein raises the DCAD without depleting chloride, which may be needed for sweating. Low protein diminishes generation of acid, heat and urea, hence spares water. Protein, 30 g/Mcal fortified with lysine and threonine, is being tested against protein, 50 g/Mcal.

- **Bowel ballast.** A bioenergetic model was developed for a 500 kg equine athlete using 36 Mcal ME/day. In the model, a hay:oats:oil (45:45:10) diet compared to hay and oats (50:50) produces about 2 Mcal less heat per day, requires about 6 L less water per day, and carries about 12 kg less bowel ballast (dead weight in the large bowel). About 4 L less water in the bowel ballast is available for absorption, and this potential disadvantage must be weighed against the advantages noted above.

- **Safety.** Horses have been carefully observed by veterinarians for 34 weeks with no sign of adverse effects when fed these complete feeds.

BEHAVIORAL AND PHYSIOLOGICAL RESPONSES OF HORSES TO INITIAL TRAINING, THE COMPARISON BETWEEN PASTURE VERSUS STALLED HORSES

E. RIVERA, S. BENJAMIN, B. D. NIELSEN, J. E. SHELLE, A. J. ZANELLA
Michigan State University, East Lansing, MI

There are evidences, for several species, that learning ability may be impaired in animals housed in social isolation or barren environments. The environment in which the animal is housed affects interaction between humans and horses. Horses housed on pasture have the ability to interact socially and are exposed to a more diverse environment but have less contact with humans. Horses kept in stalls experience human contact but often have a lack of social interaction and are exposed to more barren environments. Response to initial training may be affected by housing conditions.

Behavioral and physiological responses were monitored in 16 2-year-old Arabian horses subjected to a standardized "training procedure" (n=12) carried out by the same trainer or selected as control (C) (n=4). The horses were kept in pasture (P) (n=8) or in individual stalls (S) (n=8) for three months prior to this experiment. Twelve horses (6 P & 6 S) were subjected to training and 4 horses (2 P & 2 S) were used as control. Initial training consisted of releasing the animal into the round pen and encouraging the horse to run in one direction until the horse accepts the trainer. Circa ten minutes post-release, the horse approached the trainer and a halter was used to handle the horse. After handling, a saddle was placed on the animal's back and the horse was released in the round pen. After that the trainer mounted the horse and taught some basic commands. The training session lasted an average of 30 minutes. Control horses were released into the round pen and left to explore the environment for 30 minutes. Behavioral observation assessing the interactions between the trainer and the horse in training were collected using a video recorder. Measures of plasma cortisol were monitored by radioimmunoassay in blood samples collected by jugular puncture in training days 1, 7, 21, and 28. Samples were collected prior to training (basal), immediately post-training (pt), 15 minutes post-training (pt15) and 75 minutes post-training (pt75).

After 20 minutes of handling the trainer was able to mount and ride the horse and minimal behavioral responses were observed. There were no differences in basal cortisol levels among the three groups studied prior to training (P =181.2 ±20.9; S= 194.30 ± 57.7 and C= 182.8 ±32.1; p=. 97). Plasma cortisol levels showed significant changes after training or exposure to a novel environment in the three groups for day 1 (P: F test 3.59, p=0.03; S: F test 6.22, p=0.003 and C: F test 3.5, p=0.04). Differences between basal cortisol levels and post-training levels were not evident for control horses in days 7, 21, and 28 post-training, although they could be observed in trained horses. Effective training may be accomplished in rather a short time using humane techniques. Housing conditions appeared to have no significant effect on cortisol release during initial training in horses. These results may generate recommendations that could be given to horse owners on which housing conditions may maximize the opportunity for success of training procedures.

THE EFFECT OF EXERCISE ON THE DIGESTIBILITY OF AN ALL FORAGE OR MIXED DIET IN THOROUGHBRED HORSES

J.D. PAGAN[1], P. HARRIS[2], T. BREWSTER-BARNES[1] AND S.G. JACKSON[1]
Kentucky Equine Research, Inc., Versailles, KY
WALTHAM Centre for Pet Nutrition, Leicestershire, UK

Four conditioned Thoroughbred geldings were used in a 2 x 2 factorial design to investigate the effect of exercise and diet on apparent nutrient digestibility. The experiment consisted of 4 periods that were 4 weeks in length. During each period, the horses were fed either an all forage diet (4.54 kg alfalfa cubes + 5.45 kg alfalfa/grass hay)(FORAGE) or a mixture of forage and grain (3.63 kg sweet feed + 2.72 kg alfalfa/grass hay + 2.27 kg alfalfa cubes)(MIXED). During each period, one horse from each diet was exercised daily on a high speed treadmill (EX). During the fourth week of each period, a complete collection digestion trial was conducted. Each morning during the collection period, the EX horses performed an exercise bout on the treadmill (inclined to 3°) which consisted of a 5 min warm-up walk, 1600 m at 4 m/s, 1600 m at 7 m/s, 1600 m at 9 m/s, 1600 m at 7 m/s, 1600 m at 4 m/s and a 5 min warm-down walk. Each afternoon, the EX horses were hand walked 1600 m. The NON-EX horses were hand walked 1600 m twice daily.

The dry matter digestibility of the MIXED diet was significantly higher than the FORAGE diet (54.6% vs 62.1%) (p<.01). ADF, NDF and hemicellulose digestibility were significantly higher in the FORAGE diet (p<.05). Exercise resulted in a small but statistically significant decrease in dry matter digestibility (57.8% vs 58.9%)(p<.05). This decrease in DM digestibility was primarily from a reduction in ADF digestibility. Potassium digestibility was also significantly reduced in the EX horses (66.3% vs 74.3%)(p<.05).

A COMPARISON OF GRAIN, VEGETABLE OIL AND BEET PULP AS ENERGY SOURCES FOR THE EXERCISED HORSE

K. CRANDELL[1], J.D. PAGAN[1], P. HARRIS[2], AND S.E. DUREN[1]
[1]*Kentucky Equine Research, Inc., Versailles, KY*
[2]*WALTHAM Centre for Pet Nutrition, Leicestershire, UK*

A number of studies have evaluated the effect of adding fat to a performance horse's diet. Most have replaced a portion of the grain in the ration with fat, so that total carbohydrate intake was reduced. Little research has been conducted to evaluate the effect of substituting other energy sources for starch. Therefore, this experiment was designed to compare a traditional high grain diet with diets which provide 15% of the total caloric intake from either vegetable oil or a highly fermentable fiber source (beet pulp).

Six three-year-old Thoroughbreds (4 geldings, 2 fillies) were fed one of three diets in a replicated 3 x 3 Latin square design where each period lasted 5 weeks. Diet 1 (CONTROL) consisted of 3.62 kg of sweet feed (47% oats, 42% corn and 11% molasses) and .9 kg of a balancer pellet. Diet 2 (FAT) contained 15% of the daily DE as soybean oil and Diet 3 (FIBER) contained 15% of the daily DE as beet pulp. The oil and beet pulp replaced sweet feed so that the diets remained isocaloric on a digestible energy basis. The horses were also fed 5.43 kg of an alfalfa/whole corn plant cube daily.

During the last week of each period, the horses performed a standardized exercise test (SET) on a treadmill. The SET consisted of a warm-up (WU), which included a 2 minute walk and an 800 meter trot (~4 m/s), an 800 meter gallop at 8.5 m/s (20 mph) and a 2400 meter gallop at 11 m/s (26 mph) followed by a warm down (wd) of 800 meters at the trot (~4 m/s) and a 2 minute walk. Each horse performed the test 3 hours after eating only the grain portion of its morning meal. Heart rates were recorded during the last 15 s of each 800 meter interval. Blood samples were drawn from an indwelling jugular catheter before eating, at 30 minute intervals up to the SET, after the warm-up, at the end of every 800 meters at the gallop, after the warm down and at 15 and 30 minutes post exercise. The collected plasma was analyzed for lactate, glucose, cortisol, insulin, triglycerides and glycerol.

Blood glucose was significantly lower ($p<.05$) in the FAT horses during the three hours post feeding as compared to either the CONTROL or FIBER horses. Insulin was also significantly lower ($p<.05$) in the FAT horses both post feeding and throughout exercise. Cortisol was significantly lower ($p<.05$) in the FAT horses as compared to the CONTROL horses during exercise. Following the SET, the FAT horses drank significantly more ($p<.10$) water (7.0 liters) than either the CONTROL (4.8 liters) or the FIBER (4.7 liters) horses. Substituting 15% of DE as vegetable oil had a greater effect on metabolic response to exercise than a 15% substitution of beet pulp.

TIMING AND AMOUNT OF FORAGE AND GRAIN AFFECTS EXERCISE RESPONSE IN THOROUGHBRED HORSES

J.D. PAGAN[1] AND P. HARRIS[2]
[1]Kentucky Equine Research, Inc., Versailles, KY
[2]WALTHAM Centre for Pet Nutrition, Leicestershire, UK

There is considerable debate among horse trainers about how to feed horses before exercise. Should horses be fed or fasted before work and how should hay be fed relative to grain? Three experiments were conducted to evaluate if feeding hay with and without grain affects glycemic and hematological responses in Thoroughbred horses at rest and during a simulated competition exercise test (CET) on a high-speed treadmill.

In experiment one, six Thoroughbred horses were fed 2.27 kg of a sweet feed mix either 1) alone, 2) with 2.27 kg of hay or, 3) 2 h after receiving 2.27 kg of hay. Blood samples were taken 2 h before grain and hourly until 8 h post feeding. Blood was analyzed for glucose, insulin, lactate, PCV, and total plasma protein. Time of hay feeding affected glycemic response, plasma protein and water intake post grain feeding. Therefore, a second experiment was designed to evaluate how these changes would affect metabolic response to a CET.

In experiment two, four trained Thoroughbred horses were used in a 4 x 4 Latin square design. The four dietary treatments included: 1) overnight fast, 2) ad lib hay the night before and 2.27 kg of a sweet feed mix 2 h prior to exercise, 3) 2.27 kg of hay 3 h prior to CET and 2.27 kg of grain 2 h prior to exercise, and 4) 2.27 kg of grain 2 h prior to exercise.

The CET was performed on an inclined treadmill (3°) and consisted of a 10 min walk, 10 min trot (3.7 m/s), 2 min gallop (10.7 m/s), 10 min walk, 10 min trot (3.7 m/s), 10 min walk and 8 min canter (9 m/s). Blood samples were taken hourly before, at the end of each step during the exercise test, and 60 and 120 min post exercise. These samples were analyzed for glucose, insulin, cortisol, lactate, total plasma protein, PCV, Na^+, Cl^-, and K^+. Plasma volume was determined immediately before exercise using an indocyanine green clearance method. Ad lib hay feeding resulted in a 9% reduction in plasma volume. Fasted horses had lower blood lactate after the 8-min canter (p<0.10) compared to the other three treatments. Heart rate (HR) was significantly different between treatments. Based on these results, a third experiment was conducted to evaluate how different forage feeding regimes without supplemental grain would affect exercise response to the CET.

During experiment three, 4 horses (4 x 4 Latin square) completed the same CET as above after either 1) fasting, 2) 2.27 kg of hay 3 hours before CET, 3) ad lib hay the night before, or 4) overnight grazing in a grass paddock. No grain was fed before the CET and the same measurements were taken as in experiment two. Feeding hay or grazing affected plasma protein, PCV, electrolytes and glycemic response to exercise. The results of these experiments suggest that an overnight fast before an extended bout of exercise may be beneficial.

NUTRITIONAL ERGOGENIC AIDS IN THE HORSE - USES AND ABUSES

[1]PAT A. HARRIS & [2] R.C. HARRIS
[1]WALTHAM Centre for Pet Nutrition, Leicestershire, UK
[2]Exercise Physiology Research Group, Chichester, UK

What is an Ergogenic Aid?

If one imagines two horses with identical genetic makeup with equal physiological ability, ridden by jockeys with equal riding skill and with the same way of riding a race, being raced against each other, all things being equal the two horses should finish at exactly the same time. But what if one of the horses were given 'formula X' to improve its performance, enabling it to win outright. 'Formula X,' in this case, is an ergogenic aid. The term ergogenic comes from the Greek 'ergon' meaning work and 'genic' meaning producing. An ergogenic aid therefore can be used to describe any factor which can increase or improve work production. This could result in an increase in speed or endurance or strength.

A number of factors could have caused the improved performance in our example and have been shown to be important in man, including:
1. Psychological factors
2. Mechanical or biomechanical factors including improved equipment (shoes, track design, etc.)
3. Pharmacological agents
4. Physiological improvements (including most importantly those obtained through training)
5. Nutritional supplements.

The key to optimal sport performance has been said to be the proper production and control of energy. To this end the appropriate biomechanical, psychological and physiological training specific to the nature of the athletic event will improve the control and utilization of the various energy systems and maximize energy efficiency and production. The adaptations that occur in the body cells, tissues and organs in response to chronic exercise training are fairly specific to the imposed demands and are fairly well documented, at least in man.

In man, nutritional ergogenic aids have been suggested to be far less efficient, in general, in improving physical ability than an appropriate training regimen. This is likely to be true also for the horse. However, athletes training at levels close to their upper limit require relatively large increases in training effort in order to achieve even a small increase in performance. Very small variations in performance, whether achieved through significant increases in training effort or by addition of 'formula X' among elite athletes could make the difference between finishing first or in the middle of the pack. The comparative ease of taking 'formula X' compared with the significant increase in training effort required explains why there has been such a drive for effective nutritional ergogenic aids.

Possible ways that ergogenic aids, apart from equipment, etc., could improve performance in the horse could include (see also Harris, 1994):

1. Psychological effects
2. Improved coordination or recruitment of muscle fibers
3. Provision of a supplementary fuel source or the feeding of a feed with a higher energy content
4. Increased levels of available stored energy
5. Improved efficiency of conversion of the chemical energy of the feed, or stored energy, to mechanical energy for work
6. Improved ATP/ADP homeostasis in contracting muscle fibers
7. Decreased substrate depletion
8. Decreased end product accumulation including improved intracellular acid base regulation.

These could result in increased mechanical energy for work and/or a delayed onset of fatigue or improved neuromuscular coordination. Many substances have theoretical ergogenic properties.

Nutritional ergogenic aids may be considered to be compounds or elements that can be administered orally and have a nutritionally oriented function. This paper will concentrate on these postulated nutritional ergogenic aids, but it will not consider pharmaceuticals such as steroids or other agents that may fundamentally affect underlying metabolic and structural elements within the muscle fibres.

How Could One Design the Ideal Nutritional Ergogenic Aid?

There are three basic steps to producing an ideal nutritional ergogenic aid.

1. Identify an aspect of exercise physiology that is limiting to competition (of a particular type).
2. Identify a nutritional compound or element that will positively affect this aspect.
3. Finally and most importantly, provide evidence that the feeding or administration of the compound or element is effective in the field (as well as the laboratory); this will necessarily include evidence in the target species of (Harris, 1994):
   • absorption from the gut
   • uptake into the target tissue
   • demonstrably improved tissue function
   • improvement in performance.

Unfortunately too many substances are marketed without adequate understanding of their function, little or no evidence that the metabolic or physiological mechanism which they contribute to is limiting, and no evidence that they will affect performance in the field. The outcome may be a short term financial advantage for the manufacturer but a disappointed rider/owner and on occasions an unhappy horse. Often a compound's efficacy is based on anecdotal evidence, owner hopes or expectations, and results based on work in other species. On the other hand potential benefits should not be dismissed out of hand simply because of the difficulties in demonstrating an effect on performance. This may prevent justification of the compound or element on scientific grounds.

Why Are There Difficulties in Confirming the Efficacy of Ergogenic Aids?

There are a large number of potential reasons why there have been difficulties in confirming the efficacy of a variety of ergogenic aids. Some of these are outlined below:

1. Studies in horses are very expensive so tend to use only small numbers of animals.
2. Individual variability in response means that it may be difficult to assess scientific significance if small numbers are used. If only one out of five horses shows a marked improvement this may not show up as a statistical effect in studies, but nonetheless could be highly significant to the individual (responding) athlete in the field.
3. The effect on performance may be so small as to be masked by the normal within-subject variability intrinsic within any testing program and the sensitivity of the procedures employed, but in competition could make the difference between winning or being placed.
4. Alternatively, laboratory studies could show statistically significant effect of a compound or element on some physiological or biochemical function but which in the field is not manifest in any significant improvement in performance.
5. Extrapolation from other species is often used but not always appropriate, e.g. creatine is not well absorbed in the horse compared with the dog or man.
6. Dose response effects need to be taken into account.
7. Study design can be difficult especially regarding:
    *   randomization
    *   handlers/assessors/riders/drivers etc. being blind to treatment
    *   pattern of running over the test distance varying between test periods.
8. How do we assess performance?
9. How representative is the treadmill to the racetrack or show arena?
10. What may be suitable for a sprint race horse may be contraindicated in the endurance horse.

Results of differing studies are often contradictory and in practice there are very few credible studies available.

What About the Ethics Behind Ergogenic Aids?

In our three step example one could have added a fourth step: that the product is not banned as an illegal substance. For this we may have to establish:

1. What is allowable?
2. What should be allowable?
3. When is an ergogenic aid an advance in optimal feeding and therefore legal and acceptable?
4. When is an ergogenic aid a prohibited substance?

The IOC (International Olympic Committee) doping legislation stipulates that any physiologic substance taken in abnormal quantities with the intention of artificially and unfairly increasing performance should be construed as doping, violating the ethics of sport performance.

But what if an ergogenic aid would help for example with the welfare of horses undergoing prolonged endurance events? Should it be allowable even if it improves performance? If one takes this to its logical conclusion then the feeding of water and electrolytes to horses in endurance rides could be considered use of ergogenic aids!

What about creatine? Will there be questions over the legality of its use in horses if, in the future, a way of improving absorption and utilization is found? Creatine is not a normal component of horse diets, especially not at the levels likely to be required (although the horse synthesizes it within its own body), but then neither is fat in large amounts a normal component of horse diets. What if feeding creatine just enables the horse to be restored to its potential capacity as may arguably be the case in man?

Dietary Manipulation and the Feeding of Supplemental Fat

The horse evolved as a grazing animal which escaped predators by flight and was adapted to an almost constant supply of forage, which was predominantly digested in the hindgut. Today the horse might be expected to carry a rider and undertake fairly exhaustive repetitive work. The horse therefore often needs more energy than it would have required in the wild. The gross (chemical) energy (GE) of the feed is decreased by the energy lost with ingestion, foraging, chewing and then fermenting the feed (often included in the heat of maintenance) and via losses in the feces. The amount left is the digestible energy (DE) content of the diet. Potential energy is also lost via gas production from the fermentation processes and via urea in the urine. The conversion of the residual chemical energy, or metabolizable energy, to mechanical energy of movement is substantially less than 100%, with most energy being lost as heat. Work efficiency or Kw is only 20 - 25%. The amount of net energy available from the diet depends obviously on the feed given; for example corn has more net energy than barley and oats. Barley has around twice that of good hay which in turn is over twice that of wheat straw. Note: that barley in Grecian times was referred to as aiding performance by providing additional energy over and above forage. Is this the first ergogenic aid?

The relative energy contents for a kg of three different types of diet based on the partitioning system of Kronfeld (1996) are shown below. Traditionally energy in horse feeds is depicted as digestible energy, but the DE system tends to overestimate the energy potential of a high fiber feed compared with a highly hydrolyzable carbohydrate feed, as fiber predominantly produces volatile fatty acids, which are not as efficiently used as glucose (Harris, 1997).

Relative energy contents (MJ/kg) for three diets:

	Hay only	Hay:Oats (50:50)	Hay:Oats:Fat (45:45:10)
GE	15.6	16.3	18.5
DE	8	9.8	12
ME	7.5	9.3	11.6
NEm[1]	5.4	7.1	9.3

[1]NEm = Net energy for maintenance (The net energy will be considerably lower for work.)

Consider a 500 kg three day eventer on the cross-country day that needs approximately 60 MJ of net energy to live and compete. If it is assumed that all this energy is to be supplied by its daily ration, a horse would need to eat around 22 kg of hay which would not be possible (appetite being restricted to around 10 - 12 kg DM). This explains why, even though animals vary in their energy efficiency, it would be difficult if not impossible to maintain weight and energy output in regularly competing horses fed only on hay. If oats were substituted as part of the diet then the ration would be more realistic, but it would still be unlikely that the horse could eat all provided. However, if energy dense fat was added to the diet then it could be possible to match this energy requirement with intake (Kronfeld, 1996).

Feeding Fat:

1. reduces the amount of concentrates that need to be fed to maintain energy intake, which effectively means that often a horse can retain a healthy fiber intake despite a high energy requirement. May have behavioral advantages.
2. increases the energy density of a feed so that a horse can effectively take in more energy even if appetite decreases.
3. decreases heat load as more efficient in conversion to mechanical energy than fiber or carbohydrate. Useful under hot and humid conditions.
4. potentially helps reduce bowel ballast and possibly water requirements.

Feeding fat supplemented diets to horses has resulted in a range of effects on a variety of physiological and metabolic parameters as well as on performance. These variations may result from the variances in the study protocols and horses used in these trials. Because of the variable results, a consensus view of the benefits on performance beyond those listed above is not yet available. Long term fat supplementation in combination with appropriate training, however, has been suggested to result in the following adaptations which could result in improved performance (see also Harris, 1997; Potter et al., 1992):

1. increased mobilization of free fatty acids (FFA) and increased speed of mobilization.
2. increased speed of uptake of FFA into muscle (Orme et al., 1997) - often considered to be rate limiting.
3. a glycogen sparing effect so that fatigue is delayed and performance improved - could be especially important in endurance activities.
4. increased high intensity exercise capacity (Eaton et al., 1995).
5. increased pre-exercise muscle glycogen levels (Meyers et al., 1989; Scott et al., 1992; Hughes et al., 1995).

The profile of FFA contained within a fat supplemented diet may also be important since this may in turn influence the FFA profile of the fats stored within the body, and therefore the FFA mobilized with exercise. FFA constitute a broad spectrum of molecules, from the so-called short to long fatty acids, which may well show different rates of diffusion through cells affecting their rate of utilization with exercise. Although techniques are available for the loading of plasma with FFA with a predetermined profile (Orme & Harris, 1997), and for the investigation of their use during exercise (Orme et al., 1995), no studies have been undertaken in the horse to examine specifically the utilization of the different FFA during exercise. Medium chain triglycerides (MCT) have been proposed as a potential ergogenic aid during exercise in man. In humans MCT are emptied more rapidly from the stomach (Beckers et al., 1992), rapidly absorbed and hydrolyzed by the small intestine, and secreted directly into the systemic circulation. Furthermore, MCT do not require the acetyltransferase system to cross the inner mitochondrial wall in order to undergo oxidation. Orally supplied MCT, however, do not lead to a sparing of muscle glycogen use during moderate to intense exercise (Jeukendrup et al., 1995 & 1996; Massicotte et al., 1992) and their effect on performance is questionable.

Dietary Ergogenic Strategies for Racing

An analysis of the metabolic events occurring in the muscles during intense racing exercise indicates four areas where improved metabolic function could theoretically be beneficial to performance. These are:

1. increased availability of locally stored muscle glycogen
2. an increased rate of oxidative metabolism
3. improved intracellular acid-base regulation
4. increased phosphagen support of ATP metabolism.

I. Increased Availability of Locally Stored Glycogen

Aerobic and anaerobic utilization of carbohydrate constitutes the major fuel store for the regeneration of ATP in muscle during racing. Endurance rides of 100 km or more may bring about total depletion of the glycogen store in some muscle fibers (Snow et al., 1981; Snow et al., 1982), but over normal racing distances the decrease is probably no more than 30-40% (Hodgson et al., 1984; Harris et al., 1987; Snow et al., 1987). With normal non-exercise muscle glycogen content of around 600 mmol/kg dry muscle (dm) glucose units, glycogen is unlikely to be limiting in flat or harness racing, although over longer National Hunt distances local depletion within some muscle fibers is a possibility (Snow et al., 1981). Glycogen loss during training is accommodated easily by a normal diet containing concentrates (Snow and Harris, 1991) and is unlikely to lead to depletion prior to racing.

The use of "glycogen loading" stems from early work in man (Bergström and Hultman, 1966; Bergström et al., 1967). The normal glycogen content of rested human muscle is approximately 300 mmol/kg dm but may increase twofold by a combination of exercise and diet. This strategy is now followed ardently by human middle distance and endurance runners. There have been claims both for (Topliff et al., 1983; 1985) and against (Topliff et al., 1987; Snow et al., 1987) a similar effect in horses fed a carbohydrate supplement. However, no effect on either the rate of muscle glycogen resynthesis, or the final level reached, was seen when increased levels of starch were added to the feed (Topliff et al., 1987; Snow, 1992). Glycogen repletion, following exercise resulting in a 40% decrease in the muscle store, was approximately the same in horses fed a low carbohydrate (LC) diet consisting of hay, a normal carbohydrate (NC) diet consisting of pelleted concentrate plus hay, or a high carbohydrate (HC) diet which was the NC diet supplemented with intravenous infusions of 0.45 kg glucose (circa 0.9 g/kg bwt) on each of the first 2 days (Snow et al., 1987). However, an increased rate of glycogen resynthesis was observed by Davie et al. (1995) when 6 g/kg bwt dextrose was infused intravenously following exercise resulting in a 50% decrease in the muscle content. Muscle contents after 24 h recovery, however, were no longer significantly different with or without dextrose infusion and therefore it would appear that any advantage gained in the repletion of the muscle glycogen stores, even by this extreme procedure, is short-lived.

In general, there seems to be little or no case for adding high carbohydrate supplements to feed in preparing horses for racing, in contrast to the well described use of such supplements by human athletes to bring about "glycogen loading" of the muscles.

II. An Improved Rate of Oxidative Metabolism

Undoubtedly the best way to achieve an increase in the oxidative capacity of muscle is through training. However, a number of nutritional strategies have been used in an attempt to increase this still further.

Q10

One example is ubiquinone or Q10 but, apart from its known role in mitochondrial electron transport, there is scant evidence that it in any way limits oxidative metabolism in normal, well-fed animals. Similarly, there are no studies showing uptake from the gut and into the muscles of the horse, and no evidence of any effect upon tissue function or physical performance. Q10 has been shown in rats to be absorbed from the gut but orally supplied Q10 exhibits a restricted distribution within tissues (Zhang et al., 1995; Zhang et al., 1996). Studies in humans have generally failed to find any effect upon performance (Laaksonen et al., 1995) or the attenuation of muscle lactate formation (Porter et al., 1995) with exercise performed close to the onset of blood lactate accumulation (OBLA).

L-CARNITINE

L-carnitine is an essential cofactor for the transport of long chain fatty acids across the inner mitochondrial membrane and has been suggested to be rate limiting to fat utilization during exercise. This, however, seems unlikely in view of the very high concentration of L-carnitine that is found in tissues such as muscle (circa 6-7 mM) (Foster and Harris, 1992; Harris et al., 1995) relative to its affinity for mitochondrial bound carnitine-palmitoyl transferase. Of possibly more relevance to racing is the role of L-carnitine in carbohydrate oxidation in buffering the mitochondrial concentration of acetyl CoA (Alkonyl et al., 1975; Foster and Harris, 1987; Harris et al., 1987) and the regulation of pyruvate dehydrogenase activity (Carlin et al., 1990, Constantin-Teodosiu et al., 1991). Accumulation of acetylcarnitine provides a metabolic sink for the temporary storage of 2-carbon acetyl units formed from pyruvate and ß-oxidation of FFA, as well as preventing the local depletion of coenzyme A, the effect of which would be to inhibit functioning of the TCA cycle. The accumulation of acetylcarnitine following a warm-up exercise may account for 90% of the available L-carnitine pool in the muscle, providing a valuable source of 2-carbon units during the early stages of racing. The high concentration of L-carnitine found in muscle is clearly of advantage in respect of this role, and a further increase in concentration could in theory enhance its overall function still further. As in man, however, L-carnitine is only poorly absorbed in the horse (Harris et al., 1995) although oral supplementation with 10-60 g L-carnitine per day will effect a doubling or more of the plasma concentration (Foster et al., 1988). Despite this, prolonged oral supplementation at these levels for 58 days had no effect on the muscle content. In a further study, Harris et al. (1995) administered 10 g L-carnitine intravenously daily for 26 days. Despite plasma concentrations 30 times higher than normal immediately after infusion, and 3 times higher still after 6 h, no change was observed in the muscle content. Thus the case for manipulation of the muscle L-carnitine (by any means) is very tenuous, and this is probably true also for the human. Despite this, L-carnitine supplements continue to be popular among athletes and are often included in compound supplements claimed to enhance "fat burning." L-carnitine supplements have been available for horses for some years.

DICHLOROACETATE AND 2-CHLOROPROPIONATE

The activation of pyruvate dehydrogenase through the use of dichloroacetate (DCA) or 2-chloropropionate (2-CP) does, however, appear to be effective in increasing the oxidation rate of pyruvate during the early stages of exercise, and to attenuate the accumulation of lactate under these conditions. This has been demonstrated in man (Mercier et al., 1994, Timmons et al., 1998) and dogs (Timmons et al., 1996), following oral and intravenous administration, but has yet to be shown to be an effective method of limiting lactate under race conditions. No studies of either DCA or 2-CP have been undertaken in the horse, although anecdotal evidence indicates that the former has been used in racing. Gannon and Kendall (1982) reported a positive effect of DCA on the performance of greyhound dogs when this was administered as the diisopropylammonium salt in conjunction with N,N-dimethylglycine.

BRANCHED CHAIN AMINO ACIDS (BCAA)

Dietary supplied BCAA may affect performance by increasing the concentration of TCA intermediates (anaplerosis) available for condensation with acetylCoA enabling an increase in the turnover rate of the cycle, as well as an effect on factors contributing to central fatigue. While BCAA supplements are available to the human athlete, data on their ergogenic effect following oral ingestion is contradictory (e.g. Bigard et al., 1996; Blomstrand et al., 1996). In the study of Blomstrand et al. (1996), a positive effect on metabolism (increased alanine synthesis, reduced fall in muscle glutamate, and lower glycogen utilization during exercise) was apparent and was favorable to an increase in endurance performance. Little or no quantitative data on the uptake, metabolism and effect on performance of BCAA are currently available for the competitive horse, although once again BCAA supplements are available.

III. Improved Intracellular Acid-Base Regulation

It is inevitable during racing that lactate accumulation will occur in muscle, decreasing the intracellular pH. As the race continues, loss of adenine nucleotide may commence, signifying a breakdown in the normal mechanisms regulating ADP homeostasis at the myosin-actin cross-bridge interface. It is probable that the two events are linked, i.e. the onset of adenine nucleotide loss and pH decrease (Sewell and Harris, 1992; Sewell et al., 1992), while the consequential rise in ADP may initiate a loss of performance arising from local muscle fatigue. It is essential, if performance is to be maintained, that H^+ ions released with lactate are either transported out of the muscle fibers or neutralized by physicochemical buffering.

SODIUM BICARBONATE

Sodium bicarbonate has been used widely in racing to effect an improvement in intracellular acid-base regulation. It is cheap, easily obtained, relatively simple to administer and has proved effective in facilitating H^+ removal from the working muscle fibers. Results from different studies, however, have been contradictory, possibly reflecting the wide range of doses used and inadequate information on the changes with time in plasma bicarbonate and pH following ingestion. Greenhaff et al. (1990) reported a peak increase in plasma acid-base excess (ABE) of 6.8 mmol/l 6 h following intubation of 0.6 g $NaHCO_3$/kg bwt. A lower response was seen using 0.3 g/kg bwt, a dose used in the horse by Lawrence et al. (1987) where it resulted in improved maintenance of plasma pH during exercise. Subsequent studies (Greenhaff et al., 1991b) using 0.6 g/kg bwt showed a reduced loss of muscle adenine nucleotide to IMP during a 2- min standardized treadmill exercise test resulting in a blood lactate concentration of approximately 20 mmol/l. These results are indicative of an improvement in *intracellular* pH control during the 2 minutes of exercise. Lactate efflux from muscle into plasma appeared to be increased with bicarbonate. The effect of $NaHCO_3$ administration upon actual race performance, however, remains unresolved. Lloyd et al. (1993) observed a detrimental effect of 1.0 g/kg bwt during treadmill exercise at approximately 110% VO_2max. No effect of 0.6 g/kg bwt was observed by Greenhaff et al. (1991a) during a 1000 m field test confirming reports of Kelsö et al. (1987). However, field studies are particularly difficult to undertake and the intrinsic high within-horse variance in performance time (measurable on test and retest), as well as differences in the pattern of the exercise performed, could easily obscure any small effects upon performance. Thus, although there is a clear rationale for the use of $NaHCO_3$, and its effect in the target tissue (plasma) has been established, there is inadequate evidence of any effect on performance under field conditions.

INTRACELLULAR PHYSICOCHEMICAL BUFFERING

Sodium bicarbonate, however, provides only an indirect approach to improving intracellular acid-base status, necessitating as it does the efflux of H^+ ions out of the cell. The primary defense against intracellular H^+ increase is afforded by the cell's own physicochemical buffers made up of organic and inorganic phosphates, amino acids and proteins, and the histidine dipeptide (of which the predominant species in equine muscle is carnosine). In equine muscle carnosine may account for 30% or more of the physicochemical buffering in type II fibers (Sewell et al., 1991) where it is found in highest concentrations (Dunnett and Harris, 1995; Harris et al., 1998). Although its importance to intracellular acid-base regulation is evident, the physiology governing its synthesis and metabolism remains to be elucidated. Addition of 0.4% L-histidine to the feed over 14 days resulted in a 19.2% increase in muscle carnosine content, compared to horses fed a diet containing 0.14% L-histidine (Powell et al., 1991). The increase, however, was not statistically significant. The alternative is that carnosine synthesis is limited by the availability of ß-alanine. However, administration of this is known to result in the efflux of the beta amino acid, taurine, from tissues including heart and skeletal muscle and in humans is associated with symptoms of paresthesia, even at low doses (R.C.Harris, M.Dunnett, J.Fallowfield and J.Coakley, unpublished).

Phosphate potentially constitutes the next largest physicochemical buffer in muscle, with the majority of this in the resting state being combined with creatine to form phosphorylcreatine (PCr). PCr is a relatively weak buffer of H^+ ions at normal physiological pH. However, a much bigger buffering effect is observed only after release of the PCr bound phosphate, ultimately to form inorganic and sugar phosphates. Increasing the PCr content of resting muscle by creatine (Cr) "loading" (discussed in the next section) may (at least in the human) increase the potential buffering from this pool by as much as 10%.

IV. Increased Phosphagen Support of ATP Metabolism

The continued accumulation of lactate in working muscle fibers will ultimately exceed the capacity of the physicochemical buffers and the cell's capacity to transport H^+ ions. Intracellular pH will fall affecting both the contractile process and the mechanisms regulating ADP removal at myosin-actin cross-bridge sites. Failure to maintain normal ATP/ADP homeostasis is a major threat to the continuation of the cross-bridge cycling rate necessary to maintain work output. Although improvements in the pyruvate oxidation rate (see previous) and H^+ buffering and transport (see previous) will delay the point when this is reached, improvement in the support given to ATP/ADP homeostasis is an important last step where metabolic intervention may be used to improve performance, in particular manipulation of the PCr content, which functions in the cell as the "low ADP threshold sensor."

The comprehensive role of PCr in muscle as a buffer to ADP accumulation, and high energy phosphate transfer and integration within the cell, is summarized in Figure 1. As already noted, PCr also constitutes the largest pool of metabolically active phosphate able to contribute to acid-base regulation when this is released from PCr (with exercise leading to a net fall in PCr). The size of the PCr store at the start of exercise will be important both to buffering ADP accumulation as well as H^+ accumulation. A greater concentration of free Cr will also contribute to a faster resynthesis, in absolute terms, in PCr (Greenhaff et al., 1994) which will be especially important in events requiring intensive intermittent exercise.

Studies in man have shown that Cr administered orally is absorbed rapidly from the gut and taken up into muscle (Harris et al., 1992). Supplementation of the diet with 4 x 5g CrH_2O for 5 days resulted in a 30% increase in the muscle [Cr + PCr] store in some individuals, with a proportionate increase occurring in both PCr and Cr. A maximum content in human muscle of around 160-180 mmol/kg dm was indicated in these studies. Lower doses in humans of 2-3 g taken daily, but over a longer period, will bring about a similar increase in the muscle [PCr + Cr] content. Uptake into muscle is by means of a transporter and is greatest in vegetarian subjects who do not regularly encounter Cr in their diet. Creatine uptake appears depressed in subjects regularly taking Cr supplements (R.C. Harris, unpublished). It is facilitated locally by exercise (Harris et al., 1992), and in the non-exercised state by insulin (Haughland and Chang, 1975; Green et al., 1995). Elevation of the muscle Cr and PCr contents has been shown in numerous studies to increase the capacity for sustained or intermittent hard exercise (Balsom et al., 1993a; Greenhaff et al., 1993; Harris et al., 1993; Birch et al., 1994) and may also exert an anabolic effect increasing peak strength (Earnest et al., 1995; Vandenberghe et al., 1997). Muscle Cr elevation does not appear to enhance the capacity for prolonged submaximal exercise (Balsom et al., 1993b; Stroud et al., 1994) although it is probable that an effect would be seen with changes in pace with exercise continued to the point of marked glycogen depletion (affecting ATP/ADP homeostasis). The increased capacity with Cr loading to undertake short term exhaustive exercise is associated with a reduction in adenine nucleotide loss (Greenhaff et al., 1993; Balsom et al., 1993a).

Creatine is rapidly absorbed in the dog (Harris and Lowe, 1995) where in the feral state up to 5 g may be ingested in a single meal (calculated for a dog of 35 kg). In contrast, in the horse Cr is only poorly absorbed (Sewell and Harris, 1995). Intubation of 50 mg/kg bwt resulted in an increased plasma concentration from 40 to 100 Fmol/l after 4-6 h; the same dose resulted in an increase to 800-1000 Fmol/l in the human. Greater absorption in the horse may occur when Cr is coadministered with feed (R.C. Harris, unpublished), but is still far below that observed in man and dog. Administration of 3 x 50mg/kg bwt added to the drinking water over 13 days had no effect on the muscle [PCr + Cr] content (Sewell and Harris, 1995). Thus although there is again a clear rationale for an effect on performance by Cr and effects have been established in man (and partially in dog), adequate evidence for an effect of Cr supplementation is lacking in the competition horse.

Concluding Remarks

Several issues here will, inevitably, be viewed with some misgivings by bodies governing equestrian sports. In some areas the edges between what is and what is not acceptable are blurred by the fact that many ergogenic aids are found naturally in the body. For example, in the case of carnosine, would high levels of either of its precursors be regarded as acceptable in the knowledge that, if this were to result in an increase in muscle carnosine, this could affect performance? One of these precursors, L-histidine, occurs naturally in the free or protein bound form in feeds, but not so ß-alanine. If this defies the rules of equine competition, can we justifiably administer thiamine, which is again a metabolic regulator, in this case a cofactor of pyruvate dehydrogenase? Certainly it would be difficult to see any greater justification for the latter compared, for instance, with carnitine which functions in the same relative metabolic area and which like thiamine is found in the normal diet.

Because the horse is herbivorous, Cr is not a natural component of the equine diet, in contrast to man and the dog. There is no doubting the role of Cr and PCr in energy metabolism in muscle and other tissues (e.g. brain, heart, spermatozoa, retina) and that a low content may not only impair performance but also enhance the degradation of adenine nucleotide to IMP and ultimately to uric acid. Degradation of IMP to uric acid is one of several routes by which highly reactive free radicals are formed and which during exercise may challenge the antioxidant defenses (Mills et al., 1997). A high Cr (and carnosine content) may in this case be regarded as beneficial to the horse, helping to limit the damage resulting from intensive exercise. Finally, in humans some of the most encouraging reports of the effectiveness of Cr have been in the supplementation of the elderly. While veteran races for horses and dogs have yet to be introduced, should we not be considering investigating the use of dietary Cr in, for instance, elderly working dogs? How different should we now regard the feeding of compounds such as Cr, carnitine and carnosine, or their precursors, to the feeding of vitamins, selenium or carbohydrate supplements?

References

Alkonyi I, Kernor J and Sandor A (1975) The possible role of carnitine and carnitine acetyltransferase in the contracting from skeletal muscle. *FEBS Lett* 52, 265-268.

Balsom P D, Ekblom B, Söderlund K, Sjödin B and Hultman E (1993a) Creatine uptake and dynamic high intensity exercise. *Scand J Med Sci Sports* 3, 143-149.

Balsom P D, Harridge S, Söderlund K, Sjödin B and Ekblom B (1993b) Creatine supplementation per se does not enhance endurance exercise performance. *Acta Physiol Scand* 149, 521-523.

Beckers E J, Jeukendrup A E, Brouns F, Wagenmakers A J M and Saris W H M (1992) Gastric emptying of carbohydrate-medium chain triglyceride suspensions at rest. *Int J Sports Med* 13, 581-584.

Bergström J and Hultman E Muscle glycogen synthesis after exercise. An enhancing factor localized to the muscle cells in man. *Nature (London)* 210, 309-310.

Bergström J, Hermansen L, Hultman E and Saltin B (1967) Diet, muscle glycogen and physical performance. *Acta Physiol Scand* 71, 140-150.

Bigard A X, Lavier P, Ullmann L, Legrand H, Douce P and Guezennec C Y (1996) Branched-chain amino acid supplementation during repeated prolonged skiing exercises at altitude. *Int J Sports Nutrition* 6, 295-306.

Birch R, Noble D and Greenhaff P L (1994) The influence of dietary creatine supplementation on performance during repeated bouts of maximal isokinetic cycling in man. *Eur J Appl Physiol* 69, 268-270.

Blomstrand E, Ek S and Newsholme E A (1996) Influence of ingesting a solution of branched-chain amino acids on plasma and muscle concentrations of amino acids during prolonged submaximal exercise. *Nutrition* 12, 485-490.

Carlin J I, Harris R C, Cederblad G, Constantin-Teodosiu D, Snow D H and Hultman E (1990) Association between muscle acetyl-CoA and acetylcarnitine levels in the exercising horse. *J Appl Physiol* 69, 42-45.

Constantin-Teodosiu D, Carlin J I, Cederblad G, Harris R C and Hultman E (1991) Acetyl group accumulation and pyruvate dehydrogenase activity in human muscle during incremental exercise. *Acta Physiol Scand* 143, 367-372.

Davie A J, Evans D L, Hodgson D R and Rose R J (1995) Effects of intravenous dextrose infusion on muscle glycogen resynthesis after intense exercise. *Equine Exercise Physiology 4. Equine Vet J* Suppl 18, 195-198.

504   *Nutritional Ergogenic Aids in the Horse*

Dunnett, M and Harris R C (1995) Carnosine and taurine contents of different fibre types in the middle gluteal muscle of the Thoroughbred horse. *Equine Exercise Physiology 4. Equine Vet J*, Suppl. 18, 214-217.

Earnest C P, Snell P G, Rodriquez R, Almada A L and Mitchell T L (1995) The effect of creatine monohydrate ingestion on anaerobic power indices, muscular strength and body composition. *Acta Physiol Scand* 153, 207-209.

Eaton M D, Hodgson D R, Evans D L, Bryden W L and Rose R J (1995) Effect of a diet containing supplementary fat on the capacity for high intensity exercise. *Equine Exercise Physiology 4. Equine Vet J* Suppl 18, 353-356.

Foster C V L and Harris R C (1987) Formation of acetylcarnitine in muscle of horse during high intensity exercise. *Eur. J. Appl. Physiol.* 56, 639-642.

Foster C V L and Harris R C (1992) Total carnitine content of the middle gluteal muscle of Thoroughbred horses: normal values, variability and effect of acute exercise. *Equine Vet J.* 24, 52-57.

Gannon J R and Kendall R V (1982) A clinical evaluation of N,N-dimethylglycine (DMG) and diisopropylammonium dichloroacetate (DIPA) on the performance of racing greyhounds. *Canine Practice* 9, 6-13.

Green A L, Hultman E, Macdonald I A, Sewell D A, Greenhaff P L (1996) Carbohydrate ingestion augments skeletal muscle creatine accumulation during creatine supplementation in humans. *Am J Physiol* 271, E821-E826.

Greenhaff P L, Bodin K, Söderlund K and Hultman E (1994) The effect of oral creatine supplementation on skeletal muscle phosphocreatine resynthesis. *Am J Physiol* 266, E725-E730.

Greenhaff P L, Casey A, Short A H, Harris R C, Söderlund K and Hultman E (1993) The influence of oral creatine supplementation on muscle torque during repeated bouts of maximal voluntary exercise in man. *Clinical Science* 84, 565-571.

Greenhaff P L, Hanak J, Snow D H, Dobias P, Jahn P, Skalicky J and Harris R C (1991a) Metabolic alkalosis and exercise performance in the Thoroughbred horse. *Equine Exercise Physiol.* 3, 353-360.

Greenhaff P L, Harris R C, Snow D H, Sewell D A and Dunnett M (1991b) The influence of metabolic alkalosis upon exercise metabolism in the Thoroughbred horse. *Eur J Appl Physiol* 63, 129-134.

Greenhaff, P L, Snow, D H, Harris, R C and Roberts, C A (1990) Bicarbonate loading in the Thoroughbred horse: dose, method of administration and acid-base changes. *Equine Vet. J. Suppl 9, 83-85.*

Harris P A (1997) Energy sources and requirements of the exercising horse. *Annu. Rev. Nutr.* 17, 185 - 210.

Harris R C (1994) Naturally occurring substances. In: *Proc 10th Intl Conference Racing Analysts and Veterinarians, Stockholm, Sweden.* P Kallings, U Bondesson and E Houghton (eds) R & W Publications, Newmarket, pp79 - 84.

Harris R C, Dunnett M and Greenhaff P L (1998) Carnosine and taurine contents in individual fibers in human vastus lateralis muscle. *J Sports Science* (in press).

Harris R C, Foster, C V L and Hultman, E (1987) Acetylcarnitine formation during intense muscular contraction in humans. *J Appl Physiol* 63, 440-442.

Harris R C, Foster C V L and Snow D H (1995) Plasma carnitine concentration and uptake into muscle with oral and intravenous administration. *Equine Exercise Physiology 4. Equine Vet. J.,* Suppl. 18, 382-387.

Harris R C and Lowe J A (1995) Absorption of creatine from meat or other dietary sources by the dog. *Veterinary Record* 137, 595.

Harris R C, Marlin D J and Snow D H (1987) Metabolic response to maximal exercise of 800 and 2000 m in the Thoroughbred horse. *J Appl Physiol* 63, 12-19.

Harris R C, Söderlund K and Hultman E (1992) Elevation of creatine in resting and exercised muscle in normal subjects by creatine supplementation. *Clinical Science* 83, 367-374.

Harris R C, Viru M G, Greenhaff P L and Hultman E (1993) The effect of oral creatine supplementation on running performance during short term exercise in man. *J Physiol* 467, 74.

Hodgson D R, Rose R J, Allen J R and DiMauro J (1984) Glycogen depletion patterns in horses performing maximal exercise. *Res Vet Sci* 36, 169-173.

Hughes S L, Potter G D, Greene L W, Odom T W and Murray-Gerzik M (1995) Adaptation of Thoroughbred horses in training to a fat supplemented diet. *Equine Exercise Physiology 4. Equine Vet J* Suppl 18, 349-352.

Jeukendrup A E, Saris W H M, Schrauwen P, Brouns F, and Wagenmakers A J M (1995) Metabolic availability of medium-chain triglycerides coingested with carbohydrates during prolonged exercise *J Appl Physiol.* 79, 756-762.

Jeukendrup A E, Saris W H M, Van Diesen R, Brouns F, and Wagenmakers A J M (1996) Effect of endogenous carbohydrate availability on oral medium chain triglyceride oxidation during prolonged exercise. *J Appl Physiol* 80, 949-954.

Kelsö T B, Hodgson D R, Will E H, Bayley W M, Grant B D and Gollnick P D (1987) Bicarbonate administration and muscle metabolism during high-intensity exercise. In: *Equine Exercise Physiology 2.* J R Gillespie and N E Robinson (eds) ICEEP Publications, California, pp 438-447.

Kronfeld D S (1996) Dietary fat affects heat production and other variables of equine performance under hot and humid conditions. *Equine Vet J.* 11 (Suppl.22) 24 - 35.

Laaksonen R, Fogelholm M, Himberg J J, Laakso J and Salorinne Y (1995) Ubiquinone supplementation and exercise capacity in trained younger and older men. *Eur J Appl Physiol* 72, 95-100.

Lawrence L M, Miller P A, Bechtel P J, Kane R A, Kurz E V and Smith J S (1987) The effect of sodium bicarbonate ingestion on blood parameters in exercising horses. In: *Equine Exercise Physiology 2.* J R Gillespie and N E Robinson (eds) ICEEP Publications, California, pp 448-455.

Lloyd D R, Evans D L, Hodgson D R, Suann C J and Rose R J (1993) Effects of sodium bicarbonate on cardiorespiratory measurements and exercise capacity in Thoroughbred horses. *Equine Vet J* 25, 125-129.

Massicotte D, Peronnet F, Brisson G R and Hillarie-Marcel C (1992) Oxidation of exogenous medium-chain free fatty acids during prolonged exercise: comparison with glucose. *J Appl Physiol* 73, 1334-1339.

Mercier B, Granier P, Mercier J, Anselme F, Ribes G and Préfaut C (1994) Effects of 2-chloropropionate on venous blood lactate concentration and anaerobic power during periods of incremental intensive exercise in humans. *Eur J Appl Physiol* 68, 425-429.

Mercier B, Granier P, Mercier J, Anselme F, Ribes G and Préfaut C (1994) Effects of 2-chloropropionate on venous blood lactate concentration and anaerobic power during periods of incremental intensive exercise in humans. *Eur J Appl Physiol* 68, 425-429.

Meyers M C, Potter G D, Evans J W, Greene L W and Crouse S F (1989) Physiologic and metabolic responses of exercising horses fed added dietary fat. *J Equine Vet Sci* 9, 218-223.

Mills P C, Smith N C, Harris R C and Harris P (1997) Effect of allopurinol on the formation of reactive oxygen species during intense exercise in the horse. *Res Vet Science* 62, 11-16.

Orme, C E, Harris, R C and Marlin D J (1995) Effect of elevated plasma FFA on fat utilization during low intensity exercise. *Equine Exercise Physiology 4. Equine Vet J*, Suppl. 18, 199-204.

Orme C E and Harris R C (1997) A comparison of the lipolytic and anticoagulative properties of heparin and pentosan polysulphate in the Thoroughbred horse. *Acta Physiol Scand* 159, 179-185.

Orme C E, Harris R C, Marlin D J and Hurley J S (1997) Metabolic adaptation to a fat supplemented diet in the Thoroughbred horse. *Br J Nutr* 78, 443-458.

Powell D M, Lawrence L M, Novakofski J, Moser L R and Biel M J (1991) Effect of dietary L-histidine supplementation on muscle carnosine and buffering capacity in horses. *Proc 12th Equine Nutrition and Physiology Symposium* 143, pp 115-119.

Porter D A, Costill D L, Zachwieja J J, Krzeminska K, Fink W J, Wagner E and Folkers K (1995) The effect of oral coenzyme Q10 on the exercise tolerance of middle-aged, untrained men. *Int J Sports Med* 16, 421-427.

Potter G D, Hughes S L, Julen T R, Swinney D D L (1992) A review of research on digestion and utilization of fat by the equine. In *Eur Konf Ernahr. Pferdes., Pferdeheilkunde*, Hannover, pp 119 -23.

Scott B D, Potter G D, Greene L W , Hargis P S and Anderson J G (1992) Efficacy of a fat-supplemented diet on muscle glycogen concentration in exercising Thoroughbred horses maintained in varying body conditions. *J Equine Vet Sci* 12, 109-113.

Sewell, D A, Harris, R C and Dunnett, M (1991) Carnosine accounts for most of the variation in physico-chemical buffering in equine muscle. *Equine Exercise Physiol. 3*, 276-280.

Sewell, D A and Harris, R C (1992) Adenine nucleotide degradation in the Thoroughbred horse with increasing exercise duration. *Eur. J. Appl. Physiol.* 65, 271-277.

Sewell D A and Harris R C (1995) Effect of creatine supplementation in the Thoroughbred horse. *Equine Exercise Physiology 4. Equine Vet J* Suppl. 18, 239-242.

Sewell, D A, Harris, R C, Hanak J and Jahn P (1992) Muscle adenine nucleotide degradation in the Thoroughbred horse as a consequence of racing. *Comp Biochem Physiol.* 101B, 375-381.

Snow D H (1992) A review of nutritional aids to energy production for athletic performance. *AESM Proc.* pp 91-95.

Snow D H, Baxter P B and Rose R J (1981) Muscle fiber composition and glycogen depletion in horses competing in an endurance ride. *Veterinary Record* 108, 374-378.

Snow D H and Harris R C (1991) Effects of daily exercise on muscle glycogen in the Thoroughbred racehorse. *Equine Exercise Physiol. 3*, 299-304.

Snow D H, Harris R C, Harman J and Marlin D J (1987) Glycogen repletion following different diets. In: *Equine Exercise Physiology 2*, Gillespie, J.R. and Robinson, N.E. eds, ICEEP Publications, California, pp 701-710.

Snow D H, Kerr M G, Nimmo M A and Abbott E M (1982) Alterations in blood, sweat, urine and muscle composition during prolonged exercise in the horse. *Vet Rec* 110, 377-384.

Steenge G R, Lambourne J, Casey A, Macdonald I A, Greenhaff P L (1998) Stimulatory effect of insulin on creatine accumulation in human skeletal muscle. *Am J Physiol* 275, E974-E979.

Stroud M A, Holliman D, Bell D, Green A L, Macdonald I A and Greenhaff P L (1994) Effect of oral creatine supplementation on respiratory gas exchange and blood lactate accumulation during steady state incremental treadmill exercise and recovery in man. *Clinical Science* 87, 707-710.

Timmons J A, Constantin-Teodosiu D, Poucher S M, Worrall V, Macdonald I A and Greenhaff P L (1995) Dichloroacetate enhances skeletal muscle performance during ischemic work. *J Clin Invest* 97, 879-883.

Timmons J A, Gustavsson T, Sundberg C J, Jansson E, Hultman E, Kaiser L, Chwalbinska-Moneta J, Constantin-Teodosiu D, Macdonald I A and Greenhaff P L (1998) Substrate availability limits human skeletal muscle oxidative ATP regeneration at the onset of ischemic exercise. *J Clin Invest* 101, 79-86.

Topliff D R, Potter G D, Dutson T R, Kreider J L and Jessup G T (1983) Diet manipulation and muscle glycogen in the equine. *Proc 8th Equine Nutrition and Physiology Symposium* pp 224-229.

Topliff D R, Potter G D, Kreider J L, Dutson T R and Jessup G T (1985) Diet manipulation, muscle glycogen metabolism and anaerobic work performance in the equine. *Proc 9th Equine Nutrition and Physiology Symposium* pp 119-124.

Topliff D R, Lee S F and Freeman D W (1987) Muscle glycogen, plasma glucose and free fatty acids in exercising horses fed varying levels of starch. *Proc 10th Equine Nutrition and Physiology Symposium* pp 421-424.

Vandenberghe K, Goris M, Van Hecke P, Van Leemputte M, Vangerven L and Hespel P (1997) Long-term creatine intake is beneficial to muscle performance during training. *J Appl Physiol* 83, 2055-2063.

Zhang Y, Åberg F, Appelkvist E L, Dallner G and Ernster L. (1995) Uptake of dietary coenzyme Q supplementation is limited in rats. *J Nutr* 125, 446-453.

Zhang Y, Turunen M and Appelkvist E L (1996) Restricted uptake of dietary coenzyme Q is in contrast to the unrestricted uptake of alpha-tocopherol in rat organs and cells. *J Nutr* 126, 2089-2097.

GLYCOGEN DEPLETION AND REPLETION IN THE HORSE – POSSIBLE LIMITING FACTOR IN PERFORMANCE (REVIEW)

A.J. DAVIE, D. L. EVANS, D. R. HODGSON AND R. J. ROSE
The University of Sydney, Australia

Introduction

The importance of glycogen as a substrate and its role in both performance and recovery has been extensively studied in human athletes. Glycogen's key role is as a substrate for muscle. Contraction of muscle involves the conversion of chemical energy into mechanical work, with adenosine triphosphate (ATP), adenosine diphosphate (ADP), inorganic phosphate (P_i), magnesium (Mg^{2+}) and hydrogen (H^+) being directly involved in this process. As ATP supply within the muscle is limited, the energy for the rephosphorylation of ATP comes from creatine phosphate (PCr) and substrate breakdown (carbohydrate and fat). However, the rate of ATP production from fat oxidation is slow and limits its usage to low intensity activities (Sahlin, 1986). As the intensity increases above 30% to 50% VO_{2max}, the importance of a contribution of carbohydrates increases with the contribution of ATP from glycolysis and PCr increasing at intensities above 70% to 80% VO_{2max}. Carbohydrates for muscle energy supply can be provided both endogenously (within the muscle) or exogenously (liver and blood). The importance of the initial muscle glycogen concentration in endurance performance in humans has been reported by Bergström et al. (1967), Hargreaves et al. (1984) and Coyle (1991) and for the horse by studies of endurance rides (Snow et al., 1981; Hodgson et al., 1983), road and track components of three-day event competitions (Hodgson et al., 1984) and during a 4-hour slow trot (Lindholm, 1979).

In contrast to the findings supporting the importance of the muscle's initial glycogen concentration on endurance performance, the same support has not been coming forth for muscle glycogen and sprint performance in human studies (Costill et al., 1981; Symons and Jacobs, 1989) or high intensity exercise in horses (Topliff et al., 1983; Topliff et al., 1985; Davie, 1996). The muscle's initial glycogen concentration seems to be more important as the time period of the exercise increases with the short term high intensity exercise not so dependent on the muscle glycogen concentrations.

Glycogen Use During Exercise

The pattern of glycogen usage and selective fiber depletion in humans with varying intensities of exercise (Essén and Henriksson, 1974) has also been demonstrated for the horse (Essén et al., 1984). In low intensity activities (endurance rides of 100 km), the ST fibers display the highest degree of depletion with FT fibers showing moderate levels of depletion. This selective glycogen depletion was also reported by Davie (1996) in which horses exercising at intensities of 60% VO_{2max} for 30 min resulted in some glycogen depletion in all fiber types, with most depletion evident in the ST and FTH fibers. Essén and Henriksson (1974) reported that during a 2-hour exercise period, the glycogen concentration in FT fibers had decreased by 29% whereas the glycogen concentration of ST fibers had decreased by 86% of their resting value. Rates of glycogen utilization during low intensity exercise have been reported at 4.1 mmol/kg (dwt) (Hodgson, 1984). For high intensity exercise, Lindholm (1974) reported that when trotting speed was increased from approximately 5 m/s to 12.5 m/s the rate of glycogen usage increased from 0.3 to 14 mmol/kg/min. This rate of glycogen usage is similar to that reported by Hodgson (1984) of 15.4 mmol/kg/min during a graded exercise test and that of Snow and Harris (1991) in which they found similar glycogen depletion rates following a 1000 m and 1600 m gallop, and Snow et al. (1985) following four 620 m gallops. However, these rates are much lower than those reported by Harris et al. (1987) of 160 and 64 mmol/kg/min (dwt) during an 800 m and 1000 m gallop.

Fatigue

Exercise requires an integration of many systems, each containing varied elements, and any factor that upsets this integration could cause fatigue (Brooks and Fahey, 1985). The cause of this inability of muscle to maintain a specific level of contraction could lie in either the central nervous system, the final motor neuron, the neuromuscular junction or the muscle (Åstrand and Rodahl, 1970; Saltin and Karlsson, 1971; Brooks and Fahey, 1985; Sahlin, 1986; Enoka and Stuart, 1992).

One problem in trying to identify the cause of fatigue is that it is a multidimensional phenomenon and can vary in accordance with the activity itself, the training and physiological status of the individual and the environmental conditions (Brooks and Fahey, 1985). Fatigue may take place at a single site, in the case of a maximal lift, or it may involve many sites when there is depletion of energy supplies, accumulation of waste products, changes in pH, depletion of nervous system transmitters and dehydration (Kraemer, 1983).

The onset of fatigue is most often associated with either the accumulation of metabolic by-products such as hydrogen and inorganic phosphate or a decline in muscle glycogen concentration. In an endeavor to improve endurance performance, athletes have attempted to increase muscle glycogen storage capacity 2 to 3 times greater than normal. The issue of glycogen loading, as it applies to humans, would seem to have little practical application for the racing horse based on the degree of depletion that occurs during such events (Harris et al., 1987; Snow and Harris, 1991). Further, Topliff et al. (1983 and 1985) reported no improvement in performance with increases in muscle glycogen concentration of up to 36% above resting concentrations.

The Effects of Muscle Glycogen Concentration on Low Intensity Exercise Performance

The onset of fatigue in human endurance events has been related to muscle glycogen depletion (Bergström et al.,1967; Hargreaves et al., 1984; Coyle, 1991). Bergström et al. (1967) found a strong correlation between the initial muscle glycogen concentration and work time during endurance exercise. At exercise intensities of between 70% to 80% VO_{2max}, exhaustion coincided with the muscle's glycogen stores being depleted (Saltin and Karlsson, 1971).

Topliff et al. (1983) reported that in horses the onset of fatigue during a run at 3.0 m/s on a treadmill was earlier when the muscle glycogen concentration was reduced prior to the run. In endurance rides of 160 km, muscle glycogen depletion of more than 70% of ST fibers and substantial depletion of FT have been reported in horses (Hodgson et al., 1983). In the road and track components of three day event competitions, mean decreases in muscle glycogen concentration of 306 mmol/kg (dwt) with a mean rate of utilization of 4.1 mmol/kg (dwt) have been reported (Hodgson et al., 1984).

The effects of a reduced skeletal muscle glycogen concentration on physiological responses to low intensity treadmill exercise were examined by Davie (1996). Reductions in skeletal muscle glycogen concentrations of up to 29% had no significant effect on VO_2, heart rate and temperature during a 30-min treadmill run at 60% VO_{2max}. A decreased glycogen concentration did not affect the rate of glycolysis, confirming the findings of Topliff et al. (1983). In 7 of the 18 endurance runs, horses were unable to complete the 30 min of exercise, even though substantial quantities of muscle glycogen were still available, suggesting that muscle glycogen concentration was not the key contributing factor to fatigue. The horses' inability to complete the treadmill run also was not associated with lactate accumulation. Other factors that may have contributed to the fatigue are accumulation of ammonia or increased temperature (Greenhaff et al., 1991; McConaghy et al., 1995). It is interesting to note that in some human studies, similar results have been reported. Costill et al. (1971), Gollnick et al. (1973) and Symons and Jacobs (1989) reported that for both prolonged and short exhaustive runs, glycogen depletion was the unlikely cause of fatigue. However, Gollnick et al. (1973) stated that even though the total muscle glycogen concentration had decreased by 63%, the loss of glycogen from the FT fibers may have been sufficient to result in the inability of these fibers to function adequately. In contrast, Bergström et al. (1967) found a good correlation between the initial muscle glycogen concentration and exercise time.

The Effects of Muscle Glycogen Concentration on High Intensity Exercise Performance

In both humans and horses, the muscle's initial glycogen concentration appears to be unimportant for performance of high intensity exercise (Costill et al., 1981; Symons and Jacobs, 1989; Davie, 1996). Further, the provision of a carbohydrate supplement prior to high intensity exercise in humans also has been shown not to be beneficial. Glycogen supercompensation, which is used for performance enhancement in prolonged exercise, does not enhance high intensity exercise performance in humans (Housh et al., 1990; Madsen et al., 1990).

In contrast to human studies, Topliff et al. (1985) reported that a decrease in muscle glycogen concentration in the horse does affect its capacity for anaerobic work. This is in contrast to the work by Davie (1996) who failed to support Topliff's findings. In a study in which horses ran on a treadmill to exhaustion at 115% VO_{2max} 5 hours after a glycogen depletion run (high intensity exercise), there was no significant difference between horses performing with either a decreased muscle glycogen concentration or normal resting concentrations in reference to oxygen uptake or plasma lactate concentrations. Muscle glycogen concentration in the horse was reduced by 22% without having a significant effect on physical performance.

The lack of an effect of a reduced muscle glycogen concentration on lactate concentration during exercise is in contrast to that reported for humans. Studies in humans have shown that lactate production is reduced with a decreased muscle glycogen concentration. The basis of a reduced lactate formation with a low initial muscle glycogen concentration is the relationship between energy demand and by-product removal. If the glycogen concentrations are low to begin with, then the available substrate for glycolysis is reduced, leading to a reduced rate of glycolysis and reduced lactate production. Reductions in muscle glycogen concentration of greater than 22% may be required in horses before an effect on rates of lactate accumulation are observed.

The rate of change of the VO_2 at the commencement of high intensity exercise in this study was also not affected by the initial glycogen concentrations. There was no significant difference between horses with reduced versus normal muscle glycogen concentrations for time to peak oxygen uptake, with all horses reaching peak oxygen uptake levels between 45 and 60 seconds. This time to peak VO_{2max} supports the findings of Bellenger et al. (1994) who reported that horses reached 90% to 95% of mean steady state VO_2 by 45 seconds. The VO_2 responses are in agreement with responses reported for humans. Bergström et al. (1967) reported similar oxygen uptake responses during exercise regardless of the differing initial glycogen concentrations. However, Widrick et al. (1993) reported that in subjects performing time trials of 120-min duration, under conditions of varying initial muscle glycogen concentrations and carbohydrate feeding throughout the trials, in the first 71% of each trial the VO_2 was similar for all conditions. In the low glycogen trial VO_2 was lower at the end of the exercise period.

Key factors causing fatigue in humans parallel factors causing fatigue in the horse. In both human and horse endurance events, heat stress and substrate availability have been indicated as key factors in fatigue. In the horse substantial depletion of muscle glycogen has been reported during endurance rides, supporting the concept that substrate availability may play a key role in fatigue. Miller and Lawrence (1986) and Snow and Harris (1991) argued that the changes in muscle lactate and pH are also unlikely to be the cause of fatigue in such events. Even though some metabolic acidosis occurs, the changes are unlikely to be of sufficient magnitude to induce fatigue. Heat stress, however, has been shown to play a role in reduced performance in such events (McConaghy et al., 1995).

In sprinting events such as Thoroughbred racing, the onset of fatigue is unlikely to be muscle glycogen related. In such events the total time of performance is from 57 s for 1000 m to 3 min 20 s for 3200 m. Fatigue in such events could be either central or peripheral based with the most likely factors being changes in the chemistry of the muscle cell. Lactate concentrations as high as 31 mmol/L and muscle pH of 7.0 have been reported following a 1000 m gallop (Harris et al., 1987). The accumulations of lactate and resultant fall in cell pH have been demonstrated as factors affecting both energy production and excitation contraction in the muscle.

Muscle Glycogen Resynthesis Following Exercise

The rate of muscle glycogen resynthesis following exercise is affected by factors such as the type of fibers depleted (Terjung et al., 1974), glucose dose administered after exercise (Blom et al., 1986; Blom et al., 1987), the timing of the administration of the supplement after exercise (Ivy et al., 1988), the activity state of glycogen synthetase and the initial glycogen concentration (Danforth, 1965; Larner et al., 1967; Bergström et al., 1972; Kochan et al., 1979). The major limiting factor in glycogen synthesis appears to be the muscle's ability to synthesize glycogen, rather than the availability of intracellular substrate (Fell et al., 1982).

In humans, rates of muscle glycogen resynthesis of 23 mmol/kg/h (dwt) during the first 2 hours of recovery, following the provision of 3.0 g/kg of glucose (Ivy et al., 1988), 30.8 mmol/kg/h (dwt) following the ingestion of 2 g/kg of a 25% carbohydrate solution (Ivy et al., 1988) and 23 mmol/kg/h (dwt) with the ingestion of 0.7g/kg of glucose every 2 hours for 8 hours (Blom et al., 1987) have been reported.

In horses, muscle glycogen resynthesis rates of 12.5 mmol/kg/h (dwt) for the first 8 hours following a high carbohydrate diet plus infusion (Snow et al., 1987), 5.6 mmol/kg/h (dwt) and 7.8 mmol/kg/h (dwt) for the 20-24 hour period following a 160 km endurance ride and for the first 4 hours immediately following a treadmill exercise test (Hodgson, 1984) and 19.8 mmol/kg/h for the first 6 hours and 14.6 mmol/kg/h and 7.1 mmol/kg/h for the first 12 and 24 hours following exercise (Davie, 1996) have been reported.

Effects of Fiber Type on Muscle Glycogen Resynthesis

The rate of muscle glycogen resynthesis depends on the type of muscle fiber. Conlee et al. (1978) showed that fast twitch red fibers, which have the highest total glycogen synthetase I activity, displayed the most rapid rate of glycogen synthesis, while fast twitch white fibers, which had the slowest rate of glycogen synthesis, had the lowest total glycogen synthetase I activity. Slow twitch red muscle which had an intermediate total glycogen synthetase activity, had an intermediate rate of glycogen synthesis. This difference in rate of resynthesis between fiber types is based on differences in the enzymatic capacity of the different muscle fiber types to convert glucose to glycogen.

Effects of Diet on Muscle Glycogen Resynthesis

Davie (1996) illustrated that the administration of 3.0 g/kg of glucose polymer, as either a single or split dose, after exercise did not have a significant effect on the rate of muscle glycogen resynthesis in the 24 hours after exercise. However, blood glucose and insulin concentrations were higher in treated horses than in controls. This suggests that differences in the blood glucose concentrations were not sufficiently different to affect glycogen resynthesis, or that blood glucose *per se* is not an important factor affecting the rate of muscle glycogen resynthesis in the horse, or that the contribution of glucose, via hepatic glycogenolysis and gluconeogenesis, was adequate to maintain a sufficient level of plasma glucose for glycogen resynthesis in the control group.

In a further study investigating the effect of post-exercise administration of dextrose (d-(+)-glucose) at a rate of 6 g/kg (bwt) intravenously as a 20% solution at a mean infusion rate of 1.67±0.05 l/h for 8 hours, it was reported that there was a significant increase in resynthesis rates. The rates reported of 19.8 mmol/kg/hr are much higher than those reported previously after a routine hay/grain diet in the post-exercise period (Hodgson, 1984) or when the additional infusion of 0.45 kg glucose over 6 hours was added to a high carbohydrate diet (Snow et al., 1987).

Based on the similar rates of resynthesis between the control group in the study by Davie (1996) and those reported by Hodgson (1984), it is suggested that even when feed is withheld, the capacity for glycogen resynthesis is similar to that found during normal feeding in the first 6 hours after exercise. For this to occur, gluconeogenesis must be adequate to maintain a plasma glucose concentration sufficient for glycogen resynthesis.

A glucose dose greater than 3 g/kg (bwt) administered intravenously can have an effect on glycogen replenishment rates. This finding may have potential practical implications for horses required to compete on successive days in endurance events. The provision of glucose intravenously overcomes the potential problems of gastrointestinal disturbances and laminitis which can be associated with oral glucose administration.

Conclusion

For events lasting longer than one hour, the general consensus is that initial muscle glycogen concentration plays a significant role in performance. There is no evidence that equine skeletal muscle glycogen concentration is increased by provision of extra dietary carbohydrate during recovery periods after exercise.

References

Åstrand, P. O. and K. Rodahl. 1970. "Textbook of work physiology." McGraw and Hill, NY.

Bellenger, S. J., A. J. Davie, D. L. Evans, D. R. Hodgson, R. J. Rose. 1994. "Effects of low intensity training on the kinetics of gas exchange in Standardbred horses": *In 4th International Conference on Equine Exercise Physiology. Qld. Australia :7.*

Bergström, J., L. Hermansen, E. Hultman and B. Saltin. 1967. "Diet, muscle glycogen and physical performance." *Acta Physiol. Scand.71:140-150.*

Bergström, J. and E. Hultman. 1967. "A study of the glycogen metabolism during exercise in man." *Scand. J. Clin. Lab. Invest.19: 218-228.*

Bergström, J., E. Hultman and A. E. Roch-Norlund. 1972. "Muscle glycogen synthetase in normal subjects." *Scand. J. Clin. Lab. Invest.29:231-236.*

Blom, Per C. S., N. K. Vøllestad and D. L. Costill. 1986. "Factors affecting changes in muscle glycogen concentration during and after prolonged exercise." *Acta Physiol. Scand. 128(Suppl 556):67-74.*

Blom, Per C. S., A. T. Hostmark, O. Vaage, K. R. Kardel and S. Maehlum. 1987. "Effect of different post-exercise sugar diets on the rate of muscle glycogen synthesis." *Med. Sci. Sports Exerc. 19(5):491-496.*

Brooks, G. A. and T. D. Fahey. 1985. "Exercise Physiology: Human bioenergetics and its application." Macmillan Pub. Comp.NY.

Conlee, R. K., R. C. Hickson, W. W. Winder, J. M. Hagberg and J. O. Holloszy. 1978. "Regulation of glycogen resynthesis in muscles of rats following exercise." *Am. J. Physiol. 235(3):R145-R150.*

Costill, D. L., K. Sparks, R. Gregor and C. Turner. 1971. "Muscle glycogen utilization during exhaustive running." *J. Appl. Physiol. 31(3): 353-356.*

Costill, D. L., W. M. Sherman, W. J. Fink, C. Maresh, M. Witten and J. M. Miller. 1981. "The role of dietary carbohydrates in muscle glycogen resynthesis after strenuous running." *Am. J. Clin. Nutr. 34:1831-1836.*

Coyle, E. F. 1991. "Carbohydrate feedings: Effects on metabolism performance and recovery." *Med Sport Sci. Basel, Karger. 32:1-14.*

Danforth, W. H. 1965. "Glycogen synthetase activity in skeletal muscle." *The J. Biol. Chem. 240(2):588-593.*

Davie, A. J. 1996. " The study of energy supply for exercise in horses and factors influencing glycogen resynthesis in equine skeletal muscle." *PhD Thesis.University of Sydney. NSW. Australia.*

Enoka, R. M. and D. G. Stuart. 1992. "Neurobiology of muscle fatigue." *J. Appl. Physiol. 72(5):1631-1648.*

Essén, B. and J. Henriksson. 1974. "Glycogen content of individual muscle fibers in man." *Acta Physiol. Scand. 90:645- 647.*

Essén-Gustavsson, B., K. Karlström and A. Lindholm. 1984. "Fiber types, enzyme activities and substrate utilisation in skeletal muscles of horses competing in endurance rides." *Equine Vet. J. 16(3):197-202.*

Farris, J. W., K. W. Hinchcliff, K. M. McKeever and D. R. Lamb. 1994. "Endurance performance is enhanced by glucose supplementation during submaximal exercise." *In: Proceedings of the 4th International Conference on Equine Exercise Physiology. Australia."*

Fell, R. D., S. E. Terblanche, J. L. Ivy, J. C. Young and J. O. Holloszy. 1982. "Effect of muscle glycogen content on glucose uptake following exercise." *J. Appl. Physiol. Respirat. Environ. Exercise Physiol. 52(2):434-437.*

Gollnick, P. D., R. B. Armstrong, W. L. Sembrowich, R. E. Shepherd and B. Saltin. 1973. "Glycogen depletion pattern in human skeletal muscle fibers after heavy exercise." *J. Appl. Physiol. 34(5):615-618.*

Greenhaff, P. L., J. B. Leiper, D. Ball and R. J. Maughan. 1991. "The influence of dietary manipulation on plasma ammonia accumulation during incremental exercise in man." *Eur. J. Appl. Physiol. 63:338-344.*

Hargreaves, M., D. L. Costill, A. Coggan, W. J. Fink and I. Nishibata. 1984. "Effect of carbohydrate feedings on muscle glycogen utilization and exercise performance." *Med. Sci. Sports Exerc. 16(3):219-222.*

Harris, R. C., Marlin, Snow, D. 1987. "Metabolic responses to maximal exercise of 800 and 2,000 m in the Thoroughbred horse." *J. Appl. Physiol. 63(1):12-19.*

Hodgson, D. R. 1984. "Studies on some physiological responses to exercise in horses." *PhD Thesis. University of Sydney. NSW.Australia.*

Hodgson, D. R, R. J. Rose, J. R. Allen and J. DiMauro. 1984. "Glycogen depletion patterns in horses competing in day 2 of a three day event." *Cornell Vet. 75:366-374.*

Hodgson, D. R, R. J. Rose, J. R. Allen. 1983. "Muscle glycogen depletion and repletion patterns in horses performing various distances of endurance exercise". *In:Equine Exercise Physiology Eds: D.H Snow, S.G.B. Persson, R.J. Rose. Granta Editions, Cambridge:ICEEP Publications: 229-236.*

Hodgson, D. R. 1984. "Studies on some physiological responses to exercise in horses." *PhD Thesis. University of Sydney. NSW.Australia.*

Housh, T. J., H. A. deVries, G. O. Johnson, S. A. Evans, G. D. Tharp, D. J. Housh and R. J. Hughes. 1990. "The effects of glycogen depletion and supercompensation on the physical work capacity at the fatigue threshold." *Europ. J. Appl.Physiol. & Occup. Physiol. 60(5):391-395.*

Ivy, J. L., A. L. Katz, C. L. Cutler, W. M. Sherman and E. F. Coyle. 1988. "Muscle glycogen synthesis after exercise: effect of time of carbohydrate ingestion." *J. Appl. Physiol. 64(4): 1480-1485.*

Kochan, R. G., D. R. Lamb, S. A. Lutz, C. V. Perrill, E. M. Reimann and K. K.Schlender. 1979. "Glycogen synthase activity in human skeletal muscle: Effects of diet and exercise." *Am. J. Physiol. 236(6): E660-E666.*

Kraemer, J. 1983. "Fatigue and utilization of fatigue curves." *National Strength and Conditioning Assoc. J. Aug-Sept.:35-36.*

Larner, J., C. Villar-Palasi, N. D. Goldberg, J. S. Bishop, F. Huijing, J. I. Wenger, H. Sasko and N. B. Brown. 1967. "Hormonal and non-hormonal control of glycogen synthesis-control of transferase phosphatase and transferase I kinase." In Whelan,W.J. Control of glycogen metabolism (pp.1-17). Academic Press London and New York.

Lindholm, A. 1974. "Glycogen depletion pattern and the biochemical response to varying exercise intensities in Standardbred trotters." *S. Afr. Vet. Ass., 45(4): 341-343.*

Lindholm, A. 1979. "Substrate utilization and muscle fiber type in Standardbred trotters during exercise." In: Proceedings of the American Association of Equine Practioners. 25: 329-336.

Madsen, K., P. K. Pedersen, P. Rose and E. A. Richter. 1990. "Carbohydrate supercompensation and muscle glycogen utilization during exhaustive running in highly trained athletes." *Europ. J. Appl. Physiol. & Occup. Physiol. 61(5-6):467-472.*

McConaghy F. F., Hales J. R. S., Rose R . J., Hodgson D. R. 1995. "Selective brain cooling in the horse during exercise and heat stress." J. Appl. Physiol. 79:1849-1854.

Miller, P. A. and L. M. , & M. S. Lawrence. 1986. "Changes in equine metabolic characteristics due to exercise fatigue." *Am. J. Vet. Res. 47(10):2184-2186.*

Sahlin, K. 1986. "Metabolic changes limiting muscle performance." *In: Biochemistry of Exercise VI. International Series on Sports Science. Vol 16. Eds. B. Saltin. Pub. Human Kinetics. Illinios.*

Saltin, B. and J. Karlsson. 1971. "Muscle glycogen utilization during work of different intensities." *In:Muscle Metabolism During Exercise. (Ed) Pernow, B.; Saltin, B. Plenum Press N.Y:289-300.*

Snow, D. H., P. Baxter and R. J. Rose. 1981. "Muscle fiber composition and glycogen depletion in horses competing in an endurance ride." *Veterinary Record. 108: 374-378.*

Snow, D. H., R. C. Harris and S. P. Gash. 1985. "Metabolic response of equine muscle to intermittent maximal exercise." *J. Appl. Physiol. 58(5):1689-1697.*

Snow, D. H., R. C. Harris. 1991. "Effects of daily exercise on muscle glycogen in the Thoroughbred racehorse." In: *Equine Exer. Physiol. 3. Eds. S.G.B Persson, A. Lindholme, L. B. Jeffcott. ICEEP Publications Davis, California: 299-304.*

Snow, D. 1991. "Fatigue and exhaustion in the horse." Australian Equine Veterinarian. 9(3): 108-111.

Spencer, M. K., Z. Yan and A. Katz. 1992. "Effect of low glycogen on carbohydrate and energy metabolism in human muscle during exercise." *Am. J. Physiol. 262(Cell, Physiol 31):C975-C979.*

Symons, J. D. and I. Jacobs. 1989. "High-intensity exercise performance is not impaired by low intramuscular glycogen." *Med. Sci. Sports Exerc. 21(5):550-557.*

Terjung, R. L., K. M. Baldwin, W. W. Winder and J. O. Holloszy. 1974. "Glycogen repletion in different types of muscle and in liver after exhausting exercise." *Am. J. Physiol. 226(6):1387-1391.*

Topliff, D. R., G. D. Potter, T. R. Dutson, J. L. Kreider and G. T. Jessup. 1983. "Diet manipulation and muscle glycogen in the equine." *In: Proceedings of the Equine Nutrition and Physiology Symposium:114-120.*

Topliff, D. R., G. D. Potter, J. L. Kreider, T. R. Dutson and G. T. Jessup. 1985. "Diet manipulation, muscle glycogen metabolism and anaerobic work performance in the equine." *In: Proceedings of the 9th Equine Nutrition and Physiology Symposium: 224-229.*

Widrick, J. J., D. L. Costill, W. J. Fink, M. S. Hickery, G. K. McConell and H. Tanaka. 1993. "Carbohydrate feedings and exercise performance: effect of initial muscle glycogen concentration." *J. Appl. Physiol. 74(6) 2998-3005.*

TIME OF FEEDING CRITICAL FOR PERFORMANCE

J.D. PAGAN
Kentucky Equine Research, Inc.

Introduction

One of the most frequently asked questions regarding feeding the performance horse is when to feed before a competition. Several studies have evaluated how feeding grain before exercise affects plasma concentrations of nutrients and hormones and substrate utilization during exercise (Rodiek et al., 1991; Zimmerman et al., 1992; Lawrence et al., 1993; Lawrence et al., 1995; Stull and Rodiek, 1995; Duren et al., 1998). In each of these studies, a pre-exercise concentrate meal suppressed free fatty acid (FFA) availability and enhanced glucose uptake by muscle during exercise. Forage was not fed with the concentrate in any of these studies. Thus, it is not known whether feeding hay along with grain will alter substrate availability during exercise. Therefore, a series of experiments was conducted to first evaluate how feeding forage along with grain influences plasma variables and water intake and then to determine whether these changes affect exercise performance. Additionally, a study was conducted to determine how forage alone affects exercise response. Since time of feeding is particularly important for three-day event horses, the exercise test used was a competition exercise test (CET) performed on a high speed treadmill and designed to simulate the physiological and metabolic stresses of the speed and endurance test of a three-day event (Marlin et al., 1995).

Timing of Hay Relative to Grain Feeding

Surprisingly, the type of forage and time that it is fed relative to grain can have a large effect on fluid balance and prececal starch digestibility. Meyer et al. (1993) showed that substituting grass hay for ground alfalfa meal resulted in a decrease in the prececal starch digestibility of ground corn from 45% to 16%. He attributed this drop to changes in rate of passage and dilution of substrates and enzymes in the chyme by increased secretion of digestive juices.

KER has conducted research to determine how the timing of hay feeding relative to a grain meal affects plasma variables and water intake in Thoroughbreds. In this study, six Thoroughbred horses received 2.27 kg of orchardgrass hay at three different times relative to a 2.27 kg grain meal. Treatment 1 (hay/+4h) received hay 4 h after grain. Treatment 2 (hay/-2h) received hay 2 h before grain and Treatment 3 (hay/0h) received hay and grain together. Blood samples were taken immediately before and for 8 h post grain feeding. Insulin, glucose, lactate, total plasma protein and hematocrit were measured in each sample. Water intake was measured hourly throughout the test.

Feeding hay either before or with grain significantly reduced the glycemic response of the grain meal. Insulin production post feeding was also reduced. In addition, when hay was fed, total plasma protein (TP) became elevated in the next hour's blood sample. Interestingly, feeding only grain resulted in essentially no change in TP, even though the level of grain intake (2.27 kg) was the same that elicited a large change when hay alone was fed. Water intake was significantly influenced by time of hay feeding. Following hay feeding, water intake was greatly increased. The increase in water intake also corresponded to increased TP, suggesting that decreased plasma volume may have triggered a thirst response.

The large increase in TP seen with hay feeding probably resulted from a decrease in plasma volume due to greater saliva and digestive juice production. Plasma volume has been reported to drop by as much as 24% in response to a large meal, and this drop was accompanied by hyperproteinemia (Clarke et al., 1990). Kerr and Snow (1982) found that feeding hay, but not concentrates, caused elevations in hematocrit and TP. Meyer et al. (1985) measured the amount of saliva produced when horses ate either hay, pasture or a grain feed. When fed hay and fresh grass, the horses produced 400-480 g of saliva per 100 g of dry matter consumed. When a grain-based diet was offered, the horses produced only about half (206 g/100 g DM intake) as much saliva.

The decrease in glycemic response with hay feeding was probably a result of an increased rate of passage of the grain through the small intestine which resulted from greater volumes of fluid (saliva and drinking water) in the GI tract. Further evidence that hay feeding reduced grain digestibility in the small intestine is supplied from plasma lactate levels post-feeding. Plasma lactate increased 3 h after grain feeding when hay was fed either after or along with grain, suggesting the grain may have been fermented in the large intestine in these treatments. If feeding hay before or along with grain increases rate of passage to the point that prececal starch digestibility is compromised, then this should be discouraged since excessive starch fermentation in the hindgut can lead to a number of problems including colic and laminitis (Clarke et al., 1990).

When grain was fed alone, there was a dramatic drop in hematocrit in the 2 h post-feeding. It is not known why this occurred or if it is of physiological significance, but it certainly brings into question the relevance of resting hematocrit values when sampling times are not standardized. In fact, resting blood samples are probably of limited use anytime to assess red cell status. Persson (1979) examined resting hemoglobin values in Standardbred horses, with daily blood samples collected for 7 days. He reported up to a 30 percent variation in resting Hb values and warned that they provided no useful indication of total body Hb.

Feeding and Exercise

The above study clearly demonstrated that feeding hay and grain markedly affects glycemic response and fluid balance in resting horses. Do these changes affect the horse during exercise? To answer this question, KER conducted a second experiment in which four mature trained Thoroughbred horses were used in a 4 X 4 Latin square design to determine whether feeding grain with or without hay prior to a treadmill competition exercise test (CET) would affect substrate utilization and performance.

The four treatments tested were: (1) (FASTED) 2.27 kg of hay at 2200 h the day before but no grain or hay on the morning of CET; (2) (GRAIN only) 2.27 kg of hay at 2200 h the day before and 2.27 kg of grain 2 h before CET; (3) (GRAIN + AM HAY) 2.27 kg hay at 2200 h the day before and 2.27 kg hay 3 h before CET and 2.27 kg grain 2 h before CET; (4) (GRAIN + AD LIBITUM HAY) Ad libitum hay from 1800 h the day before and 2.27 kg grain 2 h before the CET.

The exercise test used was a competition exercise test (CET) carried out on a high speed treadmill which was designed to simulate the physiological and metabolic stresses of the speed and endurance test of a three-day event (Marlin et al., 1995). The CET was performed on an inclined treadmill (3°) and consisted of a 10 min walk *(Phase A)* (*1.4 m/s*), 10 min trot *(Phase A)* (3.7 m/s), 2 min gallop *(Phase B)* (10.7 m/s), 20 min trot *(Phase C)* (3.7 m/s), 10 min walk *(Phase C)* (1.4 m/s) and 8 min canter *(Phase D)* (9.0 m/s). Following exercise, the horses were hand walked for an additional 30 min.

Blood samples were taken hourly before exercise, during the last 30 s of each speed, and 30, 60, and 120 min after treadmill exercise. Insulin, cortisol, glucose, lactate, total plasma protein and hematocrit were measured. Body weight and plasma volume were measured immediately before exercise. Heart rate was recorded during the final 30 s at each speed throughout the CET.

A third experiment was conducted where four mature trained Thoroughbred horses (2 mares and 2 geldings) were used in a 4 X 4 Latin square design experiment to determine whether feeding forage but no grain prior to a treadmill competition exercise test would affect substrate utilization and performance.

The treatments tested were: (1) (FASTED) 2.27 kg grass hay fed at 2200 h the day before CET and no hay or grain the morning of the CET; (2) (AM HAY) 2.27 kg grass hay fed at 2200 h the day before and 2.27 kg of hay 3 h before the CET; (3) (AD LIBITUM HAY) For 7 d before the CET, horses were offered 5-6 kg of grass hay at 2200 h, 1.13 kg hay at 1200 h and 2.27 kg at 1600 h. If all of the hay was consumed during the night before the CET, the horses were offered an additional 2 kg of hay at 0600 h on the morning of the CET; (4) (GRAZING) For 7 nights before the CET, horses were housed in small grass paddocks from 1600 h to 0700 h and then they were placed in stalls and offered 2-2.5 kg grass hay. On the morning of the CET, the horses were removed from the grass paddocks at 0600 h, and they were offered no hay for 3 h before the CET. During these 7 d, the horses received no grain. The tests were conducted in late fall when the amount of forage available in these paddocks was sparse, but of high quality. The horses underwent the same CET as in Experiment 2 at 2 wk intervals. Blood sampling and analysis were also the same as in Experiment 2.

Data from Experiments 2 and 3 are combined in Figures 1a-1f. Fasted values represent four observations from Experiment 2 and four from Experiment 3 (n=8). Grain values represent all 3 grain treatments from Experiment 2 (n=12) and forage values represent all three forage treatments from Experiment 3 combined (n=12). From these figures, it is obvious that feeding grain before exercise has a large effect on glucose (Figure 1a) and insulin (Figure 1b) with a large drop in glucose occurring early in exercise, while feeding forage produced little effect on either parameter. Free fatty acids (Figure 1c) were also depressed in the grain fed horses preexercise and continued to be lower than either the fasted or forage fed horses throughout exercise. Plasma lactate (Figure 1d) appeared to be uninfluenced by diet as has been shown in other studies of feeding and exercise (Duren et al., 1998; Lawrence et al., 1995). Cortisol (Figure 1e) increased in all of the treatments throughout exercise, but appeared higher in the grain fed horses. Finally, TP was elevated both before and during exercise in the forage fed horses when compared to either the fasted or grain fed treatments (Figure 1f).

As in several other studies (Rodiek et al., 1991; Zimmerman et al., 1992, Lawrence et al., 1993; Lawrence et al., 1995; Stull and Rodiek, 1995; Duren et al., 1998), feeding grain before exercise with or without hay reduced FFA availability and increased glucose uptake into the working muscle. This would not be beneficial for horses competing in the speed and endurance phase of a three day event.

Feeding hay, either with grain or ad libitum the night before exercise, resulted in higher lactate production, heart rates, TP and hematocrit during exercise. Additionally, GRAIN + AD LIBITUM horses had elevated body weights and reduced plasma volume before exercise. Therefore, feeding hay along with grain before competition would appear to have no benefit as compared to feeding grain alone.

Feeding only forage before exercise had a much smaller effect on glycemic and insulin response to exercise than a grain meal. Additionally, feeding forage did not affect FFA availability. In the AM HAY fed horses, TP was elevated before and during exercise, and heart rate was elevated during the gallop in the AD LIBITUM HAY horses. Both of these responses in the hay fed horses were probably due to increased gut fill and a movement of water from the plasma into the gut. Even though the GRAZING horses tended to be heavier, they did not suffer from reduced plasma volume or elevated heart rates during exercise. This is probably because water was able to equilibrate between the plasma volume and gut so there was no reduction in plasma volume before exercise.

The results of these experiments indicate that feeding hay along with grain will result in a decrease of plasma volume and increase in body weight which may be detrimental to performance. Feeding grain either with or without hay 2 h before exercise will reduce FFA availability and increase glucose uptake by the working muscle. This is probably not desirable during prolonged exercise. Feeding only forage before competition does not appear to interfere with FFA availability and has no adverse effects other than possibly reducing plasma volume and increasing body weight. If forage is fed in small amounts or if time in a grass paddock is limited, then these effects will probably be minimal. Since completely withholding forage may lead to stomach ulcers (Pagan, 1997), the slight risk of reduced plasma volume and increased gut fill is more than outweighed by the potential benefit to the horse's long term health and well-being.

Figure 1a.

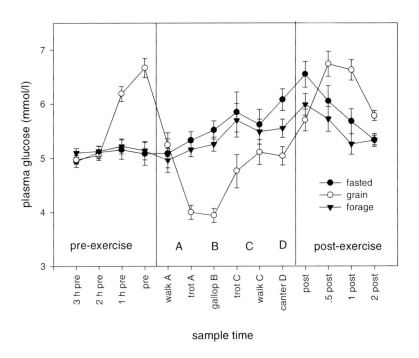

Figure 1b.

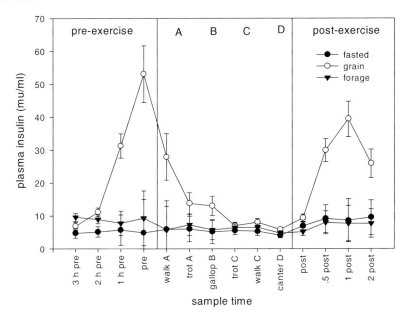

Figure 1c.

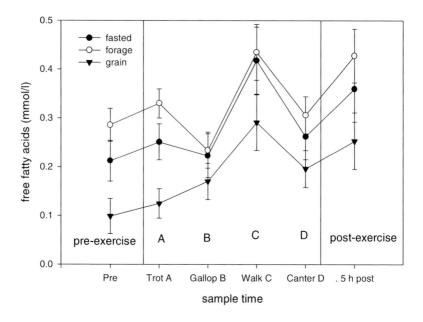

Figure 1d.

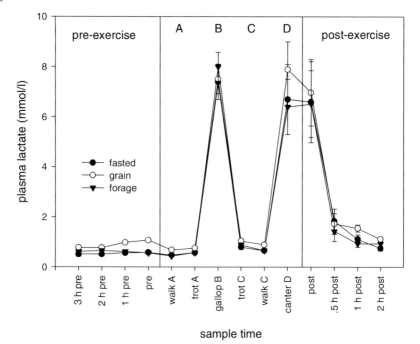

Figure 1e.

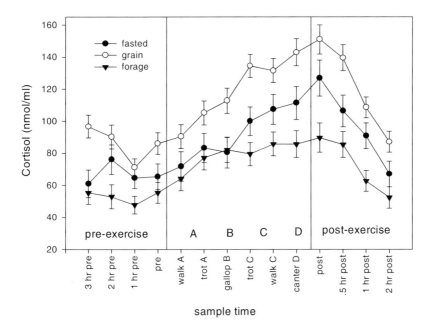

Figure 1f.

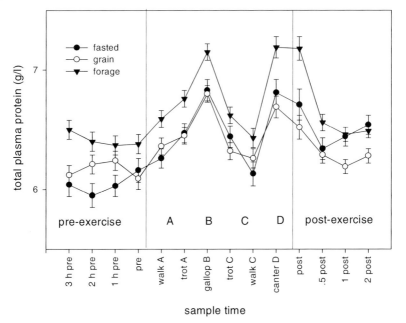

Acknowledgments

These studies were supported by a grant from the Waltham Centre for Pet Nutrition, Leicestershire, UK.

References

Clarke, L.L., Roberts, M.C. and Argenzio, R.A. 1990. Feeding and digestive problems in horses: Responses to a concentrated meal. In: Clinical Nutrition (H.F. Hintz ed.), Veterinary Clinics of North America. W.B. Saunders. Philadelphia.

Duren, S.E., Pagan, J.D., Harris, P.A. and Crandell, K. 1998. Time of feeding and fat supplementation affect exercise response in TB horses. In: Equine Exercise Physiology 5. In press

Kerr, M.G. and Snow, D.H. 1982. Alterations in plasma proteins and electrolytes in horses following the feeding of hay. Vet. Rec. 110, 538-540.

Lawrence, L.M., Soderholm, L.V., Roberts, A. M., and Hintz, H.F. 1993. Feeding status affects glucose metabolism in exercising horses. J. Nutr. 123, 2151-2157.

Lawrence, L.M., Hintz, H.F., Soderholm, L.V., Williams, J. and Roberts, A.M. 1995. Effect of time of feeding on metabolic response to exercise. Equine vet J., Suppl. 18, 392-395.

Marlin, D.J., Scott, C.M., Harris, R.C., Schroter, R.C., Mills, P.C., Harris, P.A., Orme, C.E., Roberts, C.A., Marr,C. and Barrelet, F. 1995. Physiological responses of non-heat acclimated horses performing treadmill exercise in cool (20⁰C/ 40% RH), hot/dry (30⁰/40% RH) or hot/humid (30⁰/80% RH) conditions. Equine vet J., Suppl. 22.

Meyer, H., Coenen, M., and Gurer, C. 1985. Investigations of saliva production and chewing in horses fed various feeds. Proc. 9th ENPS, East Lansing, MI, 38-41.

Meyer, H., Radicke, S., Kiengle, E., Wilke, S. and Kleffken, D. 1993. Investigations on preileal digestion of oats, corn and barley starch in relation to grain processing. In: Proceedings of 13th ENPS, Gainesville, FL, pp. 92-97.

Pagan, J.D. 1997 Gastric ulcers in horses: A widespread but manageable disease. World Equine Vet Review, 2:4, 28-30.

Persson, S.G.B. 1979. Value of haemoglobin in the horse. Nord. Vet. Med., 21:513-523.

Rodiek, A., Bonvicin, S., Stull, C. and Arana, M. 1991. Glycemic and endocrine responses to corn or alfalfa fed prior to exercise. In: Equine Exercise Physiology 3. Eds: Persson, S.V., Lindholm, A. and Jeffcott, L.B. ICEEP Publications, Davis California. Pp. 323-330.

Stull, C. and Rodiek, A. 1995. Effects of post prandial interval and feed type on substrate availability during exercise. Equine vet. J., Suppl. 18. 362-366.

Zimmerman, N.I., Wickler, S.J., Rodiek, A.V. and Hower, M.A. 1992. Free fatty acids in exercising horses fed two common diets. J. Nutr. 122, 145-150.

VITAMIN REQUIREMENTS AND SUPPLEMENTATION IN ATHLETIC HORSES

M.T. SAASTAMOINEN
Agricultural Research Centre of Finland, Animal Production Research, Finland

Many vitamins regulate glucocorticoid synthesis, thus limiting some of the negative responses associated with stress, and are therefore necessary to athletic horses. Vitamin A is required for development of epithelial cells. Vitamin A is also important in maintaining the integrity of mucous membranes, and these tissues are certainly put under considerable stress in exercising horses. According to recent studies, it is not possible to give very precise estimates for vitamin A requirements; 45-75 IU/kg body weight per day can be recommended, and 100 IU is sufficient in all circumstances. Vitamin E is an important antioxidant, and it protects against oxidative damage to tissues induced by exercise. Horses can tolerate low dietary levels of vitamin E for at least four months, but 3 to 5 mg/kg body weight per day can be recommended for horses in training and racing. Concerning vitamin D, the influence of it on calcium metabolism is not as great in horses as it is in other domestic animals under normal conditions, and deficiencies in practical diets are not likely. There are no new studies upon which to base recommendations; 5-10 IU/kg body weight per day seems to be adequate. The requirements for water-soluble vitamins can normally be met by practical diets and their synthesis in the body. Vitamin C has been studied most during recent years, mainly because it is also an antioxidant which prevents radical-induced oxidative damage. Decreased synthesis of vitamin C due to stressful conditions and infections may increase requirements. Based on studies, 30-40 mg/kg body weight per day can be recommended in the case of infections and injuries. There are no accurate requirements determined for B-complex vitamins. Needs for B_1 and B_2 are 4-5 and 2 mg/kg feed DM, respectively. Horses that are anemic, perform poorly or have low serum values may benefit from 20 µg daily supplementation of folic acid. 0.02-0.03 mg/kg body weight per day biotin is sufficient for healthy horses, but 0.03-0.05 mg/kg body weight per day can be recommended for horses with poor hooves. No dietary requirements have been determined for fat-soluble vitamin K.

THE EFFECT OF FISH OIL SUPPLEMENTATION ON EXERCISING HORSES

C. I. O'CONNOR, L. M. LAWRENCE, A. ST. LAWRENCE AND S. HAYES

University of Kentucky, Lexington, KY

Thirteen horses of Thoroughbred or Standardbred breeding were used to study the effect of dietary fish oil supplementation on exercising horses. Horses were assigned to either fish oil (FO, n=7) or corn oil (CO, n=6) treatment groups. The fish oil (Omega Protein, Hammond, LA) contained 11.3% eicosapentaenoic acid and 7% docohexaenoic acid. All horses received timothy hay and a mixed grain concentrate at rates necessary to meet their energy needs. Oil was topdressed on the concentrate daily at a rate of 324 mg/kg BW. Horses were exercised 5 d/wk for 9 wk in a program of increasing intensity. Blood samples were obtained on d 0 (before supplementation), d 28 and d 63. Following the 9-week training period horses performed a standard exercise test on a high speed treadmill. The exercise test consisted of a 5-min warm-up at 1.9 m/s, 0% grade, followed by a step test on a 10% grade at incremental speeds of 2 to 8 m/s. Blood samples were obtained during exercise and recovery. Serum cholesterol and lipids decreased during conditioning ($P < .05$) and there was a time x treatment interaction ($P < .05$). Compared to horses receiving CO, horses receiving FO had lower serum lipids and cholesterol at week 4 ($P < .05$) and lower serum triglycerides at week 9 ($P < .05$). During exercise, heart rates were lower ($P < .05$) for horses receiving FO, but no differences in plasma lactate were detected ($P > .05$). Serum cholesterol was lower ($P < .05$) in horses receiving the FO treatment throughout exercise. Serum insulin and plasma free fatty acids were lower ($P < .10$) in horses receiving FO than in horses receiving CO during the initial stages of the exercise test (warm-up to 5 m/s). Plasma glucose was lower ($P < .05$) for the FO group during exercise recovery from 6 min to 30 min post exercise. Addition of fish oil to the diet altered plasma lipid characteristics of horses and may have affected insulin sensitivity and glucose metabolism in response to exercise.

INDEX

A

Acid detergent fiber (ADF) 7, 14, 121-22, 128, 133, 135, 137, 233, 292-94, 485
Adenosine diphosphate (ADP) 502, 509
Adenosine triphosphate (ATP) 172, 201, 339-40, 497, 501-02, 509
Aerobic exercise 34, 81, 482
Aerobic metabolism 201, 477, 497
Alanine 77, 240-41, 403, 499, 501-02
Aldosterone 340-41
Alkali disease 87, 324
Alopecia 324
Aluminum (Al) 119-24, 140, 161, 165, 167, 390
Amino acid *(see also individual amino acids)* 32-34, 37, 55, 137, 153-55, 159, 172, 208, 237-43, 251-52, 283, 305, 319, 403-04, 407, 410, 412, 441, 456, 467, 499-501
Ammonia 241, 251, 283, 374, 410, 511
Anabolism 319
Anaerobic exercise 482, 512
Anaerobic metabolism 190, 201, 258
Anemia 321-22
Angular limb deformities 194, 320, 397
Anhidrosis 312
Antacid 120, 389-90
Antioxidant 33-34, 87, 90, 92, 171-72, 311, 323-24, 527
Apples 442-43
Arginine 295, 407
Arsenic (As) 89
Articular cartilage 143, 318
Ascorbate (ascorbic acid) 172-73, 312, 394
Aspartate aminotransferase 311
Azotemia 410

B

B vitamins *(see also individual vitamins)* 55, 115, 169, 172, 194-95, 305, 312
Bacteria 283, 312, 405, 409
Bagasse 463
Bahiagrass hay 239
Barley 38, 57-60, 97-100, 113, 204, 247-49, 252, 270, 276-78, 441, 457, 464, 467, 495

Beet pulp 69, 249, 255, 394-95, 409, 463, 487
Behavior 163, 373, 483-84
Bluegrass 137, 455
Bone 158, 161, 165-68, 170, 173-75, 185, 192-93, 305-06, 310-12, 318, 320, 325, 330, 332-36, 339, 353-55, 361, 374, 397-98, 461, 482
Boron (B) 119, 121-24, 140
Bran 90, 249, 255, 409-11, 425, 442, 447, 454-55, 457, 464
Brome 456
Broodmares 179, 231, 264-65, 285-90, 318, 360, 448, 458
Buffel grass 461

C

Cadmium (Cd) 89, 193
Caffeine 441
Calcitonin 173
Calcium (Ca) 5, 8-9, 14-26, 53, 115, 119-25, 133, 135, 167, 169, 173-75, 183, 185, 191-93, 196-97, 262, 264, 266, 272, 285-89, 305, 310, 319, 329-36, 342, 355-56, 359, 361-62, 367-68, 370-71, 374, 381-82, 390, 393-95, 398-99, 410, 413, 442, 454-55, 457, 461, 467, 482, 527
Cancer 92, 308
Cannon bone circumference 185, 354-55, 361
Canola 457, 465
Carbohydrate *(see also Starch)* 3, 5-7, 52, 74, 76-80, 82-84, 109, 189-90, 195, 199-201, 208, 211, 255, 305-06, 322, 395, 398-99, 414, 421-22, 497-98, 509, 513-15
Carbon dioxide 77-79, 481
Carnitine 32-33, 172, 241, 372, 498-99, 503
Carnosine 500-03
Carotene 139, 169-70, 308-10, 372, 422
Carrots 429, 442-43
Cartilage 35, 143, 161, 167, 193, 318, 322
Cassava 255
Cecum 64-69, 98-101, 251, 283, 312
Cereal grains 59-60, 154, 199, 211
Chaff 441-42, 450, 462-64, 466, 468, 470
Chloride (Cl) 119-24, 138, 193, 339-40, 342-49, 367-68, 370-71

Protein 4-5, 7-9, 14, 17, 19, 21-22, 24-25,
41, 51, 53, 55, 57-58, 64-65, 74, 77,
80-81, 84, 97, 113, 115, 117, 120-22,
124, 128, 132, 135-37, 153-55, 159,
160, 167, 172-74, 181, 185, 190-91,
200, 202-03, 205, 207-08, 220-25,
237-43, 248-54, 257, 263, 266, 268-69,
274, 276-79, 281-83, 305-06, 319,
355-56, 358, 362, 365-70, 374, 381-83,
394-95, 398-400, 404, 406, 408-10,
412-14, 421-22, 441, 443, 446, 454-57,
462, 464-66, 469, 475, 479, 481-82,
489, 502, 519-21
Proteoglycan 161
Proteolysis 404
Pulse 414
Pyridoxine 272, 308
Pyruvate 194, 241, 498-99, 501, 503

Q

Q10 (ubiquinone) 33, 498

R

Reed canarygrass 456
Renal function 309, 393-95, 407-08, 410
Respiration 324, 414, 482
Respiratory quotient 77
Retinoic acid 170
Retinol 422
Rhizoma peanut hay 239
Riboflavin 115, 272, 305, 308, 312, 360
Ribonucleic acid 407
Rice 255
Rice bran 409, 425, 455
Rice pollard 464
Ruminants 5, 7, 39, 52, 63-66, 87, 115,
119-20, 125, 128, 202, 249, 251,
268-69, 282
Ryegrass 255, 456, 462

S

Salt (sodium chloride) 40, 91, 106, 119-20,
147-48, 342-43, 347-49, 400, 408,
429, 442-43, 450, 467
Selenate 90, 323,
Selenite 90, 179, 323-24
Selenium (Se) 5, 7, 10, 87-92, 111, 138,

171, 173, 179, 194, 272, 317, 323-25,
343, 359, 365, 372, 374, 398, 413,
422, 442, 446, 462, 503
Sensitivity analysis 1-2, 6, 8-11, 421-22
Serine 240
Seteria grass 461
Silage 113, 252, 255
Silicon (Si)140, 167
Silver (Ag) 89
Small intestine 6, 16, 38, 58, 65, 68-69,
95-97, 99, 101, 105, 199-200, 232,
248, 250-52, 283, 387, 407, 409, 414,
456-57, 496, 520
Sodium (Na) 7, 10, 53, 120-24, 135, 138,
193, 262, 264, 266, 272, 285-89,
339-49, 359, 365, 367-68, 370-71,
410, 413, 443, 446, 470, 489
Sodium bicarbonate 500
Sodium chloride *(see Salt)*
Sodium selenite (salts of selenium) 87,
89-90, 179, 323
Sodium zeolite A (SZA) 167
Sorghum 37-39, 95, 248, 255, 410, 455,
464
Soy hulls 425, 463
Soyabean meal 249, 252, 276, 280-81
Soybean (soyabean) meal 139, 154, 223,
240-41, 249, 252, 276-77, 280-81,
410, 413, 422, 454-57, 465
Stable isotope tracer 73
Stalling 161, 165-66
Stallions 210, 229-30, 257, 261-62, 271,
281, 284-90, 458, 460
Standardbred 397, 461-62, 466-68
Starch *(see also Carbohydrate)* 37-38,
41, 43, 57-58, 60, 63, 66, 95-101,
105-06, 109, 113, 200-01, 232-34,
248-49, 254-55, 293, 409, 413-14,
425, 427, 456-57, 463, 481, 487, 497,
519-20
Stomach 38, 65-68, 99, 101, 109, 199,
387-89, 414, 496, 526
Straw 204, 207, 252, 271, 280-82, 290,
439-42, 450, 495
Sugar 49, 51, 398, 422, 501
Sugar beets 49, 66, 69, 255, 429, 441-42,
447, 454
Sugar cane 49
Sulfur (S) 34, 53, 89, 138, 272, 359

loss 155, 204, 231, 342, 383, 393,
  395, 399, 400, 403, 408-09,
  479-80
Wheat 38, 57-58, 60-61, 90, 99-100, 113,
  249, 255, 441, 457, 462, 464
Wheat bran 90, 248-49, 425, 442, 447, 454,
  457
Wheat middlings 139, 457
White muscle disease 88-89, 323
Withers height 155, 158-59, 185, 354-55,
  361
Wobbler syndrome 397

Y

Yearlings 6, 32, 60, 154, 156, 159, 165-66,
  185, 192, 227-28, 238, 241, 243,
  281, 311, 325, 400, 422, 468-69
Yeast 88, 90, 92, 179, 309, 319, 323-24,
  395

Z

Zinc (Zn) 5, 7, 10, 14-17, 22-26, 53, 119,
  121-24, 135, 181, 183, 185, 193, 272,
  317, 319, 322, 324-25, 343, 359, 362,
  365, 372, 379-80, 398-99, 413, 422,
  442, 446, 454-55, 457

Contributing Authors

DR. YO ASAI

Dr. Asai received his Ph.D. degree from Kyoto University in Japan. Currently, Asai is chief of the Equine Science Division of the Hidaka Training and Research Center, a part of the Japanese Racing Association. The Equine Science Division was established in 1998 and practical studies relating to growth and development of equine athletes are undertaken year-round. Prior to the opening of the Equine Science Division at the Hidaka Training and Research Center, Asai performed equine nutrition and growth research at the Equine Research Institute.

DR. KEITH C. BEHNKE

Dr. Behnke received his B.S. degree in feed technology and his M.S. and Ph.D. degrees in grain science from Kansas State University. Dr. Behnke is currently a professor at Kansas State University in the Department of Grain Science and Industry with major responsibilities in teaching and research. As a researcher, his major interest is in the area of feed processing. He has published extensively in scientific literature as well as in numerous horse publications. He is a member of a number of professional associations including the American Society of Animal Science and American Feed Industry Association.

DR. DAVID M. CALDWELL

Dr. Caldwell is a native of Indiana. He received his B.S. and Ph.D. degrees in ruminant nutrition from Purdue University. He has served as beef and horse nutritionist for a number of feed companies including Wayne Feeds, National Molasses Company, and Cargill Molasses Division. He is currently working with Westway Trading Corporation.

DR. MANFRED COENEN

Following completion of his veterinary training at the School of Veterinary Medicine Hannover, Dr. Coenen commenced postgraduate work at the Institute for Animal Breeding, Behavior and Husbandry in Mariensee where he was awarded his doctorate in veterinary medicine. In 1996, Coenen was appointed the chairman of animal nutrition and dietetics for the Institute of Animal Nutrition at the School for Veterinary Medicine Hannover.

DR. BOB COLEMAN

Dr. Coleman has recently joined the staff at the University of Kentucky as the Extension Horse Specialist in the Department of Animal Sciences. He received his B.Sc. And M.Sc. from the University of Manitoba and his Ph.D. degree from the University of Alberta. He served as the Provincial Equine Extension Specialist for Alberta Agriculture, Food and Rural Development. Dr. Coleman's research interests are in the area of feedstuff evaluation and feed management practices. In particular he has looked at the use of alfalfa products and grain processing for horses. He is the a president of the Equine Nutrition and Physiology Society.

DR. KATHLEEN CRANDELL

Dr. Crandell received her M.S. and Ph.D. degrees from Virginia Polytechnic Institute and State University where she performed research in equine nutrition, exercise physiology and reproduction. Dr. Crandell joined the staff of KER in 1996. She is located in Middleburg, Virginia and assists with technical support and consultation for KER eastern and South/Central American clientele.

DR. GARY CROMWELL

Dr. Cromwell is a professor in the Department of Animal Sciences at the University of Kentucky. Cromwell received his M.S. and Ph.D. degrees in animal nutrition from Purdue University. Many of his research efforts focus on amino acid and mineral nutrition of swine, mineral bioavailability and efficacy and safety of feed additives. More recently, he has investigated the effects of diet manipulation on nutrient excretion, odor and factors which impact the environment. Cromwell has served as president of the American Society of Animal Science and has held editorial roles for the Journal of Animal Science. He spearheaded the National Research Council Subcommittee on Swine Nutrition which prepared the tenth edition of Nutrient Requirements of Swine. Cromwell presently serves as chairman of the Committee on Animal Nutrition for the National Research Council.

DR. DEREK CUDDEFORD

Dr. Cuddeford received his B.Sc. in agricultural science from London University and his M.Sc. in animal nutrition from Aberdeen University. Dr. Cuddeford studied mineral metabolism intensively to obtain his Ph.D. from Edinburgh University. Currently, Dr. Cuddeford is a senior lecturer in the Department of Veterinary Clinical Studies at The Royal (Dick) School of Veterinary Studies at Edinburgh University. His primary research interests lie in partitioning digestion throughout the equine digestive tract, developing in vitro systems to replace in vivo methods of evaluating feeds for horses, and generating predictive methods for use in nutrition research and feed characterization. Dr. Cuddeford has written numerous popular press and scientific articles and has taught countless undergraduate and postgraduate equine nutrition courses.

DR. ALLAN DAVIE

Dr. Davie earned his Ph.D. at the University of Sydney by investigating energy supply in exercising Thoroughbreds and factors influencing muscle glycogen resynthesis in equine skeletal muscle. He remains interested in muscle physiology and maintains specific involvement with the impact of endurance exercise on strength performance, biochemical adaptations of muscle to training, and thermoregulation. Dr. Davie is a senior lecturer at the School of Exercise Science and Sports Management at Southern Cross University in Lismore, Australia.

DR. STEVE DUREN

Dr. Duren is a native of Idaho. He received his B.S. in animal sciences from the University of Idaho and his M.S. and Ph.D. degrees from the University of Kentucky where he was involved in equine nutrition and exercise physiology research. His Ph.D. research focused on blood flow distribution in fasted and fed ponies at rest and during endurance exercise. After graduating from the University of Kentucky, Dr. Duren served as an equine nutritionist for a regional feed manufacturer in central Kentucky. He joined the staff of KER in 1994 and is stationed in Idaho where he is responsible for KER's western and mid-western clientele.

DR. RAY GEOR

Dr. Geor completed his B.V.Sc. degree in his native New Zealand. Following a one-year internship in Australia and a two-year stint in large animal practice in New Zealand, he completed a residency in large animal medicine in Canada. Dr. Geor is board certified in veterinary internal medicine and has been a faculty member at the University of Minnesota and the Ontario Veterinary College at the University of Guelph. Dr. Geor joined KER in 1999 as Director of Research and Development. Geor designs and implements nutrition and exercise physiology studies, and is also instrumental in developing new products.

DR. PAT HARRIS

Dr. Harris graduated from the Cambridge University Veterinary School. After a year as House Physician at the Clinical School, she began working for the

Animal Health Trust to pursue a Ph.D. by studying aspects of equine rhabdomyolysis syndrome. In 1988 she became head of the Clinical Chemistry and Hematology Laboratories as well as being section head of the Metabolic Lameness Research group. In 1995 she joined the Waltham Centre for Pet Nutrition. Dr. Harris' main research interests are tying-up syndrome, electrolytes and bone metabolism. Dr. Harris is a past president of the British Equine Veterinary Association.

DR. HAROLD HINTZ

Dr. Hintz is a professor of animal nutrition in the Department of Animal Sciences at Cornell University's College of Agricultural and Life Sciences. Dr. Hintz has been a faculty member at Cornell University since 1967. He received his B.S. from The Ohio State University and Ph.D. from Cornell University. As a member of the equine nutrition team at Cornell, he has conducted research on energy, mineral and protein requirements of the horse. He is a member of the NRC subcommittee on the publication Nutrient Requirements of Horses.

DR. PETER HUNTINGTON

Dr. Huntington received a veterinary degree from the University of Melbourne in 1981 and then worked in equine practice in Berwick. He joined the Department of Agriculture (Victoria) as the horse specialist veterinary officer. While with the Department of Agriculture, he conducted research on equine nutrition. In addition to being a popular speaker, Huntington is a prolific writer and is the author of the popular book Horse Sense - The Australian Guide to Horse Husbandry. Dr. Huntington is an external lecturer and examiner at the University of Melbourne Veterinary School and holds numerous other appointments in the horse industry. He is currently the Director of Nutrition of KER Australasia and corresponds with clients in Australia and the Far East.

DR. DAVID S. KRONFELD

David Kronfeld is the Paul Mellon Distinguished Professor of Agriculture and Professor of Veterinary Medicine at Virginia Tech. He hails from New Zealand and has trained in veterinary science and biochemistry in Queensland, and in comparative physiology at UC-Davis. He is board certified in veterinary internal medicine and in veterinary nutrition. Dr. Kronfeld worked at the New Bolton Center of the University of Pennsylvania for 28 years, ending as Chief of Medicine. He moved to Virginia in 1988 to study pasture systems for mares and foals, and to compare diets fed to horses working on a treadmill.

DR. LAURIE LAWRENCE

Dr. Lawrence is a professor at the University of Kentucky in the Department of Animal Sciences. She received her M.S. and Ph.D. degrees from Colorado State University. Her primary research interests are in the areas of nutrition and metabolism of performance horses. In particular, she has studied the contribution and regulation of fuel sources during exercise and the factors that affect water balance. She is the author or coauthor of fifty research papers, three book chapters and more than seventy-five articles in bulletins, proceedings or popular press publications. Lawrence currently serves on the Board of Directors for the American Society of Animal Science.

MIKE LENNOX

Mr. Lennox received his B.S. degree in agriculture from the University of Guelph. Following graduation, he held several positions in the feed industry including serving as a feed mill manager and product specialist. He opened his own feed consulting and sales company in Canada and operated this business for many years. Currently, Lennox is responsible for feed formulation and quality assurance for Kentucky Equine Research clients.

DR. WILLIAM MARTIN-ROSSET

Dr. Martin-Rosset has directed the equine nutrition research program at the National Institute for Agricultural Research in France since 1970. In 1984, he and his research associates created a horse feed evaluation system (the net energy system) which included recommended allowances for energy and protein. Martin-Rosset and his colleagues revised these recommendations in 1990. These allowances have been widely published since they were unveiled and have become particularly prevalent since the 1996 European Annual Meeting for Animal Production (EAAP) which was held in Lillehammer, Norway. Martin-Rosset is currently serving as vice president of the equine nutrition branch of the EAAP.

DR. JILL McCUTCHEON

Dr. McCutcheon received her B.Sc. and D.V.M. from the University of Guelph and subsequently completed a residency in pathology and a Ph.D. in equine exercise physiology at Washington State University. In 1990, she joined the faculty of the University of Guelph. McCutcheon's research interests include muscle and bone metabolism and function as well as fluid and electrolyte regulation in the horse. In addition, McCutcheon has examined the effects of training, heat and humidity, fluid loss and electrolyte supplementation on the exercising horse.

DR. BRIAN D. NIELSEN

Dr. Nielsen received his B.S. degree at the University of Wisconsin-River Falls where he had a major in animal science with an equine emphasis and a Chemistry minor. He received his M.S. degree in equine nutrition and exercise physiology from Texas A&M University. His master's research involved feeding sodium zeolite A to young racehorses to increase training distance before breakdown. Dr. Nielsen completed his Ph.D. in May of 1996. His Ph.D. work dealt with mineral balance in young racehorses entering training. Later he began work at Michigan State University as an assistant professor in the Department of Animal Science where he works in equine exercise physiology.

DR. ED OTT

Dr. Ott received his undergraduate and Ph.D. degrees from Purdue University in Lafayette, Indiana. Following graduation, he worked as a research nutritionist for Ralston Purina. Later he moved to the University of Florida's Horse Research Center where he is a professor of animal nutrition at the Institute of Food and Agricultural Sciences. Ott is primarily involved with the equine teaching and research programs. He is a past president of the Equine Nutrition and Physiology Society and was chairman of the 1989 National Research Council subcommittee on horse nutrition.

DR. JOE PAGAN

Dr. Pagan received his B.S.A. from the University of Arkansas in animal nutrition and received his M.S. and Ph.D. degrees from Cornell University in equine nutrition and exercise physiology. Dr. Pagan's research at Cornell forms the basis for many of the energy requirements used in the 1989 version of Nutrient Requirements of Horses. After he left Cornell, Dr. Pagan was a guest researcher at the Swedish University of Agriculture Sciences where he conducted nutrition and exercise physiology research. Dr. Pagan founded Kentucky Equine Research in 1988 to be an international research, consulting and product development firm with emphasis on equine nutrition and sports medicine. Dr. Pagan serves as a consultant to feed manufacturers around the world.

DR. SARAH RALSTON

Dr. Ralston received her veterinary and Ph.D. degrees from the University of Pennsylvania. She is currently on the animal science faculty at Rutgers University and is serving as chairperson of the American College of Veterinary Nutrition. Dr. Ralston's most recent research has focused on the effects of age (young and old) on glucose and insulin metabolism in horses. Research performed by Dr. Ralston was instrumental in the development of feeds designed specifically for geriatric horses. She is also working on patented techniques to identify foals at risk of osteochondritis dissecans as well as rations to reduce the incidence in predisposed foals.

PAUL SIROIS

Mr. Sirois received his B.S. degree in animal science from Rutgers University and his M.S. degree in animal science from the University of Minnesota. He initially worked for a cooperative of independent feed manufacturers as a dairy nutritionist evaluating and developing rations for producers in New York and Pennsylvania. For the past 13 years, Sirois has been employed as the manager of Dairy One Forage Lab in Ithaca, New York. Dairy One is a domestic and international leader in providing quality feed and forage analysis.